III One Sample Confidence Intervals

Proportions   Method I $\hat{p} \pm z_{\alpha/2}\sqrt{\hat{p}(1 - \hat{p})/n}$

                Method II $\hat{p} \pm z_{\alpha/2}\sqrt{1/4n}$

Mean σ known $\overline{X} \pm z_{\alpha/2}\sigma/\sqrt{n}$

       σ unknown $\overline{X} \pm t_{\alpha/2}S/\sqrt{n}$

Variance $L_1 = (n - 1)S^2/\chi^2_{\alpha/2}$

           $L_2 = (n - 1)S^2/\chi^2_{1-\alpha/2}$

Standard Deviation $L_1 = \sqrt{(n - 1)S^2/\chi^2_{\alpha/2}}$

                   $L_2 = \sqrt{(n - 1)S^2/\chi^2_{1-\alpha/2}}$

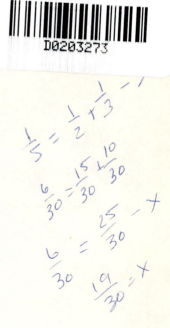

IV Two Sample Confidence Intervals

Difference in Means
Independent Samples
Variances Equal

$$(\overline{X}_1 - \overline{X}_2) \pm t_{\alpha/2}S_p\sqrt{1/n_1 + 1/n_2} \quad \text{where}$$

$$S_p^2 = \frac{(n_1 - 1)S_1^2 + (n_2 - 1)S_2^2}{n_1 + n_2 - 2}$$

Difference in Means
Independent Samples
Variances Unequal

$(\overline{X}_1 - \overline{X}_2) \pm t_{\alpha/2}\sqrt{S_1^2/n_1 + S_2^2/n_2}$ where $t_{\alpha/2}$
is based on the t distribution with

$$\nu \approx \frac{(S_1^2/n_1 + S_2^2/n_2)^2}{\dfrac{(S_1^2/n_1)^2}{n_1 - 1} + \dfrac{(S_2^2/n_2)^2}{n_2 - 1}}$$

Differences in Means
Paired Samples

$$\overline{D} \pm t_{\alpha/2}S_D/\sqrt{n}$$

Difference in Proportions   Method I

$$(\hat{p}_1 - \hat{p}_2) \pm z_{\alpha/2}\sqrt{\hat{p}_1\frac{(1 - \hat{p}_1)}{n_1} + \hat{p}_2\frac{(1 - \hat{p}_2)}{n_2}}$$

                          Method II

$$(\hat{p}_1 - \hat{p}_2) \pm z_{\alpha/2}\sqrt{1/4n_1 + 1/4n_2}$$

Continued on back endleaf

Introduction to Statistics

Introduction to

Statistics

J. S. Milton
J. J. Corbet
P. M. McTeer

Radford University

D. C. Heath and Company
Lexington, Massachusetts   Toronto

To: Amy and Neil Corbet
  Elizabeth McTeer
  Deborah and David Savage

Preface

This *Introduction to Statistics* is intended for students studying the subject at the precalculus level. Over the past several years there has been a renewed interest in the understanding and application of statistics. Students studying the material for the first time, however, find it difficult to grasp the concepts involved because most problems are word problems that require a certain amount of reasoning ability, not simply rote memorization. Each problem looks totally different; it is only with practice and experience that students begin to recognize patterns and problem types. In order to develop this recognition, students need to have available a large number of interesting examples and exercises that present the same concept in different settings. This need has been met in this text.

Chapter 1, Descriptive Methods, introduces students to concrete examples of data handling. In this chapter, we present many standard techniques for displaying and summarizing data sets. We also introduce some of the new graphical techniques of EDA, exploratory data analysis. The chapter on probability, Chapter 2, is quite complete. It can be covered in its entirety in about two weeks, or in less time if Section 2.1 is used as a reading assignment and Sections 2.6 through 2.9 are omitted. Chapter 3, Random Variables, and Chapter 4, The Binomial, Normal, and Poisson Distributions, provide the link between probability and statistics. Chapter 5, The Language of Statistical Inference, introduces the basic vocabulary underlying the areas of estimation of hypothesis testing. We discuss, in a simple context, the ideas of point and interval estimation and hypothesis testing via preset α levels and P values. Chapters 6 through 11 cover the main topics of normal theory statistics, and Chapter 12 is an introduction to distribution-free methods. Optional sections can, in all cases, be omitted. However, if included, these sections will enhance the level of presentation of the course.

Throughout the text, the data sets presented are rather small so that students will not be overwhelmed by the computational aspects of statistical analyses. However, we do not intend to imply that statisticians routinely rely on very small samples. Students are encouraged to use their hand-held calculators since our purpose is to emphasize the interpretation of statistical results rather than the mere computation of these results. Since the computer is becoming more and more important in people's lives, we include some instruction in the use of computer packages. The package chosen for this purpose is SAS® (Statistical Analysis System). This was done because of its popularity and ease of use; we do not intend to imply that it is superior to the other well-known packages such as SPSS (Statistical Package for the Social Sciences; McGraw-Hill), BMD (Biomedical Computer Programs; University of California Press), or MINITAB (Duxbury Press). The computer instruction is presented in Computing Supplements at the appropriate points in the text. These sections are optional, but students with access to a computer are encouraged to give them a try.

Problems and examples in the text are drawn from business, sports, psychology, economics, and medicine. They range from the simple to the challenging, starred exercises being somewhat more difficult than other exercises. Answers to most odd-numbered exercises can be found at the back of the book.

Each chapter ends with a vocabulary list and a set of review exercises, which should help the student solidify the concepts presented in the chapter. As additional aids to the student, both a *Solutions Guide* and a *Study Guide* are available.

We feel that this text will provide students with a knowledge of the mechanics of problem solving and also with an idea of when and how to apply statistical methods. The student who completes this text should have an adequate background for doing basic research or reading research papers in his or her field.

We wish to thank Mary Lu Walsh, Mary Le Quesne, Antoinette Schleyer, and Ted Simpson for their help and encouragement during the writing of this text. A special thanks goes to the following members of the Radford University Computing Center for their patience, care, and expertise in the production of the draft manuscripts of the text: Shirley Wilson, Brenda Grubb, Merle Hoffman, Shirley Fanning, and Louis Kent. We also want to acknowledge the following reviewers for their many helpful suggestions: John S. Bowdidge, Southwest Missouri State University; David M. Crystal, Rochester Institute of Technology; Ronald M. Davis, Northern Virginia Community College; Mark Marcucci, University of South Carolina; Andrew Matchett, University of Wisconsin-LaCrosse; Mel A. Mitchell, Clarion State College; John B. Rushton, Metropolitan State College; Larry Scott, Virginia Commonwealth University; Anne D. Sevin, Framingham State College; Galen R. Shorack, University of Washington; Kenneth V. Turner, Anderson College.

J. S. Milton

J. J. Corbet

P. M. McTeer

Contents

3   Random Variables   113

4   The Binomial, Normal, and Poisson Distributions   141

5   The Language of Statistical Inference   177

10   Categorical Data   413

11   Analysis of Variance   443

12   Some Distribution-Free Alternatives   485

Appendix A   A1

References   A31

Answers to Odd-Numbered Exercises   A33

Index   A73

Introduction: Statistics – The Tool of Decision Makers

In this day and age we are deluged with figures. Many of these figures are commonly referred to as statistics. Figures are published on a huge range of topics: We read government reports on the state of the social-security system, the arms race, and the cost of medical care. We are forced to deal daily with the effects of changes in the prime-interest rate, the cost of living, and the inflation and unemployment rates. We are influenced to change our driving habits when we hear news reports that "statistics indicate that a child not protected by a child's safety seat is five times more likely to be killed in an automobile crash than is one who uses such a restraint." To most of us, these and other "statistical" reports are mysterious and somewhat threatening. We are never quite sure how they arose, how accurate they are, or what they really mean.

The purpose of this text is to introduce you to the methods used in the field of applied statistics. It is convenient to break these methods into two broad categories called **descriptive methods** and **inferential methods**. Descriptive methods are techniques, both analytic and graphical, that are used to simply describe a data set. These techniques are presented in Chapter 1 and are used throughout the text. Inferential methods are techniques that are used to draw conclusions or to make inferences about a large group of objects based on the observation of only a portion of the members of the group. These methods make use of the concepts of probability theory. They are discussed in Chapters 5 through 12, after a brief survey of the fundamentals of probability. The methods presented in this text are those that are used to prepare many of the statistical reports that you read about in the newspaper and hear about on news reports. By gaining some insight into the manner in which statistical reports are prepared,

1

it is hoped that they will appear less mysterious and that you will be able to interpret better the figures, and assess the meaning of the findings that are placed before you.

When studying any subject for the first time, you must learn its basic vocabulary. Fortunately, there are only a few concepts that must be understood before we begin our detailed study of statistical methods. We introduce these concepts here on an intuitive level via some typical problems that can be approached statistically. The terms presented will be defined more precisely in later chapters as the need arises.

Statistical studies have one important characteristic in common. In particular, interest always centers on a target group of objects. We want to describe the characteristics of this group or draw some conclusions concerning its behavior. Unfortunately, the group is usually too large to study in its entirety. For this reason, the conclusions that we reach must be based on the observation of only a portion or a subset of its members. The terms used to describe this situation are *population* and *sample*.

The **population** in a statistical study is the group of objects about which conclusions are to be drawn.

A **sample** is a portion or subset of objects drawn from the population.

These examples should clarify the meaning of the two terms.

Example 1

The use of alcohol among teenagers is of concern to all of us. A study is to be conducted to help answer questions concerning the drinking habits of young persons in the United States in the 12- to 19-year-old age group. Questions to be answered via the data gathered are as follows: What proportion of the individuals in this age group drink on a regular basis? and If an individual is a regular drinker, what is the average age at which drinking began? The population here consists of *all* persons in the United States who are at least 12 years old but who have not yet reached their twentieth birthday. This group does exist but it is so large that it is physically impossible to study it in its entirety. We must try to answer the questions posed by selecting and interviewing only a sample of those young people who constitute the population.

Example 2

Industrial robots are being used more and more frequently in American industry. They are of particular value in the automotive industry. A study of the characteristics of a robot designed to spray-paint the hoods of automobiles is to be conducted. The primary question to be answered is, On the average, how many hoods can such a robot paint before it requires cleaning and other maintenance? In this example, the population is somewhat hypothetical. It includes *all* robots of the type described, both those that currently exist, and those that will be produced in the future. Since we cannot study objects that do not yet exist, we must draw conclusions about these robots based on a sample of those currently in use.

Example 3

A city council is considering the possibility of beginning a limited bus service for its citizens. Before taking this step, the opinion of its 25,000 adult citizens is to be sought. In particular, council members want to know whether a majority of the adult citizens favors the proposal. They also want to determine the average number of blocks that an individual is willing to walk to catch the bus so that effective routes can be developed. The population here consists of the 25,000 individuals living in this city who are considered to be adults. Since it is impossible to contact each of these persons individually, we must sample to obtain an idea of public opinion.

In a statistical study, interest centers on some particular variables associated with the population. The values assumed by these variables can change from one member of the population to another. The change is due to random or chance influences and is therefore somewhat unpredictable. The term *random variable* is used to describe such a variable. That is,

A **random variable** is a variable whose value is determined by the outcome of some chance experiment.

In Example 1, the use of alcohol among teenagers is being studied. To determine the proportion of young people in the population who drink on a regular basis, we might ask the question, Do you usually drink on at least three separate occasions per week? We can define a random variable X by agreeing to record a 1 for X if the answer to this question is yes; otherwise, we will record a 0. The variable X is random because its value can change from person to person, and the particular value that it assumes for a given individual depends on chance. We also want to determine the average age at which regular drinkers began to drink. To do this, we need to record the value of the random variable Y, the age at which drinking began, for the appropriate individuals.

Example 2 entails trying to determine the characteristics of a particular type of industrial robot. To help answer the question, On the average, how many hoods can such a robot paint before it requires cleaning and other maintenance? we will study the random variable Z, the number of hoods painted per robot. The values assumed by Z will vary from robot to robot due to the differences in workmanship, differences in environmental conditions, and other chance factors.

We leave the identification of the random variables to be studied in Example 3 as an exercise.

Random variables studied in this text fall into two broad categories. They are either *continuous* or *discrete*.

A **continuous random variable** is a random variable that, prior to the experiment, can conceivably assume any value in some interval or continuous span of real numbers.

The random variable Y, the age at which a regular drinker began to drink, is continuous. It can conceivably lie anywhere between perhaps 10 and 20 years of age, excluding the value 20 itself.

A **discrete random variable** is a random variable that assumes its values only at isolated points.

The random variable X, which assumes only the values 0 or 1, depending on whether or not the individual sampled answers yes or no to the question, Do you usually drink on at least three separate occasions per week?, is discrete. The random variable Z, the number of hoods painted by a robot before maintenance is necessary, is discrete. Its set of possible values is 0, 1, 2, ...

We will use uppercase letters such as X, Y, and Z to denote random variables; we will use lowercase letters x, y, and z to denote the observed values of these variables. For example, if a young woman reports that she usually drinks on at least three separate occasions per week and that she started drinking at age 14 years and 6 months, then we write $x = 1$ and $y = 14.5$.

As you will soon discover, statistical studies can be thought of as studies of the behavior of one or more random variables. Associated with these random variables are certain constants or numerical measures which are descriptive in nature. The average value of the variable over the entire population is such a measure; it describes the value about which the values of the random variable tend to cluster. The difference between the smallest and largest value assumed by the random variable over the entire population is another such measure; it gives us a rough idea of the degree of dispersion or spread exhibited by the random variable. Measures such as these are called *population parameters*.

A **population parameter** is a descriptive measure associated with a random variable when the variable is considered over the entire population.

Unfortunately, since we do not usually study the entire population, the actual numerical values of a specific population parameter are seldom known. We must attempt to approximate their true values based on information obtained from a sample. Before we can use information gained from a sample to answer questions concerning the population from which the sample is drawn, we must be able to describe our sample in a logical way. To do so, we make use of what are commonly called *statistics*.

A **statistic** is a descriptive measure associated with a random variable when the variable is considered only over a sample.

Example 4   In Example 1, we want to determine the average age at which a regular drinker in the 12- to 19-year-old age group began to drink. This average age is a population parameter. It is the value that we would obtain if we could do the following: locate and interview every person in the United States who is at least 12 years old but who has not yet reached his or her twentieth birthday, determine which of these individuals are regular drinkers, record the age at which each regular drinker began to drink, and then average these values. This is obviously an impossible task! Nevertheless, we can approximate the value of this population parameter in a logical way. We simply interview a sample of young people and record the age at which the regular drinkers in the sample began to drink. We then average these values. This sample average is a statistic. Its value probably is not exactly the same as that which would be obtained if we could interview everyone in the population. However, for sufficiently large samples it usually will be fairly accurate. Similarly, the proportion of teenagers in the population that drinks is a population parameter; the proportion of members of the sample who drink is a statistic.

Statistics actually serve two purposes. They describe the sample itself and they allow us to make inferences about the population from which the sample is drawn. Roughly speaking, a statistical study entails these steps:

Steps in a Statistical Study

1. Identify the population under study.
2. State the questions that you want to answer concerning the population.
3. Identify the random variables that are of interest to you and that will help you answer the questions posed.
4. Identify the population parameters that are important to you.
5. Draw a sample from the population.
6. Evaluate the statistics that will approximate the important population parameters.
7. Apply inferential techniques when needed to answer the questions posed.

The remainder of this text is spent showing you how to implement this outline.

Exercises

1. A banker is interested in studying the characteristics of the 1000 loans granted by his bank over the last year. Questions to be answered are, What proportion of the loans that we made exceeds $10,000? and, What is the average amount of the loans that we made? The answers to these questions will allow the banker to anticipate the future demand for money.
   (a) Identify the population under study.
   (b) Identify the random variables that are of interest. Are these random variables continuous or discrete?
   (c) Identify the population parameters that are important.
   (d) Does this population exist or is it somewhat hypothetical? Explain.
   (e) A sample of 100 loans is selected. It is found that 30 of these loans (.3 or 30%) exceed $10,000. Can we conclude that the proportion of loans in the population that exceeds $10,000 is *exactly* .3? Explain.

2. A public-health official is interested in describing the characteristics of the handicapped children in her district who have attended a special summer camp in the past or who might attend the camp in the future. She wants to answer the questions, What proportion of these children are afraid of the water? and, What is the average distance in miles that a child can walk before tiring? The answers to these questions will help her plan appropriate camp activities.
   (a) Identify the population under study.
   (b) Identify the random variables that are of interest. Are these random variables discrete or continuous?
   (c) Identify the population parameters that are important.
   (d) Is this population real or is it somewhat hypothetical? Explain.
   (e) A sample of 25 handicapped children is selected, and each child's parents are asked to estimate the distance that their child can walk before tiring. The average response is 1/2 mile. Can we conclude that the average distance that children in the population can walk before tiring is *exactly* 1/2 mile? Explain.

3. Identify the random variables to be studied in Example 3. Are these random variables discrete or continuous?

4. When a public-opinion poll is conducted in the United States, it is usually done by contacting approximately 2500 persons. A news report states that a recent poll indicates that 40% of the American public is satisfied with the performance of our President in the area of foreign affairs. Is this value a statistic or a population parameter? Explain.

5. At the *end* of a basketball season, it is reported that a particular team won 70% of its games, made 57% of its attempted field goals, and made 65% of its attempted free throws. Are these figures statistics in the true sense of the word or are they population parameters?

6. Identify each of these random variables as being either discrete or continuous:
   (a) the number of automobiles sold per day at a particular dealership
   (b) the weight of a newborn baby
   (c) the length of a telephone conversation
   (d) the speed of a baseball as it crosses home plate
   (e) the volume of blood lost by a patient during an operation
   (f) the number of pints of A-positive blood collected per day at a blood-mobile
   (g) the number of new accounts opened per day at a particular department store
   (h) the interest rate paid on six-month money-market certificates per week
   (i) the length of time required for a subject in a study of learning to memorize a random word list
   (j) the dress size of women who shop at a large department store

Vocabulary List and Key Concepts

descriptive methods

inferential methods

population

sample

random variable

continuous random variable

discrete random variable

population parameter

statistic

Descriptive Statistics

As mentioned in our brief introduction, the first job in analyzing a data set is usually to try to describe it in easily understandable terms. This process helps us to answer the questions posed prior to our study and it often leads us to ask other questions that are equally important.

Different fields of study may require special analytical techniques. We may need to treat data on the heat buildup on the underside of the space shuttle upon re-entry into the earth's atmosphere somewhat differently from that data gathered to ascertain public opinion concerning the electoral college system. In this book we cannot delve into the details of data analysis in any particular special area. However, there are certain general concepts that apply to the analysis of data regardless of the field of interest. In this chapter we consider some of these general concepts. In particular, we consider methods for describing a data set both graphically and analytically. We also introduce some of the tools of *exploratory data analysis* (EDA). In the words of John W. Tukey, a well-known statistician and data analyst, [22]

> We will be exploring numbers. We need to handle them easily and look at them effectively. Techniques for handling and looking—whether graphical, arithmetic, or intermediate—will be important. The simpler we can make these techniques the better—so long as they work—and work well.

The methods presented in this chapter have been found to work well in almost every discipline that requires the analysis of numerical data.

1.1 Some Detective Work: Introducing Exploratory Data Analysis

Before beginning to analyze a data set it is important to realize what the data represent. In particular, it is important to realize that each number in a data set is an *observed value of some random variable*. Sometimes we have available an observation for every object of interest to us; that is, we have data for the entire population. Usually we have data only for a sample of objects drawn from the population. The next two examples illustrate the difference between these types of data sets.

Example 1

A psychologist is working with a group of ten children who are being treated for autism, a mental disorder in which the individual avoids contact with other people and often cannot speak. The psychologist wants a profile of this group of patients. One variable to be considered is X, the age in months at which autistic behavior was first observed. These data are obtained:

    1   6   8   3   2   3   14   24   7   4

Note that each number in this data set corresponds to an observed value of the random variable X. In this case the psychologist is interested only in obtaining a profile of the ten children under treatment. Thus, the population of interest consists of these ten children alone and therefore the psychologist has data for the entire population.

Example 2

One variable being studied by geologists is X, the magnitude of a California earthquake as measured on the Richter scale. (This is a log scale, that is, 2 is ten times as strong as 1, and so on.) This scale, first published in 1935, assigns a number to each quake based on seismological data. Roughly speaking, the scale is interpreted as follows:

    1   only observed instrumentally

    2   barely felt near its epicenter

    4.5 felt to distances up to 20 miles of its epicenter; slight damage

    6+ moderately destructive

    7+ major quake

    8+ great quake

These data are obtained:

1.0	8.3	3.1	1.1	5.1
1.2	1.0	4.1	1.1	4.0
2.0	1.9	6.3	1.4	1.3
3.3	2.2	2.3	2.1	2.1
1.4	2.7	2.4	3.0	4.1
5.0	2.2	1.2	7.7	1.5

These figures should be thought of as 30 observations on the random variable X. Since scientists estimate that California is hit by thousands of shocks every year, we do not have a value for every member of the population; we are dealing with data from only a portion or sample of the population.

Strictly speaking the population in Example 2 is the set of all earthquakes that have been measured in California. The sample is the collection of 30 earthquakes upon which our measurements are based. However, it is common practice to refer to the 30 observations on X as being a sample. It is usually clear from the context of the problem whether the word *sample* refers to the objects being observed or the numerical data set associated with those objects.

When the data available are population data, any pertinent question can be answered by direct observation. There is no uncertainty concerning the characteristics of the population. However, if the data represent only a sample of observations drawn from the population, statistical methods are needed to get a clue as to the nature of the population.

The first bit of detective work to be done, regardless of the nature of the data set, is to get an idea of the distribution of the random variable being observed. That is, we want to determine where most of the values lie—whether they are widely spread, and whether they tend to form a bell, U, or some other characteristic pattern. Usually this is hard to do by looking at the numbers individually. We need a quick method for grouping or categorizing numbers that gives us an idea of their distribution. One such method is to construct a *stem-and-leaf diagram*. This procedure was introduced recently by Tukey [22] and is one of the primary tools of exploratory data analysis. A stem-and-leaf diagram consists of a series of horizontal rows of numbers. The number used to label a row is called its *stem*, and the remaining numbers in the row are called *leaves*. There are no hard and fast rules about how to construct such a diagram. Roughly, the steps followed are these:

Constructing a Stem-and-Leaf Diagram

1. Choose some convenient numbers to serve as stems. To be useful at least five stems are needed. The stems chosen are usually the first one or two digits of the numbers represented in the data set.

2. Label the rows via the chosen stems.

3. Reproduce the data graphically by recording the digit following the stem as a leaf on the appropriate stem.

4. Turn the graph on its side to get an idea of the way in which the numbers are distributed. In particular, try to answer such questions as these:
   (a) Do these data tend to cluster near a particular stem or stems or do they spread rather evenly across the diagram?
   (b) Do these data tend to taper toward one end or the other of the diagram?
   (c) If a smooth curve is sketched across the top of the diagram, does it form a rough bell? Is it flat? What is its general shape?

A few examples should clarify these ideas.

Example 3

Consider these observations on the random variable X, the magnitude of a California earthquake as measured on the Richter scale:

1.0	8.3	3.1	1.1	5.1
1.2	1.0	4.1	1.1	4.0
2.0	1.9	6.3	1.4	1.3
3.3	2.2	2.3	2.1	2.1
1.4	2.7	2.4	3.0	4.1
5.0	2.2	1.2	7.7	1.5

```
1 |                1 | 02            1 | 02409211435          1 | 02409211435
2 |                2 | 0             2 | 02723411             2 | 02723411
3 |                3 | 3             3 | 310                  3 | 310
4 |                4 |               4 | 101                  4 | 101
5 |                5 |               5 | 01                   5 | 01
6 |                6 |               6 | 3                    6 | 3
7 |                7 |               7 | 7                    7 | 7
8 |                8 |               8 | 3                    8 | 3
```

(a)   (b)   (c)   (d)

Figure 1.1
Stem-and-leaf display for the magnitude of a sample of California earthquakes as measured on the Richter scale: (a) picking the stems, (b) recording the first four data points, (c) the entire data set displayed, (d) getting an idea of the shape of the distribution.

Since the values in this data set lie between 1.0 and 8.3, let us use the digits 1, 2, 3, 4, 5, 6, 7, and 8 as stems for our diagram. We label the rows in the diagram as shown in Figure 1.1(a). We next reproduce the data graphically by recording the number appearing after the decimal point for every observation as a leaf on the appropriate stem. The manner in which this is done for the first four observations is shown in Figure 1.1(b). The stem-and-leaf display for the entire data set is shown in Figure 1.1(c).

To get an idea of the shape of the distribution, turn the book on its side and look at the smooth curve that has been drawn in Figure 1.1(d). We can see at a glance that these data tend to cluster at the lower end of the scale and taper off toward the upper end. This indicates that most of the earthquakes studied were mild and registered on the lower end of the Richter scale. If the sample studied is indicative of the population as a whole, it would be rather unusual to observe a severe earthquake in this region.

Sometimes splitting off the last digit of a number to form a single leaf, as was done in Example 3, leads to too many or too few stems to allow us to get a good idea of the shape of the distribution. When this occurs we must modify our procedure. The problem of too few stems can often be solved by labeling two or more stems the same. This idea is illustrated in Example 4.

Example 4

In a study of growth in males these data are obtained on the random variable X, the circumference in centimeters of a child's head at birth.

33.1	34.6	34.2	36.1	34.2	35.6
34.5	35.8	34.5	34.2	34.3	35.2
33.7	36.0	34.2	34.7	34.6	34.3
33.4	34.9	33.8	33.6	35.2	34.6
33.7	34.8	33.9	34.7	35.1	34.2
33.4	34.1	34.0	35.1	35.3	

33	144		33	144		33	144
33	77		33	77896		33	77896
34			34	122022332		34	122022332
34	56		34	569857766		34	569857766
35			35	12132		35	12132
35	8		35	86		35	86
36			36	01		36	01

     (a)   (b)   (c)

Figure 1.2
Stem-and-leaf display for a sample of observations on the circumference in centimeters of a male child's head at birth: (a) recording leaves on multiple stems, (b) the entire data set displayed, (c) getting an idea of the shape of the distribution.

If we construct a stem-and-leaf diagram for these data using exactly the same procedure as that of Example 3 we would have only four stems labeled 33, 34, 35, and 36. It is hard to get a good idea of the shape of a distribution with only a few stems. To overcome this problem we will label two or more stems the same. For example, since there are quite a few observations that begin with 33, we can split these into two groups by placing leaves 0, 1, 2, 3, and 4 on one row and leaves 5, 6, 7, 8, and 9 on another as shown in Figure 1.2(a). The entire stem-and-leaf display is given in Figure 1.2(b).

Many of the statistical procedures presented later are developed under the assumption that the random variable studied has at least an approximate bell-shaped distribution. The stem-and-leaf diagram is an aid in determining whether or not this assumption is reasonable. For example, it would be rather surprising to hear that the random variable X, the magnitude of a California earthquake as measured on the Richter scale, has a bell-shaped distribution. The stem-and-leaf diagram of Figure 1.1(d) does not look much like a bell! On the other hand, the diagram of Figure 1.2(c) does tend to have a rough bell shape. It would not be too surprising to hear a claim that X, the circumference of a male child's head at birth, has a bell-shaped distribution.

We have illustrated the concept of a stem-and-leaf display for rather well-behaved data sets. The reader is referred to [22] for a full discussion of this topic.

Keep in mind that this graphical method is only a means of getting acquainted with the data. Unless the data are population data, they provide us only with a clue to the actual distribution of the variable being studied.

Exercises 1.1

1. Before a new food is put on the market, manufacturers often run taste tests in selected supermarkets to assess the reaction of the public to the product. That is, shoppers in the market are invited to try the product and express their opinion. Are data gathered in this way sample data or population data?

Suppose that a majority of the shoppers who test the product indicate that they would buy the product if it is marketed. Does this guarantee that a majority of the buying public will purchase the product when it comes on the market? Explain.

2. Before completing tax reports for the year, an inventory of stock is required. To determine the value of the stock on hand a small business is closed for a few days and the price of every item on hand is determined and listed. Are these data sample data or population data? Does the owner of this business need to use statistical methods to determine the total value of the stock?

3. In the United States a census is held every ten years. Questions include such things as age, sex, race, and marital status of the respondent. Some questions on the form are asked of every respondent; others are only asked of a portion of the respondents. Thus, the census is designed to obtain both population and sample data. Do you think that the questions asked of every respondent actually generate population data in the sense that a response is obtained from every individual living in the United States? Explain.

4. These data are obtained on the interest rate for a sample of money-market funds for the previous one-week period:

13.4	14.8	14.7	15.1	14.2
12.8	13.5	12.1	14.3	14.8
15.4	14.5	13.9	16.0	13.8
14.9	15.7	14.1	13.3	14.6

Construct a stem-and-leaf diagram for these data. Use stems of 12, 13, 14, 15, and 16. Would you be surprised to hear a claim that this random variable tends to have a bell-shaped distribution? Explain. Would you be surprised to hear a claim that most of these funds paid interest in the 15%-range nation-wide? Explain.

5. A study of reaction time of motorists is conducted. The purpose of the study is to examine the change in reaction time in seconds of a motorist after having two cans of beer. These differences result: (Subtract in the order of reaction time after drinking the beer minus the initial reaction time.)

0.1	0.6	0.1	0.7	0.5	1.1
0.4	1.6	2.5	3.2	0.9	1.7
0.6	0.3	0.8	1.5	0.5	2.3
2.7	2.4	2.1	1.3	1.5	1.0
1.1	0.2	2.0	3.5	4.0	4.6

Construct a stem-and-leaf diagram for these data. Use stems of 0, 0, 1, 1, 2, 2, 3, 3, 4, and 4 and use the method demonstrated in Example 4. Do these data seem to suggest that the random variable being studied has a bell-shaped distribution? Explain. Would you be surprised to hear a claim that the reaction time of most motorists after drinking two beers is more than two seconds slower than normal? Explain.

6. In a psychological experiment in learning, 40 rats are used. They are randomly split into two groups of 20 each. Each rat is allowed to run a maze and the time required is noted. One group of rats is then taught to run the maze. All 40 rats are then allowed to run the maze and the time required is noted again. The variable studied is X, the difference in run times with subtraction in the order of first minus second trial.

   (a) If learning has taken place, which group would you expect to have the largest average difference?

   (b) The following data result. Construct a stem-and-leaf diagram for each. (Time is in seconds.)

	Trained			Untrained			
4.0	3.2	4.1	4.9	-2.1	-2.2	-1.1	-2.5
4.2	3.7	4.3	4.2	-1.2	2.0	-2.4	-0.6
4.4	3.6	3.5	4.9	1.3	-1.3	-0.2	-2.7
5.1	4.5	4.7	5.0	1.4	0.9	2.2	2.1
5.6	4.6	5.2	5.5	1.8	2.1	1.1	2.6

   (*Hint:* For the trained rats use stems 3, 3, 4, 4, and 5, 5; for the untrained rats use stems of -2, -1, -0, 0, 1, and 2.)

   (c) What is occurring physically when a negative value is obtained for X?

   (d) Would you be surprised if someone claimed that X has a bell-shaped distribution for the trained rats? for the untrained rats?

   (e) Does it appear that the values of X tend to be larger for the trained than for the untrained rats? Explain.

7. A study is conducted to help understand the effect of smoking on sleep patterns. The random variable considered is X, the time in minutes that it takes to fall asleep. Samples of smokers and nonsmokers yield these observations on X:

Nonsmokers

17.2	19.7	18.1	15.1	18.3	17.6
16.2	19.9	19.8	23.6	24.9	20.1
19.8	22.6	20.0	24.1	25.0	21.4
21.2	18.9	22.1	20.6	23.3	20.2
21.1	16.9	23.0	20.1	17.5	21.3
21.8	22.1	21.1	20.5	20.4	20.7
19.5	18.8	19.2	22.4	19.3	17.4

Smokers

15.1	20.5	17.7	21.3	16.0	24.8
16.8	21.2	18.1	22.1	15.9	25.2
22.8	22.4	19.4	25.2	18.3	25.0
25.8	24.1	15.0	24.1	21.6	16.3
24.3	25.7	15.2	18.0	23.8	17.9
23.2	25.1	16.1	17.2	24.9	19.9
15.7	15.3	19.9	23.1	23.0	25.1

(a) Construct a stem-and-leaf diagram for each of these data sets. Use the integers from 15 to 25 inclusive as stems.

(b) Would you be surprised to hear someone claim that there is no difference in the distribution of X for the two groups? Explain.

1.2   Picturing the Distribution: Histograms

In the last section we demonstrated the stem-and-leaf diagram, a relatively new method for getting an idea of where the values of a random variable will lie. This technique is especially useful for data sets that are moderate in size. When the number of observations in a data set is large, or when the values are widely dispersed, the stem-and-leaf diagram may not be very effective. In this section we consider a graphical method for picturing a distribution that has been in use for many years. The graphs produced, called *histograms*, are especially useful for large data sets.

Before we begin, let us mention one of the characteristics that we will be trying to detect graphically. We have already mentioned that it is of interest to determine whether the data indicate that the random variable being studied has a bell-shaped distribution. A bell is a symmetric curve, a curve in which the right and left sides are mirror images of one another. The presence of a bell shape is indicated by a data set whose values tend to cluster about a center point with a rather balanced tapering toward both ends. Some distributions are not symmetrical. Instead, the observed data tend to cluster off-center, creating a long tapering tail toward one end of the distribution. Such a distribution is said to be *skewed*. The direction of the skew coincides with the location of the long tail.

A distribution is said to be **skewed right** or to be **positively skewed** if it has a long tapering tail on the right. If the tail is to the left, the distribution is said to be **skewed left** or to be **negatively skewed**.

These ideas are illustrated in Figure 1.3. The techniques developed in this section can be used to help detect skewness.

To get an idea of the distribution of a random variable over a population, we study its distribution over a sample drawn from the population. A picture of the distribution of a data set should give us an idea of its general shape at a glance. Also it should allow us to make comparisons among categories by visual inspection. This can be done by constructing a graph in which areas are proportional to the number of objects within a given category. The human eye is usually a good judge of relative area when comparing objects with the same basic shape. Thus, we could use circles, triangles, squares or any other geometric figure to picture the distribution. The graphs that we will construct here make use of rectangles or vertical bars. Such a graph is called a *frequency histogram*.

A **frequency histogram** is a bar graph constructed in such a way that the area of each bar is proportional to the number of observations in the category that it represents.

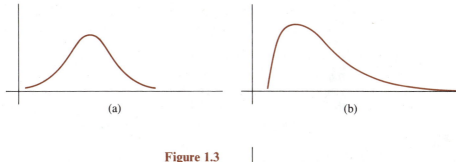

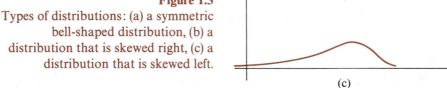

Figure 1.3
Types of distributions: (a) a symmetric
bell-shaped distribution, (b) a
distribution that is skewed right, (c) a
distribution that is skewed left.

To construct a frequency histogram we must first devise a scheme by which the data can be subdivided into categories. These categories are called *classes*. There are no universally accepted rules for doing this. However, we want to define the classes in such a way that each observation clearly falls into exactly one class. Furthermore, when we construct our histogram, each class will be represented by a vertical bar whose width corresponds to the length of the respective class. Since the area of a rectangle is found by multiplying its width by its height, it is convenient to make these widths equal. In this way, the relative area of each bar is reflected by its height alone. We can compare class frequencies by comparing the heights of their respective bars. These rules for creating classes will accomplish these goals.

Rules for Breaking Data into Classes

1. Decide on the number of classes wanted. The number chosen depends on the number of observations available. Usually 5 to 20 classes are desirable.

2. Locate the largest observation and the smallest observation.

3. Find the difference between the largest and smallest observations. Subtract in the order of largest minus smallest. This difference is called the *range* of the data.

4. Find the minimum length required to cover this range by dividing the range by the number of categories desired.

5. The actual class length to be used is found by rounding the minimum length *up* to the same number of decimal places as the data itself.

6. The lower boundary for the first class lies 1/2 unit below the smallest observation.

7. The remaining class boundaries are found by adding the class length to the preceding boundary value until all data points are covered.

These rules sound complicated but they are not! In Example 1 we illustrate their use and demonstrate the construction of a frequency histogram.

Example 1

Administrators of an industrial complex are considering the possibility of changing the pattern of work hours from an eight-hour day, five-day week to a ten-hour day, four-day week. They feel that this change might cut down on absenteeism. In order to help them make this decision, the following data were collected on the number of workers absent per day over a 6-week experimental period:

15	9	15	5	16	16
30	7	12	9	23	15
21	16	17	13	20	18
2	31	11	12	27	22
15	14	10	6	19	14

To break these data into classes we follow the seven-step procedure just outlined.

1. Since we do not have a large number of observations, let us break the data into five classes.

2. The largest observation is 31 and the smallest is 2.

3. The range of the data is $31 - 2 = 29$ units.

4. The classes must cover an interval of length 29. To do so with five classes, the minimum length of each class is given by

    $$\text{range}/5 = 29/5 = 5.8$$

5. The data are reported in whole numbers. Therefore, to find the actual length of each class, we round the minimum length 5.8 *up* to the nearest whole number, 6.

6. The lower boundary for the first class should lie 1/2 unit below the smallest observation. Since the data are integer data, we take 1 as a unit, and the lower boundary for the first class is

    $$2 - 1/2(1) = 2 - .5 = 1.5$$

7. The remaining class boundaries are found by adding 6 successively to the preceding boundary value until all data points are covered. The classes thus determined are as follows:

Class 1:	1.5– 7.5
Class 2:	7.5–13.5
Class 3:	13.5–19.5
Class 4:	19.5–25.5
Class 5:	25.5–31.5

Note that even though these classes do have common boundaries, it is impossible for an observation to fall on a boundary since the boundary

Table 1.1
Frequency distribution of absences

Class	Boundaries	Midpoint	Frequency
1	1.5– 7.5	4.5	4
2	7.5–13.5	10.5	7
3	13.5–19.5	16.5	12
4	19.5–25.5	22.5	4
5	25.5–31.5	28.5	3

values involve *one more decimal place than the data*. Hence no confusion should arise.

The midpoint of each class is needed to construct the frequency histogram. The midpoint, sometimes referred to as the *class mark*, is found by computing the arithmetic average of the class boundaries. For example, the class mark for class 1 is

$$(1.5 + 7.5)/2 = 4.5$$

Successive midpoints are found by adding the class length, 6, to the preceding midpoint. The class marks so obtained are 4.5, 10.5, 16.5, 22.5, and 28.5.

We now count the observations falling into each class. Note that the data points falling into class 1 must exceed 1.5 but be smaller than 7.5. Since the data are all whole numbers, the observations that fall into class 1 are the integers 2, 3, 4, 5, 6, and 7. By inspection of the data, we see that there are 4 such values. We do the same for each class. The results of these counts are recorded in Table 1.1 in the column labeled *frequency*. From this frequency table, we can construct the frequency histogram. The histogram, shown in Figure 1.4, consists of five vertical bars each of width 6, and each centered at its respective class mark. The height of each bar is equal to the class frequency, the *number* of observations per class.

A visual inspection of the graph reveals that the most frequently occurring class is class 3. It should not surprise you to hear that on a given day 14 to 19 workers were absent at this complex. Furthermore, even though the histogram is not perfectly symmetric, it does exhibit a rough bell shape. This is an indication

Figure 1.4
Frequency histogram for a sample of observations on X, the number of workers absent per day at a particular industrial complex

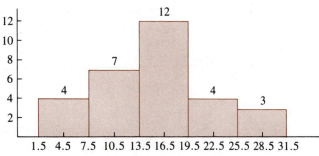

that the random variable X, the number of workers absent per day, would probably have an approximate bell shape if studied over a long period of time.

The histogram shown in Figure 1.4 is a frequency histogram. We read the actual number of observations per class from its vertical axis. Sometimes frequency information alone is misleading. For example, suppose that you are told that five employees of a particular firm were laid off last week. Is this a sign of grave economic trouble? Perhaps—perhaps not! If we are talking about laying off five out of the six employees, then there is definitely a problem. If we are talking about five of 5,000 employees, then there is probably no need to become concerned! To be completely clear in reporting our results, we should report not only the number of observations per class but also the proportion of the total that lies in each class. This proportion is called the *relative frequency*. It is found by dividing the number of observations per class by the total number of observations. When we indicate the relative frequency along the vertical axis, the resulting histogram is called a *relative frequency histogram*.

A **relative frequency histogram** is a bar graph constructed in such a way that the area of each bar is proportional to the fraction of observations in the category that it represents.

The general shape of a relative frequency histogram will be the same as that of the corresponding frequency histogram. This idea is illustrated in Example 2.

Example 2

A consumer group is studying X, the retail price of a particular prescription drug. A sample of 40 drugstores is selected and the price in dollars of a four-ounce bottle of the drug at each store is recorded. The following results are obtained:

2.00	1.98	1.48	2.99	1.20	2.06	1.98	1.20
2.50	3.02	1.75	2.05	1.71	1.10	1.82	1.80
1.75	1.17	2.25	1.90	2.03	1.89	2.15	1.96
1.87	.95	2.20	1.85	1.70	1.92	3.00	2.40
2.03	2.75	2.69	2.03	1.90	1.72	2.60	2.11

To divide these data into classes that share only a common boundary, let us use five classes, since we are still working with a small data set. Because the largest observation is 3.02 and the smallest is .95, we must cover an interval of length $3.02 - .95 = 2.07$ units, and each class must be at least $2.07/5 = .414$ units long. Rounding *up* to the same number of decimal places as the data means we use .42 as the actual class length. Since the data are reported to the nearest 1/100, we take 1/100 as a unit and hence the lower boundary for the first class is

$$.95 - 1/2(1/100) = .95 - .005 = .945$$

The remaining class boundaries are found by adding .42 successively to the preceding boundary value until all data points are covered. The classes obtained are as follows:

Table 1.2
Relative frequency distribution of drug prices

Class	Boundaries	Midpoint	Frequency	Relative Frequency
1	.945–1.365	1.155	5	5/40 = .125
2	1.365–1.785	1.575	6	6/40 = .150
3	1.785–2.205	1.995	20	20/40 = .500
4	2.205–2.625	2.415	4	4/40 = .100
5	2.625–3.045	2.835	5	5/40 = .125

Class 1:   .945–1.365

Class 2:   1.365–1.785

Class 3:   1.785–2.205

Class 4:   2.205–2.625

Class 5:   2.625–3.045

The relative frequency distribution for the data is given in Table 1.2. From this table, we can construct a relative frequency histogram by constructing bars of width .42 with each being centered at the class mark. The histogram is shown in Figure 1.5. Note that the vertical axis is labeled 0 to 1.0. This is sufficient since the fraction or proportion of observations per class will always lie between 0 and 1.0.

Figure 1.5
Relative frequency histogram for a sample of observations on X, the retail price of a particular prescription drug.

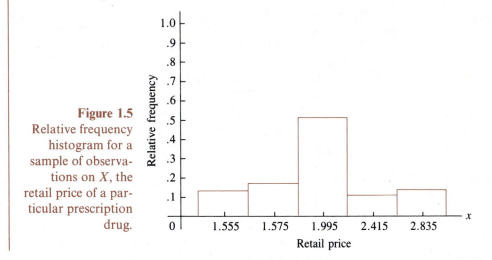

Exercises 1.2

8. A data set contains only whole number data. The largest observation is 25 and the smallest is 3. We want to divide the data into five classes.

(a) Find the range of the data.
(b) Find the minimum length needed to cover the range.
(c) Find the actual class length to be used in categorizing the data.
(d) Find the class boundaries.
(e) Find the class midpoints.

9. A data set contains data reported to the nearest tenth. The largest observation is 17.6 and the smallest is 14.1. We want to divide the data into six classes.
(a) Find the range of the data.
(b) Find the minimum length required to cover the range.
(c) Find the actual class length to be used in categorizing the data. Remember that you round the minimum length up to the nearest tenth.
(d) Find the class boundaries. Remember that here a unit is $1/10$ and half a unit is $(1/2)(1/10) = 1/20 = .05$.
(e) Find the class midpoints.

10. A data set contains data reported to the nearest hundreth. The largest observation is 59.37 and the smallest is 35.46. We want to divide the data into ten classes.
(a) Find the range of the data.
(b) Find the minimum length needed to cover the range.
(c) Find the actual class length to be used in categorizing the data.
(d) Find the class boundaries.
(e) Find the class midpoints.

11. Construct a stem-and-leaf diagram for the data of Example 1. Use stems of $0,0,$ $1,1,$ $2,2,$ and $3,3.$ Thus, for example, $0|2$ denotes 2 and $1|5$ denotes 15. Does the stem-and-leaf diagram convey the same impression of the shape of the distribution as does the histogram of Figure 1.4? Can you think of any advantages of a stem-and-leaf diagram over a histogram? Of a histogram over a stem-and-leaf diagram?

12. To decide whether or not to stay open past 5:00 P.M., a business manager collects these data on the number of customers served after 5:00 on 36 randomly selected working days.

1	3	6	2	4	6	2	1	4	6	6	1
4	9	2	3	1	5	6	2	6	5	7	11
8	4	10	12	12	15	7	5	8	7	14	4

(a) Construct a frequency distribution table using five classes.
(b) Construct a frequency histogram.
(c) The distribution appears to be skewed. In what direction is the skew?
(d) Based on the histogram, do you think that it would be unusual for the store to have fewer than four customers on a given night? Explain.
(e) Based on the direction of the skew, does it appear that the after 5:00 P.M. hours are very popular?

13. A physician is interested in studying scheduling procedures. She questions 60 patients concerning the length of time, in minutes, that they waited past their scheduled appointment time. The following observations are obtained:

60	29	34	25	31	30	−1	17	6	50
10	18	38	25	35	36	31	23	12	52
8	27	27	30	42	9	47	31	27	6
45	23	25	37	3	50	53	28	16	19
32	36	9	33	36	33	58	26	18	32
12	32	8	16	48	36	46	−5	31	59

(a) Construct a frequency distribution table using seven classes.
(b) Construct a frequency histogram.
(c) What is occurring physically when a negative value is obtained for this variable?
(d) Based on the shape of the histogram, would you feel justified in accepting the contention that this variable has a bell-shaped distribution?
(e) Based on the histogram, would you be surprised if the physician claimed that it is unusual for her patients to have to wait more than five minutes to be seen? Explain.

14. These sample data are obtained on the variable X, the length of time in hours that a jury deliberates in a criminal case involving the death penalty.

15.5	53.7	11.1	17.4	54.0
51.3	12.3	61.2	26.2	43.2
58.0	37.9	62.8	64.8	57.9
42.1	52.8	60.3	32.7	52.8
29.2	44.0	63.6	49.3	25.6
21.0	67.9	50.8	41.7	39.5
61.2	66.3	23.1	42.3	58.5
46.4	68.7	33.6	55.4	65.3
62.5	69.2	51.2	40.2	26.2
20.7	63.1	31.4	56.9	62.1

(a) Construct a frequency histogram for these data using six classes.
(b) Would you be surprised if someone claimed that this variable has a bell-shaped distribution? Explain.
(c) Based on the frequency histogram, do you think that it would be unusual for a jury to deliberate for more than 59.5 hours? Explain.

15. It is reported that during the month of August of last year the average maximum price of a gallon of unleaded gasoline across the United States was $1.35 per gallon. For publicity purposes a company wishes to claim that its average maximum was lower than this. To support its claim, company statisticians gathered these data on a sample of stations randomly selected from its chain.

1.22	1.37	1.27	1.20	1.42	1.41	1.22	1.24
1.28	1.42	1.48	1.32	1.40	1.26	1.39	1.45
1.44	1.49	1.47	1.47	1.24	1.34	1.27	1.35
1.34	1.45	1.49	1.45	1.23	1.20	1.42	1.34
1.43	1.21	1.49	1.36	1.24	1.20	1.45	
1.23	1.25	1.24	1.35	1.23	1.39	1.38	
1.46	1.48	1.26	1.36	1.22	1.46	1.39	
1.22	1.29	1.47	1.24	1.35	1.21	1.21	

(a) Construct a stem-and-leaf diagram for these data. Use as stems 1.2, 1.2, 1.3, 1.3, 1.4, and 1.4.

(b) Construct a relative frequency histogram using seven classes.

(c) Based on these graphs, do you think that it would have been unusual to have driven into one of its stations to have found that the price of a gallon of unleaded regular gasoline was more than \$1.40? Explain.

16. A highway study group is investigating the driving habits on Interstate I-81. The data obtained concerning the speed, in miles per hour, of 50 randomly selected cars and trucks observed is as follows:

39.1	35.0	48.0	47.5	40.0	42.0	45.0	51.0	67.2
55.1	52.0	56.0	51.1	49.7	45.0	44.7	46.8	68.3
52.1	54.2	55.0	54.1	50.0	52.3	53.5	49.9	
55.1	55.8	67.2	65.0	64.9	59.0	58.3	80.0	
55.0	62.3	74.1	75.0	65.0	66.0	61.0	58.0	
50.0	49.5	43.0	48.0	53.0	58.2	57.2	66.0	

(a) Construct a relative frequency histogram using six classes.

(b) Is there evidence, based upon the histogram, that the speed of the majority of the vehicles on this highway is close to the 55 mile per hour limit? Explain.

17. These observations are obtained on the random variable X, the life span in hours of the lithium batteries used in pocket calculators:

4285	564	1278	205	1850	2066	604	209
602	1379	2584	14	349	3570	99	1009
2300	478	726	510	318	737	3032	2000
582	1429	852	1000	2662	308	981	1560
701	497	3367	1402	1786	1100	35	99
1137	520	261	2778	373	414	396	83
1379	454						

(a) Construct a relative frequency histogram using seven classes.

(b) Based on the histogram, do you think that the random variable X has a skewed distribution? If so, in which direction is the skew?

(c) Do you think that it would be wise to guarantee that these batteries will last for more than 1846 hours? Explain.

1.3 Measures of Location

Recall that a population parameter is a descriptive measure associated with a random variable when it is studied over the entire population. Three parameters that measure the *center* of the distribution in some sense are of interest. These parameters, called *location parameters* or *measures of central tendency*, are the *population mean*, the *population median*, and the *population mode*. Although it is a little too soon to give you a precise technical definition of these terms, we will give you an idea of their meaning and will show you how to approximate the value of each from sample data.

To begin, imagine a large finite population together with some random variable X. The population mean, denoted by the Greek letter μ (mu), is the average value that we would get if we observed and recorded the numerical value of X for *each member of the population* and then averaged these values. For example, suppose that our population consists of every apartment for rent in New York City on a particular day, and that X denotes the monthly rent for the apartment. The population mean, μ, is the average monthly rent for these apartments. To find this mean, we would have to determine and record the monthly rent for every apartment offered and then average these values. An impossible task!

Since we usually do not have population data available, it is seldom possible to determine the exact value of μ. We must attempt to approximate its value based on sample data. We hope that the behavior of X over the parent population is reflected in the sample. Hence, we study the characteristics of the sample data in an attempt to get a feeling for the characteristics of the population. In particular, to get an idea of the average value of X over the population, we find its average value for the sample. This statistic is called the *sample mean*.

Definition 1.1

Let $x_1, x_2, x_3, \ldots, x_n$ be n observations on a random variable X. The **sample mean**, denoted by $\bar{x}$, is the arithmetic average of these values. That is,

$$\bar{x} = (x_1 + x_2 + \cdots + x_n)/n$$

The symbol $\bar{x}$ is pronounced "x bar."

Example 1

These observations are obtained on the random variable X, the monthly rent for an apartment in New York City on a particular day.

$$x_1 = \$125, \quad x_2 = \$300, \quad x_3 = \$400$$
$$x_4 = \$800, \quad x_5 = \$350$$

The sample mean is given by

$$\bar{x} = (125 + 300 + 400 + 800 + 350)/5$$

$$= 1975/5 = \$395.00$$

Note that we are *not* saying that μ, the average rent for an apartment in New York City, is \$395.00. We do not have information on every apartment for rent in the city! We are saying that $\bar{x}$, the average rent for the apartments in our sample, is \$395.00. If the apartments in our sample reflect those in the population, then $\bar{x}$ gives us a fairly good idea of the true value of μ.

Before proceeding, let us pause to introduce a notational shorthand that is used extensively in the field of statistics. In particular, note that it is awkward to have to list all the terms of a sum as we did in Example 1. We need a symbol that can be translated as add. The symbol used is Σ, the capital Greek letter sigma. Using this notation, the sample mean in general is given by

$$\bar{x} = (x_1 + x_2 + \cdots + x_n)/n = \Sigma x/n$$

The use of this notation is illustrated in the next example.

Example 2

In an experiment in learning, each subject is given two minutes to learn a random word list. The subject is then asked to list as many of the words as he or she can remember. The random variable studied is X, the number of words remembered. A sample of 10 subjects yields these data:

$$x_1 = 8 \qquad x_2 = 2 \qquad x_3 = 4 \qquad x_4 = 9 \qquad x_5 = 7$$
$$x_6 = 2 \qquad x_7 = 12 \qquad x_8 = 5 \qquad x_9 = 5 \qquad x_{10} = 7$$

The mean of this sample is given by

$$\bar{x} = \Sigma x/n$$

$$= (8 + 2 + 4 + 9 + 7 + 2 + 12 + 5 + 5 + 7)/10$$

$$= 61/10 = 6.1$$

Note that the sample mean is not necessarily equal to any one of the sample values.

Occasionally it is necessary to combine two or more sample means in a logical way. Probably the first thing that comes to mind is to average the averages. The danger of doing this without careful thought is illustrated in the next example.

Example 3

Suppose that a hospital administrator wants to approximate the average number of cases treated in the emergency room per night so that this information can be used in staff planning. A sample of five nights yields a sample mean of $\bar{x}_1 = 3$; a more extensive study based on a 100-night study period yields $\bar{x}_2 = 15$. If we average these values we get $(3 + 15)/2 = 9$. A little thought will show that this procedure does not make much sense. Note that the sample sizes are $n_1 = 5$ and $n_2 = 100$. The second sample is 20-times larger than the first! However, by averaging the two sample means we ignored this difference. Each sample mean was given the same importance or weight. To overcome this problem, let us compute what is called a *weighted mean*. This procedure adjusts for differences in sample sizes by multiplying each sample mean by its respective sample size and then dividing by the total number of observations available. In this case, the weighted mean is given by

$$\frac{n_1 \bar{x}_1 + n_2 \bar{x}_2}{n_1 + n_2} = \frac{5(3) + 100(15)}{5 + 100}$$

$$= 1515/105 = 14.43$$

This value is quite different from that obtained by ignoring sample sizes and just averaging $\bar{x}_1$ and $\bar{x}_2$.

It is easy to extend the idea presented in Example 3 to more than two samples. To do so, let $\bar{x}_1, \bar{x}_2, \bar{x}_3, \ldots, \bar{x}_k$ denote the means of k samples of sizes n_1, $n_2, n_3, \ldots, n_k$, respectively. The weighted mean is given by

$$\bar{x}_{\text{wt}} = \frac{n_1 \bar{x}_1 + n_2 \bar{x}_2 + n_3 \bar{x}_3 + \cdots + n_k \bar{x}_k}{n_1 + n_2 + n_3 + \cdots + n_k}$$

Since we are weighting by size, the effect of this is to combine or pool the k samples and then find the usual sample mean for the combined sample.

The population median is another useful measure of central tendency. It is essentially a half-way point. It is a number with the property that roughly half of the observations fall on or below the value while the others fall on or above it. For example, the median rent for apartments available for rent in New York City on a particular day is that rent such that half the apartments cost this value or less, while the rest are as expensive or more so. To approximate the half-way point of a population, we find the half-way point of a sample drawn from the population. That is, we find the *sample median*.

Definition 1.2

Let $x_1, x_2, x_3, \ldots, x_n$ be a sample of observations arranged in the order of smallest to largest. The **sample median** for this collection is given by the middle observation if n is odd. If n is even, the sample median is the average of the two middle observations.

Example 4

A manager of a clothing store wants to approximate the median age of both his male and female customers so that he can take this into account when ordering stock. Since he cannot survey every customer who will ever enter his store he must approximate these values from sample data. These data are obtained (age is in years):

Female (x)			Male (y)	
15	42	35	17	40
17	60	30	29	72
24	20	12	37	
27				

To find the median for each sample, we first arrange the data in order from smallest to largest.

Female (x)		Male (y)
12	$27 \leftarrow x_6$	17
15	30	29
17	35	$37 \leftarrow y_3$
20	42	40
$x_5 \rightarrow 24$	60	72

Since the number of observations for females is even, namely, $n = 10$, the sample median is the average of the two middle observations. That is,

$$\text{sample median} = (x_5 + x_6)/2$$

$$= (24 + 27)/2$$

$$= 25.5 \text{ years}$$

Since the number of observations for males is odd, namely, $n = 5$, the sample median is the middle observation. That is,

$$\text{sample median} = y_3 = 37 \text{ years}$$

If the data set is small, it is easy to arrange it in order and locate the sample median by inspection. However, for large data sets it is convenient to have a rule by which we can determine the position of the sample median in our ordered list. This rule is given by

$$\text{position of the sample median} = (n + 1)/2$$

Applying this rule to our previous example, we see that the position of the sample median for the data for female customers is

$$(n + 1)/2 = (10 + 1)/2 = 11/2 = 5.5$$

We take this to mean the average of the fifth and sixth observations, as before. The position of the sample median for the data for male customers is

$$(n + 1)/2 = (5 + 1)/2 = 6/2 = 3$$

The sample median is taken to be the third observation in our ordered list. If we had a data set of size $n = 728$, we could immediately conclude that the sample median is in position

$$(n + 1)/2 = (728 + 1)/2 = 729/2 = 364.5$$

That is, the sample median would be the average of the 364th and 365th observations in the ordered list.

The next example shows that in some cases the sample median gives a better description of the center of location than does the sample mean.

Example 5

Consider the following set of observations representing the market value of 10 houses on a block:

$32,000	$41,000	$28,500	$36,000	$30,500
$35,000	$32,500	$30,000	$27,000	$400,000

What kind of neighborhood is this? If we attempt to answer this question by considering the "average" value of a house on the block, we will probably conclude that the neighborhood is fairly well-off, since the mean for this sample is $69,250. This figure is obviously misleading, and the confusion is caused by the one figure $400,000, which is totally out of line with the others. Such a figure is called an *outlier*. If we answer the question by considering the median of the sample, we will get a more accurate picture, since this value is $32,250. The presence of the outlier does not seriously affect the sample median. The neighborhood is probably a more moderate neighborhood. We do not intend to imply that the sample mean is not to be trusted, but there is more involved here than meets the eye. That extra ingredient is variability, which will be discussed in the next section.

The last measure of central tendency is the **mode**. For a finite population, the population mode is the value of X that occurs most often. For instance, the mode in our rent example is the rent most frequently charged across the city. The mode of a sample is the value that occurs most often in the sample. The drawback to this measure is that there might not be a unique mode. There might be no single number that occurs more often than any other. For this reason, the mode is not a particularly useful descriptive measure.

Before closing this section let us remark that we have used small samples to illustrate these basic concepts. We do not intend to imply that samples as small as those presented are common in statistical studies. Most studies involve very large populations. Getting an accurate picture of the population characteristics usually requires extensive data collection and a major investment in time and effort.

Exercises 1.3

18. Consider these sets of observations on the random variables X and Y:

	x			y	
2	1	7	2	0	3
3	5	9	8	1	3

(a) Find Σx and Σy.

(b) Later we will need to square each observation in a data set and then add these squares. That is, we will need to evaluate sums of the form

$$x_1^2 + x_2^2 + \cdots + x_n^2 = \Sigma x^2$$

Use the above data sets to find Σx^2 and Σy^2.

(c) Find $\bar{x}$ and $\bar{y}$.

(d) Find the sample median for each data set.

(e) Does either data set have a unique mode? If so, what is its value?

19. Find the position of the sample median for data sets of the following sizes:

(a) $n = 710$

(b) $n = 813$

(c) $n = 1051$

(d) $n = 2000$

20. Find the sample mean and the sample median for each of these data sets:

	I				II			
2	1	4	8	2	1	0	9
5	10	12	6	4	3	2	7
7	3	0					

21. Find the median for the sample of Example 1. Is it safe to say that the median rent for all apartments available for rent in New York City on the day in question is $350? Explain.

22. A community is interested in beginning adult education classes. To determine the need, samples of 20 women and 15 men are selected. The number of years of formal education for each individual is found. The following data sets result:

	Women				*Men*	
8	8	12	14	8	16	7
10	18	12	7	12	16	9
16	18	14	8	12	16	20
12	12	16	12	14	16	20
7	12	16	20	14	12	12

(a) Compute the sample means for the women and men. Denote these by $\bar{x}_1$ and $\bar{x}_2$, respectively.

(b) Find the average value of $\bar{x}_1$ and $\bar{x}_2$. That is, find $(\bar{x}_1 + \bar{x}_2)/2$.
(c) Find the weighted mean for $\bar{x}_1$ and $\bar{x}_2$.

$$\text{Is } \bar{x}_{wt} = (\bar{x}_1 + \bar{x}_2)/2?$$

(d) Under what circumstances would you expect $\bar{x}_{wt}$ to equal $(\bar{x}_1 + \bar{x}_2)/2$?
(e) Is the value of $\bar{x}_1$ obtained necessarily exactly equal to the average number of years of formal education for the women in the community from which the sample was drawn? Explain.
(f) Would you be surprised if someone claimed that the average number of years of formal education for the entire community is more than 18 years? Explain.
(g) Find the sample median for each of the data sets.

23. A business manager is interested in learning more about the clientele. For two days an informal survey on the amount of money spent by each individual is conducted. These data are obtained:

I Female, 25 and over		II Female, under 25		III Male, 25 and over		IV Male, under 25	
10.98	29.80	5.98	35.00	90.00	16.25	10.02	18.35
12.03	53.00	2.03	36.20	75.00	18.21	17.98	6.95
27.00	52.00	8.02	17.25	5.98	37.50	22.50	12.03
26.50					110.00	7.32	

(a) Compute the sample mean for each data set; call these means $\bar{x}_1, \bar{x}_2, \bar{x}_3$ and $\bar{x}_4$.
(b) Compute the weighted mean for the purchases made by female customers. Is the weighted mean equal to the arithmetic average of the sample means for data sets I and II?
(c) Compute the weighted mean for purchases made by male customers. Is the weighted mean equal to the arithmetic average of the sample means for data sets III and IV? If so, why does this happen?
(d) Find the overall mean for all purchases by combining the four data sets into one and finding the sample mean for the new data set.
(e) Find the weighted mean for the four data sets and compare your answer to that of part (d).
(f) The manager thinks that to stay in business the average sale must exceed $25. Does it appear from the data that this business will succeed? Do these data guarantee that the average sale is such that the business will succeed? Explain.
(g) Find the overall median for all purchases by combining the four data sets.

24. These data are obtained on the sales per day in thousands of dollars at two franchises of the same hamburger chain:

	Franchise A				Franchise B		
0.9	2.5	1.4	2.6	0.9	0.7	4.5	0.6
1.7	3.2	3.6	1.3	0.9	4.6	0.3	5.3
2.9	3.8	2.7	5.9	1.4	3.1	5.2	4.2
3.5	0.7	4.8	4.7	2.5	5.1	1.7	5.0

(a) Find the sample mean for each franchise. Based on these statistics, does it appear that the average sales per day for franchise A is about the same as that for franchise B? Explain.

(b) Find the sample median for each franchise. Based on these statistics, does it appear that the median sales per day for franchise A is about the same as that for franchise B? Explain.

(c) Is it safe to say that the distribution of sales per day is the same for these two franchises? Be careful! Look at the stem-and-leaf diagrams for the two data sets.

25. A sociologist is studying the cultural habits of two primitive societies. One random variable under consideration is the age of the woman at the time of her first marriage. These data are obtained:

	Society A				Society B			
14	22	15	14	14	14	15	22	19	26
20	14	15	16	17	21	25	26	27	32
25	24	24	18	14	25	18	30	26	35
21	13	14	35	16	27	24	18	31	20
26	19	30	13		20				

(a) Find the sample mean for each data set. Based on these statistics, would you be surprised to hear a claim that the average age of the woman at the time of her first marriage is the same in society A as it is in society B? Explain.

(b) Find the sample median for each data set. Does it appear that the median age of the woman at the time of her first marriage is the same in these two societies? Explain.

(c) Construct a stem-and-leaf diagram for these data sets. Does either of these diagrams suggest a skewed distribution? If so, in what direction is the skew?

1.4 Measures of Variability

Recall that, in a statistical study, we are attempting to describe the behavior of a random variable X over a population by observing the way it behaves over a sample drawn from that population. As we saw in the last section, we can use the sample mean and sample median to get an idea of the center of location of X. However, X is a random variable. We expect its value to vary from one member of the population to another. Three parameters are used to describe variability within a population. These are the *population range*, *population variance*, and *population standard deviation*. The population range is the difference between the largest and smallest values of X when considered over the entire population. The population variance is a measure of the variability of X about its mean, μ. This parameter, denoted by σ^2, has the property that its value is large if X tends to assume values far from its mean; its value is small if the values of X tend to cluster near μ. The population standard deviation, denoted by σ, is the square root of its variance. Although it is too soon to give you a technical definition of the term variance, you will see what this parameter measures from the examples presented.

We approximate the population range by considering the *sample range*. This statistic is defined in a logical way.

Definition 1.3

> The **sample range** is the difference between the largest and smallest sample values. Subtraction is in the order largest minus smallest.

Note that this is the same statistic that we used earlier when breaking a data set into classes. Although the sample range is easy to compute, it is of limited value in detecting variability. To see that this is true, consider the following example.

Example 1

To compare the grades of her spring semester students with those of her fall semester students, an instructor draws a sample from each of these populations and reports the following results for the scores made on the final exam:

Spring (x)	*Fall (y)*
sample size: 23	sample size: 26
average grade: $\bar{x} = 75$	average grade: $\bar{y} = 75$
median grade: 75	median grade: 75
range: 50 with grades from 50 to 100	range: 50 with grades from 50 to 100

It appears that the final exam scores are virtually identical. However, the instructor further reveals the grade distributions for the two groups shown in Figure 1.6.

50–50–50–50–50–50	50
65–65–65	65–65
70–70	70–70–70
75	74–74–74–74
80–80	75–75–75–75–75–75
85–85–85	76–76–76–76
100–100–100–100–100–100	80–80–80
	85–85
	100

Figure 1.6
Distribution of grades for
two different classes.

There is obviously a marked difference in the dispersion of scores that we could not detect from the sample range. What is this difference, and how can it be detected? Notice that during the fall a majority of the grades clustered near the mean grade of 75, while during the spring a substantial number of students received grades that were quite far from 75. In general terms, the spring semester grades deviated more on the average from the mean score of 75 than those recorded in the fall. This difference in the grade distribution was not detected by the sample range, the sample mean, or the sample median.

The characteristic that is not being reflected by the sample statistics that we have studied thus far is the variance of the random variable X. To get an idea of σ^2, the manner in which X varies about μ in the population, we must look at how it varies about $\bar{x}$ in the sample. That is, we must define the *sample variance*. Perhaps an obvious way to define this statistic is to consider the average of the differences between the observations and the sample mean. That is, we propose

$$\Sigma(x - \bar{x})/n$$

as a measure of the variance within the sample. Does this measure the variability as desired? A numerical example will show that it does not and will also suggest a way to remedy the situation.

Example 2

The weather bureau reported temperatures in degrees Celsius for a one-week period in a southern town:

$$2 \quad 5 \quad 10 \quad -10 \quad 8 \quad 0 \quad 20$$

The mean for these data is $\bar{x} = 5$. Is there variability in the data? By inspection the answer is a resounding yes! However, if we compute the value of the proposed measure of variability we get

$$\Sigma(x - \bar{x})/7 = [(2 - 5) + (5 - 5) + (10 - 5) + (-10 - 5) + (8 - 5)$$
$$+ (0 - 5) + (20 - 5)]/7$$
$$= (-3 + 0 + 5 - 15 + 3 - 5 + 15)/7$$
$$= 0$$

There is definitely a problem! The proposed measure indicates that there is no variability in the data when this is obviously not the case.

The problem with the proposed measure is easy to spot from Example 2. We have allowed positive differences to cancel negative ones by addition. One way to overcome this problem is to square each difference before averaging. This yields

$$\Sigma(x - \bar{x})^2/n$$

as our measure of variability. Note that if a given observation x lies close to $\bar{x}$, then both $(x - \bar{x})$ and $(x - \bar{x})^2$ are small. Thus, data sets in which most of the observations lie close to $\bar{x}$ will yield a small value for this measure; those with a substantial number of observations falling far from $\bar{x}$ will yield a large value. This is what we want! However, there is a problem. It can be shown that this statistic tends to underestimate σ^2, the variance of the population from which the sample is drawn. To adjust for this, we need to decrease the size of the denominator slightly. We do so by dividing $\Sigma(x - \bar{x})^2$ by $(n - 1)$ rather than n. This results in the following definition for the term *sample variance*.

Definition 1.4

Let $x_1, x_2, x_3, \ldots, x_n$ be a set of n observations. The **sample variance** for these observations, denoted by s^2, is given by

$$s^2 = \Sigma(x - \bar{x})^2/(n - 1)$$

Example 3

These data are obtained on the length (in minutes) of a long-distance telephone call made after 11:00 P.M. and before 8:00 A.M.

| 10 | 20 | 6 | 12 | 15 | 8 | 4 | 9 | 3 | 13 |

For these data $\bar{x} = 10$ minutes. The sample variance is given by

$$s^2 = \Sigma(x - \bar{x})^2/9 = [(10 - 10)^2 + (20 - 10)^2 + \cdots + (13 - 10)^2]/9$$

$$= 244/9 = 27.1$$

The computation required to determine the value of s^2 from the definition is rather time-consuming, although not difficult. Most pocket calculators now on the market are able to compute the value of $\bar{x}$ and s^2 for you. All you have to do is enter the data! If you don't own such a calculator, the following formula will simplify the calculations needed to find s^2.

Computational formula for s^2

$$s^2 = \frac{n\Sigma x^2 - (\Sigma x)^2}{n(n - 1)}$$

To illustrate the use of this computational shortcut, let us reconsider the data of Example 3.

Example 4

To compute s^2 for the data of Example 3 using the computational shortcut, we need only to compute Σx and Σx^2. These are given by

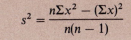

$$\Sigma x = x_1 + x_2 + x_3 + \cdots + x_{10}$$

$$= 10 + 20 + 6 + 12 + 15 + 8 + 4 + 9 + 3 + 13$$

$$= 100$$

$$\Sigma x^2 = x_1^2 + x_2^2 + x_3^2 + \cdots + x_{10}^2$$

$$= 10^2 + 20^2 + 6^2 + \cdots + 9^2 + 3^2 + 13^2$$

$$= 1244$$

Thus

$$s^2 = \frac{n\Sigma x^2 - (\Sigma x)^2}{n(n-1)}$$

$$= \frac{10(1244) - 100^2}{10(9)}$$

$$= 27.1$$

As expected, this value agrees with that obtained directly from the definition of s^2.

The last measure of variability we consider is the sample standard deviation, s, defined as the nonnegative square root of the sample variance. That is

Definition 1.5   Let $x_1, x_2, \ldots, x_n$ be a sample of n observations. The **sample standard deviation**, denoted by s, is given by

$$s = \sqrt{s^2}$$

To compute s, we first compute s^2 and then find its square root as illustrated in Example 5.

Example 5   To find the sample standard deviation of the temperatures

$$2° \quad 5° \quad 8° \quad 0° \quad 10° \quad 20° \quad -10°$$

we first find s^2. For these data

$$\Sigma x = 2 + 5 + 8 + 0 + 10 + 20 + (-10) = 35$$

$$\Sigma x^2 = 2^2 + 5^2 + 8^2 + 0^2 + 10^2 + 20^2 + (-10)^2 = 693$$

Thus

$$s^2 = \frac{n\Sigma x^2 - (\Sigma x)^2}{n(n-1)}$$

$$= \frac{7(693) - (35)^2}{7(6)}$$

$$= 86.33$$

and

$$s = \sqrt{86.33} = 9.29 \text{ degrees Celsius}$$

Since a large sample variance will imply a large sample standard deviation, the natural question to ask is, Why bother with both of them? We will see some mathematical reasons for wanting to consider both s^2 and s in Chapter 6. However, there is a practical reason for considering both that is easy to see by reviewing our previous example. When computing s^2, we looked at terms of the form $(x - \bar{x})^2$ where x and $\bar{x}$ are in degrees Celsius. We find ourselves working with a rather peculiar object called a *squared degree*; that is, the sample variance is 86.33 squared degrees. What, physically speaking, is a squared degree? Nothing, really! It is a physically meaningless unit. However, the sample standard deviation is

$$s = \sqrt{s^2} = \sqrt{86.33 \text{ (degrees)}^2} = 9.29 \text{ degrees Celsius}$$

and we are again talking about degrees, meaningful physical units. An advantage of the sample standard deviation over the sample variance as a measure of dispersion is that it involves the same physical units as the data.

We close this section with a warning. Be careful when you see the term *sample variance* used in other books. Some authors prefer division by n; others prefer division by $(n - 1)$ as we do. Also, if you are using a pocket calculator to find the variance of a data set, be sure to read your user's manual. Some calculators divide $\Sigma(x - \bar{x})^2$ by n, whereas others divide it by $(n - 1)$. The reason for this is simple. If the data set that you are analyzing is population data, then division by n yields the *exact* value of σ^2. If the data set represents data for only a sample drawn from a late population, then division by n yields an *approximate* value for σ^2 that tends on the average to be a bit too small. To adjust for this, we divide by $(n - 1)$ when dealing with sample data. Your calculator is programmed to do either calculation but it is up to you to decide which divisor is appropriate. Beware!

Exercises 1.4

26. Drug abuse takes many forms. There is some concern that doctors tend to overprescribe drugs for older persons in this country. A survey of persons over the age of 65 living in a particular district is conducted by a public health worker. The random variable studied is X, the number of drugs that have been prescribed for each individual during the last six months. These observations result from a sample of size 10.

    0   3   8   1   4   6   9   1   0   8

    (a) Find the mean and median for this sample.
    (b) Find the sample range.
    (c) Find s^2 using Definition 1.4.

(d) Find Σx^2. Use this to find s^2 using the computational formula. Did you get the same answer as you got in part (c)?

(e) Find the sample standard deviation. What physical unit is associated with s?

27. Consider these observations on the random variable X, the number of jobs available on a daily basis at a local employment center.

$$5 \quad 2 \quad 7 \quad 3 \quad 6 \quad 8 \quad 0 \quad 1 \quad 2 \quad 4 \quad 2$$

(a) For these data find Σx and $\bar{x}$.

(b) Use Definition 1.4 to find s^2.

(c) Find Σx^2 and then find s^2 using the computational formula. Compare your answer to that of part (b).

(d) Find the sample standard deviation. What unit is associated with s?

28. Many people think that alcohol is a stimulant. It is in fact a depressant. At a concentration of 0.1 percent most sensory and motor functions become impaired. In a study of alcohol consumption subjects are allowed to drink three bottles of beer. Then their blood alcohol concentration is determined. These data result:

.10	.09	.12	.08	.09
.08	.09	.09	.10	.06
.11	.08	.07	.12	.13

(a) Find the mean and median for this sample.

(b) Find the sample range.

(c) Find Σx and Σx^2

(d) Find the sample variance and the sample standard deviation.

29. The following data concern the length of active sentence, in years, received by a group of 20 men and a group of 15 women, each convicted of a similar crime.

Men (x)				Women (y)		
5.5	6.0	3.0	1.0	1.0	1.5	2.0
1.5	1.5	2.5	2.5	0.5	0.0	1.7
4.0	1.7	2.5	7.0	2.0	3.0	1.5
2.0	2.5	3.5	6.5	2.5	1.5	1.8
0.0	2.5	4.5	5.5	0.0	0.5	1.0

(a) Construct a stem-and-leaf diagram for each data set.

(b) Based on the charts of part (a), which data set do you think has the larger mean? the larger median? the larger variance? the larger standard deviation? Verify your answers by finding the sample mean, sample median, sample variance, and sample standard deviation for each group.

(c) Do you think that these data tend to support the statement that sentencing for men is different than for women for this particular crime? Explain.

30. It is often the intent of nurses and doctors to inject a drug into the gluteus muscle of the patient. The standard needles used to do so are 1.4 inches long.

A study was recently conducted to determine whether or not the drug was actually being injected into the muscle as desired. These data were obtained based on samples of 100 men and 100 women in the 30- to 40-year age group. The random variable is the thickness, in inches, of the fat layer covering the gluteus muscle.

Men (x)	Women (y)
$\Sigma x = 150$	$\Sigma y = 260$
$\Sigma x^2 = 249$	$\Sigma y^2 = 775$

(a) Find $\bar{x}$ and $\bar{y}$. Considering the length of the needle being used for the injections, does there appear to be a problem?
(b) Find the sample variance and standard deviation for each group. What physical unit is associated with the sample standard deviation?
(c) Judging from the sample results, which group seems to have the greater variability in the thickness in the fat layer covering the gluteus muscle?

31. Based on the data of Figure 1.6, which group of scores should have the smaller sample variance? Verify your answer by computing s^2 for each group.

32. These data are obtained on the sales per day in thousands of dollars at two franchises of the same hamburger chain:

Franchise A				Franchise B			
0.9	2.5	1.4	2.6	0.9	0.7	4.5	0.6
1.7	3.2	3.6	1.3	0.9	4.6	0.3	5.3
2.9	3.8	2.7	5.9	1.4	3.1	5.2	4.2
3.5	0.7	4.8	4.7	2.5	5.1	1.7	5.0

(a) Construct a stem-and-leaf diagram for each data set. Based on these diagrams, which sample do you think has the larger variance?
(b) Verify your answer to part (a) by finding the sample variance for each franchise.
(c) Suppose that we survey the two franchises at two randomly selected future dates. Would you be surprised to see daily sales over $4000 one day and under $1000 the next at franchise A? at franchise B? Explain.

33. A sociologist obtains these data on the age in years of a woman at the time of her first marriage.

Society A					Society B				
14	22	15	14	14	14	15	22	19	26
20	14	15	16	17	21	25	26	27	32
25	24	24	18	14	25	18	30	26	35
21	13	14	35	16	27	24	18	31	20
26	19	30	13		20				

(a) Construct a stem-and-leaf diagram for each sample.

(b) Which sample do you think has the higher variance?

(c) Verify your answer by finding the value of s^2 for each group.

1.5   More Exploratory Data Analysis (Optional)

In this section we introduce another graphical technique that is useful in exploring and summarizing a data set. The graph produced is called a *box-and-whisker plot* [22]. It provides a visual summary of five key numbers associated with a data set. These are the smallest observation, the lower hinge, the median, the upper hinge, and the largest observation. You are already familiar with the term median. The lower hinge is the point with the property that approximately 25% of the observations in the data set lie below this value; approximately 25% of the observations lie above the upper hinge. To find the lower hinge, we first find the median location. Recall that this is given by $(n + 1)/2$ where n is the sample size. The position of the lower hinge (lh) is given by

lower hinge

$$\text{location} = \frac{\text{median location (rounded down to the nearest integer)} + 1}{2}$$

The location of the upper hinge (uh) can be determined once the lower hinge is found. Approximately 1/2 of the observations will lie between these two hinges. Finding the hinges is not difficult, as we shall soon demonstrate. The box-and-whisker plot provides a quick visual method of determining the nature of a data set. In particular, it allows one to detect symmetry and to detect situations in which one or both tails of the distribution contain extremely large or extremely small values. Box-and-whisker plots are also useful in making a quick comparison among two or more data sets.

Example 1

To see how to construct a box-and-whisker plot, consider again the data on the magnitude of a California earthquake as measured on the Richter scale. (See Example 2 of Section 1.1.) The data are first ordered smallest to largest as shown:

1.0	1.2	1.9		2.2	3.1	5.0
1.0	1.3	2.0		2.3	3.3	5.1
1.1	1.4	2.1	median	2.4	4.0	6.3
1.1	1.4	2.1		2.7	4.1	7.7
1.2	1.5	2.2		3.0	4.1	8.3

Since the number of observations is even, the median is the average of the two middle observations. That is, the median is the average of the fifteenth and sixteenth observations in our ordered list. In this case the median is 2.2. The *median location* is halfway between the fifteenth and sixteenth observations or at position 15.5. The lower hinge location is

$$\text{lh location} = \frac{\text{median location (rounded down to nearest integer)} + 1}{2}$$

$$= \frac{15.5 \text{ (rounded down to nearest integer)} + 1}{2}$$

$$= (15 + 1)/2$$

$$= 8$$

That is, the lower hinge is the eighth observation in the ordered data set. The upper hinge is found by counting down eight observations from the largest. These values are indicated below.

1.0	1.2	2.0		2.2	3.1	5.0
1.0	1.3	1.9		2.3	3.3	5.1
1.1	1.4 (lh)	2.1	median	2.4	4.0 (uh)	6.3
1.1	1.4	2.1		2.7	4.1	7.7
1.2	1.5	2.2		3.0	4.1	8.3

To construct a box-and-whisker chart, we plot the median and hinges on a number scale, drawing small line segments through these points parallel to one another as shown in Figure 1.7(a). The hinge segments are joined by vertical lines to form a box as shown in Figure 1.7(b). To the box we attach vertical line segments which extend from the hinges to the extremes as shown in Figure 1.7(c). These segments form the "whiskers." Note that this graph vividly displays the lack of symmetry in the data. The extremely short lower whisker indicates that

Figure 1.7
Box-and-whisker chart for the magnitude of a sample of California earthquakes as measured on the Richter scale: (a) locating the median and the hinges, (b) constructing the box, (c) completing the chart by constructing the whiskers.

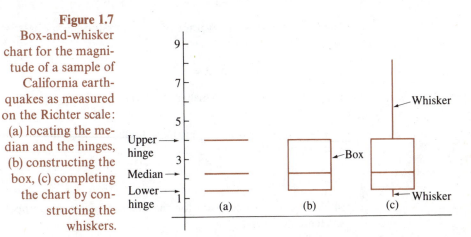

the data set contains no unusually small values; the relatively long upper whisker implies the presence of at least one value that is substantially larger than the others.

It is possible that the location of the lower hinge as computed from our formula may not be an integer. For example, we might apply the formula to obtain 3.5 as the location of the lower hinge. If this occurs, we assume the hinge to be halfway between the third and fourth observations.

Example 2

These data are obtained on the change in price over a one-week period of 35 randomly selected stocks listed on the American Stock Exchange. The data have been arranged in order from smallest to largest.

−2 7/8	−1 1/4	−3/4	−1/4	−1/8	0	5/8
−1 5/8	−1 1/8	−3/4	−1/4	−1/8	1/8	1
−1 1/2	−1	−1/2	−1/8	−1/8	1/8	1 3/8
−1 1/2	−7/8	−1/2	−1/8	−1/8	1/8	1 3/8
−1 3/8	−3/4	−3/8	−1/8	0	3/4	1 3/8

The median location is at $(n + 1)/2 = (35 + 1)/2 = 36/2 = 18$. The eighteenth member in the ordered list is $-1/8$. The location of the lower hinge is given by

$$\text{lh location} = \frac{\text{median location (rounded down to nearest integer)} + 1}{2}$$

$$= (18 + 1)/2$$

$$= 19/2 = 9.5$$

We take the lower hinge to be the average of the ninth and tenth observations in the ordered list. In this case the lower hinge is given by

$$\text{lh} = [-7/8 + (-3/4)]/2 = -13/16$$

The upper hinge is the average of the ninth and tenth observations counting from the largest down. Here,

$$\text{uh} = [1/8 + 0]/2 = 1/16$$

The complete box-and-whisker chart is shown in Figure 1.8.

Plots of this sort are informative and are beginning to appear frequently in scientific literature.

Exercises 1.5

34. Verify the values given for the median, lower, and upper hinges for each of these data sets:

I			II			
2	9	17	7	10	18	30
3	10	18	8	10	21	
5	12	20	8	12	22	
7	15	23	9	15	25	

median = 11	median = 12
lh = 6	lh = 9
uh = 17.5	uh = 21

Sketch a box-and-whisker plot for each set. Which data set is more symmetric?

35. Sketch a box-and-whisker plot of the data of Example 5 of Section 1.4, both with and without the outlier. Comment on the effect of the outlier on the plot.

36. Echocardiography is a medical technique that permits the strength of heart contractions to be measured without invasion of the body. Results obtained

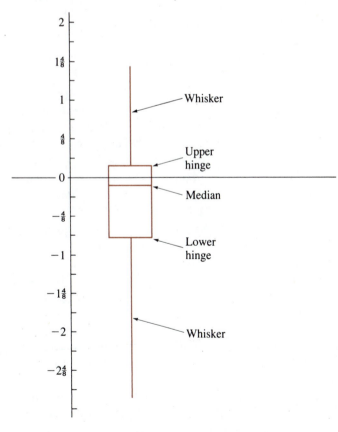

Figure 1.8
Box-and-whisker plot for the change in price over a one-week period for a sample of stocks listed on the American Stock Exchange.

with echocardiography were compared to those obtained with ventriculography, a more invasive but established method. By ventriculography, the hearts of patients in the sample were classified as normal, hypokinetic (weakly contracting), or akinetic (very weakly contracting). Measurements of the excursion (distance moved when contraction occurred) of the posterior wall of the heart were then made in each patient by echocardiography. These data resulted:

Posterior wall excursion (cm)

Normal		Hypokinetic	Akinetic
1.00	1.32	0.85	0.82
1.07	1.33	0.87	0.87
1.09	1.38	0.89	0.90
1.09	1.38	0.90	1.00
1.12	1.38	1.00	1.00
1.12	1.42	1.06	1.08
1.21	1.45	1.07	1.18
1.21	1.50	1.10	
1.23	1.51	1.12	
1.24	1.61	1.14	
1.25	1.61	1.51	
1.25	1.73		
1.29			

(a) For each data set find the median, upper hinge, and lower hinge.

(b) For each data set construct a box-and-whisker plot. Compare the medians of the three groups. Do any of the data sets appear to contain extremely large or extremely small values based on the whisker lengths?

37. Sketch box-and-whisker plots for the data given in exercise 33 of Section 1.4. Compare these plots to the stem-and-leaf diagrams that you already have for these data. Do you see that these plots convey the same basic information in a different form?

38. Sketch box-and-whisker plots for the data given in exercise 34 of Section 1.4. Compare these plots to the stem-and-leaf diagrams that you already have for these data.

39. Based on the box plot of Figure 1.7, try to draw a curve that roughly represents the distribution of the magnitude of a California earthquake as measured on the Richter scale. Check your sketch by constructing a stem-and-leaf diagram for the data. Does the stem-and-leaf diagram suggest the shape that you anticipated?

40. Based on the box plot of Figure 1.8, try to draw a curve that roughly represents the distribution of the change in price for the stocks in question. Check your sketch by constructing a frequency histogram for the data of Example 2. Use these classes:

Class 1: $-47/16 - -35/16$
Class 2: $-35/16 - -23/16$
Class 3: $-23/16 - -11/16$
Class 4: $-11/16 -   1/16$
Class 5: $1/16 -   13/16$
Class 6: $13/16 -   25/16$

Does the histogram suggest the shape that you anticipated?

1.6   Handling Grouped Data (Optional)

Raw data are seldom given in research journals or in business reports. Rather, a summary of the data is displayed. This summary sometimes is in the form of a frequency table or a histogram. It is helpful to be able to approximate the sample mean and sample variance from grouped data of this sort. In this way we can approximate the mean and variance of the population from which the sample is drawn even when the original data are not available to us. The method for doing so is illustrated in our first example.

Example 1   A study commission is looking into the possibility of beginning a limited bus service in a small town. The data of Table 1.3 give a summary of the responses to the question, If the system is put into effect, how many round trips do you expect to make per week? To use these data to approximate $\bar{x}$ and s^2 thereby approximating μ and σ^2, we make use of the class midpoints. Since the classes are not extremely wide, the midpoint of each class serves as a good approximation for each of the values in the class. Thus, in approximating the sample mean, each of the 40 observations in class 1, whose actual values are unknown to us, is replaced by the number 1; each of the 350 observations in class 2 is replaced by the number 4; this procedure is continued for the remaining three classes. To approximate $\bar{x}$, we add these values and divide by 1000, the total number of persons answering the question. Thus

$$\bar{x} \approx [1(40) + 4(350) + 7(400) + 10(175) + 13(35)]/1000$$

$$= 6445/1000$$

$$= 6.445$$

Table 1.3

Class	Boundaries	Midpoint	Frequency
1	$-.5-$ 2.5	1	40
2	2.5– 5.5	4	350
3	5.5– 8.5	7	400
4	8.5–11.5	10	175
5	11.5–14.5	13	35

Based on these data, the citizens expect to make an average of 6.445 round trips per week.

The above example illustrates how to approximate $\bar{x}$ from grouped data. For future reference, we summarize the technique used. In the formula given, m represents the class midpoint, f the class frequency, and n the sample size. Summation is over all classes.

$$\bar{x} \approx \Sigma m \cdot f/n$$

The sample variance, s^2, can be approximated in a similar manner. To see how this is done, recall that the computational formula for s^2 is

$$s^2 = \frac{n\Sigma x^2 - (\Sigma x)^2}{n(n-1)}$$

To approximate s^2, we again replace each observation by its class midpoint. This results in the following formula:

$$s^2 \approx \frac{n\Sigma m^2 \cdot f - (\Sigma m \cdot f)^2}{n(n-1)}$$

This idea is demonstrated in Example 2.

Example 2   For the data of Table 1.3,

$$\Sigma m \cdot f = 1(40) + 4(350) + 7(400) + 10(175) + 13(35)$$

$$= 6445$$

$$\Sigma m^2 \cdot f = 1^2(40) + 4^2(350) + 7^2(400) + 10^2(175) + 13^2(35)$$

$$= 48655$$

Substituting, we see that

$$s^2 \approx \frac{n\Sigma m^2 \cdot f - (\Sigma m \cdot f)^2}{n(n-1)}$$

$$= \frac{1000(48655) - (6445)^2}{1000(999)}$$

$$= 7.124$$

Taking the square root of s^2, we see that the sample standard deviation is given by

$$s \approx \sqrt{s^2} = \sqrt{7.124} = 2.669 \text{ round trips}$$

Exercises 1.6

41. Table 1.4 gives data on the amount of the loans in thousands of dollars granted to applicants at two branches of a particular bank over a one-month period:

Table 1.4

Class	Boundaries	Midpoint	Frequency
Branch A			
1	.5– 3.5		12
2	3.5– 6.5		18
3	6.5– 9.5		20
4	9.5–12.5		22
5	12.5–15.5		28
Branch B			
1	.5– 3.5		5
2	3.5– 6.5		20
3	6.5– 9.5		50
4	9.5–12.5		20
5	12.5–15.5		5

(a) Find the class midpoints.

(b) Sketch a frequency histogram for each.

(c) Do these branches seem to have the same distribution for loans? Explain.

(d) Does either distribution appear to be skewed? If so, in which direction is the skew?

(e) Based on the histogram, do you think that the sample means for these branches are close in vaiue? Verify your answer by approximating the sample mean for each data set.

(f) Based on the histogram, which branch do you think exhibits higher variability in the amount of loans granted? Verify your answer by approximating the sample variance for each group.

(g) Approximate the sample standard deviation for each group.

42. Table 1.5 gives summary data on the IQ of 180 children suffering from Down's Syndrome, a congenital condition characterized by moderate to severe mental retardation. Approximate $\bar{x}$, s^2, and s from these data. Would it surprise you to hear someone claim that the average IQ of such children is 70? Explain.

Table 1.5

Class	Boundaries	Midpoint	Frequency
1	10.5– 20.5		4
2	20.5– 30.5		34
3	30.5– 40.5		0
4	40.5– 50.5		70
5	50.5– 60.5		43
6	60.5– 70.5		19
7	70.5– 80.5		7
8	80.5– 90.5		2
9	90.5–100.5		1

Table 1.6

Class	Boundaries	Midpoint	Frequency
1	13.5– 624.5		23
2	624.5–1235.5		7
3	1235.5–1846.5		9
4	1846.5–2457.5		1
5	2457.5–3068.5		4
6	3068.5–3679.5		1
7	3679.5–4290.5		5

43. Table 1.6 gives summary data on the lifespan, in hours, of the lithium batteries used in a particular type of pocket calculator. Approximate the average
lifespan of such a battery. Also approximate the standard deviation for these
data. Do you think that it would be safe to guarantee batteries of this sort for
at least 1000 hours? Explain.

COMPUTING SUPPLEMENT A

As can be seen, the calculations necessary to summarize even a relatively small data set can become tedious and time-consuming. There are many types of hand-held calculators on the market that are programmed to do statistical calculations. In recent years several computer systems for data analysis have been developed. Among the systems currently in use are SPSS (Statistical Package for the Social Sciences: McGraw-Hill), BMD (Biomedical Computer Programs: University of California Press), MINITAB (Pennsylvania State University Press), and SAS® (Statistical Analysis System: SAS Institute Inc., Cary, N.C.). To use any of these systems one needs little background in computer science.

We present here a very brief introduction to SAS programming to give the reader some experience with computer packages. We do not mean to imply that this package is necessarily superior to the others. Once experience is gained with one package it is not difficult to adjust to the others as they are similar in nature. We introduce SAS by presenting two sample programs. These programs can be modified to analyze other data sets. Programs are written assuming that data is entered with the program. The reader is referred to a consultant at his or her own installation to determine how to access SAS.

Example 1

These data are obtained on the variable X, the retail price of a particular prescription drug. (See Example 2 of Section 1.2.)

2.00	1.98	1.48	2.99	1.20	2.06	1.98	1.20
2.50	3.02	1.75	2.05	1.71	1.10	1.82	1.80
1.75	1.17	2.25	1.90	2.03	1.89	2.15	1.96
1.87	0.95	2.20	1.85	1.70	1.92	3.00	2.40
2.03	2.75	2.69	2.03	1.90	1.72	2.60	2.11

For these data let us find the mean, variance, standard deviation, and range. Let us also construct a relative frequency histogram.

The first step in using SAS is to create a SAS data set. This is done by a series of statements that name the data set (the DATA statement), describe the arrangement of the data lines (the INPUT statement), and signal the beginning of the data itself (the CARDS statement). SAS statements may begin in any column of print and must always end with a semicolon.

The name chosen for the data set should be a one-word name of eight or fewer letters. It is helpful to choose a name that is related to the data. To name the data set, set the SAS keyword DATA, followed by the name chosen. In our example, we will name the data set "drug." We type the words

```
DATA DRUG;
```
(on a single line)

The next statement is the input statement. This statement names the variables and describes the order in which their values appear on the data lines. Variable names should also be eight letters or less in length. Here we have only one variable, price. We type

```
INPUT PRICE;
```

The computer now knows that each line contains the value of only one variable, whose name is "price."

The input statement is followed by a statement that signals that the data now follows. This statement is

```
CARDS;
```

The data follows *immediately* with only one observation typed per line. When all data have been entered, a line containing *only* a *semicolon* should be typed. This signals the end of the data. Thus far we have

DATA DRUG;	names the data set
INPUT PRICE;	names the variables in order
CARDS;	signals that the data follows
2.00	data (one observation per card)
2.50	
.	
.	
.	
2.11	
;	signals the end of data

The data are now in a SAS data set called "DRUG." We now tell the computer what to do with the data. This is done by using one or more procedure statements. Each procedure statement begins with the keyword PROC followed by the name of the procedure. We shall be using the MEANS procedure. The keyword is the word MEANS, followed by the keywords for those summary measures desired. Some of the keywords are as follows:

MAXDEC = n	(*n* is an integer from 0 to 8 that specifies the number of decimal places to be used to print results)
MEAN	(mean)
STD	(standard deviation)

```
RANGE                                        (range)
VAR                                          (variance)
N                                            (number of observations)
```

To find the summary measures called for, we type

```
PROC MEANS MAXDEC = 2 MEAN VAR STD RANGE;
```

The relative frequency histogram called for can be obtained by using the CHART procedure. The statements required are

```
PROC CHART;
VBAR PRICE/TYPE = PERCENT;
```

This first statement calls for a chart to be made; the second asks for a vertical bar chart for the variable price. The type is to be a relative frequency histogram. To obtain a frequency histogram type only

```
VBAR PRICE;
```

One may include a title for the chart, if desired. This statement uses the keyword TITLE(*n*) where *n* is the line number where the title is to be printed. For example, the statements

```
TITLE1     DRUG PRICE;
TITLE2     STUDY;
```

would result in the words **DRUG PRICE** being printed on line 1 of each page of output and the word **STUDY** on line 2. The title statement should be placed *after* the PROC statement with which it is associated. Our entire program in shown below:

```
DATA DRUG;
INPUT PRICE;
CARDS;
2.00
2.50
  ⋮
2.11
  ;
PROC MEANS MAXDEC = 2 MEAN VAR STD RANGE;
TITLE1    DRUG PRICE;
```

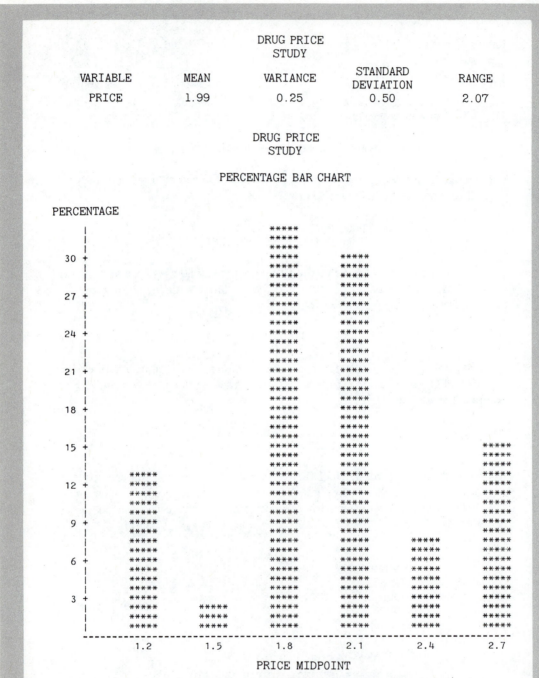

Figure 1.9

```
TITLE 2    STUDY;
PROC CHART;
VBAR PRICE/TYPE = PERCENT;
```

To adjust this program to handle another data set, change the name of the data set and the variable to reflect the new data. Also change the title. The output of this program is shown in Figure 1.9 on page 52.

Example 2

The procedure UNIVARIATE can be used to construct both stem-and-leaf displays and box-and-whisker plots. It also produces the summary measures given by the MEANS procedure as well as other measures. Its use is illustrated via the data on California earthquakes of Section 1.1. The program required is as follows:

Statement	*Purpose*
`DATA  QUAKE;`	names the data set
`INPUT SIZE @ @;`	names the variable; @@ indicates that more than one data point may be typed on each line
`CARDS;`	signals that the data follow

`1.0  1.2  2.0  3.3`
`1.4  5.0  8.3  1.0`
⋮
	data points; at least one space between each

`1.3  2.1  4.1  1.5` `;`	signals the end of the data
`PROC UNIVARIATE PLOT;`	asks for stem-and-leaf and box-and-whisker plots to be constructed along with the summary measures automatically produced by the procedure

`TITLE 1    PLOTS;`
`TITLE 2    OF;`
`TITLE 3    QUAKE DATA;`
	titles the output

The output is as shown in Figure 1.10 on page 54.

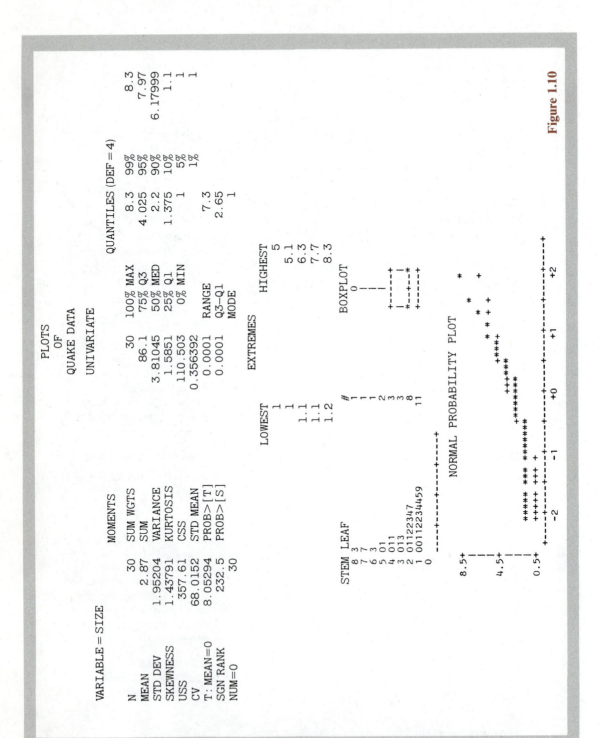

Figure 1.10

Vocabulary List and Key Concepts

stem-and-leaf diagram   sample median

skewed right (positively skewed)   mode

skewed left (negatively skewed)   sample range

frequency histogram   sample variance

relative frequency histogram   sample standard deviation

sample mean

Review Exercises

In 1970 the average cost of an initial office visit to a physician was $14.23. The standard deviation was one dollar. These data represent a sample of observations on X, the current cost of such a visit adjusted to 1970 dollars.

16.14	15.71	16.23	17.44	16.88
15.79	15.10	15.82	15.89	16.99
14.08	16.30	14.88	18.05	17.22
17.22	14.39	16.04	17.56	15.32
16.18	15.26	16.73	16.03	18.94

Use these data to answer questions 44–49.

44. Construct a stem-and-leaf diagram for these data. Use stems of 14, 15, 16, 17, and 18. Let the third digit of each number be the leaf. Does the distribution appear to be approximately bell shaped?

45. Find the sample mean for these data. Based on this statistic, can we conclude without a doubt that μ, the mean cost of such a visit currently, is higher than the old figure of $14.23? Explain.

46. Find the sample variance for these data.

47. Find the sample standard deviation for these data. What physical unit is associated with s? Based on this value, can we say without a doubt that σ, the standard deviation of X today, exceeds one? Explain.

48. Find the sample median for these data and explain briefly what this number means.

49. Do these data have a unique mode?

Diabetes is a disease that often strikes in middle age. These data are obtained on X, the age at onset of the disease in a sample of 30 randomly selected diabetics.

35.5	30.5	40.1	59.8	47.3	44.5	48.9	36.8	52.4	36.6
39.8	42.1	39.3	26.2	55.6	33.3	40.3	65.4	60.9	45.1
51.4	46.8	42.6	45.6	52.2	51.3	38.0	42.8	27.1	43.5

Use these data to answer questions 50–57.

50. Construct a relative frequency histogram for these data based on six classes.

51. Would you be surprised to hear a physician claim that this random variable has a bell-shaped distribution? Explain.

52. Find $\bar{x}$.

53. Find s^2.

54. Find s. What physical unit is associated with s?

*55. Approximate $\bar{x}$ using the grouped-data approach demonstrated in Section 1.6. Compare this figure to the exact value of $\bar{x}$ found in exercise 52. At this point do we know the exact value of μ, the true average age at the onset of diabetes in the population? Explain.

*56. Approximate s^2 using the grouped-data approach and compare your answer to that obtained in exercise 53.

*57. Approximate s using the grouped-data approach and compare your answer to that obtained in exercise 54.

Use the following data and answer exercises 58–61:
A study of the effects of smoking on sleep patterns is conducted. The random variable observed is the time in minutes that it takes to fall asleep. These data are obtained:

Smokers		Nonsmokers		
69.3	52.7	28.6	38.5	36.0
56.0	34.4	25.1	30.2	37.9
22.1	60.2	26.4	30.6	13.9
47.6	43.8	34.9	31.8	
53.2	23.2	29.8	41.6	
48.1	13.8	28.4	21.1	

58. Find the sample medians for smokers and nonsmokers respectively.

*59. Find the lower hinges and upper hinges for each group.

*60. Construct box-and-whisker plots for each group.

*61. Comment on the characteristics of these data sets as revealed by the box-and-whisker plots.

* Starred exercises are optional.

2

Probability

We have seen that it is often possible to get at least a rough idea of the characteristics of a population by examining a sample selected from that population. However, since we have not observed every member of the population, there is always some doubt as to the accuracy of the conclusions that are drawn. To make decisions based on quantitative data in such a way that the degree of accuracy can be assessed, one must have some understanding of **probability**. For this reason, we consider now the basic concepts underlying this area of mathematics.

2.1 Intuitive Probability

Most people, when asked "Do you know anything about probability?" promptly answer with a resounding no. This, however, is not usually the case. An intuitive notion of the terminology and interpretation of probabilities has become almost essential to functioning successfully in today's society. For instance, if you are like many Americans, one of the first things you do in the morning is turn on the TV or radio to get a weather report. You will hear such phrases as "there is a 45% chance of rain today" or "there is a 5% chance of rain today." Many weather reporters have gone so far as to coin the phrase "POP" (probability of precipitation) in making their reports. What do these numbers mean? If the POP is 0, should you take an umbrella to class? If the POP is 100%, should you plan an outdoor sporting event? Most people would agree that the answer to each of these questions is no, because even without having formally studied probability theory the public has an intuitive notion about how these numbers should be interpreted. The consensus is that numbers near 0 or percentages near 0% indicate that the event in question is not expected to occur; numbers near 1 or percentages near 100% indicate that the event is expected to occur. What should one do if the POP is 50%? This, in fact, depends entirely on one's mental outlook. An optimist would not prepare for rain, whereas a pessimist would take an umbrella to class. The problem is, of course, that numbers near 1/2 or percentages near 50% indicate that the chance that the event will occur is about the same as the chance that the event will not occur. Hence, it is hard to interpret the number.

This example illustrates the manner in which we interpret probabilities.

Interpreting Probabilities

1. Probabilities are numbers between 0 and 1 inclusive. They give us an idea of whether or not a physical event will occur. For ease in interpretation, these numbers can be expressed as percentages between 0 and 100%.

2. Probabilities near 0 indicate that the event in question is not likely to occur. They do not mean that the event will not occur, only that it is considered to be rare.

3. Probabilities near 1 indicate that the event in question is likely to occur. They do not mean that the event absolutely will occur, only that the event is considered to be a common occurrence.

4. Probabilities near 1/2 indicate that the event in question has about the same chance of occurring as it has of failing to occur.

The obvious question is, How does one go about assigning probabilities to physical events? Answering that question thoroughly is the topic of this chapter.

Many problems involving probability do not require a great deal of mathematical training to solve. They require instead an intuitive understanding of the question being asked and common sense.

Example 1

Each time a major sporting event occurs, money changes hands in the form of legal and illegal wagers. The payoff for such a wager depends on the perceived probability of a particular participant winning the contest in question. For example, to determine the payoff on a wager that the Dallas Cowboys will defeat the Washington Redskins, one first has to approximate the probability that this will occur. This is done by compiling available information about both teams and then expressing an opinion about the correct probability.

This example shows that informed personal opinion is an accepted method for assigning probabilities. This method, called **personal probability**, is used widely in "one-shot situations," situations in which the event under study has not occurred before and will not occur again. The advantage of this method of assigning probabilities is that it can be used at any time. However, a drawback is that its accuracy is dependent upon the accuracy and completeness of the available information and the ability of the individual making the assignment to assess the material intelligently.

Example 2 illustrates a method for assigning probabilities that requires no extensive information gathering and no experimentation. Via this method, probabilities are assigned **a priori**, before performing the experiment.

Example 2 | A box contains 25 lithium batteries for use in pocket calculators. Four of these batteries are weak but this fact cannot be detected visually. If we select one battery from the box at random, what is the probability that the one chosen is weak? If we draw a battery at random, then we are just as likely to choose one battery as any other. That is, the 25 choices that we can make are intuitively equally likely. Since there are four weak batteries in the box, any one of which can be selected, the probability of drawing a weak battery is 4/25. Note that we arrive at this answer without having to physically select a battery from the box.

In this example, we used an intuitively acceptable and powerful principle that comes into play in a wide variety of problems. Namely, if all possible outcomes of an experiment are assumed to be **equally likely**, then to determine the probability that a particular event A will occur, we need count only two things:

1. The number of ways in which the experiment itself could proceed denoted by $n(S)$
2. The number of ways in which the event A could occur denoted by $n(A)$

The probability that A will occur is then given by

$$\text{probability of } A = P[A] = n(A)/n(S)$$

This assignment of probabilities is termed **classical probability** or the **a priori** approach. It is widely used, especially with games of chance such as craps, poker, roulette, and other casino games. It requires that all possible outcomes of the experiment be equally likely, but this condition is often satisfied and hence does not severely limit the applicability of the method. When applicable, it results in a completely accurate statement of the probability that the event in question will occur.

A third method for assigning probabilities to events does require some experimentation. For this reason, we say that these probabilities are assigned **a posteriori**, after the experiment has been conducted a number of times.

Example 3 | It has been reported that the probability of the birth of twins is 1/96. How is this number obtained? It may be the personal opinion of an informed expert. However, more likely, this figure is found by examining medical records and observing the number of live births and the number of live twin births. On the basis of this information it has apparently been observed that

$$P[\text{twins}] \approx 1/96$$

That is, approximately 1 out of every 96 observed live births results in the birth of twins. Based on past experience, our best answer to the question, What is the probability that the next birth will result in twins? is 1/96.

In this example, we have been using an intuitively reasonable method for approximating probabilities in situations in which direct observation is possible and in which the classical approach to probability does not appear feasible. In

Figure 2.1
Copyrighted by the Chicago Tribune. Used with permission.

repetitive situations, we may reasonably conclude that the probability that an event A will occur is approximately given by

$$P[A] \approx \frac{\text{number of times event } A \text{ has occurred}}{\text{number of times the experiment was performed}}$$

This method of approximating probabilities is called the **relative frequency** or **a posteriori** approach to probability. This approach is usually more accurate than the personal approach and is therefore preferable to that approach in those instances in which it can be used.

There are three methods for assigning probabilities—personal, classical, and relative frequency. Each method has advantages and disadvantages. Some problems clearly call for one approach over another, whereas others may be solved by each of the methods. Figure 2.1 illustrates a mixed approach to the assignment of probabilities.

Exercises 2.1

In exercises 1–5, which approach(es) to probability should be used to answer the question posed? Be ready to defend your choice.

1. A man wishes to place a wager on the outcome of the World Series. The game will be played by the Braves and the Yankees, and he wishes to consider the question, What is the probability that the Braves will win?

2. A woman contracted German measles while pregnant, and she is afraid that her child may have suffered mental or physical damage. She asks her doctor, "What is the probability that my child will be damaged as a result of my illness?"

3. There has been discussion that regular use of marijuana leads to heroin addiction; this is an argument against the legalization of marijuana. What is the probability that a regular user of marijuana will become addicted to heroin?

4. A game of chance is being played at a county fair. The rules are as follows: A duck is selected from a moving line of plastic ducks. If the duck has an even number on the bottom, the player wins; otherwise he loses. What is the probability that he will win on a single play of the game?

5. In Monopoly, a player is allowed to roll a pair of fair dice and moves the playing piece ahead the number of blocks shown on the dice. If, for example, a sum of 7 is rolled, the piece is allowed to move forward 7 blocks. If doubles is rolled, the player is allowed to roll again. What is the probability that on a given throw, doubles is rolled?

In exercises 6–17, find, or approximate, the probability requested. Identify the method that you use as being either classical or relative frequency.

6. Toss a fair coin once. Find the probability that it lands with heads showing.

7. A single fair die is tossed once. Find the probability of rolling an odd number; an even number; a number less than 3.

8. Fifty microprocessor chips produced by a particular company are selected and tested. It is found that four are defective. What is the probability that the next chip produced by this company will be defective?

9. A baseball player has obtained 20 hits in his last 65 times at bat. What is the probability that he will get a hit at his next time at bat?

10. A loaded coin is tossed 100 times. It lands heads 75 times. What is the probability that it will land heads on the next toss?

11. A hospital has 50 units of blood labeled "A positive" available. Unknown to the staff, four of these are mislabeled and are actually "A negative." One unit of blood is selected at random. Thus, each unit has the same chance of being chosen as any other. Find the probability that a mislabeled unit is selected.

12. Five hundred tickets have been sold for a lottery. You hold exactly one ticket. The winner is selected by placing all tickets in a box and drawing the winner at random. What is the probability that you will win?

13. Ten thousand tax forms are filed in a small town. Of these, 1000 have errors in arithmetic. A form is selected randomly and checked for accuracy. What is the probability that it will contain an arithmetic error?

14. A basketball player has hit 82% of her free throws in the past. She is fouled and receives a free throw. What is the probability that she will make the shot?

15. A bag contains 150 jelly beans. Ten of these are licorice. A child reaches into the bag and selects one jelly bean at random. What is the probability that a licorice bean is selected?

16. A grocery shelf contains 10 cans of mushrooms. Three of these are from a lot that is being recalled because of a botulism scare. Before the recall is announced, a shopper purchases one can chosen randomly from the shelf. What is the probability that the can selected is from the lot that is being recalled?

17. An ordinary deck of playing cards consists of 52 cards. There are four suits: hearts, diamonds, clubs, and spades. Hearts and diamonds are red whereas clubs and spades are black. Each suit has 13 cards labeled 2, 3, 4, 5, 6, 7, 8, 9, 10, jack, queen, king, and ace. The cards are shuffled and one card is drawn at random. What is the probability of drawing a black card? a red card? a diamond? an ace? a face card? (jack, queen, or king)

18. Suppose that the Redskins and the Cowboys are to meet in the playoffs. Since there are two teams involved and one of them is the Redskins, a person reasons that the probability that the Redskins will win the game is 1/2. What is wrong with this reasoning?

19. Each of the 50 states in the United States sends a delegation to the House of Representatives. A state delegate is selected at random to serve on a committee. It is argued that since there are 50 states, the probability is 1/50 that a person chosen is from Texas. Criticize this argument.

20. Toss a pair of coins 64 times and count the number of times that you get two heads, two tails, and one of each. Using the relative frequency approach, approximate:

$P[\text{two heads}]$, $P[\text{two tails}]$, $P[\text{one of each}]$.

Based on your results would you be surprised to hear someone claim that these three events are equally likely?

2.2   Sample Spaces and Events

So far, we have been considering probability from an intuitive standpoint. We have used such words as *outcome*, *experiment*, and *event*, assuming that you have an idea of what these terms mean in a physical context. Let us take a moment to clarify their meaning when used in a mathematical context.

An **experiment** is any physical action or process that is observed and the outcome noted.

Experiments that are of real interest are those whose outcome cannot be predicted with certainty. Tossing a coin, dealing a poker hand, firing a missile, administering a dose of a drug, or just getting up in the morning classify as experiments in this context. In order to analyze such an experiment, one must investigate two things: (1) the possibilities and (2) the probabilities. It is clear that these concepts should be considered in the order stated; we cannot talk about what is probable without first considering what is possible. Thus, our first definition concerns the question, What is possible in a given experiment? This question is answered by specifying what is called a *sample space* for the experiment.

Definition 2.1

A **sample space** for an experiment is a set S of possible outcomes for the experiment. An element of S is called a **sample point**.

There are several points to note concerning this definition. For a set to serve as a sample space for an experiment, it must satisfy one important condition. In particular, each physical outcome of the experiment must correspond to *exactly* one sample point. We do not want to leave anything out and we do not want anything represented more than once! Furthermore, sample spaces are not unique; there may be more than one sample space associated with a given experiment. The use of the phrase "a sample space" rather than "the sample space" is an intentional attempt to emphasize this point. However, some sample spaces may be more desirable than others. Usually, a sample space based on precise information is preferable to one based on broad generalizations. Also, a sample space in which each sample point occurs with the same probability is preferred if such a sample space exists. These points are clarified in the next example.

Example 1

A card is drawn at random from an ordinary deck. Let us consider some sets that may serve as sample spaces.

$S_1 = \{$red, black$\}$

$S_2 = \{$club, diamond, heart, spade$\}$

$S_3 = \{x : x$ is a card in the deck$\}$

$S_4 = \{$honor (10, jack, queen, king, ace), not an honor$\}$

$S_5 = \{$face card (jack, queen, king), numbered card$\}$

$S_6 = \{$face card, ace, nonface card$\}$

Note that S_1, S_2, S_3, and S_4 all satisfy the definition of the term *sample space*. Note also that S_1, S_2, and S_3 satisfy the additional condition that each sample point occurs with the same probability. S_6 fails to be a sample space since the physical outcome that the ace of spades is drawn corresponds to the sample point "ace" and to the sample point "nonface card." S_5 fails to be a sample space since the outcome "ace of spades" corresponds to no sample point of S_5.

Consider an experiment and some physical event associated with the experiment. For example, consider the experiment of rolling a fair die once. A logical sample space for this experiment is

$$S = \{1, 2, 3, 4, 5, 6\}$$

The physical event that a number less than 4 is rolled, corresponds to the set $\{1, 2, 3\}$. When we identify the set of sample points that corresponds to the occurrence of a physical event we also call that set an *event*. This idea leads us to the next definition.

Definition 2.2

Let S be a sample space for an experiment. Any subset A of S will be referred to as an **event**. We write $A \subseteq S$.

Two events or subsets of S are of particular interest. These are S itself and the empty set, $\varnothing$. The sample space corresponds to a physical event that is certain to occur. For this reason, S is called the **sure event** or **certain event**. The empty set corresponds to physical events that are impossible. We therefore refer to $\varnothing$ as the **impossible event**.

Example 2

In many games of chance, such as Monopoly, craps, Parcheesi, and backgammon, a pair of fair dice is rolled. To determine the probabilities of the occurrence of events in these games, it is necessary to consider a detailed sample space for the experiment. The simplest way to form such a sample space is to think in terms of rolling dice, one red and the other white. We may then write S as a set of pairs of numbers (x, y), with x being the number on the red die and y the number on the white. Thus, S is as shown in Table 2.1. Typical physical events associated with this experiment are

A: a sum of 7 is rolled C: a sum of 13 is rolled

B: a sum of 12 is rolled D: each die shows a number
                                         smaller than 7

The subsets of S that correspond to the occurrence of events A and B are

$$A = \{(3, 4), (4, 3), (5, 2), (2, 5), (6, 1), (1, 6)\}$$

$$B = \{(6, 6)\}$$

Note that since it is physically impossible to roll a sum of 13 with a pair of ordinary dice, event C is the impossible event; that is, $C = \varnothing$. Similarly, since each die is certain to show a number smaller than 7, event D is the sure event; that is, $D = S$.

Table 2.1

Sample Space for the Two-Dice Experiment (x = red die; y = white die).

$$
S = \begin{array}{cccccc}
(1, 1) & (1, 2) & (1, 3) & (1, 4) & (1, 5) & (1, 6) \\
(2, 1) & (2, 2) & (2, 3) & (2, 4) & (2, 5) & (2, 6) \\
(3, 1) & (3, 2) & (3, 3) & (3, 4) & (3, 5) & (3, 6) \\
(4, 1) & (4, 2) & (4, 3) & (4, 4) & (4, 5) & (4, 6) \\
(5, 1) & (5, 2) & (5, 3) & (5, 4) & (5, 5) & (5, 6) \\
(6, 1) & (6, 2) & (6, 3) & (6, 4) & (6, 5) & (6, 6)
\end{array}
$$

As you can see, we sometimes consider more than one event in a single experiment. It is often necessary to form a new event from others via the set operations of *union, intersection*, or *complementation*. You are probably familiar with these terms, but let us review their meaning.

The **union** of two events A and B is denoted by $A \cup B$, and consists of all sample points that are in event A, or in event B, or in both A and B.

The **intersection** of events A and B is denoted by $A \cap B$. It consists of all sample points that are in both A and B; that is, $A \cap B$ is the set of sample points that are common to A and B.

The **complement** of event A, denoted by A', is the set of all sample points that are *not* in set A.

These ideas are illustrated in the next example.

Example 3

Consider the experiment of identifying an individual's ABO blood type taking into consideration the Rh factor. The sample space for this experiment is

$$S = \{A^+, A^-, B^+, B^-, AB^+, AB^-, O^+, O^-\}$$

Let A denote the event that an individual's blood contains the antigen A and let P denote the event that the Rh factor is positive. The sample points corresponding to these events are

$$A = \{A^+, A^-, AB^+, AB^-\}$$

$$P = \{A^+, B^+, AB^+, O^+\}$$

The event that an individual's blood contains the A antigen *or* the Rh-positive is denoted by $A \cup P$. The sample points that constitute this new event are

$$A \cup P = \{A^+, A^-, AB^+, AB^-, B^+, O^+\}$$

The event that an individual's blood contains the antigen A *and* is Rh-positive is denoted by $A \cap P$. In this case

$$A \cap P = \{A^+, AB^+\}$$

The event P′ corresponds to the fact that an individual does *not* have Rh-positive blood. Thus,

$$P' = \{A^-, B^-, AB^-, O^-\}$$

One question that we usually ask about two events A and B is, Can these events occur at the same time? If they cannot, we say that the events are mutually exclusive. In our previous example, the events P and P' are mutually exclusive. They cannot occur at the same time. It is impossible to have and not have Rh-positive blood at the same time! Note that $P \cap P' = \varnothing$. This is not a coincidence. Mutually exclusive events are always characterized by the fact that their intersection is empty. These ideas are summarized in the next definition.

Definition 2.3

Two events A and B are **mutually exclusive events** if they cannot occur at the same time. That is, A and B are mutually exclusive if $A \cap B = \varnothing$.

Pictorial representations of sample spaces, events, and their relationships to one another are useful in solving probability problems. In these representations, called *Venn diagrams*, we use a rectangle to represent the sample space and closed curves within the rectangle to represent events. We illustrate the idea by considering the Venn diagrams for the events discussed in Example 3.

Example 4

In Example 3, our sample space consists of all possible ABO blood types for human beings, taking into consideration the Rh factor. We picture S as a rectangle, as shown in Figure 2.2(a). The event A, "an individual's blood contains the A antigen," is a subset of S. This is depicted by drawing a closed curve within the rectangle as in Figure 2.2(b). Events A and P, "the Rh factor is positive," are not mutually exclusive. The closed curve representing P does overlap that of A as shown in Figure 2.2(c). The event $A \cup P$, "an individual's blood contains the A antigen *or* is Rh-positive," is shown by the shaded region of Figure 2.2(d). The event $A \cap P$, "an individual's blood contains the A antigen *and* is Rh-positive," is shown by the shaded region of Figure 2.2(e). The event P', "an individual does *not* have Rh-positive blood," is depicted by the region outside of the region P. This complement is shown in Figure 2.2(f).

The ability to translate from symbols and pictures to English and from English to symbols and pictures is essential! The main points to remember are

Translating Probability Statements

1. The English word NOT involves set complements. In probability problems, interest centers on the probability that a given event will not occur.

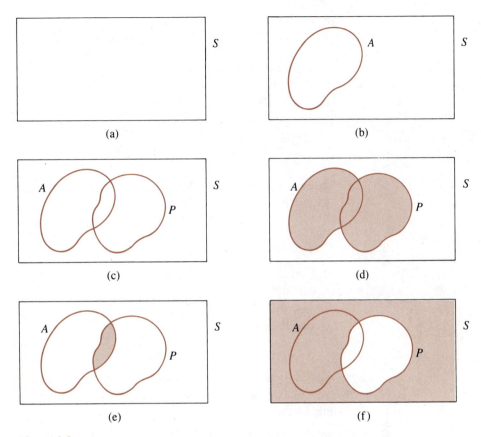

Figure 2.2
S, the sample space: (a) a rectangle is used to depict the sample space, (b) $A \subseteq S$, (c) A and P are not mutually exclusive, so the closed curves must overlap, (d) shaded region depicts $A \cup P$, (e) shaded region depicts $A \cap P$, (f) shaded region depicts P'.

2. The English words AND or BUT involve set intersections. Probability problems involving these words in their statement concern themselves with the probability that both or all of the events will occur.

3. The English word OR involves set union. Probability problems with this word in their statement concern the probability that one or the other, or perhaps both, of the events will occur.

Exercises 2.2

21. An experiment consists of rolling a pair of fair dice once. Use Table 2.1 to list the sample points that constitute each of the events in parts (a)–(f):
(a) A: "A sum less than 6 is rolled."

    (b) B: "Doubles is rolled."
    (c) $A \cup B$: "A sum less than 6 *or* doubles is rolled."
    (d) $A \cap B$: "A sum less than 6 *and* doubles is rolled."
    (e) A': "A sum greater than or equal to 6 is rolled."
    (f) $A' \cap B$: "A sum greater than or equal to 6 *and* doubles is rolled."
    (g) Are the events A and B mutually exclusive? Explain.

22. The sample space for the experiment of identifying an individual's ABO blood type taking into consideration the Rh factor is

$$S = \{A^+, A^-, B^+, B^-, AB^+, AB^-, O^+, O^-\}$$

The events A, "an individual's blood contains the A antigen," and P, "the Rh factor is positive," are given by

$$A = \{A^+, A^-, AB^+, AB^-\}$$

$$P = \{A^+, B^+, AB^+, O^+\}$$

    (a) Find the sample points that constitute the event B, "an individual's blood contains the B antigen."

In parts (b)–(i), find the sample points that constitute the event given and describe the event in words:

    (b) A' (c) $A \cap P'$ (d) $A \cap B$ (e) $(A \cap B)'$
    (f) $A' \cap B'$ (g) $A \cup B$ (h) $(A \cup B) \cap P$ (i) $A' \cap B' \cap P'$

23. A sociologist is interested in studying divorce and intends to study a group of married couples over a long period of time. The following events are of interest:

    P: "The parents of at least one member of the couple were divorced while that member was living at home."

    C: "The couple divorces."

    (a) What physical events correspond to the following?

        $P \cup C$ $P \cap C$ $P \cap C'$ $P' \cap C'$

    (b) Express the following event in set notation: "the parents of neither member of the couple divorced while the member was living at home, but the couple divorced."
    (c) Do you think that events P and C are mutually exclusive? Explain.
    (d) Construct Venn diagram representations for each of the events listed in part (a).

24. A businessperson is interested in the behavior of the stock market and the rate of inflation. In order to study this situation the following events are pinpointed:

    A: "The stock market shows an increasing trend."

    B: "The inflation rate increases."

(a) Describe each of these events in words:

$$A \cap B \qquad A \cup B \qquad A' \cap B \qquad A \cap B'$$

(b) Express the following event in set notation: "the stock market does not show an increasing trend and the inflation rate does not increase."
(c) Do you think that A and B are mutually exclusive? Explain.
(d) Construct Venn diagram representations for each of the events listed in part (a).

25. A study is conducted to investigate these three events:

P: "The child is premature."

M: "The child's mother smokes."

B: "The child has a birth defect."

(a) Describe the physical event that corresponds to each of the following:

$$P' \qquad P \cap M \cap B' \qquad P' \cap M' \cap B \qquad (P \cup B) \cap M'$$

Express each of the events in parts (b)–(e) in set notation.
(b) "The child's mother does not smoke."
(c) "The child's mother does not smoke but the child is born prematurely."
(d) "The child is premature, and the mother smokes, and there is a birth defect."
(e) "The child is not premature, the mother does not smoke, and there is no birth defect."
(f) Do you think the events P and M are mutually exclusive? Explain.

26. When a computer goes down, the cause may be a hardware problem (H) or an operator error (E). These events are not mutually exclusive. Explain what this means in a practical sense. Draw a Venn diagram representation of H and E that makes it clear that the events are not mutually exclusive.

2.3   Some Rules of Probability

In the previous sections, we introduced the basic terminology of probability theory. We discussed the ideas of a sample space S, an event, the sure or certain event, the impossible event, and mutually exclusive events. We also considered the classical and relative frequency approaches to problems involving chance. To make full use of these concepts in solving complex problems, it is necessary to pause and consider the mathematical structure underlying probability theory. That is, we need to consider briefly some of the properties of probabilities and some of the rules that govern their behavior.

The first three properties that we state are logical intuitively and are accepted without proof. If you are familiar with the term *axiom* from your study of plane geometry, then you can think of these properties as being the **axioms of probability**.

Basic Properties of Probability

1. $P[S] = 1$
2. $P[A] \geq 0$ for any event A
3. $P[A_1 \cup A_2 \cup A_3 \ldots] = P[A_1] + P[A_2] + P[A_3] + \cdots$ for every finite or infinite collection of mutually exclusive events.

The first two properties are not difficult to understand since they are used intuitively by most people even if they have had no formal training in probability. Property 1 simply states that the probability assigned to a certain event, one that is sure to occur, is 1. Property 2 asserts that probabilities cannot be negative. Property 3 is referred to as the **countable additivity** property of probabilities. It simply gives one the right to compute the probability of the union of a collection of mutually exclusive events by adding together the probabilities associated with the individual events. The idea is illustrated in Example 1.

Example 1 A broker feels that the probability that a given stock will go up in value during the day's trading is .3 and the probability that it will go down in value is .1. What is the probability that it will go up or down? This question can be answered easily by noting first that the two events

A_1: "The stock goes up in value."

A_2: "The stock goes down in value."

constitute a finite collection of mutually exclusive events. The events are mutually exclusive since the closing price of the stock cannot be both above and below its starting price simultaneously. We are thus asked to find

$$P[A_1 \cup A_2]$$

By Property 3,

$$P[A_1 \cup A_2] = P[A_1] + P[A_2] = .3 + .1 = .4$$

These three basic properties can be used to derive a series of rules that help us calculate the probabilities of various events. These rules are not hard to understand. Their derivations are outlined as exercises.

Rule 1 $P[\varnothing] = 0$

This rule states a fact that we pointed out earlier. Namely, that the probability assigned to the impossible event is zero.

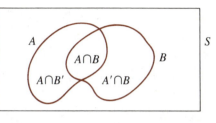

Figure 2.3
$A \cap B \subseteq A$ and $A \cap B \subseteq B$.

Rule 2

$$P[A'] = 1 - P[A] \qquad \text{for any event } A$$

This rule states that the probability that an event A will *not* occur is equal to 1 minus the probability that it will occur. For example, if the probability that a heart transplant will be successful is given as 2/3, then the probability that it will not be successful is $1 - 2/3 = 1/3$.

Property 3 allows us to compute the probability of the union of two or more events provided those events are mutually exclusive. We now develop a rule, called the **addition rule**, by which the probability of the union of two events can be found even if the events are *not* mutually exclusive. To derive this rule, let us consider the Venn diagram shown in Figure 2.3.

Note that $A \cap B \subseteq A$ and also $A \cap B \subseteq B$. Thus, if we add the probabilities associated with events A and B, as we have done previously, the probability associated with the event $A \cap B$ will be included twice. This mistake can be corrected by subtracting the term $P[A \cap B]$ from the sum $P[A] + P[B]$. In this way, we obtain the addition rule.

Rule 3

Addition Rule

$$P[A \cup B] = P[A] + P[B] - P[A \cap B]$$

Example 2 illustrates the use of this rule.

Example 2

A student referendum is held to determine student opinion concerning the construction of a recreation complex on campus. Unknown to the administration, 50% of the student body actually favors the construction (F). When the referendum is held, only 40% of the students vote (V). Overall, 32% of the student body vote in favor of the construction ($V \cap F$). If a student is selected at random from among the student body, what is the probability that he or she either voted *or* was in favor of the building? Since percentages may be interpreted as being probabilities, we are given that $P[F] = .5$, $P[V] = .4$ and $P[V \cap F] = .32$. We are asked to find $P[V \cup F]$. By the addition rule,

$$P[V \cup F] = P[V] + P[F] - P[V \cap F]$$

$$= .4 + .5 - .32$$

$$= .58$$

Venn diagrams are especially useful in displaying information in an easily accessible form. Pertinent questions often can be answered by inspection without having to worry about using a formula at all. To illustrate this idea, we reconsider the information concerning the student referendum given in Example 2.

Example 3

Let V denote the event that a student voted in the referendum and let F denote the event that a student favors the construction of the recreation complex. We are given that $P[F] = .5$, $P[V] = .4$ and $P[V \cap F] = .32$. To use this information fully we construct a Venn diagram that apportions the probability within the sample space S among the mutually exclusive events $V \cap F$, $V \cap F'$, $V' \cap F$, and $V' \cap F'$. The apportionment is done by the following steps:

Step 1   By Property 1, $P[S] = 1$. [See Figure 2.4(a).]

Step 2   Note that $P[V \cap F] = .32$. [See Figure 2.4(b).]

Usually, finding the probability of the intersection of the events is the first thing to do when using a Venn diagram. Here it was given in the problem.

Step 3   Note that $P[V] = .40$. Hence 40% of the probability involved should be accounted for within event V in the diagram, and 32% of this probability has already been located in $V \cap F$. Hence, $.40 - .32 = .08$ must be the probability associated with the event $V \cap F'$. [See Figure 2.4(c).]

Step 4   Similarly, since $P[F] = .50$, $P[F \cap V'] = .50 - .32 = .18$. [See Figure 2.4(d).]

Step 5   Finally, since $P[S] = 1$, $P[F' \cap V'] = 1 - P[F \cup V] = 1 - .58 = .42$. [See Figure 2.4(e).]

The probability that a student either voted for or favored the construction of the building, but not both, is denoted by $P[(V \cap F') \cup (F \cap V')]$. This probability can be found by inspecting Figure 2.4(e), and adding together the probabilities associated with the two shaded regions shown. In particular,

$$P[(V \cap F') \cup (F \cap V')] = .08 + .18 = .26$$

Note that the addition rule, in fact, relates the probability of the union to the probability of the intersection. By rewriting the rule in the alternative form

$$P[A \cap B] = P[A] + P[B] - P[A \cup B]$$

it can be used to find the probability of an intersection.

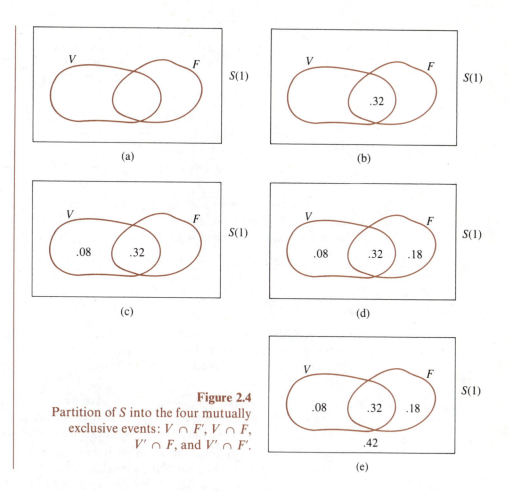

Figure 2.4
Partition of S into the four mutually
exclusive events: $V \cap F'$, $V \cap F$,
$V' \cap F$, and $V' \cap F'$.

Example 4

A recent study shows that 15% of the used cars on the market have had their
odometers set back (A). Furthermore, 5% of the used cars available have been
involved in an accident (B). The study also shows that 18% have had their
odometers set back *or* have been in an accident. Find the probability that a
randomly selected used car has had its odometer set back *and* has been in an
accident. The key word in the question is *and*. We are looking for $P[A \cap B]$. We
are given that

$$P[A] = .15$$
$$P[B] = .05$$
$$P[A \cup B] = .18$$

Using the addition rule in alternative form, we can conclude that

$$P[A \cap B] = P[A] + P[B] - P[A \cup B]$$
$$= .15 + .05 - .18$$
$$= .02$$

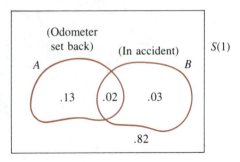

Figure 2.5
Partition of *S* according to the
accident status and odometer reading
on a used car.

The Venn diagram representation for these data is given in Figure 2.5. From the diagram we see that the probability that a used car has had its odometer set back but has not been in an accident is given by

$$P[A \cap B'] = .13$$

Exercises 2.3

27. An experiment consists of rolling a pair of fair dice. Find the probability of rolling a sum of 7; a sum of 11. Are these events mutually exclusive? Find the probability of rolling a sum of 7 *or* 11. What rule or property are you using to find this probability?

28. An experiment consists of drawing a single card from an ordinary deck. Find the probability that the card is an ace; that it is a king. Are these events mutually exclusive? Find the probability that the card is an ace *or* a king.

29. When a couple, each with one recessive (blue) and one dominant (brown) gene for eye color, parent a child, the probability that the child will have blue eyes is 1/4. What is the probability that the child's eyes will not be blue?

30. The Mets are involved in the World Series. Suppose that the probability that they will win at least one game is .7. Find the probability that they will win no games in the series.

31. Let *A* and *B* be events such that $P[A] = .6$, $P[B] = .3$ and $P[A \cap B] = .1$. Find each of these probabilities:
    (a) $P[A']$
    (b) $P[B']$
    (c) $P[A \cup B]$
    (d) $P[A' \cap B]$
    (e) $P[A \cap B']$
    (f) $P[A' \cap B']$

32. Let *A* and *B* be events such that $P[A] = .4$, $P[B] = .7$ and $P[A \cup B] = .9$. Find each of these probabilities:
    (a) $P[A']$
    (b) $P[B']$

(c) $P[A \cap B]$

(d) $P[A' \cap B]$

(e) $P[A \cap B']$

(f) $P[A' \cap B']$

33. Let A and B be events such that $P[A' \cap B] = .1$, $P[A \cap B'] = .3$ and $P[A \cup B] = .7$. Find each of these probabilities:
    (a) $P[A \cap B]$
    (b) $P[A' \cap B']$

Use Venn diagrams to help answer the questions asked in exercises 34–39.

34. A juvenile court study shows that 60% of all children seen in its court come from low-income families (L), 50% come from broken homes (B), and 40% are involved in both situations ($L \cap B$). If a child is selected at random from the children seen by this juvenile court, find the probability that:
    (a) the child is from a low-income family or a broken home
    (b) the child is from a low-income family but not a broken home
    (c) the child is not from a broken home
    (d) the child is from neither a broken home nor a low-income family

35. A blood-bank worker in a college town reports that 80% of the blood donated is from volunteer donors (V), 50% is from college students (C), and 40% is from persons who are college students and volunteer donors. Find the probability that a randomly selected donor is:
    (a) a college student or a volunteer donor
    (b) a college student but not a volunteer donor
    (c) neither a college student nor a volunteer donor
    (d) not a college student
    Are the events V and C mutually exclusive? Explain.

36. In a market-research project designed to ascertain the effectiveness of advertising, a business executive found that of all people surveyed, 35% had heard the firm's ads (H); 30% had actually patronized the company (P); and 50% had heard the firm's ads or had patronized the company ($H \cup P$). Asssume that these percentages reflect those of the general public. If a person is selected at random and questioned, find the probability that he or she:
    (a) has heard the ads and has patronized the business
    (b) has heard the advertising but has not patronized the business
    (c) has not heard the ads and has not patronized the business

37. In a recent controversy in a large religious organization, two proposals were put forth. One would permit the ordination of women to the priesthood, and the other would revise the book of prayer currently in use. It was estimated that among the lay people involved, 60% opposed the ordination of women (the rest were in favor of it) and that 20% favored the revision of the prayer book (the rest opposed it). It was felt also that 50% favored the ordination of women or favored the prayer book revision. Find the probability that a randomly selected individual from this group will:

(a) favor both the ordination of women and the revision of the prayer book

(b) favor exactly one proposal

(c) oppose both proposals

38. During a recent downward trend in the economy, 10% of the work force in a small industrial town was unemployed, 15% was eligible for food stamps, and 0.5% fell into both categories. If a worker is selected at random in this town, find the probability that he or she

(a) is unemployed and eligible for food stamps

(b) is unemployed or eligible for food stamps

(c) is unemployed but not eligible for food stamps

39. In a study of voting patterns among United States senators, records show that 70% voted for a tax cut, and 37% voted for an increase in military spending, while 77% voted for a tax cut or an increase in military spending. If a senator is selected at random and questioned, find the probability that he or she:

(a) voted for both measures

(b) voted for the increase in military spending but not for the tax cut

(c) voted for neither measure

*40. Show that $P[\varnothing] = 0$. (*Hint*: Write the sample space as $S = S \cup \varnothing$ and apply properties 3 and 1.)

*41. Show that $P[A'] = 1 - P[A]$. (*Hint*: Write the sample space as $S = A \cup A'$ and apply properties 3 and 1.)

2.4   Conditional Probability

Here we introduce the notion of conditional probability. The name given to this topic suggests what is to be done. We wish to compute the probability that event A will occur conditioned on or if, or given that some event B has already occurred. For example, we might wish to compute the probability that a poker player has a full house if we know that he holds two aces; the probability that an incumbent United States president will be returned to office, given the fact that he has operated under a balanced budget while in office; the probability that a business firm will succeed if its nearest competitor fails. Note that the key words are the words *if* and *given that*. The general situation encountered in practice, which will call for a conditional approach, is described below.

There is an experiment to be studied. A sample space S for the experiment has been constructed, an event A has been identified, and its probability, $P[A]$, has been determined. We then receive additional information that some event B has occurred. We wish to determine the effect, if any, that this new information has on our choice of $P[A]$. That is, we wish to determine the probability that A will occur given that B has already occurred. We call this probability the *conditional probability of A, given B* and denote it by $P[A \mid B]$. Note that even though there are two events involved, there is only one whose actual occurrence is in doubt, namely A. We know that B has occurred. Hence, $P[A \mid B]$ represents only one number.

There are two methods for handling conditional probability problems: common sense or the formal definition of $P[A \mid B]$. We will have occasion to use each.

Example 1

A single fair die is rolled once. What is the probability that a number less than 4 is rolled? This question is not new. To answer it, we simply consider the sample space

$$S = \{1, 2, 3, 4, 5, 6\}$$

in which each sample point is equally likely and occurs with probability 1/6. Relative to S, the event A "the number is less than 4" occurs with probability $3/6 = 1/2$. That is, $P[A] = 1/2$. Suppose that we now receive information that the number rolled is odd. Now what is the probability that the number is less than 4? Common sense tells us that some of the sample points of S are no longer feasible and that a new sample space for the experiment is given by

$$S_1 = \{1, 3, 5\}$$

Note that each sample point in S_1 is equally likely and occurs with probability 1/3. Relative to S_1, event A occurs with probability 2/3. Denoting the event "the number rolled is odd" by B, we have shown that $P[A \mid B] = 2/3$. Since $P[A] \neq P[A \mid B]$, the receipt of the information that B has occurred does have an effect on the probability of A.

Our method of solution above is straightforward: We simply consider the original sample space S, throw out of it all sample points not consistent with our new information B, form a new sample space S_1, and recompute the probability of event A relative to S_1. This procedure works well when outcomes are equally likely and relatively small in number.

We cannot expect that every question involving conditional probability will be solvable by the above method. However, the following definition allows us to handle complex problems with ease. We shall show that the previous intuitive method of solution is, in fact, just an application of this definition.

Definition 2.4

Let A and B be events such that $P[B] \neq 0$. The **conditional probability** of A, given B, denoted by $P[A \mid B]$, is given by

$$P[A \mid B] = \frac{P[A \cap B]}{P[B]} = \frac{P[\text{both events occur}]}{P[\text{given event occurs}]}$$

Note that there is one mathematical restriction on the events involved, namely that $P[B] \neq 0$. This restriction is necessary since we cannot divide by 0. This does not limit the physical application of the definition, since B represents an event that is assumed to have occurred. Hence B is not impossible and $P[B] \neq 0$. Note that the probabilities here are all computed relative to the same sample space S. The definition automatically takes care of the sample space reduction that was done in our first example. To see that this is true, let us reconsider the question asked in Example 1.

Example 2

A single fair die is rolled once. Let A denote the event that the number obtained is less than 4; let B denote the event that an odd number is rolled. For this experiment,

$$S = \{1, 2, 3, 4, 5, 6\}$$

Let us compute $P[A \mid B]$ from the definition, keeping in mind that all probabilities are computed relative to the sample space S.

$$P[A \mid B] = \frac{P[A \cap B]}{P[B]} = \frac{P[\text{number is less than 4 and odd}]}{P[\text{number is odd}]}$$

$$= \frac{2/6}{3/6} = 2/3$$

Note that this is the same result as was obtained previously.

The next example illustrates the use of Definition 2.4 in solving a problem for which no explicit sample space is available.

Example 3

Economists propose several ways to slow the rate of inflation and stimulate the economy. Congressional reaction to two of these measures is sought. It is found that 50% of all members of Congress favor a tax cut (T), 30% favor a budget cut (B), and 25% favor both of these measures ($B \cap T$). If a member of Congress is selected randomly and interviewed, what is the probability that the individual favors a budget cut, given that he or she favors a tax cut? If the individual does not favor a budget cut, what is the probability that he or she does favor a tax cut? We are given that $P[T] = .55$, $P[B] = .30$, and $P[B \cap T] = .25$. We are asked to find $P[B \mid T]$ and $P[T \mid B']$. Definition 2.4 can be applied immediately to find $P[B \mid T]$. In particular,

$$P[B \mid T] = \frac{P[B \cap T]}{P[T]} = \frac{.25}{.55} = \frac{5}{11}$$

Finding $P[T \mid B']$ requires a little more thought. We know that

$$P[T \mid B'] = \frac{P[T \cap B']}{P[B']}$$

We also know that $P[B'] = 1 - P[B] = 1 - .30 = .70$. However, we do not know $P[T \cap B']$. To find this probability, we construct the Venn diagram shown

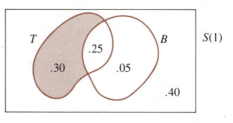

Figure 2.6
Partition of S according to opinion
concerning tax and budget cuts.

in Figure 2.6. The shaded region denotes the event $T \cap B'$. Its probability is seen
to be .30. Substituting, it is easy to see that

$$P[T \mid B'] = \frac{P[T \cap B']}{P[B']} = \frac{.30}{.70} = \frac{3}{7}$$

Conditional probability problems are not hard to solve once the question has
been translated from English into symbols, but be careful—it is not usually true
that $P[A \mid B] = P[B \mid A]$! The order in which you write the events does matter, so
read the problem very carefully to be sure that you correctly identify the given
event.

Exercises 2.4

42. A single die is rolled once. A: the number is less than 3; B: the number is
even; and C: the number is divisible by 3. Find:
    (a) $P[A]$ (b) $P[B]$ (c) $P[A \cap B]$ (d) $P[A \mid B]$
    (e) $P[C]$ (f) $P[A \cap C]$ (g) $P[A \mid C]$ (h) $P[B \cap C]$
    (i) $P[C \mid B]$ (j) $P[B \mid C]$ (k) Is $P[B \mid C] = P[C \mid B]$?

43. An experiment consists of drawing a single card from an ordinary well-
shuffled deck. Let H: the card is an honor (ace, king, queen, jack, ten), B: the
card is black. Find:
    (a) $P[H]$ (b) $P[B]$ (c) $P[H \cap B]$
    (d) $P[H \mid B]$ (e) $P[B \mid H]$ (f) Is $P[H \mid B] = P[B \mid H]$?

44. A family has three children. The set of possible birth orders for these children, identifying each child as being either male (M) or female (F) is

$$S = \{MMM, MMF, MFM, FMM, FFM, FMF, MFF, FFF\}$$

    (a) Assume that the sample points listed are equally likely. Find the prob-
    ability that exactly two of the three children are male.
    (b) If it is known that the first two children born are male, what is the
    probability that there are exactly two male children in the family?

45. Consider the experiment of rolling a pair of fair dice once. Let A denote the
event that the sum is even, B denote the event that the sum exceeds 8, and C
the event that doubles is rolled. Find:
    (a) $P[A]$ (b) $P[A \cap B]$ (c) $P[B]$ (d) $P[A \mid B]$
    (e) $P[B \mid A]$ (f) $P[C]$ (g) $P[A \cap C]$ (h) $P[A \mid C]$
    (i) $P[C \mid A]$ (j) $P[B \cap C]$ (k) $P[B \mid C]$ (l) $P[C \mid B]$

46. A market-research study showed that 25% of the businesses surveyed advertise in the local paper (A) and that 80% showed a profit for the year (Y). It also showed that 22% advertised in the local paper and showed a profit. Based on these data, approximate the probability that a randomly selected business
    (a) showed a profit, given that it advertised in the local paper
    (b) did not show a profit, given that it advertised in the local paper
    (c) advertised in the local paper, given that it showed a profit
    (d) advertised in the local paper, given that it did not show a profit

47. The probability that a fugitive will be placed on the FBI's Ten Most Wanted List is .02, the probability that a fugitive will be captured is .80, and the probability that a fugitive will be placed on the list or captured is .81.
    (a) Find the probability that a fugitive will be placed on the list and captured.
    (b) Find the probability that a fugitive will not be captured.
    (c) Find the probability that a fugitive will be captured, if placed on the Ten Most Wanted List.
    (d) Find the probability that a fugitive will not be captured, if placed on the Ten Most Wanted List.

48. In sampling student opinion, it is found that 40% of the students surveyed favor the resumption of the military draft for young men (M); 35% favor a draft of young women for service in some capacity (not necessarily military) (W); and 55% favor neither proposal ($W' \cap M'$). If these figures are taken to be a representative of general student opinion, what is the probability that a randomly selected student will favor a draft of young women, if it is known that he or she favors a military draft for young men? What is the probability that a student will favor a military draft of young men, if it is known that the student is not in favor of any kind of draft for women?

49. Health officials estimate that 15% of the adult population has hypertension (H). Surveys indicate that 75% of all adults do not think that they personally have the disease (N). It is estimated that 6% of all adults have hypertension but do not think they have the disease ($H \cap N$). If an adult patient thinks that he or she does not have hypertension, what is the probability that the disease is, in fact, present? If the disease is present, what is the probability that the patient will suspect its presence?

50. The **false positive rate** of a medical test is defined to be the probability that the test indicates the presence of the disease when in fact the disease is not present. That is,

$$\text{false positive rate} = P \left[\begin{array}{l|l} \text{test results are} & \text{subject does not} \\ \text{positive} & \text{have the disease} \end{array} \right]$$

    (a) Ideally, should the false positive rate be large or small? Explain.
    (b) A study of a new method for detecting kidney disease is conducted. Forty-nine percent of the subjects were reported as positive. Using

another detection method it was found that 39% of the subjects actually had the disease. It was also found that 17% of the subjects did not have the disease, but the new test indicated its presence. Use these data to approximate the false positive rate for this new procedure.

(c) If you take the new test and the result is positive, is this clear evidence that the disease is present? Explain.

(d) How do you think that the **false negative rate** for a medical test is defined?

(e) From a patient's point of view, is it more important to have a low false positive rate or a low false negative rate? Explain.

51. Use the data of exercise 34 to find the probability that a randomly selected child is
   (a) from a low-income family given that he or she is from a broken home
   (b) not from a low-income family given that he or she is from a broken home
   (c) from a broken home if he or she is from a low-income family

52. Use the data of exercise 35 to find the probability that a randomly selected blood donor was
   (a) a volunteer given that he or she was a college student
   (b) a college student given that he or she was not a volunteer
   (c) not a volunteer given that he or she was a college student

53. Use the data of exercise 36 to find the probability that a randomly selected individual has
   (a) patronized the business given that he or she had heard the firm's ads
   (b) patronized the business given that he or she had not heard the firm's ads

2.5   Independence and the Multiplication Rule

We have already considered one relationship that might exist between two events. In particular, two events may or may not be mutually exclusive. We now consider another possible relationship between events. Namely, we define what is meant when we say that one event is independent of another. The notion of conditional probability will help us develop this definition in a mathematical sense. Most people have an intuitive idea of the meaning of the phrase "these events are independent." To see how to express this concept mathematically, let us consider the following pairs of events:

A: "A person is less than 4 feet tall."
B: "A person weighs over 150 pounds."

H: "Your college cafeteria serves hot dogs for lunch."
L: "The space shuttle makes a safe landing."

Which of these is a pair of independent events? Common usage of the word *independent* leads us to say that A and B are not independent, but that H and L are independent. It is obvious that height does affect weight and hence that A and B are not to be considered independent events. It is equally as obvious that the lunch menu at your college cafeteria has no influence on the space shuttle and hence that H and L should be thought of as being independent. Intuitively, we are saying this:

> Two events are considered to be **independent events** if the occurrence or nonoccurrence of one has no influence on the occurrence or nonoccurrence of the other.

To understand the mathematical implications of this idea, let us assume that the probability that your college cafeteria will serve hot dogs for lunch is .70. That is, $P[H] = .7$. We now ask, if the space shuttle makes a safe landing, what is the probability that your college cafeteria will serve hot dogs for lunch? That is, What is $P[H|L]$? Since H and L are independent, it is evident that $P[H|L] = .7$ as before. The information concerning the space shuttle is completely irrelevant and hence does not influence the figure for $P[H]$. This observation leads us quite naturally to the definition of the term *independent events*.

Definition 2.5

> Let A and B be events such that $P[B] \neq 0$. Events A and B such that $P[A|B] = P[A]$ are said to be **independent events**.

Note that this says that two events are independent if the probability of A, given B, is the same as the probability assigned to A *before* receipt of the information that B has occurred. This definition can be used to test events for independence.

Example 1

Streams located near power plants and other industrial plants that release wastewater into the water system may suffer thermal or chemical pollution. Studies indicate that 5% show signs of both types of pollution, 45% are chemically polluted, and 35% are thermally polluted. Are the events C: "a stream is chemically polluted," and T: "a stream is thermally polluted" independent? To see this, note that

$$P[C|T] = \frac{P[C \cap T]}{P[T]} = \frac{.05}{.35} = \frac{5}{35} = .14$$

Note also that $P[C] = .45$. Since $P[C|T] \neq P[C]$, we can conclude that these forms of pollution are not independent.

Definition 2.5 can be used to find a formula for computing the probability that two events that are assumed to be independent will occur simultaneously. That is, we can derive a formula for $P[A \cap B]$, when A and B are independent. To do so, note that

$$P[A|B] = \frac{P[A \cap B]}{P[B]} \qquad \text{holds for any events } A \text{ and } B$$
$$\text{as long as } P[B] \neq 0$$

$$P[A|B] = P[A] \qquad \text{holds if } A \text{ and } B \text{ are independent}$$

Since we have two expressions, each equal to $P[A \mid B]$, they must be equal to each other. That is, we can conclude that if A and B are independent and $P[B] \neq 0$, then

$$\frac{P[A \cap B]}{P[B]} = P[A]$$

Solving this equation for $P[A \cap B]$, we obtain the following rule for finding the probability of the joint occurrence of two independent events:

Rule 4

Multiplication Rule for Independent Events

If A and B are independent, then $P[A \cap B] = P[A] \cdot P[B]$.

This rule is often taken as the definition of the term *independent events*. We illustrate its use in Example 2.

Example 2

In the United States approximately 46% of the population has type O blood. Approximately 39% has a negative Rh factor. If a person is selected at random, what is the probability that he or she will have type O-negative blood? Genetic studies have shown that an individual's blood type is independent of the Rh factor. Let A: the blood group is O; B: the Rh factor is negative. We can conclude that:

$$P[\text{type O and negative}] = P[A \cap B] = P[A] \cdot P[B]$$

$$= (.46)(.39) = .179$$

That is, approximately 17.9% of all individuals in the United States have type O-negative blood.

The idea of independence is extended to include more than two events in a natural way. In particular, suppose that we have a collection of events that are independent in the sense that the occurrence of one has no effect on the occurrence of any of the others. To find the probability that all of these events will occur, we simply multiply their individual probabilities. This idea is illustrated by considering Example 3. Strange as it may seem, the situation described in this example actually did occur a few years ago!

Example 3

Shortly after the crash of two 747 jumbo jets, it was reported that a large number of people had played a weird hunch and wagered on the number 7470 in the Massachusetts lottery. The winning number in this lottery is generated by selecting four digits in order in such a way that at each step in the selection process each digit from 0 to 9 has probability 1/10 of being chosen. What is the probability that the winning number will be 7470? We need only note that, due to the method of generating the winning number, the events

A_1: "The first digit is 7."

A_2: "The second digit is 4."

A_3: "The third digit is 7."

A_4: "The fourth digit is 0."

are independent. Hence,

$$P[7470 \text{ wins}] = P[A_1 \cap A_2 \cap A_3 \cap A_4]$$
$$= 1/10 \cdot 1/10 \cdot 1/10 \cdot 1/10 = 1/10000$$

The number 7470 did win the lottery! It was also reported that so many people played this hunch that the normal winning payoff of around $6000 was reduced to approximately $2000.

Independence is often assumed in cases where it does not necessarily hold. The following example, taken from *Three Mile Island*, [19], by Mark Stephens, illustrates the sometimes disastrous consequences of making this incorrect assumption!

Example 4

WASH 1400, a 1975 Atomic Energy Commission study of the probability of reactor accidents and their results, estimated that a meltdown and breach of containment would kill 27,000 people, injure another 73,000 and cause $17 billion in property damage.

Of course, the WASH 1400 study, led by Dr. Norman Rasmussen of MIT, also assessed the probability of such a total meltdown at one in 10 million per year. That is, with the 72 commercial power reactors operating today in the United States, there is less than one chance in 10,000 of such a serious accident happening in the next decade.

And yet, Unit 2 stood glowing in its steam in some defiance of Dr. Rasmussen's calculations. There was a flaw in those figures that had so reassuringly said that we need not expect a meltdown in this century—a flaw that was taken advantage of by Unit 2.

The methodology of WASH 1400 makes use of "event trees": sequences of actions that would be necessary for accidents to take place. These event trees did not assume any interrelation between events—that they might be caused by the same error in judgement or as part of the same mistaken action.

The statisticians who assigned probabilities in the writing of WASH 1400 said, for example, that there was a one-in-a-thousand risk of one of the auxiliary feedwater control valves—the twelves—being closed. And if there is a one-in-a-thousand chance of one valve being closed, the chance of both valves being closed is one-thousandth of that, or a million-to-one. But both of the twelves were closed by the same man on March 26—and one had never been closed without the other.

The odds continued to change when an operator both opened a let-down line to drain the cooling system and turned off the high pressure injection that was sending water to the reactor core. To Dr. Rasmussen, they were *independent events, each to be assigned a probability, the product of which would be the real risk of an operator letting the reactor run dry.* But both events happened together. They were the logical outcome of an operator's thinking he was overfilling the cooling system. It was one mistake, not two, and was a thousand times more likely to occur.

The type of optimistic thinking that was the foundation of WASH 1400 pervaded the nuclear industry. So small were the chances of a major accident and so fail-safe the emergency systems, that certain accidents and situations didn't even have to be considered in emergency planning.[1]

If we pause to consider exactly what situations we can now handle, there is one obvious question that remains unanswered—How do we compute the probability of the simultaneous occurrence of two events that are not independent? What is $P[A \cap B]$ for nonindependent events? To answer this question, note that if A and B are events such that $P[B] \neq 0$ then

$$P[A \mid B] = \frac{P[A \cap B]}{P[B]}$$

Multiplying each side of this equation by $P[B]$ yields the multiplication rule for nonindependent events.

Rule 5

Multiplication Rule for Nonindependent Events

Let A and B be events such that $P[B] \neq 0$. Then

$$P[A \cap B] = P[A \mid B] \cdot P[B]$$

This rule is easy to use. Again, the trick is to read the problem carefully so that you will translate the information given from English into symbols correctly. The next example demonstrates the idea.

Example 5

Approximately 46% of the individuals in the United States have type O blood. It is estimated that during World War II, 4% of those with type O blood were mistyped as being type A. What is the probability that a randomly selected individual has type O blood, but is typed as being type A? To answer this question let O: the individual has type O blood; A: the individual is typed as A. We are given that $P[O] = .46$; $P[A \mid O] = .04$. We are asked to find $P[O \cap A]$. By the multiplication rule,

$$P[O \cap A] = P[A \mid O] \cdot P[O] = (.04)(.46) = .018$$

[1] From *Three Mile Island* by Mark Stephens. Copyright © 1980, by permission of Random House, Inc., pp. 26–27.

Exercises 2.5

54. Let A and B be events such that $P[A] = .5$, $P[B] = .6$, and $P[A \cup B] = .7$.
    (a) Find $P[A \cap B]$.
    (b) Find $P[A \mid B]$.
    (c) Are A and B independent events? Explain.

55. Let A and B be events such that $P[A] = .5$, $P[B] = .6$, and $P[A \cap B] = .3$.
    (a) Find $P[A \cup B]$.
    (b) Find $P[A \mid B]$.
    (c) Are A and B independent events? Explain.

56. In the United States, 9% of the population has type B blood and 61% of the population has a positive Rh factor. What is the probability that a randomly selected individual has B-positive blood?

57. If a man and a woman, each with one recessive (blue) and one dominant (brown) gene for eye color, parent a child, then the probability that the child will have brown eyes is 3/4. If this couple has two children, what is the probability that they will both have brown eyes? That they will both have blue eyes? That one will have blue eyes and the other brown eyes?

58. A woman who is a carrier of classical hemophilia has a probability of 1/2 of passing the disease on to each son. What is the probability that her first child will be a boy and will have hemophilia? If she has exactly three sons what is the probability that all three will have hemophilia? That none will have the disease? That at least one will be affected?

59. It is estimated that 25% of all females are blond (natural or otherwise); 10% of all cars on the market are yellow; 20% of all males under 25 wear a beard and 30% wear a mustache; and 0.1% of all dogs are Lhasa Apsos. A drugstore was robbed, and a witness claimed that a young man under 25 with a beard and a mustache was seen leaving the store. He was observed getting into a yellow car driven by a woman with blond hair; a Lhasa Apso was in the front seat of the car. The police later arrested a couple matching this description and charged them with the crime. What is the probability that by chance a couple would meet this description? What are you assuming in order to compute this probability? Do you feel that the assumption is completely warranted?

60. **Sampling with Replacement**: In selecting a set of n objects from a finite collection of N objects, $n < N$, the sampling may be done either with or without replacement. That is, an object may be selected, observed, and returned to the original collection to, perhaps, be selected again; or an object may be selected and retained. The former method insures independence among draws; the latter does not.
    (a) Two cards are drawn in succession *with* replacement from an ordinary deck of 52 cards. What is the probability that the second card will be a jack, given that the first is a jack?

(b) The cards of part (a) are drawn without replacement. What is the probability that the second will be a jack, if the first is a jack?

*61. During World War II, blood was collected from bloodmobiles in batches of 100 pints (100 pints were collected and then pooled). It was later found that 67% of such pools were contaminated with serum-hepatitus virus. What is the probability that a person selected at random who contributed to one of the batches was a virus carrier? (*Hint*: Let q be the probability that such an individual did not contribute virus to the pool. Use this to find the probability that the pool is virus-free.) Assume independence among individuals.

62. The reliability of a system or a device is defined as the probability that a system or device performs adequately for a specified period in a given environment. Consumers and producers alike are interested in products that have high reliability. Warranties are based on the producer's determination of the reliability of the product. For example, a one-year auto warranty implies that a manufacturer has confidence that the car will perform adequately for at least a year with high probability.

(a) A simplified version of the Apollo system used to land men on the moon is shown in Figure 2.7. In this system (independent components connected in series) the entire system functions if and only if all components shown operate properly. That is, the system fails if any component fails. Denote the reliability of a component k by R_k.

(i) Find an expression for the reliability of the entire system.

(ii) If the reliability of each component pictured is .999, what is the system reliability?

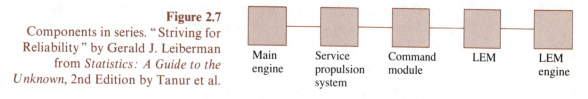

Figure 2.7
Components in series. "Striving for Reliability" by Gerald J. Leiberman from *Statistics: A Guide to the Unknown*, 2nd Edition by Tanur et al.

Main engine   Service propulsion system   Command module   LEM   LEM engine

(b) Suppose that the Apollo system is constructed in such a way that the main engine A_1 has an independent backup engine A_2 as shown in Figure 2.8

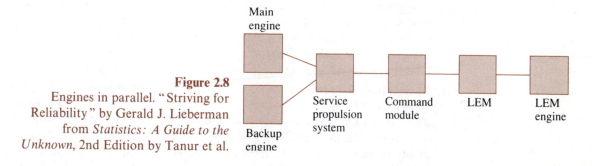

Figure 2.8
Engines in parallel. "Striving for Reliability" by Gerald J. Lieberman from *Statistics: A Guide to the Unknown*, 2nd Edition by Tanur et al.

Main engine

Backup engine

Service propulsion system   Command module   LEM   LEM engine

In this system (independent components with main engines in parallel) the main engine component fails if and only if both A_1 and A_2 fail.

  (i) Find an expression for the reliability of the main engine component.

  (ii) If both engines A_1 and A_2 have reliability .98, find the reliability of the main engine system. Compare this figure with the single-engine reliability of part (a).

  (iii) Find the reliability of the above system if each of the remaining components has reliability .999.

63. It is estimated that 50% of Americans are overweight and that 20% suffer from high blood pressure. It is also thought that 50% of all persons suffering from high blood pressure are overweight. If an individual is selected at random from the American public, what is the probability that the person will be overweight and suffering from high blood pressure? What is the probability that the person will not be overweight but suffering from high blood pressure?

64. Some 75% of all registered Democrats vote a straight party ticket, whereas only 60% of all registered Republicans do so. Of all registered voters, 38% claim to be Republicans, 52% claim to be Democrats, and the rest are independent.

  (a) What is the probability that a ballot selected at random will be a straight Republican ticket and will have been turned in by a registered Republican?

  (b) What is the probability that a randomly selected ballot will not be straight Republican but will have been turned in by a registered Republican?

*65. In medical studies, researchers are interested in the false positive rate and the false negative rate of certain screening tests. These rates are defined as follows:

α: false positive rate

α: $P[\text{disease found} \mid \text{disease is not present}]$

β: false negative rate

β: $P[\text{disease not found} \mid \text{disease is present}]$

An ideal test is one in which both rates are low. Suppose it is known that the false positive rate of the test for a certain disease is 5% and the false negative rate is 10%. Suppose, for a given group of subjects tested, 20% are reported as positive. If a person is selected at random from the group tested, what is the probability that the person actually has the disease? (*Hint*: $P[\text{disease found}] = P[\text{found and present}] + P[\text{found but not present}]$.)

*66. **Randomized Response Method for Getting Honest Answers to Sensitive Questions:** This is a method used to guarantee an individual that answers to sensitive questions will be anonymous, thus encouraging a truthful response. This method is, in effect, an application of the formula for finding the probability of an intersection. It operates as follows. Two questions A and B are posed, one of which is sensitive and the other not. The probability of receiving a yes to the nonsensitive question must be known. For example, one could ask

A: Does your Social Security number end in an odd digit? (Nonsensitive)

B: Have you ever intentionally cheated on your income taxes? (Sensitive)

We know that $P[\text{answer yes} \mid \text{answered } A] = 1/2$. We wish to approximate $P[\text{answer yes} \mid \text{answered } B]$. The subject is asked to flip a coin and answer A if the coin comes up heads and answer B if it is tails. In this way, the interviewer does not know which question the subject is answering. Thus a yes answer is not incriminating. There is no way for the interviewer to know whether the subject is saying, "Yes, my Social Security number ends in a odd digit," or "Yes, I have intentionally cheated on my income taxes." The percentage of subjects in the group answering yes is used to approximate $P[\text{answer yes}]$.

(a) Use the fact that the event "answer yes" is the union of the event "answer yes and answered A" with the event "answer yes and answered B" to show that $P[\text{answer yes} \mid \text{answered } B]$ equals

$$\frac{P[\text{answer yes}] - P[\text{answer yes} \mid \text{answered } A]P[\text{answered } A]}{P[\text{answered } B]}$$

(b) If this technique is tried on 100 subjects and 60 answered yes, find the approximate probability that a person randomly selected from the group has intentionally cheated on income taxes.

*67. In a study of high school students, each subject was asked to roll a die and then flip a coin. If the coin came up heads, the subject was to answer question A below and if tails, question B.

A: Did the die land on an even number?

B: Have you ever smoked marijuana?

In a group of 50 subjects, 35 answered yes. Use this information to approximate the probability that a student randomly selected from this group has smoked marijuana.

2.6   Bayes' Rule (Optional)

A well-known result in mathematics is **Bayes' rule,** named for the Reverend Thomas Bayes (1761). This rule is used to compute a conditional probability. It comes into play when the information given does not permit us to apply the definition of conditional probability directly. The statement of Bayes' rule is rather frightening at first glance! We will show you how to use it via a numerical example before attempting a formal statement of this important result.

Example 1

It is estimated that there is a 20% chance that unemployment will increase by more than 1% next year. If this increase does occur, then there is a 90% chance that Congress will enact a federally funded job program; otherwise the probability of such a program being funded is only 30%. Suppose that the job program was funded by Congress. What is the probability that unemployment rose by more than 1%? To answer this question, let

U: unemployment increases by more than 1%

J: Congress enacts a job program

We are given that

$$P[U] = .20 \qquad P[J\,|\,U] = .90 \qquad P[J\,|\,U'] = .30$$
$$P[U'] = .80 \qquad P[J'\,|\,U] = .10 \qquad P[J'\,|\,U'] = .70$$

We are asked to find $P[U\,|\,J]$. We know that

$$P[U\,|\,J] = \frac{P[U \cap J]}{P[J]}$$

Unfortunately, neither of these probabilities is given to us. However, they can each be computed using methods developed earlier. In particular, by Rule 5, the multiplication rule for nonindependent events, we know that

$$P[U \cap J] = P[J\,|\,U] \cdot P[U]$$

$$= (.9)(.2)$$

$$= .18$$

We also know that either the job program is enacted and unemployment increases or else the job program is enacted and unemployment does not increase. That is,

$$P[J] = P[J \cap U] + P[J \cap U']$$

By applying Rule 5 to each term on the right side of this equation, we see that

$$P[J] = P[J\,|\,U] \cdot P[U] + P[J\,|\,U'] \cdot P[U']$$

$$= (.9)(.2) + (.3)(.8)$$

$$= .18 + .24 = .42$$

Combining these results, we see that

$$P[U\,|\,J] = \frac{P[U \cap J]}{P[J]}$$

$$= \frac{.18}{.42}$$

$$= 18/42$$

In this example, we actually used Bayes' rule to answer the question posed. For future reference, let us outline the characteristics that are present in a problem in which Bayes' rule is applicable and the general method used to solve the problem.

Applying Bayes' Rule

1. There are n mutually exclusive events A_1, A_2, ..., A_n whose union is S. (In our example, $A_1 = U$ and $A_2 = U'$.)

2. We receive information that some event E has occurred. (In our example, we received information that the job program was funded. Here, $E = J$.)

3. We want to find the probability that one of the original events will occur, given that event E has occurred. That is, we want to find $P[A_i | E]$. (In our example we wanted to find $P[A_1 | E] = P[U | J]$.)

4. We are not given $P[A_i \cap E]$ or $P[E]$, and so we cannot use the definition of conditional probability to answer the question posed directly. (In our example we were not given $P[A_1 \cap E] = P[U \cap J]$ or $P[E] = P[J]$.)

5. We must compute $P[A_i \cap E]$ and $P[E]$ using rule 5, the multiplication rule for nonindependent events. (In our example, we used rule 5 to find $P[A_1 \cap E] = P[U \cap J]$ and $P[E] = P[J]$.)

6. We substitute the values found for $P[A_i \cap E]$ and $P[E]$ into the definition of conditional probability to solve the problem. (In our example, we substituted $P[A_1 \cap E] = P[U \cap J]$ and $P[E] = P[J]$ into the expression $P[U | J] = P[U \cap J]/P[J]$ to answer the question posed.)

Bayes' rule summarizes the results of the computations indicated in items 5 and 6 into one convenient formula. This formula is as follows:

Rule 6

Bayes' Rule

Let A_1, A_2, A_3, ..., A_n be a collection of mutually exclusive events such that $A_1 \cup A_2 \cup A_3 \cup \cdots A_n = S$. Let E be an event such that $P[E] \neq 0$. Then for each $i = 1, 2, 3, ..., n$

$$P[A_i | E] = \frac{P[E | A_i] \cdot P[A_i]}{P[E | A_1]P[A_1] + P[E | A_2]P[A_2] + \cdots + P[E | A_n]P[A_n]}$$

The next example illustrates the use of Bayes' rule in a situation in which the sample space is divided into more than two mutually exclusive events.

Example 2   A manufacturer has available three machine operators. The first and most experienced operator, A, produces defective items only 1% of the time, whereas the other two operators, B and C, have defective rates of 5% and 7%, respectively.

The experienced operator is on the job 50% of the time; B works 30% of the time, and C, 20%. A defective item is produced. What is the probability that it was produced by operator A? Let

A_1: operator A works

A_2: operator B works

A_3: operator C works

E:　a defective part is produced

We are given the following probabilities:

$$P[E|A_1] = .01 \qquad P[E|A_2] = .05 \qquad P[E|A_3] = .07$$
$$P[A_1] = .50 \qquad P[A_2] = .30 \qquad P[A_3] = .20$$

We are asked to find $P[A_1|E]$. By Bayes' rule, this probability is

$$P[A_1|E] = \frac{P[E|A_1] \cdot P[A_1]}{P[E|A_1]P[A_1] + P[E|A_2]P[A_2] + P[E|A_3]P[A_3]}$$

$$= \frac{(.01)(.50)}{(.01)(.50) + (.05)(.3) + (.07)(.2)}$$

$$= \frac{.005}{.005 + .015 + .014} = \frac{.005}{.034} = 0.1471$$

That is, if we know that a defective item is produced, there is only about a 15% chance (14.71%) that the experienced operator A was at work.

Exercises 2.6

68. Use the information given in Example 1 to find the probability that unemployment did not rise by more than 1% given that the job program was funded.

69. Use the information given in Example 2 to find the probability that operator B was at work given that a defective part was produced. What is the probability that operator C was at work given that a defective part was produced?

70. The probability that a woman exposed to German measles will contract the disease is .2. If she is exposed to risk while pregnant, the probability is .1 that her child will suffer from a particular birth defect; otherwise, the probability of such a defect is only .01. Let D denote the event that the mother contracts the disease while pregnant and let B denote the event that her baby is born with the birth defect.
    (a) Express all the probabilities given in symbolic form.
    (b) A child is born with this defect. What is the probability that the mother contracted German measles while pregnant with this child?
    (c) A child is born without the defect. What is the probability that the mother contracted German measles while pregnant with this child?

71. The probability that July (winter) wheat futures in Chicago will fall by more than 10 cents per bushel from December 1 to January 1 is .7, if the precipitation in Kansas is 2 inches or greater during this period. Otherwise this probability is only .2. The probability of precipitation of 2 inches or more in Kansas during this period is 5%. Suppose that July wheat futures fall by 12 cents during the given one-month period. What is the probability that Kansas had less than 2 inches of rain or snow during December?

72. A screening test for cancer has a low false positive rate and a low false negative rate, in that, with careful use, one finds the disease in 95% of the patients having it; and in only 5% of the patients not having it. Assume that 4% of a certain population has the disease. What is the probability that a person who reacts positively to the test will, in fact, have the disease?

73. A government report gives these data on the drinking habits of men and women in the United States:

	Men	Women
Abstainers	23%	40%
Heavy	21%	5%
Moderate	46%	37%
Infrequent	10%	18%

Assume that half of the population is male.
(a) An individual is selected at random and found to be an abstainer. What is the probability that the individual is male?
(b) An individual is selected at random and found to be a heavy drinker. What is the probability that the individual is female?

74. It is estimated that 10% of all federal prisoners have a positive self-image, 40% have a neutral self-image, while the rest have a negative self-image. The probability of rehabilitating a prisoner with a negative self-image is .1. With a neutral self-image this probability is .4, and with a positive one it is .8.
(a) A prisoner is rehabilitated. Find the probability that the original self-image was negative.
(b) A prisoner is not rehabilitated. Find the probability that the original self-image was negative.

75. One controversial problem of interest to psychologists is the "cab problem." The statement of this problem is as follows:

A cab was involved in a hit-and-run accident at night. Two cab companies, the Green and the Blue, operate in the city. You are given the following data:

(i) A witness identified the cab as a Blue cab. The court tested his ability to identify cabs under the appropriate visibility conditions. When presented with a sample of cabs (half of which were Blue, half were Green) the witness made correct identifications in 80% of the cases and erred in 20% of the cases.

QUESTION: What is the probability that the cab involved in the accident was Blue rather than Green?[1]

Let B denote the event that the cab is Blue and let I denote the event that the cab is identified as Blue. Use Bayes' rule to answer the question by finding $P[B \mid I]$.

2.7 Counting Sample Points (Optional)

As we have noted, the first step in solving a probability problem is to determine what is possible. That is, we must think about a sample space for the experiment. The sample space can be constructed by listing the sample points, if the number of possibilities is small. When the number of possibilities is moderate in size, it is helpful to have a systematic method for determining or counting the elements of S. In this section we consider a convenient method for listing sample points called a **tree diagram**. We then consider the multiplication principle, a method for counting sample points that does not require that they be listed.

Tree diagrams are easy to construct. They are applicable when the experiment can be visualized as taking place in stages. To see how to draw a tree, consider the experiment of tossing a nickel and a dime simultaneously. What is a sample space for this experiment? It is easy to list the elements of S as ordered pairs in a manner similar to that used in the two-dice problem by agreeing that the first member of each pair represent the face appearing on the nickel, and the second that on the dime. We let H represent heads and T tails. Thus,

$$S = \{(H, H) (H, T) (T, H) (T, T)\}$$

For this example, the list is easy to obtain.

However, suppose that a nickel, a dime, a quarter, and a penny are tossed. Then the elements of S are logically elements of the form (H, H, T, H) or (H, T, T, H), where the results on each coin are listed in the order in which the coins are mentioned. If one attempts to list all the elements of S randomly, the task is difficult. It is hard to keep track of what has already been listed, and it is also difficult to know when the list is complete. To overcome both problems a tree diagram can be used.

To construct a tree diagram, we first ask, How many stages are involved in the experiment? In this case there are four, because we are tossing four coins, a nickel, a dime, a quarter, and a penny. The nickel can land in two ways, either heads (H) or tails (T). This is indicated in the tree as shown in Figure 2.9(a). After the nickel lands, the dime can also land either heads or tails. This is illustrated in Figure 2.9(b). Proceeding in a similar fashion with the quarter and penny, we obtain the entire tree, shown in Figure 2.9(c). The sample space can be read from

[1] "Base rates in Bayesian inference: Signal detection analysis of the cab problem," Michael Birnbaum, *American Journal of Psychology*, Vol. 96, 1, pp. 85–94, 1983. Copyright © 1983 by the Board of Trustees of the University of Illinois.

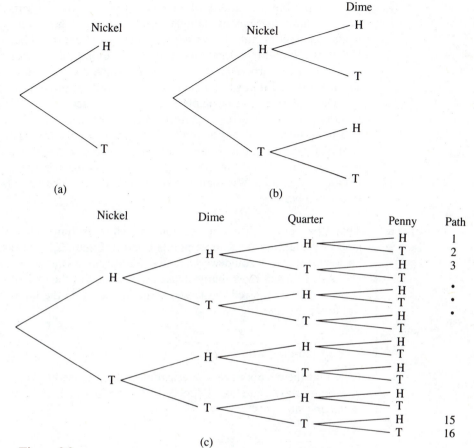

Figure 2.9
Tree diagram for the toss of four coins: a nickel, a dime, a quarter, and a penny.

the tree by reading the "paths" through the tree. A path is traced by placing your pencil at the beginning of the tree and passing it through four points without picking it up, as indicated in Figure 2.9(c). Thus

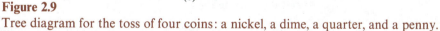

Since each path through the tree generates a sample point, it is easy to see that there are 16 sample points in S.

In the classical approach to probability, one assumes that each possible physical outcome of the experiment is equally likely, and, hence, that there exists a sample space for the experiment such that each sample point occurs with the same probability. In the classical context, the probability of an event is given by

$$P[A] = \frac{\text{number of sample points in } A}{\text{number of sample points in } S} = \frac{n(A)}{n(S)}$$

To compute the probability of an event A in the classical setting, it is necessary only to know how many sample points exist and how many of them satisfy condition A; it is not always necessary to know exactly what these points are. Hence, we should investigate methods for counting the number of ways in which experiments can proceed and events occur which do not require complete listings. The first method that we shall consider is the **multiplication principle**.

The multiplication principle is a counting procedure that is especially useful in situations in which the experiment involved can be visualized as taking place in a finite number of stages. (Note that tree diagrams are also applicable in this context.) The principle is more difficult to state verbally than it is to use in practice; when given the opportunity, most people use the principle correctly without much ado. We shall illustrate its use with examples before stating the rule formally.

Example 1

Toss a pair of fair dice, one red and one white. In how many ways can the dice land? By counting the sample points listed in Figure 2.2 it is easy to see that the answer is 36. How could we have arrived at this answer without having made a list? Ask yourself, How many stages are involved in the experiment? The answer is two: stage 1: the red die lands; stage 2: the white die lands; write down one slot for each stage:

$$\underline{} \quad \underline{}$$
red die   white die

Note that these slots have a physical meaning attached to them. Ask yourself, In how many ways can stage 1 proceed? The answer is 6, and a 6 should be placed in the first slot:

$$\underline{6} \quad \underline{}$$
red die   white die

Now ask yourself, After stage 1 has occurred, in how many ways can stage 2 proceed? Again, the answer is 6 as indicated below:

$$\underline{6} \quad \underline{6}$$
red die   white die

The multiplication principle then says that altogether the entire experiment can proceed in $6 \cdot 6 = 36$ ways. This answer agrees with our previous result.

Example 2

In how many ways can a nickel, dime, quarter, and penny land when tossed? We know from the tree of Figure 2.9 that the answer is 16. Using the multiplication principle, we think of the experiment as a four-stage process and write down four slots, one corresponding to each coin.

$$\underline{} \quad \underline{} \quad \underline{} \quad \underline{}$$
nickel   dime   quarter   penny

Each coin can land in two ways. Hence there are

$$\frac{2}{\text{nickel}} \cdot \frac{2}{\text{dime}} \cdot \frac{2}{\text{quarter}} \cdot \frac{2}{\text{penny}} = 16 \text{ ways}$$

in which the coins can land.

Example 3   A student wishes to select a science course from among biology, chemistry, and physics; a social science course from among sociology, government, psychology, and geography; and a physical education course from among golf, tennis, bowling, archery, and basketball. In how many ways can the student set up a three-course schedule? There are

$$\frac{3}{\text{science}} \cdot \frac{4}{\text{social science}} \cdot \frac{5}{\text{physical education}} = 60 \text{ ways}$$

in which this three-stage process can proceed. If one course is randomly selected from each of the three categories, what is the probability that psychology will not be chosen and an outdoor sport (golf, tennis, or archery) will be chosen? Since courses are being selected randomly, each of the 60 possible schedules is equally likely. The number of schedules satisfying the stated restrictions is

$$\frac{3}{\text{science}} \cdot \frac{3}{\text{social science}} \cdot \frac{3}{\text{physical education}} = 27 \text{ ways}$$

Using the classical approach, the desired probability is 27/60.

We now state the multiplication principle formally.

Rule 7

Multiplication Principle

Assume that an experiment takes place in k stages and that stage 1 can occur in n_1 ways; after it has occurred in one of these ways, stage 2 can occur in n_2 ways; after it has occurred in one of these ways, stage 3 can occur in n_3 ways, and so forth. Then altogether the experiment can proceed in $n_1 \cdot n_2 \cdot n_3 \cdots n_k$ ways.

The correct use of the multiplication principle requires an ability to visualize how an experiment can be broken into stages and how each stage can then proceed. Here are a few general guidelines to keep in mind:

1. Watch out for repetition versus nonrepetition. For example, when writing a Zip code, digits can be repeated. The number 24141 is a perfectly legitimate Zip code. However, when lining up children in a row, repetition is not allowed since it is physically impossible for one child to stand in two places simultaneously. Whether or not repetition is allowable depends entirely on

"If each answer raises three more questions,
we'll have 2,940 . . . and if each answer to those
raises three more, we'll have 8,820 . . . and if . . ."

Figure 2.10
Copyrighted by the Chicago Tribune. Used with permission.

the context of the problem. Unless stated otherwise, or physically prohibited, we assume that repetition is allowable.

2. Fill slots with stated restrictions (trouble spots) first, if possible. For example, if we were interested in counting the number of Zip codes that end in an even digit, then we would fill the last digit first since it has a restriction placed upon it.

3. Watch out for the use of subtraction. Occasionally one encounters problems characterized by the fact that the total number of ways for an experiment to proceed, $n(S)$, is easy to compute; the number of ways in which event A can proceed, $n(A)$, is to be calculated, but is hard to compute directly; the number of ways in which A' can proceed, $n(A')$, is easy to compute. Thus, $n(A)$ can easily be found by subtraction. That is,

$$n(A) = n(S) - n(A')$$

For example, in the game of Yahtzee, five fair dice are rolled. It is easy to see that $n(S) = 6 \cdot 6 \cdot 6 \cdot 6 \cdot 6 = 6^5 = 7776$. Suppose that we are interested in the event A: at least two dice show different numbers. To find $n(A)$ directly is virtually impossible. However, $n(A') = 6$ where A' is the event that all dice show the same number. Hence, $n(A) = n(S) - n(A') = 7770$.

These exercises involve tree diagrams and the multiplication principle. Look for the similarities that exist between the exercises and the examples just presented so that you can begin to develop the ability to recognize problem types.

Exercises 2.7

76. A stock market analyst classifies each stock as being either a blue chip stock or not. Each stock is also classified as to whether it goes up, down, or is unchanged in the day's trading. Construct a tree diagram for the two-stage process involved in fully classifying a stock. List the sample points generated by the tree.

77. Each patient seen at a public-health clinic is classified according to sex, whether or not he or she has a history of diabetes in the immediate family, and the status of the patient relative to diabetes. In classifying the history use three categories: yes (Y), no (N), unknown (U). In classifying the status of the patient, use three categories: diabetic (D), borderline (B), not diabetic (ND). Construct a tree for this three-stage classification process. List the sample points generated by the tree.

78. In humans, the four blood types A, B, AB, and O are recognized. Construct a tree to represent the possible blood types for a married couple. Assume that there are no restrictions due to some distant blood relationship and ignore the Rh factor.

79. When bags of cookies are mechanically filled, there is some variation in the number of cookies in each bag. To be acceptable, a bag must contain at least 20 cookies. To monitor its filling process, four bags of cookies are selected at random from a conveyor belt and checked. Let A denote the selection of an acceptable bag and U the selection of an unacceptable bag.
    (a) Construct a tree diagram for this four-stage process.
    (b) List the sample points in *S*.
    (c) The process is considered to be in control if at least three of the four bags are acceptable. List the sample points that comprise the event *I* "the process is in control."
    (d) List the sample points that comprise the event *O* "the process is out of control."
    (e) Are the events *I* and *O* mutually exclusive?

80. A family has three children. If each child is identified only by its sex, how many birth orders are possible? Assuming that each of the birth orders is equally likely, what is the probability that at least one child will be a boy?

81. Two commercials are scheduled during a 10-minute TV program. The first commercial is chosen from those listed in group I and the second from group II.

I	*II*
beer	Weight Watchers
candy	Alcoholics Anonymous (public service)
aspirin (drug)	highway safety (public service)
cold remedy (drug)	toothpaste
antacid (drug)	

(a) In how many ways can the commercials be run?

(b) In how many ways can the commercials be run if the first must be a drug ad and the last cannot be a public service ad?

(c) If an ad is randomly selected from each group, what is the probability that the candy ad will be run with the Weight Watchers ad, or that the beer ad will be run with a public service ad, thus creating a rather incompatible set of commercials?

82. An ice cream shop advertises that it offers over 1,000 possible sundaes for its customers. It has available the following ingredients: six kinds of ice cream, chopped almonds, chocolate sauce, salted peanuts, cashews, butterscotch sauce, cherries, pineapple, strawberries, and whipped cream. A sundae consists of one type of ice cream, one type of nut, plus any other ingredients desired. Is the ad being truthful? (*Hint:* This is an eight-stage process. When making a decision concerning the extra ingredients you have two choices: you can take the ingredient or not take it as your taste dictates.)

83. Post-office boxes may be opened by dialing in succession three letters from the letters A through J. Only one sequence of three letters opens the box.

(a) How many possible three-letter sequences can be formed?

(b) If one randomly dials letters on this type box, what is the probability that the box will open on the first attempt?

(c) If it is known that series of this type always end with a vowel and never have repeated letters, how many three-letter sequences are feasible?

(d) If a thief has available the information in part (c), what is the probability that he or she can open the box in a single try?

84. An auditor is spot-checking the records of a large business firm for accuracy. Each transaction has an invoice identifier of the form

——— ———   ——— ——— ———

where the first two positions are letters and the last three are digits. The auditor does not examine all records, only those records with invoice identifiers beginning with A, D, H, L, P, or T and ending in an even digit, with *no digits being repeated*. Letters can be repeated. What is the maximum number of records to be checked?

85. The basic storage unit of digital computers is a *bit*. A bit is either on (denoted by 1) or off (denoted by 0) at any given time. Computer pictures make use of picture elements called *pixels*. A pixel of two bits can code four gray levels by designating them as 00, 01, 10, and 11. How many gray levels can be designated using a four-bit pixel? an eight-bit pixel?

86. In trying to break a computer code, the five-letter password has been narrowed to the letters A, B, Q, Z, T, S, and X. If it is known that the password contains no repeated letters, what is the maximum number of passwords that could be tried before the correct sequence of letters is found?

87. A psychologist is working with a chimpanzee in a word-recognition experiment. The word *banana* is spoken. If the chimp makes a correct response

when it hears the word, then it receives a reward. The experiment consists of using the word five times and noting the response, either success or failure, each time.

(a) In how many ways can the experiment proceed?

(b) Assume that, on each of the five trials, the chimp is guessing. Thus, it is just as likely to succeed as to fail at each stage so that each of the possible outcomes of the experiment is equally likely. What is the probability that no two consecutive stages will be the same?

88. A man and a woman, each with one recessive (blue) and one dominant (brown) gene for eye color, parent a child. The child inherits one gene randomly from each parent. If at least one inherited gene is for brown eyes, the child will have brown eyes.

(a) Construct a tree diagram to represent the possible gene type for the child. Let B denote the inheritance of a gene for brown eyes and b the inheritance of a gene for blue eyes. Treat the experiment as a two-stage process with stage 1 being the inheritance of a gene from the father, and stage 2 the inheritance of a gene from the mother.

(b) Identify the eye color of the child for each of the four gene types found in part (a).

(c) Is this couple just as likely to have a blue-eyed child as they are a brown-eyed child?

(d) The couple in question parents six children. How many birth orders are possible identifying each child by eye color only?

(e) Do you think that each of the birth orders of part (d) is equally likely? Explain.

89. The four ribonucleotides, adenine (A), uracil (U), guanine (G), and cytosine (C), provide the principal substance of RNA, the intermediate information-carrying molecule involved in translating the DNA genetic code. They do so by forming "words," which are sequences of three of the ribonucleotides, not necessarily all different. For example, the sequence UUU denotes the amino acid phenylalanine; the sequence AUG denotes methionine.

(a) How many "words" can be formed?

(b) How many of these "words" have no repeated nucleotides?

(c) How many of these "words" contain at least two identical nucleotides?

(d) What is the probability that a randomly selected "word" will begin with adenine and contain no identical nucleotides?

2.8   Counting Arrangements of Objects: Permutations (Optional)

Counting problems fall into two broad categories. They entail *permutations* or *combinations*. In this section we consider methods for counting permutations. In the section that follows this one we consider the problem of counting combinations. Before we begin, it is essential that you be able to distinguish a permutation from a combination. We begin by defining these terms.

Definition 2.6   A **permutation** is an arrangement of objects in a definite order.

Notice that

1. The word *permutation* is a noun; it is the name given to an *arrangement* of objects.
2. The key words in the definition are the words *order* and *arrangement* (which implies order).

Definition 2.7   A **combination** is a selection of objects without regard to order.

Notice that

1. The word *combination* is also a noun; it is the name given to a collection of objects chosen *without regard to the order in which they are chosen.*
2. The key words in the definition are the words *without regard to order* and *selection.*

How does one read a counting problem and decide whether or not it is a permutation problem or a combination problem? Just ask yourself, Does the order in which the actions are taken or the objects chosen make any difference? That is, is order important? If the answer is yes, then you are dealing with a permutation; otherwise, you should think in terms of combinations. Watch for the words *arrange* and *select* in the statement of the problem; they are clues to the problem type and also to the method of solution.

Example 1

1. Spell your first name. Is this a permutation of letters or a combination of letters? Obviously it is a permutation, since the order in which the letters are written certainly matters.
2. A *venire* is a group of persons from which a jury is selected. A particular venire consists of 30 individuals from which a jury of 12 is to be chosen. Does this jury constitute a permutation of people or a combination? It is a combination since we are interested only in who is selected. We are *not* interested in the order in which jurors are chosen.

Usually, the question to be answered in a permutation problem is, How many permutations are possible? Tree diagrams and lists can be used to answer the question. There are also several permutation formulas that apply to simple problems. However, the most often used method is the multiplication principle. It is the method that should come to mind once you have decided that order is important.

Example 2 | A starting gate for a horse race has 10 gates, and 10 horses are entered in the race. In how many ways can the horses be arranged in the gates? Since we are interested in the order in which the horses are arranged, this is a permutation problem. The experiment consists of a 10-stage process with each stage being the placement of a horse in a gate. By the multiplication principle, the number of permutations or arrangements of horses is $10 \cdot 9 \cdot 8 \cdot 7 \cdot 6 \cdot 5 \cdot 4 \cdot 3 \cdot 2 \cdot 1 = 3,628,800$. You would not want to try a tree this time!

Note that it is awkward to have to list all the integers from 1 to 10 to denote the product in the last example. Since products of this type arise often in counting problems, we need a shorthand notation to indicate a product that begins at some positive integer, n, decreases by one with each factor, and stops at 1. Such notation is called *factorial notation*.

Definition 2.8

> Let n be a positive integer. By *n factorial*, denoted by $n!$, we mean
>
> $$n \cdot (n-1) \cdot (n-2) \cdots 3 \cdot 2 \cdot 1$$
>
> By $0!$ we mean 1. That is, $0! = 1$

The use of this notation is illustrated in the next example. Check your calculators! Many calculators have a built-in program to evaluate factorials.

Example 3 | To evaluate the quotient $10!/(3!\,7!)$ we first expand each of these factorials via Definition 2.8. Factors common to the numerator and denominator are then canceled as shown:

$$\frac{10!}{3!\,7!} = \frac{10 \cdot 9 \cdot 8 \cdot 7 \cdot 6 \cdot 5 \cdot 4 \cdot 3 \cdot 2 \cdot 1}{(3 \cdot 2 \cdot 1) \cdot (7 \cdot 6 \cdot 5 \cdot 4 \cdot 3 \cdot 2 \cdot 1)} = 120$$

Although the multiplication principle is used to solve most permutation problems, there is one type of problem involving order for which a formula is convenient. In particular, there are instances in which one is concerned with order but in which the objects being arranged are indistinguishable from one another. In these cases the multiplication principle may not apply. To learn how to handle such a problem consider the following example.

Example 4 | How many permutations are there of the letters in the word KOOK? Let us answer this question by listing the possibilities:

KOOK   KOKO   OKOK   KKOO   OOKK   OKKO

Evidently the answer is 6. That is, the number of permutations of four objects, two of one type and two of another is 6. Note that the answer 6 can be expressed as

$$6 = 4!/(2!\,2!)$$

The numerator is 4! because there is a total of four objects being permuted. The factor 2! appears twice in the denominator because there are two types of objects involved, namely two Ks and two Os.

This technique is generalized to provide a convenient formula for finding the number of permutations of objects, some of which are indistinguishable from one another.

Rule 8

Permutations of Indistinguishable Objects

Let n objects be divided into k categories with objects within categories being indistinguishable. Let n_j represent the number of objects in the jth category, $j = 1, 2, \ldots, k$ and $n_1 + n_2 + \cdots + n_k = n$. The total number of distinguishable arrangements for these objects is given by

$$\frac{n!}{n_1! \, n_2! \, n_3! \cdots n_k!}$$

This formula is especially useful when the number of permutations becomes too large to list.

Example 5

How many permutations are there of the letters of the word *bookkeeping* (reportedly the only word in the English language with three consecutive sets of double letters)? That is, how many permutations are there of 11 objects, two of a first type, two of a second type, two of a third type, and the rest distinct? Applying the above theorem the answer is seen to be

$$11!/(2!\,2!\,2!\,1!\,1!\,1!\,1!\,1!\,1!) = 4,989,600$$

Exercises 2.8

Each of the exercises 90 to 95 involves either a permutation or a combination. Identify each as to problem type.

90. An ardent poker player is interested in finding the probability that when five cards are dealt at random, four will be aces.

91. A man at the Big A has bet on the first race. He has wagered that Native Dancer will come in first, and Slew will be second. He is interested in the number of finishes that will allow him to win his wager.

92. A student is interested in her class standing, since she feels that this has an effect on her chances of admission to law school.

93. A group of students is to be selected at random from the student body to serve on a committee to study registration procedures.

94. Each week the Associated Press lists the top 20 teams in basketball.

95. Twenty students try out for the school volleyball team. Only 12 will be selected for the team.

96. Evaluate each of these expressions:
    (a) 6!   (b) 8!   (c) 0!   (d) 8!/(3! 5!)   (e) 4!/(4! 0!)
    (f) 6!/(3! 2! 1!)   (g) 10!/(5! 3! 2!)

97. How many arrangements are possible for a math, a science, a psychology, and an art book on a shelf with spaces for four books? With spaces for only three books?

98. In how many ways can the letters A, B, Q, Z, T, S, and X be arranged to form a computer password if
    (a) the password has seven letters with no repetition?
    (b) the password has five letters with no repetition?

99. An employer has eight employees, all of whom have comparable credentials. In a recession these workers will be laid off one at a time as economic conditions require.
    (a) In how many orders can the eight workers be let go?
    (b) Three workers are laid off before the economy recovers. In how many orders could the layoffs have occurred? Assume that you are one of the eight workers. If layoffs are randomly conducted so that each of the permutations is equally likely, what is the probability that you will *not* be laid off?

100. In how many ways can the notes shown in Figure 2.11 be permuted to form an eight-note tune?

Figure 2.11

101. A medical researcher has nine subjects available for use in an experiment. Two drugs, A and B, are to be tested against a placebo, P. Each is to be used on exactly three subjects with assignment of treatment to subject to be made randomly. A typical arrangement of treatments would be: ABBPAP-BAP. How many such assignments are possible?

102. A string of Christmas tree lights has space for 15 lights. Three colors are to be used: red, white, and blue. Six are to be red, five blue, and four white. In how many ways can these lights be arranged on the string? Some of the arrangements counted above would not be very pleasing to the eye. For example, six of the arrangements have all the reds together, all the blues together, and all the whites together. Suppose we wish to be a bit more artistic than that, and decide to begin and end the string with red, mass the blue in the middle, and then arrange the remaining reds and whites in the spaces left. How many such arrangements are now possible?

2.9   Counting Selections of Objects: Combinations (Optional)

We have already discussed the definition of the term combination. Recall that a combination is a selection of objects without regard to order. We develop here a formula to count the number of combinations that exist in a given situation.

Example 1

A new doughnut store has opened in town. The store advertises 27 varieties of doughnuts. J. K. goes to the shop and picks out five varieties to sample: lemon, orange, cherry, blueberry, and vanilla. He discovers that he has only enough funds to purchase three doughnuts. In how many ways can he select three different varieties from the five listed? Note that this problem is not a permutation problem since J. K. has no interest in which variety is selected first, second, or third. He is interested only in the final result. To answer the question, let us simply list the possibilities:

lemon, orange, cherry	lemon, blueberry, vanilla
lemon, orange, blueberry	orange, cherry, blueberry
lemon, orange, vanilla	orange, cherry, vanilla
lemon, cherry, blueberry	orange, blueberry, vanilla
lemon, cherry, vanilla	cherry, blueberry, vanilla

The answer, evidently, is that there are ten combinations of three doughnuts that can be selected from five varieties. Notationally, we write $_5C_3 = 10$. The left subscript denotes the total number of objects available, the right subscript the number to be selected. The letter C represents the word *combination*.

Example 2

Six persons have volunteered to take part in a psychological experiment. Only four are needed. In how many ways can we select four subjects from among the six available? That is, what is $_6C_4$? Again, since the numbers are small, we can answer the question by labeling the volunteers A, B, C, D, E, F and then listing the possibilities:

A, B, C, D	A, B, E, F	B, C, D, E	A, B, C, E	A, C, D, E
B, C, D, F	A, B, C, F	A, C, D, F	B, C, E, F	A, B, D, E
A, C, E, F	B, D, E, F	A, B, D, F	A, D, E, F	C, D, E, F

Thus, there are 15 combinations of six volunteers selected four at a time. That is, $_6C_4 = 15$.

Obviously, as the number of objects from which to choose and the number selected increases, the listing method of counting combinations becomes impractical. For example, to use listing to determine the number of ways in which a five-card poker hand could be dealt from an ordinary deck would require a listing of all possible sets of five cards selected from 52. A monumental task! Is there a pattern to the numbers involved? If so, can the pattern be shown to hold for any number of objects n and any size collection r? We claim that this can be done.

Note that $_5C_3 = 10$, and that 10 could have been written as $5!/(3!\,2!)$. Note also that $_6C_4 = 15$, and that 15 could have been written as $6!/(4!\,2!)$. What is the pattern? In each case, the number in the numerator is the total number of objects from which to choose, factorial ($n!$). In each case, two numbers are involved in the denominator: the number of objects to be selected, factorial ($r!$) and the difference between the total number and the number being selected, factorial $[(n - r)!]$. This pattern is stated in general terms in Rule 9.

Rule 9

Combinations Formula

The number of combinations of n distinct objects selected r at a time, denoted by $_nC_r$ or $\binom{n}{r}$ is given by:

$$_nC_r = \binom{n}{r} = \frac{n!}{r!\,(n - r)!}$$

This formula is important. It can be used in conjunction with the multiplication principle to solve some interesting probability problems, as demonstrated in the next example.

Example 3

What is the probability that a randomly dealt five-card poker hand will contain exactly two aces? To find this probability, we must count two things, the number of five-card poker hands possible and the number of these that contain exactly two aces. The former is found directly from Rule 9 and is given by

$$\binom{52}{5} = \frac{52!}{5!\,47!} = \frac{52 \cdot 51 \cdot 50 \cdot 49 \cdot 48 \cdot 47!}{5 \cdot 4 \cdot 3 \cdot 2 \cdot 1 \cdot 47!} = 2{,}598{,}960$$

We view the dealing of a hand containing exactly two aces as a two-stage process. In stage 1, we select two aces from the four available. This can be done in $_4C_2 = 6$ ways. In stage 2, we select three cards from the 48 cards in the deck that are *not* aces. This can be done in $_{48}C_3 = 17{,}296$ ways. By the multiplication principle, this two-stage process results in $6 \cdot (17{,}296) = 103{,}776$ hands that contain exactly two aces. Using classical probability,

$$P\begin{bmatrix} \text{exactly} \\ \text{two aces} \end{bmatrix} = \frac{103{,}776}{2{,}598{,}960}$$

Exercises 2.9

103. Evaluate:

(a) $_9C_4$ (b) $_8C_0$ (c) $_7C_7$ (d) $\binom{8}{2}$ (e) $\binom{6}{3}$ (f) $\binom{4}{0}$

104. An instructor asks 12 questions on a test. The student is asked to select and answer 10 of these questions. In how many ways can the questions be chosen?

105. A pizza shop offers 15 toppings for its basic pizza. A customer wants to select four of these toppings. In how many ways can the selection be made?

106. How many juries of size 12 can be selected from a venire of 30 potential jurors? If you are a member of the venire, what is the probability that you will *not* be selected?

107. A manufacturer of engine blocks ships the product in lots of size 100. Before accepting a shipment, the buyer randomly selects five engines from the lot for testing. He cannot test every block in the shipment because the test entails cutting the block in half. He will accept the shipment only if no defective blocks are found. A certain lot actually contains 10 defective blocks. What is the probability that none of the defective ones will be selected for testing?

108. Five hundred mice are born in the laboratories of a biological supply house. Four have a birth defect that can be detected only by killing the mouse. A research team purchases 20 mice for use in an experiment. The experiment will be ruined if any of the defective mice are included. What is the probability that no defective mice will be included? What is the probability that exactly one defective mouse will be included?

109. Commercial blood banks buy blood from anyone willing to sell it. It has been determined from past experience that approximately 10% of all blood so collected is contaminated with hepatitis virus. A hospital receives a shipment of 100 pints of A-positive blood from one such commercial source. Assume that 10 of these are contaminated. If five transfusions of A-positive blood are given and the five pints used are selected at random from the shipment, what is the probability that no one will be exposed to hepatitis from this source? What is the probability that exactly two patients will be exposed to this disease? (Assume that each pint is administered to a different patient.)

Vocabulary List and Key Concepts

probability	sample space
personal probability	sample point
a priori	event
equally likely	sure event (certain event)
classical probability	impossible event
a posteriori	union
relative frequency	intersection
experiment	complement

mutually exclusive events

axioms of probability

countable additivity

addition rule

conditional probability

false positive rate

false negative rate

independent events

multiplication rule for independent events

multiplication rule for nonindependent events

randomized response method

Bayes' rule

tree diagram

multiplication principle

permutations

combinations

factorial notation

permutations of indistinguishable objects

Review Exercises

110. It has been observed that the methane reading in a particular coal mine exceeds the safety level about once in every 50 days of operation. What is the approximate probability that on a randomly selected day the level will exceed the safety level, thus forcing an evacuation of the mine? What approach to probability are you using to answer this question?

111. A random digit generator on a pocket calculator is designed so that the digits 0, 1, 2, 3, 4, 5, 6, 7, 8, and 9 are equally likely to appear each time the generator is activated. What is the probability of obtaining three zeros in succession with such a generator?

For exercises 112–118, use the following data: Assume that 85% of all defendants who are brought to trial are in fact guilty of the offense with which they are charged. Assume that 75% of all defendants who are brought to trial are found to be guilty. Assume also that 90% of all defendants are guilty, or are found to be guilty, of the offense with which they are charged. A defendant is selected at random.

112. Find the probability that the defendant is innocent of the offense with which he or she is charged.

113. Find the probability that the defendant is guilty and is found to be guilty.

114. Find the probability that the defendant is innocent but nevertheless is found to be guilty.

115. Are the events "the defendant is guilty" and "the defendant is found to be guilty" mutually exclusive? Explain.

116. Find the probability that a guilty person will be found to be not guilty.

117. Find the probability that an innocent person will be found to be guilty.

118. Are the events "the defendant is guilty" and "the defendant is found to be guilty" independent? Explain.

For exercises 119–126, use the following data: A roulette wheel has 38 slots, each of the same size. Two of the slots numbered 0 and 00 are green. The other 36 slots are numbered 1 to 36 with half being red and the other half being black. When the wheel is spun, a small ball comes to rest in one of the slots, thus determining the winning number and color.

119. What is the probability that any single specific number will win?

120. The numbers 0 and 00 are called *house numbers*. What is the probability that a house number will win?

121. What is the probability that a red number will win?

122. What is the probability that a red or a black number will win?

123. If the wheel is spun twice, what is the probability that the ball will land on red both times?

124. If the wheel is spun twice, what is the probability that the second number will be black, given that the first is red?

125. On a single spin are the events "the number is green" and "the number is red" mutually exclusive?

126. What is the probability that the number 39 will be a winner?

*127. Let A and B be independent events. Show that events A and B' are also independent. [*Hint:* Note that $A = (A \cap B) \cup (A \cap B')$ and that the events $A \cap B$ and $A \cap B'$ are mutually exclusive.]

For exercises 128–130, use the following data: Experience shows that the probability that a skin test will indicate the presence of an allergy to a particular substance when in fact the patient is not allergic to the substance under study is 0.10. The probability that the test will not detect the allergy in patients sensitive to the substance is 0.15. Some 30% of the persons tested are allergic to the substance. A patient is selected at random.

128. What is the false positive rate for the test?

129. What is the false negative rate for the test?

130. What is the probability that the patient is allergic to the substance, given that the test indicates an allergy?

131. A car salesman feels that the probability that he can sell a car to a male customer is .4 and the probability that he can sell a car to a female customer is .3. Seventy percent of his customers are male. Find the probability that he can sell a car to his next customer.

*132. A desk contains three drawers. Drawer one contains two gold pieces, drawer two contains two silver pieces, and drawer three contains one of each. A drawer is selected at random and a coin is selected at random from the drawer. The coin is silver. Find the probability that the other coin in the drawer is also silver. (*Hint:* Find P[drawer 2 | coin is silver].)

*133. An appliance dealer has seven brands of TV sets from which to select four.
    (a) In how many ways can the selection be made?
    (b) One of the brands is made in Japan. How many of the selections will include this brand? How many will exclude this brand?
    (c) If the four brands are selected at random, what is the probability that the brand made in Japan will not be included?

*134. A computer program has five subroutines that can be inserted into the program in any order. How many arrangements are possible for these subroutines?

*135. A rat is allowed to run a maze 10 times. A run is recorded as being a success if the rat makes no wrong turns while making its way through the maze; otherwise, it is recorded as a failure. It is later reported that there were seven successes and three failures. In how many orders can such a sequence of runs have occurred?

Random Variables

We defined the term random variable in the Introduction. In Chapter 1 we looked at some methods for describing the behavior of a random variable based on information obtained from a sample. Chapter 2 entailed a discussion of some of the basic ideas of probability. In this chapter we link the ideas of probability and random variables; we discuss how to use probability to help describe the behavior of a random variable.

3.1 The Basic Definition

Before we begin, let us review some of the basic definitions given in the Introduction.

Definition 3.1

A **random variable** is a variable whose value is determined by chance.

We are dealing with an entity that varies in value and whose behavior is governed by chance. Random variables are usually denoted by capital letters. It is helpful to choose a letter that reminds us of the meaning of the random variable being studied. The following example provides a collection of random variables for illustrative purposes.

Example 1

1. An experiment consists of rolling a pair of fair dice. The random variable X is defined by

    X = sum of the spots on the two dice

2. A roulette wheel has 38 compartments numbered 0, 00, 1, 2, ..., 36. The numbers 0 and 00 are green, and the numbers 1 through 36 are split evenly

113

between black and red. The compartments are each of the same size. The wheel is spun and a small ball placed inside comes to rest in one of the compartments as the wheel stops. If the wheel is perfectly balanced, the probability of the ball landing on any specific number is 1/38. A gambler plays roulette and always places a wager on green. The gambler continues to play until a win occurs. The random variable W is defined by

W = number of trials necessary to obtain a win

3. The morning shift at a local police station begins at 7:30 A.M. and past experience indicates that at least one call will be logged by 8:00 A.M. The random variable T denotes the time at which the first call of the day is received.

4. An experiment consists of observing the random variable L, where L is the length of time that it takes for a computer program to run.

Each of the above examples has associated with it a random variable. Each variable X, W, T, and L can change in value from trial to trial, and chance is certainly a factor. Before each experiment, we cannot say with certainty what the outcome will be; after each experiment, the variable will have assumed a numerical value determined by chance.

To define a random variable, we give a rule that allows us to associate a real number with each outcome of our experiment. Thus, it is correct to say that a random variable is a function in the usual mathematical sense. It is a function that assigns a real number to each element of the sample space for the experiment. Recall that there are two basic types of random variables of interest in beginning courses. These are *discrete random variables* and *continuous random variables*. With practice it is not difficult to distinguish the two. It is essential to do so later since methods of computing associated probabilities depend on the type of variable involved.

Definition 3.2

A **discrete random variable** is a random variable that assumes its values only at isolated points.

To determine whether or not a random variable X is discrete, we ask the question, What are the possible values for X? If we can count the possibilities, then X is discrete. If we can begin to count the possibilities, but soon realize that the set of possible values for X is unending, then X is also discrete. In a mathematical sense, we are saying that a random variable is discrete if it can assume a finite or a countably infinite number of possible values. This idea is illustrated in the next example.

Example 2

1. The random variable

   $$X = \text{sum of the spots on two dice}$$

   is discrete since X can assume only the values 2, 3, 4, 5, 6, 7, 8, 9, 10, 11, and 12. The number of possibilities is finite, namely 11.

2. The random variable

   $$W = \text{the number of trials necessary to win in roulette}$$

   is discrete since W can assume any of the possible values 1, 2, 3, 4, 5, 6, 7, ... This collection of possibilities is countably infinite.

Sometimes when we ask, What are the possible values for X? we are forced to admit that, conceivably, X can assume any value in some interval of real numbers. When this occurs, X is not discrete because any interval contains an uncountable number of possible values. In this case, we say that X is *continuous*.

Definition 3.3

A **continuous random variable** is one that can assume any value in some interval or intervals of real numbers, and the probability that it assumes any specific value is zero.

Example 3

1. The random variable

   $$T = \text{time that the first call of the day is received at a local police station}$$

   is continuous. If we attempted to list the possibilities, we would find it impossible to do so. We might begin by posing the list

   $$7:30 \qquad 7:40 \qquad 7:50 \qquad 8:00$$

   This list is clearly incomplete since the first call might arrive at 7:35. Thus, we pose the following possibilities:

   $$7:30 \qquad 7:35 \qquad 7:40 \qquad 7:45 \qquad 7:50 \qquad 7:55 \qquad 8:00$$

   This is an improvement but still incomplete, since 7:31 is not included. Regardless of how the list is amended, we find it still inadequate. Between any two real numbers there exists another real number that is a possible value for T; time is measured *continuously*. Hence, to describe the set of values for T, we are forced to admit that T can conceivably assume any value in the continuous interval of real numbers between, say, 7:30 and 8:00. That is, the value of T will be in the interval (7:30, 8:00).

2. The random variable

   $$L = \text{the length of time that it takes for a computer program to run}$$

is continuous. Let us assume that it is reasonable to expect that the value of L is less than four minutes. That is, the value of L lies in the interval $(0, 4)$.

The definition of continuous random variable requires that the random variable assume any specific value with probability 0. What does this mean? To answer this question, let us ask, What is the probability that the first call to a police station arrives at exactly .123179823965 seconds before 8:00? The probability of hitting this exact number is extremely small. For all practical purposes, it is 0. Any real number has an infinite decimal expansion, thus the a priori probability of a random variable assuming any stated value is 0. Keep this fact in mind when dealing with the computation of probabilities associated with continuous variables.

Note that counting problems usually give rise to discrete random variables; measuring problems generate continuous random variables.

Exercises 3.1

Identify each random variable as being either discrete or continuous. If the random variable is discrete, list its possible values.

1. A scientist has a cage containing 20 white rats. Fifteen are male and five are female. She reaches in and randomly selects five rats for use in her experiment. The random variable is X, the number that are female.

2. A single die is tossed. The random variable is Y, the number rolled.

3. An experiment consists of measuring T, the length of time that it takes a rescue squad to respond to an emergency call.

4. A man prepares a bag of Halloween candy for trick-or-treaters. The bag contains 5 Tootsie Pops; 6 pieces of bubble gum; 10 Sugar Daddies; 8 boxes of Red Hots; and 1 box of Sweet Tarts. A witch, a ghost, and a child dressed as a bag of trash are the first three visitors; each in turn reaches into the bag and selects at random one item. The random variable is the number (N) of Tootsie Pops selected.

5. A person on a high-protein diet is interested in the random variable X, the amount of weight lost in a week's time.

6. An experiment consists of rolling a pair of fair dice, one red and one white. There are two random variables: H, the highest number occurring on either die; and D, the difference of the numbers appearing where subtraction is done in the order of red minus white.

7. An economist is interested in the random variable C, the number of persons filing for unemployment benefits each day.

8. A physician studies L, the length of remission in leukemia patients who have achieved their first remission.

9. The proprietor of a hamburger franchise is interested in the volume of Coke sold per day.

10. A physician is interested in the amount of liquid drug to prescribe for patients.

11. A psychologist is interested in TV commercials directed toward children. The random variable is the number of commercials shown between 7 and 11 on Saturday mornings that depict adults in a negative role.

12. A criminologist is interested in the random variable N, the number of arrests made by the FBI in a district each month.

13. A highway-safety group is interested in studying the random variable S, the speed of a car or truck as a checkpoint is passed.

14. A businessman is interested in the random variable O, the total overtime accumulated by his employees each week.

3.2   Computing Probabilities

When dealing with a random variable X, it is not enough to ask, What are the possible values for X? We also need to consider the probabilistic behavior of the variable. That is, we want to answer such questions as:

If a pair of fair dice is rolled once, what is the probability that the sum of the spots will be greater than nine?

If a gambler plays roulette until a win occurs, wagering each time on green, what is the probability that the number of trials needed will be more than three?

What is the probability that the first call of the day at a police station will arrive between 7:35 A.M. and 7:45 A.M?

If a computer program is run, what is the probability that it will take between 1 and 2 minutes?

To answer these and similar questions, we must consider the probability distribution for the random variable. That is, we need to find a table, formula, or curve that allows us to determine both the possibilities and the probabilities associated with the variable. Because of the basic differences in the two types of variables, the discrete case must be handled separately from the continuous one. We begin by defining what is called the *probability function* or the *probability density function* for a discrete random variable.

Definition 3.4	Let X be a discrete random variable. The **discrete probability function**, f, for X is given by $$f(x) = P[X = x] \qquad \text{for } x \text{ real}$$

Note that the random variable is denoted by capital X and the observed numerical value of X by a lowercase x. The probability function in the discrete case is a table, or an equation, which gives the possible values for X, together with the probability that X assumes those values.

Since probabilities cannot be negative, a probability function cannot assume negative values. The probability associated with a sample space is 1. Thus, if we add the values of f over all possible values of X, the total should be 1. In fact, these two properties completely characterize the probability function of a discrete random variable.

Properties That Identify a Probability Function for a Discrete Random Variable

1. $f(x) \geq 0$ for each real number x.
2. $\Sigma f(x) = 1$.

Note that by $\Sigma f(x)$, we mean simply to add over all possible values of X. If a real number x is not a possible value for the random variable X, then it occurs with probability zero ($f(x) = 0$) and is of no interest. In expressing probability functions throughout this text, $f(x)$ is assumed to be 0 for all values of x not specifically mentioned.

Example 1

An investor has five stocks which she follows each day. The random variable being studied is X, the number of stocks that increase in value each day. The probability function for this variable is shown in Table 3.1. Note that, to be a probability function, the probabilities in the second row must sum to 1. Thus, the probability that no stocks increase in value, $f(0)$, is .34. To find the probability that, on a given day, a majority of the stocks will increase in value, we need to find $P[X \geq 3]$. From the table,

$$P[X \geq 3] = P[X = 3] + P[X = 4] + P[X = 5]$$

$$= f(3) + f(4) + f(5)$$

$$= .10 + .05 + .01 = .16.$$

Note that

$$P[X > 3] = P[X = 4] + P[X = 5]$$

$$= f(4) + f(5) = .05 + .01 = .06.$$

The point to notice is that $P[X \geq 3] \neq P[X > 3]$. In the discrete case care must be taken since excluding an endpoint that occurs with nonzero probability does affect the result.

Table 3.1

x	0	1	2	3	4	5
$P[X = x] = f(x)$	$\cdots$	.30	.20	.10	.05	.01

Table 3.2

x	2	3	4	5	6	7	8	9	10	11	12
$P[X = x] = f(x)$	1/36	2/36	3/36	4/36	5/36	6/36	5/36	4/36	3/36	2/36	1/36

Example 2

The game of craps is popular at Las Vegas. It is reported that although it takes 11 months to cure a pair of dice for the tables, it takes only an average of two days to wear them out. The rules of the game are as follows: A player rolls the dice. If a 7 or an 11 is rolled, the player immediately wins. If a 2, 3, or 12 is rolled, the player immediately loses. If any other number is rolled, then this number is called the player's *point*, and the player must continue to roll until either a 7 is rolled, in which case the player loses, or else the point is rolled and the player wins. What is the probability that a player will win on the first roll of the dice? To answer this and other questions concerning craps, we need the probability function for the experiment of rolling a pair of fair dice with the variable being X, the sum of the spots on the two dice. This function can be read from Table 2.1 and is given in Table 3.2. From the table it is easy to see that

$$P[\text{win on first roll}] = P[\text{roll 7 or 11}]$$
$$= P[\text{roll 7}] + P[\text{roll 11}]$$
$$= f(7) + f(11)$$
$$= 6/36 + 2/36 = 8/36$$

Let us now consider the notion of a probability function in the continuous case. That is, we want to develop a function that allows us to compute probabilities associated with a continuous random variable X. It is obvious that the definition used in the discrete case will not work, since, if X is continuous, $P[X = x] = 0$ for all x. The characteristics needed for a function to be a probability function for a continuous random variable are given in Definition 3.5.

Definition 3.5

Let X be a continuous random variable. A function f such that

1. $f(x) \geq 0$ for each real number x
2. the total area bounded by the graph of f and the horizontal axis is 1
3. the probability of observing X with a value between any two points a and b is equal to the area of the region bounded by the graph f, $x = a$, $x = b$, and the horizontal axis

is called a **continuous probability function** for the random variable X.

Note that conditions 1 and 2 are analogous to conditions 1 and 2 for the probability function in the discrete case. Note also that condition 3 can be written as

$$P[a < X < b] = \text{area under the graph of } f \text{ between } a \text{ and } b$$

If a and b are real numbers, then, since $P[X = a] = P[X = b] = 0$, we can write

$$P[a < X < b] = P[a \leq X \leq b] = P[a \leq X < b] = P[a < X \leq b]$$

That is, in the continuous case, adding an endpoint to an interval, or deleting one, does not affect the numerical probability involved. This is not true in the discrete case, since endpoints there may occur with nonzero probability.

Example 3   The graph of the probability function for T, the time at which the first call of the day is received at a local police station is shown in Figure 3.1(a). The time span from 7:30 A.M. to 8:00 A.M. is identified with the interval 0 to 30 minutes. The graph is of constant height, c. What is the value of c? To be a probability

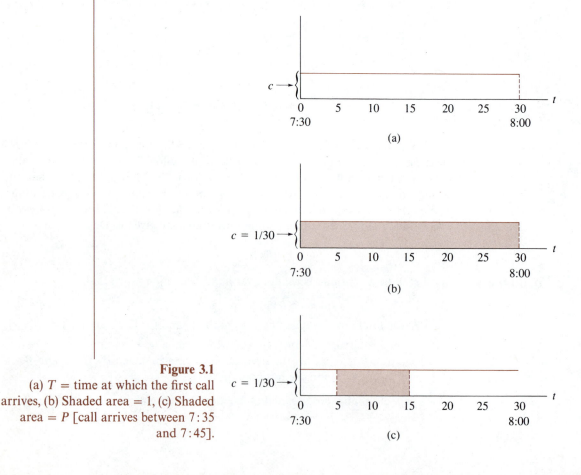

Figure 3.1
(a) T = time at which the first call arrives, (b) Shaded area = 1, (c) Shaded area = P [call arrives between 7:35 and 7:45].

function c must be chosen so that the area of the rectangle shown in Figure 3.1(b) is 1. The area of a rectangle is the product of its length by its height. Here, the length is 30 and so we choose c so that $30c = 1$ or $c = 1/30$. To find the probability that the first call of the day arrives between 7:35 A.M. and 7:45 A.M., we find the area of the region shown in Figure 3.1(c). Since this region is also a rectangle, its area is its length, 10, multiplied by its height, 1/30. The desired probability is 10/30. That is,

$$P[5 < T < 15] = 10/30.$$

Note that we can also say that

$$P[5 \leq T \leq 15] = 10/30.$$

The difference in these two questions is that the latter includes the possibilities that the first call will arrive at exactly 7:35 A.M. or 7:45 A.M. Graphically we are now including the dashed lines shown in Figure 3.1(c). Since lines have no width, they have no area. We have added no area by including these endpoints and hence have added nothing to the probability obtained previously.

A random variable whose probability function is constant as in Example 3 is said to have a **uniform distribution**. This is the simplest of all continuous random variables. Unfortunately, most continuous random variables have probability functions whose graphs are not straight lines. When this occurs, it is usually not possible to find areas geometrically. In this text, continuous probability functions that are employed are such that extensive tables of associated areas have been compiled. These tables give the probability that a random variable X assumes values less than or equal to some specified value x and is called a **cumulative distribution table**. Our job will be to learn how to use these tables to find probabilities.

How does one decide what sort of probability function is associated with a particular continuous random variable X? Sometimes past experience with the variable serves as a guide. However, when dealing with a random variable for the first time, we must ascertain its shape from a sample of observations on X. The stem-and-leaf charts and relative frequency histograms studied in Chapter 1 are used for this purpose. Often we can determine a reasonable shape for the probability function from these graphs.

Exercises 3.2

15. Find the table for the probability function for the random variable Y, the number obtained on the toss of a single die. Use the table to find $P[Y \leq 3]$; $P[Y < 3]$. Are the results the same?

16. Find the table for the probability function of the random variable H, the highest number occurring on either die when a pair of fair dice is tossed. Use this table to find $P[H > 4]$; $P[2 \leq H \leq 4]$; $P[H = 7]$.

17. Find a table for the probability function of the random variable D, the difference in the numbers appearing when a pair of fair dice, one red and one white, is rolled. (Subtract in the order of red minus white.) Use this table to find $P[D < 0]$; $P[D = 0]$; $P[D > 3]$.

18. Use Table 3.2 to find the probability of losing on the first roll in the game of craps. What is the probability of obtaining a point on the first roll in the game of craps? If you rolled losing rolls four times in succession would you have reason to suspect that the dice were loaded in favor of losing combinations? Explain on the basis of the probability involved.

19. A recent report indicates that the number of children in families receiving public assistance is lower than in the past. The probability function for this variable is shown in Table 3.3.
    (a) Find $P[X = 4]$.
    (b) Find $P[X \leq 2]$ and $P[X < 2]$. Use these values to find $P[X = 2]$ and compare your answer to that found in Table 3.3.
    (c) Find the probability that a randomly selected family on public assistance has between 1 and 3 children, inclusively.

Table 3.3

x	0	1	2	3	4
$f(x)$	.10	.15	.27	.30	?

20. An employer has four positions to fill. There are eight equally qualified applicants, four women and four men. The names are to be placed in a box and four are to be drawn at random. Table 3.4 gives the probability function for X, the number of women selected. (This table can be verified using the combinatorial methods of Chapter 2.)
    (a) What is the probability that exactly two men and two women are selected?
    (b) What is the probability that a majority of those selected are men?
    (c) Would you be surprised if at most one woman was employed? If no women were employed? Explain on basis of the probabilities involved.

Table 3.4

x	0	1	2	3	4
$f(x)$	1/70	16/70	36/70	16/70	1/70

21. Find the first four entries in the table for the probability function for the random variable W, the number of trials needed to obtain the first win when wagering successively on green in roulette. Use this table to find the probability that the number of trials needed to win will be more than three.

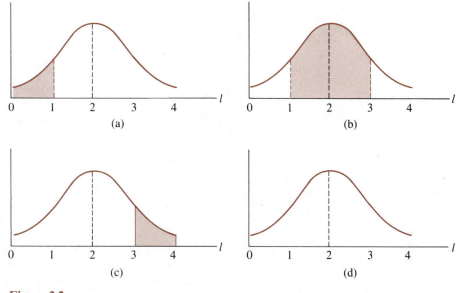

Figure 3.2
L = length of time in minutes required to run a program.

22. Consider the random variable L, the length of time it takes to hear a case in small claims court. Assume that the probability function for L is a constant, c, over the interval 20 to 100 minutes.
    (a) Find the value of c that makes this a probability function.
    (b) Find $P[L \leq 30]$ and $P[L < 30]$. Are these results the same?
    (c) Find $P[40 < L < 55]$.
    (d) Find the probability that it will take more than an hour to hear a case.
    (e) Find the probability that it will take exactly one hour to hear a case.

23. Consider the random variable L, the length of time that it takes in minutes for a computer program to run.
    (a) What probability is represented by the shaded region in Figures 3.2(a), (b), (c), and (d)?
    (b) Shade the region in Figure 3.2 corresponding to $P[2 < L < 3]$.
    (c) The curves shown in Figure 3.2 are symmetric curves centered at two minutes. What is the probability that it will take at most two minutes for a given program to run?
    (d) If $P[1 < L < 2] = 3/8$, what is the numerical value of each of the probabilities represented in Figures 3.2(a), 3.2(b), 3.2(c), and 3.2(d)?

24. Let D be the amount of a liquid drug that a physician should prescribe for a patient to obtain maximum benefit. Assume that the graph of the probability function is as shown in Figure 3.3. For parts (a)–(e) determine which region(s) in the figure correspond to each of the probabilities.

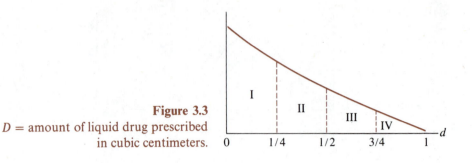

Figure 3.3
D = amount of liquid drug prescribed
in cubic centimeters.

(a) $P[D \leq 1/4]$
(b) $P[D < 1/4]$
(c) $P[1/4 < D < 3/4]$
(d) The probability that at least 1/2 cc should be prescribed.
(e) The probability that at most 3/4 cc should be prescribed.
(f) What is the numerical value of the area corresponding to regions I, II, III, and IV together?

25. Let O be the overtime accumulated by employees of a small business firm during a week. Assume that the probability function for O, given in Figure 3.4, is symmetric about five hours.
    (a) Shade the area corresponding to $P[O < 2.5]$. If this area has a value of 1/10, what is the probability that the amount of overtime accumulated during a given week will be at least 2.5 hours?
    (b) What is the probability that during a week's time the amount of accumulated overtime will be more than 7.5 hours?
    (c) What is the probability that the amount of accumulated overtime during a week's time will be exactly 5 hours?
    (d) What is the probability that the amount of overtime accumulated during a given week will be between 2.5 and 7.5 hours, inclusive?
    (e) Consider a work year of 50 weeks. For how many weeks would you expect the amount of accumulated overtime to lie between 2.5 and 7.5 hours? If overtime pay amounts to $5 per hour, what are the minimum and maximum amounts necessary in the budget to cover these weeks?

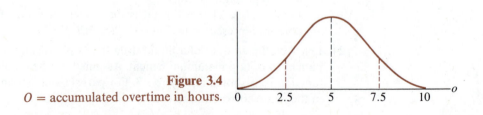

Figure 3.4
O = accumulated overtime in hours.

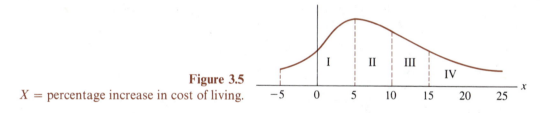

Figure 3.5
X = percentage increase in cost of living.

26. Let X denote the percentage increase in the cost of living from January of one year to January of the next. The probability function for X is shown in Figure 3.5.
    (a) What is occurring if X assumes a negative value for a given year?
    (b) Shade the area corresponding to the probability that the cost of living will decrease. If this area has a value of $1/100$, what is the probability that the cost of living will increase?
    (c) What probability is represented by the area of region II? of regions III and IV together?
    (d) Find $P[X = 15\%]$.

3.3   Measures of Location and Variability

Recall that, associated with any random variable X, there exist *parameters* or constants that characterize the random variable over the entire population under study. The most common of these are the *mean* (μ), the *variance* (σ^2), and the *standard deviation* (σ). These parameters are theoretical in nature, and their exact numerical values cannot be found from a sample. In Chapter 1, we introduced these parameters from an intuitive standpoint only. We consider here their technical definitions. We also consider how to use the probability function to determine their exact values.

To define the parameters mentioned, we consider briefly what is meant by the *expectation for X* or the *expected value for X*. Intuitively, the expected value of a random variable X, denoted $E[X]$, is the long-run theoretical average value for X. Admittedly, this definition is a bit vague. However, a simple illustration demonstrates its meaning.

Example 1   Consider the experiment of rolling a fair die once. Let X be the number obtained. If this experiment were repeated over and over again, and the results recorded, the observed values of X would fluctuate among the numbers 1, 2, 3, 4, 5, and 6. In other words, we would generate a sequence $x_1, x_2, x_3, x_4, x_5, \ldots, x_n, \ldots$ of observations on X. A typical sequence might be

$$1, 3, 2, 5, 2, 1, 1, 6, 5, 4, 2, 3, 6, 4, \ldots$$

We use this sequence to generate a sequence of arithmetic averages:

$$x_1, \frac{x_1 + x_2}{2}, \frac{x_1 + x_2 + x_3}{3}, \frac{x_1 + x_2 + x_3 + x_4}{4}, \ldots$$

In this example we obtain

$$1, \frac{1+3}{2}, \frac{1+3+2}{3}, \frac{1+3+2+5}{4}, \frac{1+3+2+5+2}{5}, \dots$$

or

$$1, 2, 2, 2.75, 2.6, 2.33, 2.14, 2.63, 2.89, 3.0, 2.91, 2.92, 3.15, 3.21, \dots$$

As the number of observations increases, this sequence of averages tends toward a specific number, the long-run theoretical average value or the expected value of X.

It is possible to approximate $E[X]$ by means of the above procedure. If we have available a large number of observations, their arithmetic average is usually close in value to $E[X]$. However, if we want to know the exact value of $E[X]$, we must know the probability function for X.

Example 2

The probability function f for the random variable X just discussed is given in Table 3.5. $E[X]$ can be found from this table by inspection. We first note that the probabilities in the table are symmetric in the sense that in the long run we expect to roll about as many ones as sixes. The numbers one and six average to 3.5. We also expect to roll about as many twos as fives. Two and five also average to 3.5. Similarly, the threes and fours average to 3.5 and occur with equal probability. Common sense points to a value of 3.5 for $E[X]$. Note that $E[X]$ need not be a possible value for X.

Table 3.5

x	1	2	3	4	5	6
$f(x)$	1/6	1/6	1/6	1/6	1/6	1/6

What can we do if the probability function is not symmetric as above? In this case, we must turn to the mathematical definition of expected value. To see how to define $E[X]$ in general, let us reconsider our die-roll experiment. Imagine rolling the die 36 times. Ideally, we should obtain each of the numbers 1, 2, 3, 4, 5, and 6 six times. The ideal average value of X is

$$\frac{\text{sum of spots}}{36} = \frac{6(1) + 6(2) + 6(3) + 6(4) + 6(5) + 6(6)}{36}$$

$$= 1(1/6) + 2(1/6) + 3(1/6) + 4(1/6) + 5(1/6) + 6(1/6)$$

$$= 3.5$$

Note the pattern on the right side of this equation. We are simply taking each possible value of X, then multiplying it by its theoretical probability, and then summing the results. The fact that this procedure works in the single-die experiment is not a coincidence! This idea is, in fact, taken as the definition of the term $E[X]$ for a discrete random variable.

Definition 3.6

Let X be a discrete random variable with probability function f. Then

$$E[X] = \Sigma x f(x)$$

The idea of expected value plays an important role in actuarial science where it is used to help determine insurance premiums. Example 3 is a simplified illustration of how this is done.

Example 3

An insurance company insures its policyholders against the following dangers: fire, tornado, flood, and theft. After research, they conclude that the probabilities of the occurrence of these disasters are .1, .01, .05, and .14, respectively. The company pays $100, $5000, $1000, and $75, respectively, to policyholders experiencing losses from these causes. They charge $150 per year for coverage. Should the company expect to make a profit on these policies? To answer this, let Y = company profit per policy. Note that if the company must pay a fire victim, Y assumes a value of $50. Similarly, a tornado claim by a policy holder yields a Y value of $-\$4850$, a company loss. The probability function for Y is shown in Table 3.6.

$$
\begin{aligned}
E[Y] &= \Sigma y f(y) \\
&= 50(.1) + (-4850)(.01) + (-850)(.05) + 75(.14) + 150(.70) \\
&= \$29.50
\end{aligned}
$$

This means that over a long period of time, the company expects an average profit of $29.50 per policy per year.

Table 3.6

y	$50	$-\$4850$	$-\$850$	$75	$150
$f(y)$	.1	.01	.05	.14	.70

We often want to find the expected value of some random variable that is expressed in terms of X, as well as the expected value of X itself. For example, we might want to find $E[X + 2]$, $E[X^2]$, or $E[(X - 3)^2]$. The random variables $X + 2$, X^2, and $(X - 3)^2$ are called *functions of* X. Let $g(X)$ denote some function of X. We find $E[g(X)]$ in a manner similar to that used to find $E[X]$. In particular,

$$E[g(X)] = \Sigma g(x) f(x)$$

This idea is illustrated in Example 4.

Example 4   Consider the single-die experiment. What is the theoretical average value of the square of the number rolled? That is, what is $E[X^2]$? Letting $g(X) = X^2$, we see that

$$E[g(X)] = \Sigma x^2 f(x)$$

$$= 1^2(1/6) + 2^2(1/6) + 3^2(1/6) + 4^2(1/6) + 5^2(1/6) + 6^2(1/6)$$

$$= 91/6 = 15.17$$

To compute the expected value of X in the continuous case requires calculus. However, we can visualize $E[X]$ geometrically as being the *balance point* for the graph of the probability function for X. That is, if we imagine taking the region bounded by the graph of f and the horizontal axis and cutting this region out of a piece of thin rigid metal, then $E[X]$ is the point at which the region could be expected to balance on a knife edge.

Example 5   Consider the random variable X, the length of time required to admit a patient to a particular hospital. The probability function for X is given in Figure 3.6. Due to the symmetry about the line $x = 20$, we know that $E[X] = 20$ minutes. That is, if a large number of admittances were observed and the length of time required for each recorded, we could expect the long run average time required to be 20 minutes.

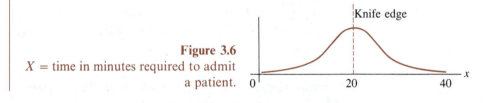

Figure 3.6
X = time in minutes required to admit
a patient.

Note that if the graph of the probability function in the continuous case is a symmetric curve, $E[X]$ will be the axis of symmetry. If the curve is not symmetric, the $E[X]$ can only be approximated by this method.

When used in a statistical context, the expected value of a random variable X is referred to by another name. In particular it is called the **mean** of the variable and is denoted by the Greek letter mu (μ). That is, the parameter that we called the *population mean* earlier is really just the expected value of X. This means that μ is another way to denote $E[X]$; these symbols are used interchangeably.

Is knowledge of the mean of a random variable sufficient to describe the variable completely? The following example will show that this parameter, though useful, does not say much about the behavior of the variable by itself.

Example 6   Consider two towns: one a so-called company town where most members of the community work for a manufacturing plant and live in modest company

Table 3.7

x	5	15	55
$f(x)$	.4	.5	.1

y	13	14	15	16	17
$f(y)$	.025	.05	.85	.05	.025

housing; the other a moderately wealthy suburb of a metropolitan area. Let X and Y denote the family income in thousands of dollars in each town, respectively. We assume that the probability functions for these variables are given in Table 3.7. Note that the mean incomes in the two towns are given by $\mu_X = E[X] = \$15,000$ and $\mu_Y = E[Y] = \$15,000$. The average incomes in both towns are the same, and yet the towns are obviously different in nature. What is the difference? Just this: In the company town, the actual family income differs from the mean income by at least $5000. However, in the suburb with high probability (.85) there is no difference between the observed and the mean incomes. The random variable X appears to deviate from the mean more than Y.

To measure the differences of the sort just encountered, we introduce the notion of variance. Recall that we indicated earlier that the population variance, σ^2, is a measure of the variability of X about its mean, μ. We can now make this definition more precise. In particular, we will measure the variability of X about μ by considering the theoretical average value of the square of the differences between X and μ. This leads to the technical definition of the term *variance* given in Definition 3.7.

Definition 3.7

Let X be a random variable. The **variance** of X, denoted by Var X or σ^2, is

$$\text{Var } X = \sigma^2 = E[(X - \mu)^2]$$

Note that Var X and σ^2 are different notations for the same thing, the long run theoretical average value of the square of the difference between a variable and its mean. If, with high probability, the variable assumes values close to its mean, it has a small variance; if, with high probability, the variable assumes values far from its mean, it has a large variance. Thus, variance is a measure of variability, or fluctuation, of a random variable about its mean, or theoretical average value as desired. Variance is a quantity mainly useful for comparative purposes. When dealing with two similar variables, it is useful to know which exhibits the greater variability. To actually compute the variance we must know the probability function. The computation of σ^2 in the discrete case is illustrated by computing the variance in income for the two towns of Example 6.

Example 7   To compute the variance in incomes for the two towns we use Table 3.7. Recall that $\mu_X = \mu_Y = 15$. Recall also that inspection of Table 3.7 led us to conclude that the random variable X fluctuates more than Y.

$$\text{Var } X = E[(X - \mu_X)^2] = E[(X - 15)^2]$$
$$= (5 - 15)^2(.4) + (15 - 15)^2(.5) + (55 - 15)^2(.1)$$
$$= 100(.4) + 0(.5) + 1600(.1)$$
$$= 200$$

$$\text{Var } Y = E[(Y - \mu_Y)^2] = E[(Y - 15)^2]$$
$$= (13 - 15)^2(.025) + (14 - 15)^2(.05) + (15 - 15)^2(.85)$$
$$+ (16 - 15)^2(.05) + (17 - 15)^2(.025)$$
$$= 4(.025) + 1(1.05) + 0(.85) + 1(.05) + 4(.025)$$
$$= .3$$

As expected, Var $X >$ Var Y.

In the continuous case, it is useful to think of variance as being a *shape parameter*. Example 8 demonstrates how σ^2 affects the shape of the graph of a probability function for a continuous random variable.

Example 8   Let

$X =$ time it takes to complete a basketball game in NCAA (college play)

$Y =$ time it takes to complete a basketball game in the NBA

Assume that $\mu_X = \mu_Y = 90$ minutes, and that the probability functions for X and Y are as shown in Figure 3.7. Note that since $\mu_X = \mu_Y = 90$, the probability functions are centered at exactly the same point. Which variable has the larger variance? Note that the probability of observing X with a value as far from the mean as 30 minutes is the area of the slashed region; the corresponding probabil-

Figure 3.7
$X =$ time in minutes required to complete an NCAA game, $Y =$ time in minutes required to complete an NBA game, Var $X >$ Var Y.

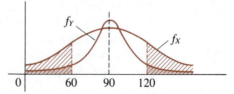

ity for Y is the area of the darker region. Clearly, the probability of observing X far from its mean of 90 minutes is higher than the corresponding probability for Y. Hence, X exhibits more variability than Y, or Var $X >$ Var Y. This is geometrically reflected in that the graph of f_X is flatter and more spread out than the graph of f_Y. In this sense, variance is a shape parameter.

The definition of the variance of X can be used to compute variances as illustrated. However, this definition is not usually used because it is tedious, requiring many subtractions and squarings. The following formula often provides an easier method for computing σ^2:

The computational shortcut for finding the variance of X is

$$\sigma^2 = E[X^2] - (E[X])^2$$

Example 9

To compute the variances in income for the two towns of Example 6 we use Table 3.7 to find $E[X^2]$ and $E[Y^2]$. Remember that $E[X] = E[Y] = 15$.

$$E[X^2] = \Sigma x^2 f(x) = 5^2(.4) + 15^2(.5) + (55)^2(.1)$$

$$= 25(.4) + 225(.5) + 3025(.1)$$

$$= 10.0 + 112.5 + 302.5 = 425$$

$$E[Y^2] = \Sigma y^2 f(y) = 13^2(.025) + 14^2(.05) + \cdots + 17^2(.025)$$

$$= 169(.025) + 196(.025) + \cdots + 289(.025)$$

$$= 225.3$$

$$\text{Var } X = E[X^2] - (E[X])^2 = 425 - 15^2 = 200$$

$$\text{Var } Y = E[Y^2] - (E[Y])^2 = 225.3 - 15^2 = .3$$

Note that these values agree with those obtained using the definition of variance directly as in Example 7.

The second measure of variability that we consider is the **standard deviation**. It is the nonnegative square root of the variance and is denoted by σ. Thus, in Example 7, the standard deviation for X is $\sigma_X = \sqrt{\text{Var } X} = \sqrt{200}$; the standard deviation for Y is $\sigma_Y = \sqrt{\text{Var } Y} = \sqrt{.3}$. Since the standard deviation measures variability in exactly the same way as does the variance, it is natural to ask why we bother with both of these measures. Recall that we posed the same question in Chapter 1 with respect to the sample variance and the sample standard deviation. The answer here is the same as that discovered earlier. Consider the question, What physical unit is associated with variance? If we examine Example 7, we see that X and Y are both reported in thousands of dollars. Thus, since the variables are squared in computing variance, the unit associated with variance is "squared dollars." What is a squared dollar? This really makes no sense physically! For this reason, variance is usually reported as a unitless number. However, since the standard deviation is the square root of the variance, the unit attached will always be the same as that of the original variable. It will usually be physically meaningful.

Exercises 3.3

27. The probability function for the random variable X, the number of cases heard in a small claims court per day, is given in Table 3.8.
    (a) Find $E[X]$.
    (b) Find $E[X^2]$.
    (c) Find the variance and standard deviation for X.
    (d) What physical unit is associated with σ?

    Table 3.8

x	0	1	2	3	4	5	6	7
$f(x)$	.01	.02	.05	.10	.14	.19	.47	.02

28. The probability function for X, the number of courses in which a student is enrolled the last semester of the senior year is given in Table 3.9.
    (a) Find the mean for X.
    (b) Find $E[X]$.
    (c) Find $E[X^2]$.
    (d) Find the variance and standard deviation for X.
    (e) What physical unit is associated with σ?

    Table 3.9

    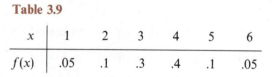

x	1	2	3	4	5	6
$f(x)$	.05	.1	.3	.4	.1	.05

29. Table 3.10 gives the probability function for C, the number of bad checks passed in a small town per week.
    (a) Find μ.
    (b) Find $E[C^2]$.
    (c) Find the variance and standard deviation for C.
    (d) What physical unit would naturally be attached to σ^2? Is this unit meaningful? What unit is associated with σ?

    Table 3.10

    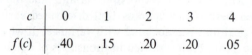

c	0	1	2	3	4
$f(c)$	.40	.15	.20	.20	.05

30. Table 3.11 gives the probability function for R, the number of cases of rabies reported per year in the United States.
    (a) By inspection, what do you think is the mean for R?
    (b) Find the mean for R. How good was your guess in part (a)?
    (c) Find $E[R^2]$.
    (d) Find the variance and standard deviation for R.
    (e) What physical unit is associated with σ?

    Table 3.11

r	0	1	2	3	4	5
$f(r)$	.22	.24	.30	.15	.08	.01

31. A woman purchases apartment dweller's insurance at $125 per year; her furnishings are insured up to a maximum of $10,000. The insurance company from past experience feels that the probabilities of claims of $500, $1000, $5000, $10,000 are, respectively, .1, .02, .01, .001. Can the company be expected to make a profit of $10 per policy?

32. A health-insurance policy costs $300 per year. It pays a maximum benefit of $5000. Past experience indicates that the probabilities of claims of $100, $200, $500, $1000, and $5000 are .30, .20, .10, .08, and .02, respectively. What is the expected profit per policy to the company? Can the company operate profitably under these conditions? The company wants to clear $20 a year per policy, and it costs $10 a year to maintain records on each account. To accomplish this goal, how much should be charged for each policy?

33. The overall probability of winning in craps is 1952/3960. In an even-money bet, a player bets a dollar in the hopes of winning a dollar. Thus, each play results in either a one-dollar profit to the player or a one-dollar loss. What is the expected profit to the player in this situation? Based on this expectation, can a player expect to make money in the long run by shooting craps?

34. One dice game that was popular in the Old West was Chuck-a-Luck. In this game a player bets on one of the numbers 1, 2, 3, 4, 5, or 6. Three dice are rolled. If the number appears exactly once, the player receives a profit equal to the amount wagered; if the number appears on exactly two of the dice, the profit is twice the wager; if the number appears on all three dice, the profit is three times the player's wager. The probability function for X, the profit per roll on a one-dollar wager is given in Table 3.12. (This table can be verified using the methods of Chapter 2.) Find the expected profit per roll to the player. In the long run, can you expect to make money playing Chuck-a-Luck? Explain.

    Table 3.12

x	1	2	3	-1
$f(x)$	75/216	15/216	1/216	125/216

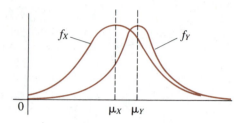

Figure 3.8
X = hourly wage for females,
Y = hourly wage for males.

35. Figure 3.8 shows the graphs of the probability functions for the random variables X, the hourly wage paid to female workers; and Y, the hourly wage paid to male workers, within a particular industry. The workers are engaged in comparable tasks.
    (a) Which variable has the larger mean?
    (b) Which variable has the larger variance?
    (c) Can you think of some possible practical explanation for these differences?

36. If $E[X] = 10$, is it possible for $E[X^2]$ to be 25? Explain. (*Hint*: Consider Var X.)

*37. Three rules govern the behavior of **expected values**. These rules are helpful in proving some of the theoretical results that you will see later in the text. The rules are as follows:

Rules for Expected Values

Let X and Y be random variables and c any real number. Then

  1. $E[c] = c$

(The expected value of a constant is that constant.)

  2. $E[cX] = cE[X]$

(Constants can be factored out of expectations.)

  3. $E[X + Y] = E[X] + E[Y]$

(Expectations for sums can be split into pieces.)

Although we will not attempt a formal proof of these rules, they are not difficult to understand. The first simply says that if one is dealing with a random variable that always assumes the same value, namely c, its long run average value is that value itself. The second says that if one systematically multiplies each observed value by some constant c, the effect is to multiply the long run theoretical average of X by that constant. The third rule allows one to find the theoretical average of the sum of two variables by finding each expectation separately and then adding the results.

Example

Let X and Y be random variables with $E[X] = 10$ and $E[Y] = 5$. Then

$$E[2X - 3Y + 7] = E[2X] + E[-3Y] + E[7] \qquad \text{rule 3}$$
$$= 2E[X] + (-3)E[Y] + E[7] \qquad \text{rule 2}$$
$$= 2E[X] + (-3)E[Y] + 7 \qquad \text{rule 1}$$
$$= 2(10) + (-3)(5) + 7 \qquad \text{substitution}$$
$$= 20 - 15 + 7$$
$$= 12$$

Let X and Y be random variables with means of 6 and 3, respectively. Use the rules for expectation to find each of the following:
(a) $E[3X]$ (b) $E[X - 1]$
(c) $E[2X + 1]$ (d) $E[-2Y]$
(e) $E[3Y - 2]$ (f) $E[2X - Y]$
(g) $E[Y + 3X]$ (h) $E[3X + 2Y + 1]$

*38. Analogous to the three rules for expectation allowing us to simplify complex expressions, there are three rules for variance. These rules will be of use in justifying the statistical procedures in later chapters. One condition that was not necessary when considering the rules for expectation is required here, that is, that X and Y be *independent random variables*. In the intuitive sense, this term is similar in meaning to the term *independent events*. We are dealing with two variables characterized by the fact that knowledge of the value of one variable gives no clue as to the value of the other. This condition does not seriously restrict the applicability of the rules, since in most statistical situations independence is a basic assumption.

Rules for Variance

Let X and Y be independent random variables and let c be any real number. Then

1. $\text{Var } c = 0$
2. $\text{Var } cX = c^2 \text{ Var } X$
3. $\text{Var}(X + Y) = \text{Var } X + \text{Var } Y$

Example

Let X and Y be independent random variables with

$$E[X] = -5 \qquad E[Y] = 3$$
$$E[X^2] = 36 \qquad E[Y^2] = 25$$

For these variables,

$$\text{Var } X = E[X^2] - (E[X])^2 = 36 - (-5)^2 = 9$$

$$\text{Var } Y = E[Y^2] - (E[Y])^2 = 25 - (3)^2 = 16$$

Also

$$
\begin{aligned}
\text{Var}[2X - 3Y + 9] &= \text{Var}[2X] + \text{Var}[(-3Y)] + \text{Var}[9] && \text{rule 3}\\
&= 2^2 \text{ Var } X + (-3)^2 \text{ Var } Y + \text{Var } 9 && \text{rule 2}\\
&= 4 \text{ Var } X + 9 \text{ Var } Y + 0 && \text{rule 1}\\
&= 4(9) + 9(16) && \text{substitution}\\
&= 180
\end{aligned}
$$

Let X and Y be independent random variables with means 6 and 3, respectively. Let $E[X^2] = 40$ and $E[Y^2] = 29$.
(a) Find σ_X^2 and σ_X.
(b) Find σ_Y^2 and σ_Y.
(c) Use the rules of variance to find each of the following:

Var[3X]	Var[3Y − 2]
Var[X − 1]	Var[2X − Y]
Var[2X + 1]	Var[Y + 3X]
Var[−2Y]	Var[3X + 2Y + 1]

*39. Let X be a random variable with mean 5 and variance 9. Let Y be a random variable with mean 7 and variance 16.
(a) Find $E[(X - 5)/3]$ and $E[(Y - 7)/4]$.
(b) Find $\text{Var}[(X - 5)/3]$ and $\text{Var}[(Y - 7)/4]$.
(c) The results in parts (a) and (b) are not coincidental. Can you make a general statement that describes the pattern that appears?
(d) Use the rules for expectation and variance to show that if X is a random variable with mean μ and standard deviation σ, then

$$E[(X - \mu)/\sigma] = 0 \text{ and } \text{Var}[(X - \mu)/\sigma] = 1$$

*40. **Chebyshev's Inequality** This inequality, derived by the Russian probabilist P. L. Chebyshev (1821–1894), emphasizes the fact that the variance and the standard deviation measure the variability of a random variable about its mean. In particular it states that

The probability that any random variable that lies within k standard deviations of its mean is at least $1 - 1/k^2$.

For example, if we know that X has mean 5 and standard deviation 2, we can conclude that the probability that X lies between 1 and 9 ($k = 2$ standard deviations) is at least $1 - 1/2^2 = .75$.

(a) Many rape victims are reluctant to report the crime to law-enforcement officers. Among those reported, let X denote the elapsed time between the attack and the filing of the report. Assume that X has a mean of 24 hours and a variance of 16. Based on these values, would it be unusual for a report to be filed between 12 and 36 hours after the attack? Would it be unusual for a report to be filed as late as 48 hours after the attack? Explain.

(b) In business, time is money. Consider the random variable X, the time that elapses between the receipt of a telephone order and the shipment of that order to the consumer. Assume that, for a particular business, X has a mean of 36 hours with a variance of 9. Would it be unusual for this firm to take two or more days to ship an order? Would it be unusual for the firm to ship the order within 30 hours of its receipt? Explain.

Vocabulary List and Key Concepts

random variable	cumulative distribution table
discrete random variable	mean
continuous random variable	variance
discrete probability function	standard deviation
continuous probability function	expected value
uniform distribution	Chebyshev's Inequality

Review Exercises

41. Let X denote the length of time that it takes to fill an order at a fast-food restaurant. Is X discrete or continuous? What is the probability that $X = 5$?

42. In a grafting experiment, 10 plants are used. X is the number of plants for which the graft is a success. The probability function for X is given in Table 3.13.

   Table 3.13

x	0	1	2	3	4	5	6	7	8	9	10
$f(x)$	?	.04	.12	.21	.24	.20	.11	.04	.01	.02	.004

   (a) Find $f(0)$.
   (b) Find $P[X = 5]$.
   (c) Find $P[X \le 5]$.
   (d) Find $P[X < 5]$.
   (e) Find $P[X > 5]$.
   (f) Would it be unusual for more than 8 grafts to be successful? Explain based on the probability involved.
   (g) Find $E[X]$.

(h) Find $E[X^2]$.

(i) Find Var X.

(j) Find σ and indicate what physical unit is associated with it.

43. In placing a wager on an event A, a bettor agrees to pay \$$a$ if the event does not occur provided he or she receives \$$b$ if the event does occur. Let X denote the profit to the bettor. The probability function for X is given in Table 3.14.

Table 3.14

x	b	$-a$
$f(x)$	$P[A]$	$P[A']$

(a) Find $E[X]$.

(b) The wager is said to be *fair* if $E[X] = 0$. Show that the wager is fair if the monetary payoffs are proportional to the probability of winning and losing the bet. That is, the wager is fair if

$$\frac{\text{amount bettor pays}}{\text{amount bettor receives}} = \frac{a}{b} = \frac{P[A]}{P[A']}$$

(c) A man wants to wager that he can roll a 5 on a single toss of a fair die. What should the monetary payoffs be in order for this wager to be fair? If he bets \$1 and wins, how much should he receive? If he bets \$2 and wins, how much should he receive?

(d) The ratio a/b is called the *odds in favor of event A*. These odds are usually written in the form $a : b$. Suppose that the odds in favor of a particular horse winning a race are $1 : 10$. A bettor places a \$100 bet on the horse and the horse wins. How much money will the bettor win?

(e) Find the odds in favor of the number 7 coming up in a single spin of a roulette wheel. (See Example 1 of Section 1.) How much money should a bettor receive if she bets \$1 on this number and the number comes up?

(f) Find the odds in favor of rolling a sum of 7 on a single roll of a pair of fair dice.

44. A random variable assumes only the values 2 and 5. Its mean is $17/4$. Find $f(2)$ and $f(5)$. (*Hint*: Let $f(2) = p$ and $f(5) = 1 - p$. Find $E[X]$ and   solve for p.)

45. A stockbroker is interested in the random variable X, the profit per stock obtained by her clients. Past experience indicates that the probability function for this random variable is as shown in Table 3.15. On the average, are her clients making money on their investments? Explain.

Table 3.15

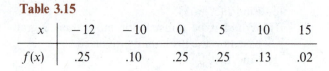

x	-12	-10	0	5	10	15
$f(x)$	.25	.10	.25	.25	.13	.02

46. A random variable has mean 3 and standard deviation 4. Find $E[X^2]$.

47. Let $f(x) = (x - 1)/2$ where $x = 0, 1, 2, 3$. Could this function serve as a probability function for a discrete random variable? Explain.

48. Figure 3.9 gives the probability function for the random variable X, the speed of a baseball in miles per hour as it crosses home plate. The curve is symmetric.
    (a) Find $E[X]$.
    (b) Find the probability that a randomly selected pitch will cross the plate at a speed in excess of 84 mph.
    (c) Find the probability that a randomly selected pitch will cross the plate at a speed exactly 90 miles per hour.
    (d) What probability is represented by the shaded region in the diagram?
    (e) Based on the diagram, would it be unusual for a randomly selected pitch to exceed 90 miles per hour as it crosses the plate? Explain.

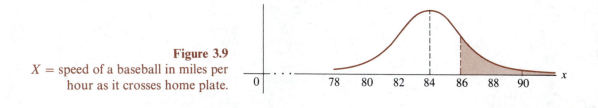

Figure 3.9
$X =$ speed of a baseball in miles per hour as it crosses home plate.

The Binomial, Normal, and Poisson Distributions

In Chapter 3 we considered the general properties of both discrete and continuous random variables. Here we present three particular distributions that have found wide application in statistical research. The first, the **binomial** or **Bernoulli** distribution, was named after James Bernoulli (1654–1705). It is important in that it provides the foundation for making inferences on proportions. The second distribution, the **normal** or **Gaussian distribution**, was discovered in 1733 by De Moivre. Pierre Laplace studied the distribution in detail in 1774. However, the discovery of the normal distribution was erroneously attributed to Gauss (1777–1855), who first referred to it in a paper in 1809. The third, the **Poisson distribution**, is widely used in the study of business and industrial processes. It was discovered by the French mathematician Simeon Denis Poisson. The binomial and Poisson distributions are discrete; the normal is continuous.

4.1 Binomial Model

We begin by considering three random variables that appear to be quite different. As we shall see, they have many common characteristics.

Example 1

1. A student takes a five-question, multiple-choice exam. Each question has four choices for answers, only one of which is correct. The student forgot to study for the exam, and guesses on each question. Let C be the number of correct answers obtained.

2. Under normal circumstances, an otherwise healthy six-year-old child who has not had mumps has a probability of .1 of contracting the disease after

141

exposure to a child with an active case. In a first-grade class 15 susceptible children are exposed to the disease. Let M be the number of children who actually contracts mumps from this source.

3. Twenty voters are selected at random from a large group of eligible voters and polled about their opinions on an upcoming bond for higher education. The actual proportion of eligible voters favoring the proposal is .7. Let B be the number of voters polled who favor the bond issue.

What do the three situations above have in common? There are essentially four points to be noted:

1. Each can be viewed as an experiment consisting of a fixed and known number of identical trials, n.

In case 1 of the example, a trial consists of answering a multiple-choice question; $n = 5$. In case 2, a trial consists of observing a child who is exposed to mumps and noting whether or not the child contracts the disease; $n = 15$. In case 3, a trial consists of polling a randomly selected voter and ascertaining an opinion concerning the bond issue; $n = 20$.

2. The outcome of each trial can be classified as being either a success or a failure.

Usually, success is defined to be the characteristic that is counted. In case 1 of the example, success is getting a correct answer; in case 2, it is observing a child who contracts mumps; in case 3, success is finding a voter who favors the bond issue.

3. The outcome of one trial has no effect on the outcome of any other. That is, the trials are independent and the probability of success p remains the same from trial to trial.

Since the student is guessing on each question, these conditions are satisfied in case 1 of the example: $p = 1/4$. Assuming that immunity in one individual has no influence on immunity in another, case 2 involves independent trials: $p = .1$. Case 3 of the example presents a debatable problem. Opinion polling usually involves sampling without replacement, which implies dependence. Thus, strictly speaking, the probability of success will change from trial to trial. However, if the group of eligible voters is large, as is the usual case in statistical applications, then the removal of a few individuals from the group will not seriously alter the success rate. We may conclude that for all practical purposes we have independence and a constant probability of success of $p = .7$.

4. The random variable of interest is the number of successes obtained in the n trials.

Conditions 1 through 4 above are the general assumptions underlying the binomial model. We summarize these properties for easy reference.

Properties of Binomial Experiments

1. The experiment consists of a fixed number of identical trials. The number of trials is denoted by n.

2. The outcome of each trial can be classified as being either a success or a failure. Trials with this property are called *Bernoulli trials*. Success is taken to be the characteristic being counted.

3. The trials are independent. Therefore, the probability of success, p, remains the same from trial to trial.

4. The random variable X being studied is the number of successes obtained in the n trials. We say that X is binomial with parameters n and p.

Before considering the probability function for a binomial random variable, let us review two ideas presented in Section 2.8. In particular, we need to recall the meaning of the term *n factorial*, and the formula for counting the number of arrangements of objects when some of the objects are indistinguishable from one another.

Definition 4.1	Let n be a positive integer. By ***n* factorial**, denoted by $n!$, we mean $$n(n-1)(n-2) \cdots 3 \cdot 2 \cdot 1$$ Zero factorial, denoted by $0!$, is defined to be 1.

Example 2

$$5! = 5 \cdot 4 \cdot 3 \cdot 2 \cdot 1 = 120$$

Consider a collection of n objects in which x are of one type and the rest of another type. The number of ways to arrange these objects in such a way that they form recognizably different patterns is given by

$$\binom{n}{x} = \frac{n!}{x!\,(n-x)!}$$

The details of the derivation of this formula are given in Section 2.8. The next example illustrates its use.

Example 3

A car salesman talks to 10 clients during the course of a day. Each contact is recorded as being either a success or a failure. For a particular day, he records 3 successes and 7 failures. Letting s denote success and f denote failure, a typical arrangement of contacts is as follows: *sfffsfsfff*. How many such arrangements are possible? Here we have $n = 10$ objects; $x = 3$ are of one type, namely, successes; the rest, $n - x = 10 - 3 = 7$, are of another type, namely failures. By the formula for counting recognizably different patterns, there are

$$\binom{n}{x} = \frac{n!}{x!\,(n-x)!}$$

or

$$\binom{10}{3} = \frac{10!}{3!\,7!}$$

$$= \frac{10 \cdot 9 \cdot 8 \cdot 7 \cdot 6 \cdot 5 \cdot 4 \cdot 3 \cdot 2 \cdot 1}{3 \cdot 2 \cdot 1 \cdot 7 \cdot 6 \cdot 5 \cdot 4 \cdot 3 \cdot 2 \cdot 1}$$

$$= 120$$

ways in which the contacts can be recorded.

To answer questions concerning the behavior of the random variables C, M, and B of Example 1, we must first obtain their probability functions. To do so, let us consider a similar problem in which the number of trials is small. If we can recognize a pattern in the solution, we can generalize the pattern and obtain the desired probability function.

Example 4   A business has a telephone-operated computer terminal tied to a central computer located at an industrial firm. The probability that the line is open at any given time is .75. During the day the firm tries to gain access to the main computer three times. The random variable X is the number of times that the line is open. To find the probability function for X, consider the chart in Table 4.1. In the chart, s denotes success in getting a line and f denotes failure to get a line.

From the table, it can be seen that the possible values for X are 0, 1, 2, and 3. Since the trials are independent, the probability associated with each sample point is found by multiplying the probabilities of success (3/4) and failure (1/4) as shown. The probability function for X is given in Table 4.2. To express this as an

Table 4.1

Sample Points	Value of X	Probability of Sample Point
sss	3	$(3/4)\,(3/4)\,(3/4) = (3/4)^3\,(1/4)^0$
ssf	2	$(3/4)\,(3/4)\,(1/4) = (3/4)^2\,(1/4)^1$
sfs	2	$(3/4)\,(1/4)\,(3/4) = (3/4)^2\,(1/4)^1$
sff	1	$(3/4)\,(1/4)\,(1/4) = (3/4)^1\,(1/4)^2$
fss	2	$(1/4)\,(3/4)\,(3/4) = (3/4)^2\,(1/4)^1$
fsf	1	$(1/4)\,(3/4)\,(1/4) = (3/4)^1\,(1/4)^2$
ffs	1	$(1/4)\,(1/4)\,(3/4) = (3/4)^1\,(1/4)^2$
fff	0	$(1/4)\,(1/4)\,(1/4) = (3/4)^0\,(1/4)^3$

Table 4.2

x	0	1	2	3
$f(x)$	$1(3/4)^0\,(1/4)^3$	$3(3/4)^1\,(1/4)^2$	$3(3/4)^2\,(1/4)^1$	$1(3/4)^3\,(1/4)^0$

equation, we need to look for patterns. Part of the pattern is obvious. Each time, the exponent on the term 3/4 is x and the exponent on the term 1/4 is $3 - x$. Hence the general form for $f(x)$ is

$$f(x) = k \cdot (3/4)^x (1/4)^{3-x}$$

What is k? Here k is the number of sample points that involves the specific value of X. However, a sample point is just an arrangement of three letters, x of which are s's and the rest $(3 - x)$ of which are f's. Hence, we may use the formula for finding the number of recognizably different arrangements to conclude that

$$k = \binom{3}{x} = \frac{3!}{x! \, (3 - x)!}$$

Thus, the equation for the probability function for the random variable X is

$$f(x) = \binom{3}{x}\left(\frac{3}{4}\right)^x\left(\frac{1}{4}\right)^{3-x} \qquad x = 0, 1, 2, 3$$

To find the general form for the probability function for any binomial random variable, we need only to realize that the reasoning used in Example 4 can be used regardless of the number of trials or the probability of success involved in the experiment. Thus, for a binomial random variable with n trials and probability of success p, the probability function is given by

Binomial Probability Function

$$f(x) = \binom{n}{x} p^x (1 - p)^{n-x} \qquad x = 0, 1, 2, 3, 4, \ldots, n$$

Since the form of this function is independent of the actual numerical values of n and p, our main problem is recognizing that in a given situation the assumptions underlying the binomial model are satisfied. Once this is done, we need only identify n and p in order to be able to write the equation for the probability function. This equation can then be used to calculate desired probabilities.

Example 5 Each of five questions on a multiple-choice exam has four choices, only one of which is correct. A student is attempting to guess the answers. The variable C is the number of questions answered correctly. C is binomial with $n = 5$ and $p = 1/4$. What is the probability that a student will get exactly three answers correct? What is the probability that the student will get, at most, three answers correct? What is the probability that the student will get at least four correct? All these questions can be answered by means of the probability function for C,

$$f(c) = \binom{5}{c}\left(\frac{1}{4}\right)^c\left(\frac{3}{4}\right)^{5-c} \qquad c = 0, 1, 2, 3, 4, 5$$

$$P\begin{bmatrix} \text{exactly 3} \\ \text{correct} \end{bmatrix} = P[C = 3] = \binom{5}{3}\left(\frac{1}{4}\right)^3\left(\frac{3}{4}\right)^2 = 10\left(\frac{1}{64}\right)\left(\frac{9}{16}\right) = \frac{90}{1024} = .0879$$

$$P\begin{bmatrix} \text{at most 3} \\ \text{correct} \end{bmatrix} = P[C \le 3] = \sum_{c=0}^{3}\binom{5}{c}\left(\frac{1}{4}\right)^c\left(\frac{3}{4}\right)^{5-c} = \binom{5}{0}\left(\frac{1}{4}\right)^0\left(\frac{3}{4}\right)^5$$

$$+ \binom{5}{1}\left(\frac{1}{4}\right)^1\left(\frac{3}{4}\right)^4 + \binom{5}{2}\left(\frac{1}{4}\right)^2\left(\frac{3}{4}\right)^3 + \binom{5}{3}\left(\frac{1}{4}\right)^3\left(\frac{3}{4}\right)^2$$

$$= 1 \cdot 1(243/1024) + 5(1/4)(81/256)$$

$$+ 10(1/16)(27/64) + 10(1/64)(9/16)$$

$$= 1008/1024 = .9844$$

$$P[\text{at least 4 correct}] = P[C \ge 4] = 1 - P[C < 4]$$

$$= 1 - P[C \le 3] = 1 - .9844 = .0156$$

A general formula for the probability function for a binomial random variable X with parameters n and p exists, so it is reasonable to expect that there is a general formula for the **mean and variance of a binomial random variable** X based on these parameters. Such is the case, as seen in the following theorem.

Theorem 4.1

If X is binomial with parameters n and p, then $E[X] = np$ and Var $X = np(1 - p)$.

The result given in the above theorem is not mathematically trivial; however, the fact that $E[X] = n \cdot p$ is not surprising. Consider the problem of sampling 20 voters for their opinions on a bond issue. If each voter has a 70% chance of favoring the issue and we independently sample 20 voters, we would naturally expect 70% of those sampled to favor the issue. That is, the expected number in favor is $20(.70) = 14$. This intuitive fact is exactly what is shown by Theorem 4.1. Furthermore, we can conclude that the variance in the number of voters in favor of the issue is given by

$$\sigma^2 = \text{Var } X = n \cdot p(1 - p)$$

$$= 20(.7)(.3)$$

$$= 4.2$$

The standard deviation for X is

$$\sigma = \sqrt{\text{Var } X} = \sqrt{4.2} = 2.049 \text{ voters}$$

Exercises 4.1

In exercises 1–6 a random variable is proposed. Decide whether the variable is exactly binomial, approximately binomial, or neither. In the first two situations, identify the numerical values of n and p and find $E[X]$ and Var X.

1. A craps shooter shoots 10 games of craps. X is the number of games won on the first roll.

2. A craps shooter decides to shoot craps until he or she wins. Y is the number of trials necessary to obtain the first win.

3. Fifteen untrained rats are each allowed to run a maze once. L is the number of rats that turn left at the first branch of the maze. (Assume that an untrained rat turns left and right with equal probability.)

4. Assume that 60% of the general public favors abolition of the electoral college system. Fifty citizens are polled at random and asked their opinion. Y is the number in favor of the abolition of the system.

5. A bowler is attempting to learn how to convert the 1–6 split. This split is shown in Figure 4.1. She sets up and bowls for this split 100 times. S is the number of splits converted. (Your answer to this question depends on whether or not you think enough learning can take place to change p enough to matter as the bowler progresses from trial 1 to trial 100.)

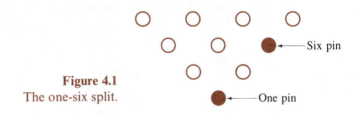

Figure 4.1
The one-six split.

6. Ten cards are drawn from an ordinary deck, one at a time, with replacement. A subject is asked to guess the suit on each draw. G is the number of correct guesses. If the drawing were done without replacement, would your answer change? Explain.

7. In a music-appreciation class a student is expected to be able to distinguish the music of Bach, Beethoven, Mozart, and Brahms. As part of the final exam, the instructor plays five small musical segments and asks the student to identify the composer of each. One student has a tin ear and resorts to guessing. Let X be the number that the student gets correct.
(a) Find $E[X]$; Var X; σ.
(b) Find the expression for the probability function.

(c) Use the probability function of part (b) to find $P[X = 0]$ and $P[X > 4]$.

(d) The instructor assumes with this test that if a student gets four or five correct, then the student learned to distinguish the composers. Is this assumption justified? Explain, based on the probability involved.

8. Assume that 40% of all car buyers are now interested in small cars. During a day, a dealership handles 20 customers. Let X be the number of customers who are interested in small cars.

(a) Find $E[X]$; Var X; σ.

(b) Find the expression for the probability function.

(c) Use $f(x)$ to compute $P[X \leq 4]$.

9. Albino rats are used to study the regulation of hormones. Ten animals are injected with a drug that inhibits the body's synthesis of protein. The probability that a rat will die during the experiment is .2.

(a) Find the expected value of X, the number of rats that die during the experiment.

(b) Find Var X and σ.

(c) Find the expression for the probability function.

(d) Use $f(x)$ to find the probability that no rats die during this experiment.

(e) What is the probability that at least one rat will die?

4.2   Computing Binomial Probabilities

It is obvious that even though the calculations required in dealing with a binomial random variable are not difficult, they do become time-consuming as n becomes larger. The binomial model occurs often in practical problems. This has led mathematicians to compute and table the *cumulative distribution* for commonly encountered values of n and p. The cumulative distribution gives the probability that a random variable X assumes values less than or equal to a specified value t. We denote the cumulative distribution function for X by F. Thus for any real number t,

$$F(t) = P[X \leq t]$$

Table I of Appendix A gives values of the cumulative distribution function for binomial random variables with $n = 5, 10, 15, 20$; and $p = .1, .2, .25, .3, .4, .5, .6, .7, .8,$ and $.9$. We illustrate the use of Table I in Example 1.

Example 1   Consider again the probability of guessing correctly when taking a five-question, multiple-choice test. To find $P[C \leq 3]$ from Table I of Appendix A, note that we are asking for $F(3)$. We need only enter Table I with $n = 5$, $p = 1/4 = .25$ and find the probability associated with $t = 3$. Reading from Table I, we obtain the number .9844. Thus,

$$F(3) = P[C \leq 3] = .9844$$

Note that this corresponds to the result obtained in Example 5 of the previous section. Other types of probabilities can also be found from Table I. For example,

$$P[C \geq 4] = 1 - P[C < 4]$$
$$= 1 - P[C \leq 3]$$
$$= 1 - F(3)$$
$$= 1 - .9844 = .0156$$

To find $P[C = 3]$ from Table I, we must first rewrite the question in terms of the cumulative distribution function. This is done easily as follows:

$$P[C = 3] = P[C \leq 3] - P[C < 3]$$
$$= P[C \leq 3] - P[C \leq 2]$$
$$= F(3) - F(2)$$
$$= .9844 - .8965$$
$$= .0879$$

Example 2   Assume that the probability of contracting mumps after exposure to the disease is .1. What is the probability that a majority of 15 susceptible youngsters will contract the disease after being exposed to it? To answer this equation we must find $P[M \geq 8]$. From Table I of Appendix A,

$$P[M \geq 8] = 1 - P[M < 8]$$
$$= 1 - P[M \leq 7]$$
$$= 1 - F(7) = 1 - 1 \approx 0$$

Note that for $n = 15$ and $p = .1$, $F(t) = 1$ for $t \geq 7$. This means that the probabilities associated with these values are so small that for practical purposes they may be considered to be 0.

What is the probability that the number of children contracting mumps from this exposure will be between two and five, inclusive? Here we are asking for

$$P[2 \leq M \leq 5] = P[M \leq 5] - P[M < 2]$$
$$= P[M \leq 5] - P[M \leq 1]$$
$$= F(5) - F(1)$$
$$= .9978 - .5490 = .4488$$

Note that if a question is not directly translatable into a statement of the form $P[M \leq t]$ for some t, it must be rewritten in terms of statements of this form in order to use Table I of Appendix A.

Exercises 4.2

10. In a survey to determine the effectiveness of a political campaign, 20 voters are randomly selected and polled. Each is asked to identify, without guessing, the office for which the candidate is running. If a majority of those polled can do so, it will be concluded that a majority of the population can identify the candidate; otherwise, it will be concluded that the candidate has a recognition problem and steps will be taken to improve the situation.

    (a) If only 30% of the population can identify the candidate, what is the probability that a majority of those sampled can do so? In this case, the candidate will come to an erroneous conclusion. What are some practical consequences of making this error?

    (b) If 60% of the population can identify the candidate, what is the probability that, at most, 10 of those sampled can do so? In this case an error will also be made. What are some practical consequences of making this error?

11. A simplified programming language uses "words," each consisting of a sequence of 10 digits, either 0 or 1 (for example, 0110101001 is a typical "word"). In transmission, the probability of a digit reversal (0 read as a 1, or vice versa) is .01. Digits are read independently. Let X denote the number of digit reversals per word transmitted.

    (a) Find $E[X]$; $Var[X]$.

    (b) Show that the probability that a given word is transmitted correctly is approximately .9.

    (c) A message is sent that consists of five words. What is the probability that the entire message will be transmitted correctly?

12. A study was run to investigate the association between a mother's smoking during pregnancy and later development of cancer in the child. The mothers of 20 children, each child suffering from a specific cancer, were each matched with four other women of similar backgrounds who had given birth at about the same time as the mother of the patient. Thus we have available 20 matched "quintuplets." In each quintuplet, there was exactly one smoker. If no association exists between smoking and the development of cancer, the probability that the smoker is the patient's mother is 1/5. Let X denote the number of quintuplets observed in which the patient's mother smokes. Note that X is binomial with $n = 20$ and $p = 1/5$.

    (a) If there is no association between smoking and the disease, what is $E[X]$?

    (b) When the experiment was run, it was discovered that there were, in fact, eight quintuplets in which the patient's mother smoked. Find the probability of seeing a result this extreme if in fact there is no association between the risk factor (smoking) and the disease. That is, find $P[X \geq 8]$.

    (c) On the basis of your answer to part (b) do you think that a link has been established between the risk factor and the disease?

13. A psychological experiment involving twins is run. Ten cards are laid face down on a table. Each card is either red, yellow, blue, or green. One twin looks at the card, concentrates on its color, and then tries to communicate mentally the color to the other twin in another room. This twin tries to identify the color of the card. Let X denote the number of cards correctly identified.
    (a) What is the probability of a correct response on a given trial if, in fact, the second twin is randomly guessing the color?
    (b) Find the expected number of correct responses if the second twin is guessing.
    (c) When the experiment is run, six correct responses are obtained. Find the probability of getting a result as extreme as this strictly by chance. That is, find $P[X \geq 6]$.
    (d) On the basis of your answer to part (c), do you think that the twins have some ability to communicate mentally?

14. A manufacturer of moist dog food wants to claim that more than 80% of all dogs prefer moist food to dry. To verify this claim the manufacturer places a bowl of dry food and an identical bowl of moist food side by side in a run. A dog is allowed to enter the run and the food type selected is observed. The experiment is conducted on a total of 20 dogs. Let X be the number of dogs that select moist food.
    (a) If 80% of all dogs prefer moist food, we would expect 16 of the 20 dogs in the experiment to select moist food. We should believe the manufacturer's claim if the number that selects moist food is larger than this expected value. Let us agree to believe the manufacturer if $X \geq 19$. Find the probability that we will believe the claim when, in fact, it is not true and $p = .8$.
    (b) Find the probability that we will not believe the claim when, in fact, it is true and $p = .9$.
    (c) The results described in parts (a) and (b) both lead to errors being made. What are some of the practical consequences, from a business standpoint, of making each of these errors?

15. In a psychological experiment, 20 subjects are used. Each is presented with the word *doctor* and asked to state whether the word is felt to be *male* or *female*. Let X denote the number of subjects who answer "male." If no stereotyping occurs, then the probability p of this response should be $1/2$. When the experiment is run, 15 subjects answer "male." Find the probability of obtaining a result this extreme if in fact $p = 1/2$. That is, find

$$P[X \geq 15 \,|\, p = 1/2]$$

What practical conclusion does this result suggest?

16. The makers of brand A television sets want to claim that their set has a clearer picture than the set of the nearest competitor. To test this claim 15

persons are allowed to view the unidentified sets placed side by side and are asked to pick the one with the clearer picture. The company marketing analyst feels that if 9 or more persons pick brand A, he has the statistical evidence he needs to support the claim. Do you agree with him? Support your answer by computing the probability that 9 or more could pick brand A strictly by chance, when no discernible difference exists between the sets.

17. If two people, each with genes for both brown and blue eyes, parent a child, the probability that the child will have blue eyes is 1/4. A person with genes for both brown and blue eyes is said to be heterozygous with respect to this trait. The parents themselves will each be brown-eyed. If each of their ten natural children has brown eyes, is there reason to suspect that the parents are not both heterozygous with respect to eye color? Explain, based on the probability of this occurring.

4.3   The Normal Distribution

In the last sections we studied an important family of discrete random variables, namely the binomial distribution. Here we consider the normal distribution, a frequently encountered family of continuous variables. We begin by defining a normal random variable in terms of its probability function. Most of the results presented here require calculus for their proofs, hence no rigorous proofs are attempted.

Definition 4.2   A random variable X is said to be distributed *normally* with mean μ and variance σ^2 if its probability function f is given by

$$f(x) = \frac{1}{\sigma\sqrt{2\pi}} \exp\left[-\frac{1}{2}\left(\frac{x-\mu}{\sigma}\right)^2 \right]$$

$$-\infty < x < \infty$$

$$-\infty < \mu < \infty$$

$$\sigma > 0$$

$$\pi \approx 3.1416$$

$$\exp \approx 2.7183$$

Do not be overly concerned about the above definition. It is included only to show you that probability functions are not always given by simple straight lines! We need only be aware of the general properties possessed by a curve with an equation of this form.

We summarize the properties of a normal curve as follows:

Properties of Normal Curves

1. The graph of the probability function for any normal random variable is a symmetric, bell-shaped curve centered at the mean as shown in Figure 4.2. Note that, as is always true with continuous random variables, μ is a location parameter. It determines the exact location of the peak of the bell along the horizontal axis.

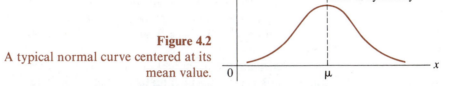

Figure 4.2
A typical normal curve centered at its mean value.

2. The dips in the curve, called *points of inflection*, are located one standard deviation to either side of the mean. That is, the points of inflection are located at $\mu \pm \sigma$ as shown in Figure 4.3. Note that, as expected, σ is a shape parameter. Small values of σ are associated with steep bells, whereas large values of σ generate rather flat bells. This idea is illustrated in Figure 4.4.

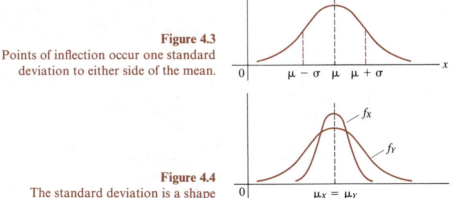

Figure 4.3
Points of inflection occur one standard deviation to either side of the mean.

Figure 4.4
The standard deviation is a shape parameter.

3. Since a normal curve is the graph of the probability function for a continuous random variable X, the total area under the curve is 1; the probability that X assumes any specified value is 0; and $P[a < X < b]$ is calculated by finding the area under the graph of f between points a and b.

Example 1

Let X denote the number of hours that a child spends per week watching television. Assume that X is normal with mean 8 and variance 4. We are assuming that the equation for the probability function for X is

$$f(x) = \frac{1}{2\sqrt{2\pi}} \exp\left[-\frac{1}{2}\left(\frac{x-8}{2}\right)^2\right]$$

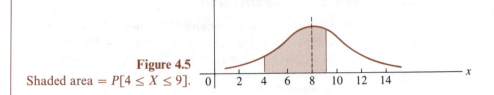

Figure 4.5
Shaded area $= P[4 \leq X \leq 9]$.

The graph of this probability function is shown in Figure 4.5. Note that the curve is centered at $\mu = 8$ and is symmetric about the line $x = 8$. Note also that since by assumption $\sigma^2 = 4$, $\sigma = 2$. Thus, the dips in the curve occur at 8 ± 2, or at the points 6 and 10.

To calculate the probability that a randomly selected child spends between 4 and 9 hours per week watching television, we have to find the area of the shaded region of Figure 4.5. A word of explanation is in order. The equation given in the definition of the normal random variable implies that a normal random variable can assume any real value. This is not true here, since it is clearly impossible to watch television for less than 0 or more than 168 hours during a single week. When we assume that X is distributed normally with mean 8 and variance 4, we are assuming that, over the set of physically possible values of X, the above curve yields acceptable probabilities. We assume that by the time the curve reaches 0 or 168 it has become so close to the horizontal axis that, for all practical purposes, the area under the curve below 0 or above 168 is 0. Thus the probability of observing values in these regions is 0.

There are infinitely many normal random variables, each completely characterized by stating the numerical value of its mean and variance. Is it necessary to make a table of the cumulative distribution function for selected values of μ and σ^2, as was the case for the binomial distribution? Thankfully, the answer is no. We can show that by algebraic methods any question concerning a normal random variable can be converted to an equivalent question concerning a normal random variable with mean 0 and variance 1. Such a variable is referred to as being *standard normal*. Hence, we need only work with the cumulative distribution function for the standard normal random variable.

Definition 4.3

A normal random variable is said to have a **standard normal distribution** if it has mean 0 and variance 1. It is denoted by Z.

The cumulative distribution function F for Z is given in Table II of Appendix A. Note that graphically $F(z)$ gives the area under the graph of the probability function for Z to the *left* of and including the point z.

There are basically two types of questions that we must be prepared to answer.

1. Given a specified value z, what is the probability that Z assumes a value less than or equal to z? That is, what is $P[Z \leq z]$?

2. Given a specified probability α, what value of z has the property that $P[Z \leq z] = \alpha$?

We shall first consider questions of the former type.

Example 2

1. Suppose we want to find $P[Z \leq -2.56]$. Since $P[Z \leq -2.56] = F(-2.56)$, we need only locate the number -2.56 in Table II of Appendix A. The first digits are found in the column headed z, with the third digit in the column headed .06. Reading from Table II $P[Z \leq -2.56] = .0052$. The area corresponding to this probability is shown in Figure 4.6.

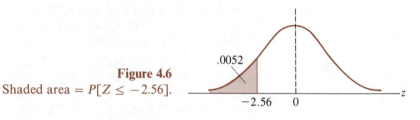

Figure 4.6
Shaded area $= P[Z \leq -2.56]$.

2. To find $P[Z > 1.52]$ we first express this probability in terms of the cumulative distribution function:

$$P[Z > 1.52] = 1 - P[Z \leq 1.52] = 1 - F(1.52)$$

From Table II, $F(1.52) = .9357$. Hence,

$$P[Z > 1.52] = 1 - .9357 = .0643$$

This probability is pictured in Figure 4.7.

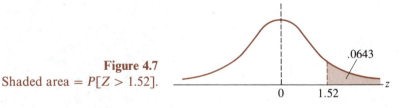

Figure 4.7
Shaded area $= P[Z > 1.52]$.

3. From Table II, $F(1.62) = .9474$ and $F(-1.53) = .0630$. Hence,

$$P[-1.53 \leq Z \leq 1.62] = P[Z \leq 1.62] - P[Z < -1.53]$$
$$= P[Z \leq 1.62] - P[Z \leq -1.53]$$
$$= F(1.62) - F(-1.53)$$
$$= .9474 - .0630$$
$$= .8844$$

Figure 4.8 shows this probability graphically.

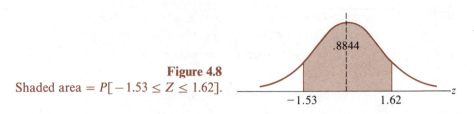

Figure 4.8
Shaded area $= P[-1.53 \le Z \le 1.62]$.

In each case in Example 2, we were given a z point and asked to find a probability. Example 3 demonstrates the use of Table II to go in the opposite direction. We will be given a probability and asked to find the corresponding z point.

Example 3

1. Find the point z such that $P[Z \le z] = .2946$. This point is pictured in Figure 4.9(a). We go to Table II of Appendix A and look in the table for the number .2946. This number is found in row $-.5$ and column .04. Hence $z = -.54$.

2. Find the point z such that $P[Z > z] = .3264$. Figure 4.9(b) shows this point. We must first realize that $P[Z > z] = 1 - P[Z \le z]$. Thus we are looking for the point z such that $1 - P[Z \le z] = .3264$. This is equivalent to asking for the point such that $P[Z \le z] = .6736$. We now look in Table II and find that $z = .45$.

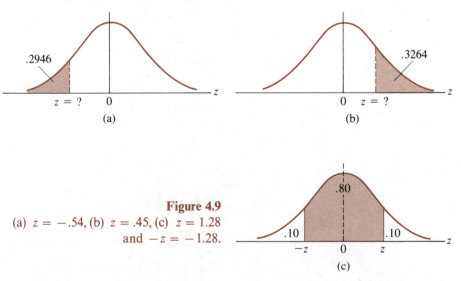

Figure 4.9
(a) $z = -.54$, (b) $z = .45$, (c) $z = 1.28$ and $-z = -1.28$.

3. Find the point z such that $P[-z \le Z \le z] = .80$. These points are shown in Figure 4.9(c). From the diagram it is evident that we are looking for the point z such that $P[Z \le z] = .90$. We look in Table II and find the number that is closest to .9000. That number is .8997 and corresponds to a value of $z = 1.28$.

We shall use the Z table extensively in the work to come. These exercises should provide you with the practice needed to learn to read the table.

Exercises 4.3

18. Let X be normal with mean 5 and variance 16.
    (a) Find $E[X]$; Var X; $E[X^2]$; σ.
    (b) Find the expression for the probability function for X.
    (c) Where are the points of inflection for the graph of the probability function located?

19. Let X be normal with mean 8 and variance 4.
    (a) Find $E[X]$; Var X; σ.
    (b) Find the expression for the probability function for X.
    (c) Where are the points of inflection for the graph of the probability function located?

20. Let Z be standard normal. Use Table II of Appendix A to find the following:
    (a) $P[Z \leq 1.75]$ (b) $P[Z < 1.75]$
    (c) $P[Z > -1.32]$ (d) $P[-1.6 \leq Z \leq 2.37]$
    (e) $P[Z = 2.1]$ (f) $P[Z \leq 4]$
    (g) $P[Z \leq -4]$ (h) $P[Z \leq -1.29]$
    (i) $P[Z < -1.29]$ (j) $P[Z > -1.29]$
    (k) $P[Z = -1.29]$ (l) $P[-1.31 \leq Z \leq 1.84]$
    (m) $P[1.05 \leq Z \leq 1.92]$ (n) $P[Z > 4.5]$

21. Use the standard normal table to find:
    (a) the point z such that $P[Z \leq z] = .9946$
    (b) the point z such that $P[Z > z] = .1685$
    (c) the point z such that $P[Z \leq z] = .10$
    (d) the point z such that $P[Z > z] = .10$
    (e) the point z such that $P[-z \leq Z \leq z] = .90$
    (f) the point z such that $P[-z \leq Z \leq z] = .95$
    (g) the point z such that $P[-z \leq Z \leq z] = .99$

4.4 Computing Normal Probabilities

It is unreasonable to expect that every normal random variable that you will encounter in practice will have mean 0 and variance 1. In fact, most normal random variables are not standard naturally. Hence we cannot use the standard normal table to compute probabilities associated with the variable directly. We must first rewrite the question in a form that will allow us to answer it via Table II of Appendix A.

The following theorem, called the *standardization theorem*, states the algebraic transformation required to convert a question concerning a general normal random variable into an equivalent question pertaining to the standard normal variable. The theorem was partially proved in exercise 39 in Chapter 3, using the rules for expectation and variance. This theorem provides the foundation for all future computations involving the normal curve.

Theorem 4.2

Standardization Theorem

Let X be a normal random variable with mean μ and variance σ^2. Then the random variable

$$\frac{X - \mu}{\sigma}$$

is standard normal.

Perhaps it is easier to remember this theorem by remembering that to standardize a normal random variable, you must subtract its mean and then divide by its standard deviation. Examples 1 and 2 show how this theorem is used.

Example 1

A new diet is being advertised in which only liquid protein and bread are eaten. The random variable of interest is X, the amount of weight lost by the end of the first week. Assume that X is normal with a mean of 12 pounds and a variance of 9. What is the probability that a person who goes on this diet will lose less than 8 pounds the first week? Mathematically, we are to find $P[X < 8]$. Geometrically, we are being asked to find the area of the shaded region in Figure 4.10(a). To do

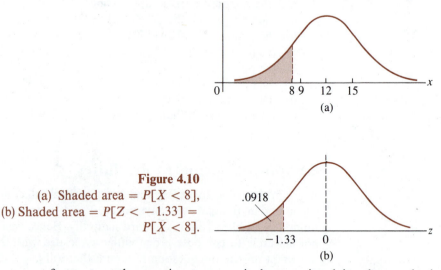

Figure 4.10
(a) Shaded area = $P[X < 8]$,
(b) Shaded area = $P[Z < -1.33] = P[X < 8]$.

so we must first convert the question to an equivalent one involving the standard normal curve by subtracting the mean of 12 and dividing by the standard deviation of 3. Thus,

$$P[X < 8] = P\left[\frac{X - 12}{3} < \frac{8 - 12}{3}\right]$$

$$= P[Z < -1.33]$$

The desired probability is read from Table II of Appendix A, and is shown in Figure 4.10(b).

Note that the areas of shaded regions in both diagrams are equal and have a numerical value of .0918.

Example 2

The exhaust emitted from automobiles contains hydrocarbons which contribute to air pollution. Let X denote the number of grams of hydrocarbons emitted by an automobile per mile. Assume that X is normally distributed with a mean of 1 gram and a standard deviation of .25 gram. To find the probability that a randomly selected automobile will emit between .75 and 1.50 grams per mile, we must standardize X by subtracting its mean and dividing by its standard deviation. We see that

$$P[.75 < X < 1.50] = P\left[\frac{.75 - 1}{.25} < \frac{X - 1}{.25} < \frac{1.50 - 1}{.25}\right]$$

$$= P[-1 < Z < 2]$$

$$= P[Z \le 2] - P[Z \le -1]$$

From the standard normal table, $P[Z \le 2] = .9772$ and $P[Z \le -1] = .1587$. Substituting, we see that

$$P[.75 < X < 1.50] = .9772 - .1587$$

$$= .8185$$

In Examples 1 and 2, we were given a value of X and asked to find a probability. Examples 3 and 4 demonstrate how to work in reverse. We will be given a probability and will be asked to find the corresponding value of X.

Example 3

One random variable monitored by diabetics is the fasting blood glucose level. Assume that this random variable is normally distributed with a mean of 106 mg/100 mL and a standard deviation of 8 mg/100 mL. Find the point x with the property that 25% of all diabetics have a fasting glucose level below this point. The desired point is shown in Figure 4.11. We are looking for the point x such that

$$P[X < x] = .25$$

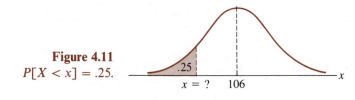

Figure 4.11
$P[X < x] = .25$.

Standardizing X by subtracting its mean and dividing by its standard deviation, we are looking for the point x such that

$$P\left[\frac{X-106}{8} < \frac{x-106}{8}\right] = .25$$

$$P\left[Z < \frac{x-106}{8}\right] = .25$$

From the standard normal table, we see that

$$P[Z < -.67] = .25$$

Since $-.67$ and $(x-106)/8$ both have area .25 to the left, these points are equal. We find x by solving the equation

$$(x-106)/8 = -.67$$

In this case, $x = 8(-.67) + 106 = 100.64$. Approximately 25% of all diabetics have a fasting blood glucose level below 106.64 mg/100 mL.

Example 4

There is concern that, because of pollutants in the water, shrimp caught off the southeastern United States coast are no longer safe for human consumption. To monitor this situation, let us assume that 100-pound samples of shrimp are tested periodically. The variable X denotes the percentage of harmful substance found. Assume that under acceptable conditions X is normal with a mean of 2% and a standard deviation of .5%. Each week a 100-pound sample is caught and tested. If X lies in the upper 5% of the distribution, then it is necessary to monitor the waters daily. Above what value must X lie in order to trigger daily monitoring? We are dealing with the curve of Figure 4.12. We want to find the value x such that $P[X > x] = .05$. To find this number, we must first standardize and then make use of Table II of Appendix A.

$$P[X > x] = P\left[\frac{X-.02}{.005} > \frac{x-.02}{.005}\right] = P\left[Z > \frac{x-.02}{.005}\right]$$

We want this probability to equal .05. Hence, we want

$$P\left[Z \le \frac{x-.02}{.005}\right] = .95$$

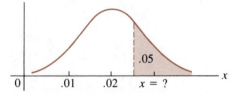

Figure 4.12
$X = $ percentage of harmful substance in a 100-pound sample of shrimp.

Let $(x - .02)/.005 = z$. We thus look for the point z such that $P[Z \leq z] = .95$. From Table II, we see that $P[Z \leq 1.64] = .9495$, and $P[Z \leq 1.65] = .9505$. Let us compromise and let $z = 1.645$. Hence, we obtain the equation

$$1.645 = \frac{x - .02}{.005}$$

Solving this equation for x, we see that $x = 1.645(.005) + .02 = .0282$. Practically speaking, if the percentage of harmful materials is found to be 2.82% or higher, it is necessary for daily monitoring to begin.

Exercises 4.4

22. New car makers stress gas mileage by quoting EPA estimates of mileage in their advertisements. Assume that, for a particular model, the variable X, the number of miles per gallon obtained on the highway, is normally distributed with mean 31 and variance 16.
    (a) Would you be surprised if you bought one of these cars and got less than 25 miles per gallon? Explain, based on the probability of this occurring.
    (b) What is the probability that a car of this type would get more than 39 miles per gallon?

23. A psychologist is experimenting with assertiveness training. Subjects are given preliminary tests to ascertain their attitude toward certain stressful situations. After a three-day training period the subjects are retested. If the course has been effective, the scores should increase. Let X denote the increase in score and assume that X is normal with mean 20 and variance 36.
    (a) Find the probability that the training will be ineffective for a given randomly selected individual. That is, find $P[X \leq 0]$.
    (b) Find the probability that a person's score will increase by more than 30 points.
    (c) Find the probability that a person's score will increase between 15 and 25 points.

24. From past experience it is felt that the random variable X, the age of the mother at the birth of her first child, is normally distributed with a mean of 20 years and a variance of 9 world-wide. Find the probability that a randomly selected mother has her first child:
    (a) before age 16
    (b) after age 30
    (c) between the ages of 16 and 24
    (d) between the ages of 14 and 26

25. There are two check-out counters at the supermarket. Suppose the amount of time (in minutes) we must wait in line at the first counter, X, is normal with mean 10 and variance 2. The time that we must wait at the second counter, Y, is normal with mean 12 and variance 16. A shopper has only 6 minutes left on a parking meter. Which line should the shopper choose to have the highest probability of getting through the line in 5 minutes or less?

26. A business firm considers the number of years of formal education, X, a prime factor in screening applicants for employment in its plant. Assume from past experience that X is normal with a mean of 10 years and a standard deviation of 3 years.
    (a) If an applicant falls into the lower 10% educationally, the applicant is automatically eliminated from consideration for employment. What cutoff point in terms of number of years of formal education is being used?
    (b) If an applicant falls into the upper 20% educationally, then the applicant is considered to be overeducated for the job and is eliminated from consideration. What is the upper cutoff point in terms of years of formal education that is being used by the business firm?

27. Under usual conditions a particular drug is known to have a pH that is normally distributed with mean 9 and variance 1/4. If the pH is either too high or too low, the drug is unacceptable, since it may cause harmful side effects or be ineffective. For this reason, if a sample of the drug lies in the lower or upper 15% of the distribution, it is not acceptable. Within what pH range must a sample of the drug fall to be acceptable?

28. An American fitness council concluded a 5-year study on adults from 20 to 29 years of age. One result of this study was that the length of time, X, for a mile run for one of the subjects was normally distributed with $\mu = 380$ seconds and $\sigma = 40$ seconds. What percentage of this group could run a mile in under 5 minutes? The council decided that the slowest 20% are not in good shape. What is the fastest time in which a person from this group can run a mile?

*29. **Empirical Rule**   Recall that Chebyshev's Inequality (exercise 40 in Chapter 3) guarantees the probability that any random variable X that lies within two standard deviations of its mean is at least .75. If X is normal, this probability is much higher. Verify the following statements which are commonly known as the *empirical rule*. Let X be normal with mean μ and variance σ^2. Then
    (a) $P[-\sigma < X - \mu < \sigma] \approx .68$
    (b) $P[-2\sigma < X - \mu < 2\sigma] \approx .95$
    (c) $P[-3\sigma < X - \mu < 3\sigma] \approx .99$

*30. One way to immunize persons sensitive to bee stings is to treat them with extracts of bee venom. However, recent research has shown that the extracts manufactured currently vary widely in strength. Let X denote the amount of antigen in a sample. An antigen is a substance that, when introduced into the body, stimulates the production of an antibody. Assume that X is normally distributed with a mean of 16 micrograms per milliliter and a standard deviation of 7 micrograms per milliliter.
    (a) Between what two values are 95% of the samples expected to lie? Don't use a calculator!
    (b) Would it be unusual to find a sample with an antigen level of less than 2 micrograms per milliliter? Explain based on the probability of this occurring. Don't use a calculator!

4.5   Approximating the Binomial Distribution

Recall that in the binomial model, the random variable X denotes the number of successes obtained in n independent trials with the probability of success p remaining the same from trial to trial. Table I of Appendix A gives the cumulative distribution for X for selected values of n and p. There are situations arising quite naturally that satisfy the assumptions of the binomial model for which the value of n or p is not listed in Table I. These problems can be handled by recourse to the binomial probability function

$$P[X = x] = f(x) = \binom{n}{x} p^x (1 - p)^{n-x}$$

However, computing probabilities by this formula for large n requires that numerous computations be performed; it is extremely time-consuming. We need a method for quickly and accurately approximating probabilities when the binomial table cannot be used. This method is the **normal approximation to the binomial distribution**.

Example 1   Suppose we are interested in studying local public opinion concerning the United States President's performance in foreign affairs. We intend to sample 100 individuals from a locality and ask each the question, Do you feel that the President is doing a good job in the area of foreign affairs? Suppose that, in fact, 10% of the people in the population feel that the answer is yes. What is the probability that, at most, five people in the sample will answer yes? What is the probability that at least 12 people will answer yes? What is the probability that between 8 and 12, inclusively, will answer yes?

If we allow X to denote the number of individuals in the sample who answer yes, X can be thought of as binomial with the number of trials, n, being 100, and the probability of success (obtaining a yes answer) being .10. We wish to use the methods of Section 4.2 to answer these questions. However, $n = 100$ is too large for use with the available binomial table, Table I of Appendix A. We need to find a way to overcome this difficulty.

To see logically how binomial probabilities can be approximated, let us consider a series of binomial variables. In particular, we will examine four binomial variables, each with probability of success $p = .3$, with values of n being 5, 10, 15, and 20. The probability functions (obtained from Table I of Appendix A) and graphs of the respective probability functions are shown in Figure 4.13. The crucial point to be noted from the graphs of these probability functions is that, as the value of n increases, the diagrams assume an approximate bell shape. It takes only a little imagination to visualize a smooth bell curve that closely fits the last block diagram in the figure.

This suggests that perhaps binomial probabilities, represented by one or more blocks in the above diagram, can be reasonably approximated by means of suitably selected areas under the appropriate normal curve. Which normal curve is appropriate? Common sense indicates that the normal curve used in the

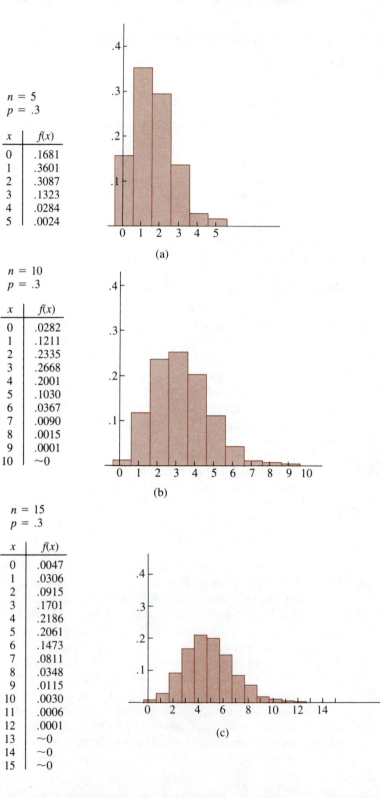

$n = 5$
$p = .3$

x	$f(x)$
0	.1681
1	.3601
2	.3087
3	.1323
4	.0284
5	.0024

(a)

$n = 10$
$p = .3$

x	$f(x)$
0	.0282
1	.1211
2	.2335
3	.2668
4	.2001
5	.1030
6	.0367
7	.0090
8	.0015
9	.0001
10	~0

(b)

$n = 15$
$p = .3$

x	$f(x)$
0	.0047
1	.0306
2	.0915
3	.1701
4	.2186
5	.2061
6	.1473
7	.0811
8	.0348
9	.0115
10	.0030
11	.0006
12	.0001
13	~0
14	~0
15	~0

(c)

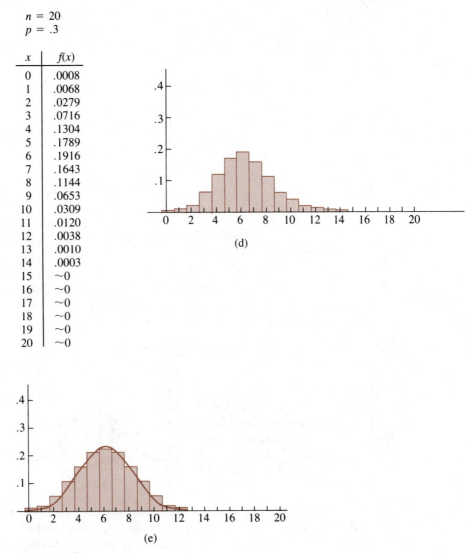

n = 20
p = .3

x	f(x)
0	.0008
1	.0068
2	.0279
3	.0716
4	.1304
5	.1789
6	.1916
7	.1643
8	.1144
9	.0653
10	.0309
11	.0120
12	.0038
13	.0010
14	.0003
15	~0
16	~0
17	~0
18	~0
19	~0
20	~0

(d)

(e)

Figure 4.13
Graph of binomial probability functions: (a) $n = 5$ and $p = .3$, (b) $n = 10$ and $p = .3$, (c) $n = 15$ and $p = .3$, (d) $n = 20$ and $p = .3$, and (e) bell curve that approximates the binomial probability function.

approximation should have the same center of location and the same spread as the binomial variable being approximated. That is, the normal variable used should have the same mean and the same variance as the binomial variable, namely np and $np(1 - p)$, respectively. It has been found through empirical studies that this method of approximation is good whenever $p \leq .5$ and $np > 5$, or $p > .5$ and $n(1 - p) > 5$.

Example 2

Let X be binomial with $n = 20$ and $p = .3$. The probability function for X is shown in Figure 4.13(d). Suppose we want to find $P[X \le 5]$. This can be done by adding together the areas of blocks centered at 0, 1, 2, 3, 4, and 5 as shown in Figure 4.14. Thus

$$P[X \le 5] = .0008 + .0068 + .0279 + .0716 + .1304 + .1789 = .4164$$

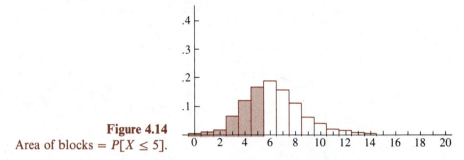

Figure 4.14
Area of blocks $= P[X \le 5]$.

Since $p = .3 \le .5$ and $np = 20(.3) = 6 > 5$, we can expect to be able to approximate closely the above probability using a normal variable Y, with mean $np = 6$ and variance $n(p)(1 - p) = 20(.3)(.7) = 4.2$. The probability function for Y is shown in Figure 4.15.

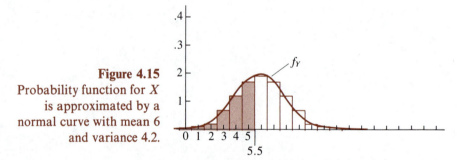

Figure 4.15
Probability function for X
is approximated by a
normal curve with mean 6
and variance 4.2.

The one point to be noted here that helps to improve our approximation is that we want to include the areas of the blocks centered at 0, 1, 2, 3, 4, and 5. How can this best be done? If we consider only the area under the normal curve less than or equal to 5, we will quite obviously be leaving out half of the area in the block centered at 5, since that block extended from the points 4.5 to 5.5. This would introduce an unnecessary and avoidable error in the calculation. To obtain the best possible approximation, we should approximate the areas within the blocks centered at 0, 1, 2, 3, 4, and 5 by the area under the normal curve Y less than 5.5. That is,

$$P[X \le 5] \approx P[Y \le 5.5]$$

This point can be seen graphically in Figure 4.16.

From this point on, the problem is a routine problem of standardization. We need only standardize Y by subtracting its mean of 6 and dividing by its standard

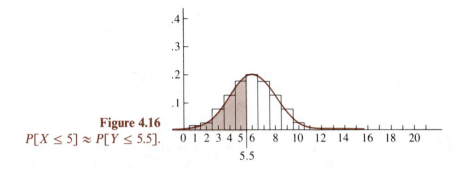

Figure 4.16

$P[X \le 5] \approx P[Y \le 5.5]$.

deviation of $\sqrt{4.2} = 2.05$. We then look up the resulting probability in the standard normal table (Table II of Appendix A). Using this procedure,

$$P[X \le 5] \approx P[Y \le 5.5]$$

$$= P\left[\frac{Y - 6}{2.05} \le \frac{5.5 - 6}{2.05}\right]$$

$$= P\left[Z \le \frac{-.5}{2.05}\right] = P[Z \le -.24]$$

$$= .4052$$

Note that this is not exactly what we got using the binomial table, but the difference between the two results is only .0112. The approximation is fairly good. This problem was worked both ways only for comparative purposes. If one has available binomial tables with the desired values of n and p, and can find the exact probability, the normal approximation should not be used.

Example 3   Consider the opinion poll of Example 1. The variable of interest was the number of persons in a sample of 100 answering yes to the question, Do you feel that the President is doing a good job in the area of foreign affairs? We assume that X is binomial with $n = 100$ and $p = .1$. To find the probability that at most 5 people answer yes we need to compute $P[X \le 5]$. This probability is approximated by a normal random variable Y with mean $n \cdot p = 100(.1) = 10$ and variance $np(1 - p) = 100(.1)(.9) = 9$. Thus

$$P[X \le 5] = P[Y \le 5.5] = P\left[\frac{Y - 10}{3} \le \frac{5.5 - 10}{3}\right]$$

$$= P\left[Z \le \frac{-4.5}{3}\right] = P[Z - 1.5] = .0668$$

That is, if in fact 10% of the people feel the President is doing well, there is only about a 6.68% chance that in a sample of 100 people 5 or fewer would support the President.

What is the probability that at least 12 answer yes? That is, what is

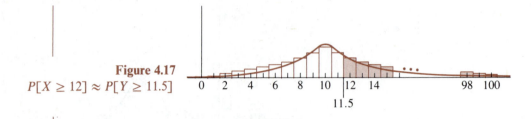

Figure 4.17
$P[X \geq 12] \approx P[Y \geq 11.5]$

$P[X \geq 12]$? To answer this question consider the diagram of Figure 4.17. To approximate $P[X \geq 12]$ as closely as possible, we should obviously consider the area under the normal curve from 11.5 on. That is,

$$P[X \geq 12] \approx P[Y \geq 11.5]$$

The problem is once again a routine standardization problem.

$$P[X \geq 12] \approx P[Y \geq 11.5]$$

$$= P\left[\frac{Y - 10}{3} \geq \frac{11.5 - 10}{3}\right]$$

$$= P\left[Z \geq \frac{1.5}{3}\right]$$

$$= P[Z \geq .5]$$

$$= 1 - P[Z \leq .5] = 1 - .6915 = .3085$$

That is, if in fact 10% of the persons in the population favor the President's policy, there is only about a 30.85% chance that in a sample of 100 people 12 or more supporters are found.

Exercises 4.5

31. Let X be binomial with $n = 20$ and $p = .5$.
    (a) From Table I of Appendix A, find $P[X \leq 5]$.
    (b) Use the normal approximation to approximate $P[X \leq 5]$ and compare your results to part (a).
    (c) From Table I, find $P[X = 5]$.
    (d) Use the normal approximation to approximate $P[X = 5]$. (*Hint*: $P[X = 5] \approx P[4.5 \leq Y \leq 5.5]$.)

32. In studying birthdays, it is assumed that one is just as likely to be born on Sunday as on any other day. To examine this theory, 49 individuals are to be randomly selected and their birthdays determined. What is the probability that at least 10 were born on Sunday? What is the probability that exactly 7 were born on Sunday?

33. When men landed on the moon for the first time, 70% of the TV-viewing public watched the coverage of the landing. In a random sample of 200

viewing units, what is the probability that more than 150 were watching coverage of this event?

34. The probability that an automobile accident is due to equipment failure is .1. What is the probability that between 7 and 10 (inclusive) of the next 100 accidents will be due to equipment failure?

35. A headache remedy is said to be 80% effective in curing headaches caused by simple nervous tension. If this remedy is used on 100 patients suffering from nervous tension, what is the probability that between 75 and 90 (inclusive) will obtain relief?

36. A politician claims that 70% of his constituents favor the legalization of marijuana. A random sample of 200 voters from his district is obtained. If the politician is right, what is the probability that fewer than 125 favor this legislation?

37. A recent report claims that fewer than 1/3 of all patients who experience their first epileptic seizure will have a second seizure within three years. To verify this claim, the progress of 150 patients is followed after their first seizure.
    (a) If the probability of a second seizure within three years, p, is 1/3, how many of these 150 patients are expected to experience a second seizure?
    (b) Let us agree to believe the claim if and only if X, the number of patients experiencing a second seizure, is 40, at most. What is the probability that we will make the error of believing that $p < 1/3$ when, in fact, $p = 1/3$?
    (c) What is the probability that we will make the error of not believing the claim when, in fact, it is true and $p = 1/4$? (*Hint*: Find $P[X > 40 | p = 1/4]$.)

38. Research indicates that the probability that it will be necessary to deliver a child by the Caesarean method is .05. Of the next 100 babies delivered at a particular hospital, how many Caesarean deliveries would you expect? Would you be surprised if at least 7 had to be delivered in this manner? if at least 10 had to be Caesarean deliveries? Explain, based on the probability involved in each case.

4.6   The Poisson Distribution (Optional)

In this section we consider discrete random variables that follow what is called the *Poisson distribution*. It comes into play in situations in which discrete events are being observed in some continuous interval of time or space. The random variable of interest is the number of occurrences of the event in the observation period. A few examples should make the idea clear.

Example 1   A nuclear-power plant releases detectable amounts of radioactive gases twice a month on the average. The plant is observed for a three-month period. Let X denote the number of times that radioactive gases are released during this time period. Here, we are interested in the event "radioactive gases are released." This event is discrete in the sense that it will occur at some isolated or discrete instant in time. We are attempting to observe and count the number of occurrences of this event in a continuous interval of time of length three months. Thus X is a Poisson random variable; it is a random variable that counts the number of occurrences of a discrete event in a continuous interval of time.

Example 2

The age of a zircon can be estimated by counting the number of uranium fission tracks on a polished surface of the zircon. The average number of tracks per square centimeter on a particular zircon is five. We examine a two-square centimeter sample of this zircon. Let X denote the number of fission tracks observed. Here, the discrete event is the occurrence of a fission track in a particular location. We are attempting to observe and count the number of occurrences of this event in a continuous interval of area of size two square centimeters. Thus X is a Poisson random variable; it is a random variable that counts the number of occurrences of a discrete event in a continuous spacial interval.

When we observe the occurrence of discrete events in a continuous interval, we are observing what is called a **Poisson process**. Each Poisson process is characterized by one parameter which we denote by λ. This parameter gives the *average* number of occurrences of the event in a unit interval. For instance, in Example 1, the average number of releases of radioactive gases is two per month. Here, the basic unit of measurement is a month and $\lambda = 2$. In Example 2, the average number of fission tracks is five per square centimeter. The measurement unit is a square centimeter and $\lambda = 5$. When we observe a Poisson process, we do so over a specified interval of time or space. We denote the size of this interval by s. In Example 1, we observe the power plant over three months and so $s = 3$. In Example 2, the zircon is examined over an area of two square centimeters so that $s = 2$. The probability function for the Poisson random variable X, the number of occurrences of the event in an interval of size s, is expressed in terms of λ and s. Although its derivation is beyond the scope of this text, it can be shown that it takes this form:

Poisson Probability Function

$$f(x) = \frac{e^{-\lambda s}(\lambda s)^x}{x!} \qquad x = 0, 1, 2, \ldots, \text{ where } e \approx 2.72$$

Working with this probability function directly is a little more difficult than working with other discrete probability functions that we have studied. If you have a calculator that will enable you to find powers of e easily then there is no problem; otherwise, computations get a bit messy. However, since this probability function comes into play in many settings, its values have been found and a cumulative table has been compiled for a large number of values of λs. Table III of Appendix A is one such table. In this table, the values of λs are listed as column headings; the values of X are listed as row headings. Remember that a cumulative table gives the probability of observing a value of X less than or equal to that specified. Example 3 illustrates the use of this table.

Example 3   Consider the nuclear-power plant described in Example 1. Since $\lambda = 2$ and $s = 3$, we are dealing with a Poisson random variable with $\lambda s = 6$. Suppose we want to find the probability that there will be at most two releases of radioactive gases during the three-month observation period. We want to find $P[X \le 2]$. This probability is calculated by summing the probability function for $x = 0, 1,$ and 2. Substituting, we see that

$$P[X \le 2] = P[X = 0] + P[X = 1] + P[X = 2]$$
$$= e^{-6}6^0/0! + e^{-6}6^1/1! + e^{-6}6^2/2!$$

By direct computations,

$$P[X \le 2] = .0025 + .0149 + .0446 = .0620$$

To obtain this probability from Table III, we look in the column headed $\lambda s = 6$ and the row headed 2. We read the number .062. The cumulative table can also be used to find other sorts of probabilities. For example, let us find the probability that there will be between 8 and 10 releases (inclusive) during a three-month period. To do so, note that

$$P[8 \le X \le 10] = P[X \le 10] - P[X < 8]$$
$$= P[X \le 10] - P[X \le 7]$$

From Table III,

$$P[X \le 10] = .957 \quad \text{and} \quad P[X \le 7] = .744$$

Substituting, the desired probability is

$$P[8 \le X \le 10] = .957 - .744 = .213$$

Before closing, let us note that there are general formulas that can be used to find the mean and variance for a Poisson random variable. These formulas require knowledge only of the numerical values of λ and s. They are:

Mean and Variance for a Poisson Random Variable

Consider a Poisson process with parameter λ observed over an interval of size s. The mean of X, the number of occurrences of the event during the observation period, is given by $E[X] = \lambda s$. Its variance is given by $\text{Var } X = \lambda s$.

Example 4

Consider the zircon described in Example 2. If we examine a two-square centi-meter segment of the material, would it be unusual to observe at least 20 fission tracks? Before answering this question, note that the average number of tracks in an area of this size is

$$E[X] = \lambda s = 5 \cdot 2 = 10$$

Since 20 is double this average value, intuition leads us to suspect that seeing at least 20 tracks is unusual. To verify that this is true, we use Table III to find $P[X \geq 20]$. This probability cannot be read directly. However, we know that

$$P[X \geq 20] = 1 - P[X < 20] = 1 - P[X \leq 19]$$

From Table III, $P[X \leq 19] = .997$. Substituting we see that

$$P[X \geq 20] = 1 - .997 = .003$$

As suspected, it would be unusual to see at least 20 fission tracks in a two-square centimeter segment of this zircon.

Exercises 4.6

39. Let X be a Poisson random variable with $\lambda s = 8$. Use Table III to find each of these probabilities:
    (a) $P[X \leq 7]$ (c) $P[X \geq 7]$ (e) $P[4 \leq X \leq 9]$
    (b) $P[X < 7]$ (d) $P[X = 7]$ (f) $P[5 \leq X \leq 12]$

40. A Poisson process with parameter $\lambda = 4$ is observed over an interval of size $s = 3$.
    (a) Find the mean of X.
    (b) Find the variance of X.
    (c) Find the standard deviation of X.

41. On the average, a copier jams twice a day. Let X denote the number of times the machine jams over a seven-day period.
    (a) What is the mean of X?
    (b) What is the variance of X?
    (c) Find the probability that the machine jams fewer than 8 times during this period; more than 12 times during this period; between 14 and 18 times (inclusive) during this time.
    (d) If the machine jams more than 23 times, is there reason to suspect that perhaps the reported average is in error? Explain by computing $P[X > 23]$ when $\lambda = 2$.

42. Suppose that the white blood count of a healthy individual averages 6000 per cubic millimeter of blood. A .001 cubic millimeter drop of blood is taken and examined. Let X denote the number of white cells found. In a healthy indi-vidual, what is the expected value of X? If at most one white cell is found, is there evidence of a white-cell deficiency? Explain by finding the probability that $X \leq 1$ in a healthy individual.

43. Airplanes leave a busy airport at the rate of 4 per minute. What is the average number of departures over a 30-second period? What is the probability that no planes depart during a particular 30-second period? What is the probability that at least one plane departs during this time period?

44. A particular mental patient averages two bouts of serious depression per six-month period. What is the probability that this patient will go for three months without an episode of depression? If the patient goes for one year with no serious depression, is there evidence that the patient is improving in this respect? Explain by finding the probability of this occurring if $\lambda = 2$.

45. A business firm has a toll-free telephone-ordering service. Orders arrive at the rate of 1 per hour. What is the average number of orders received per 8-hour day? What is the probability that at most 6 orders will be received during a given day? Would it be unusual to receive no orders during a given day? Explain.

*46. **Normal Approximation to the Poisson Distribution** Poisson probabilities can be approximated using a normal curve in a manner similar to that used to approximate binomial probabilities. The normal random variable used is assumed to have mean λs and variance λs. For example, to approximate Poisson probabilities associated with a Poisson process with $\lambda = 8$ over an observation interval of size $s = 5$, we use a normal random variable with $\mu = \sigma^2 = \lambda s = 40$. The half-unit adjustment demonstrated in Section 4.5 also should be used here.

(a) Let $\lambda s = 20$. Approximate each of these probabilities:
(i) $P[X \le 15]$ (ii) $P[X \ge 25]$ (iii) $P[X = 20]$
(b) Let $\lambda s = 8.5$. Approximate each of these probabilities:
(i) $P[X \le 10]$ (ii) $P[X \ge 11]$ (iii) $P[X = 7]$

*47. A bank auditor finds that a particular bookkeeping department averages one serious error per week. Over a 52-week period, approximate the probability that at most 35 errors will be made.

*48. As cars pass along an assembly line, a robot does spot welding. When working properly, the robot averages three poor-quality welds per hour. During an 8-hour shift, what is the average number of poor-quality welds produced? Approximate the probability of obtaining at most 20 such welds during a given shift. If at least 40 such welds are discovered during a particular shift, is there reason to suspect that there is a malfunction in the robot? Explain by finding $P[X \ge 40]$ when the robot is working properly.

Vocabulary List and Key Concepts

binomial model

Bernoulli trial

normal or Gaussian distribution

Poisson distribution

n factorial

$\binom{n}{x}$

**mean and variance of a binomial
      random variable**

standard normal distribution

standardization theorem

empirical rule

**normal approximation to the binomial
      distribution**

Poisson process

Review Exercises

Use the following information to answer exercises 49–55:

A study of air-traffic controllers shows that the probability of correctly identifying a signal is .9. Assume that the identification of one signal has no effect on that of any others. During a half-hour period 100 signals are received.

49. Let X denote the number of signals identified correctly. Argue that X is binomially distributed, and identify the values of n and p.

50. Write the expression for the probability function for X.

51. Find $E[X]$, Var X and σ.

52. What is the probability that fewer than 90 signals will be identified correctly?

53. What is the probability that more than 95 signals will be identified correctly?

54. What is the probability that between 85 and 95 signals (inclusive) will be identified correctly?

55. If more than 18 signals are incorrectly identified, would you question the accuracy of the figure .9? Explain, based on the probability that this could occur by chance when $p = .9$.

56. A binomial random variable for which $n = 1$ is called a *point binomial random variable*. Let X be a point binomial random variable with $p = .7$. Find $E[X]$, Var X, and σ.

Use the following information to answer exercises 57–62:

When a female who is a carrier of color blindness and a male who is normal with respect to this trait parent a child, the probability is 1/4 that the child born will be a color-blind male. One such couple has 5 children.

57. Let X denote the number of color-blind males born to this couple. Argue that X is binomial and identify the values of n and p.

58. Find $E[X]$ and Var X.

59. Find the expression for the probability function for X.

60. Find the probability that none of the children will be color-blind males.

61. Find the probability that at least one child will be a color-blind male.

62. Find the probability that two or three of the children will be color-blind males.

Use the following information to answer exercises 63–70:

Let X denote the score obtained by a particular gymnast on her floor routine. Assume that X is normally distributed with mean 9.0 and standard deviation .25.

63. Find the equation for the probability function for X.

64. At what values of x are the points of inflection located?

65. Find the probability that a given performance will receive a score less than 8.7.

66. Find the probability that a given performance will receive a score of more than 9.3.

67. Find the probability that the score for a given performance falls between 8.72 and 9.35.

68. Find the probability that the score for a given performance falls between 8.75 and 9.25. Don't use your calculator!

69. Above what point do her scores lie 5% of the time?

70. Below what point do her scores lie 10% of the time?

71. A new insurance plan is being investigated by officials of a particular labor union. To get an idea of the opinion of their membership, they poll 20 randomly selected members by phone. They think that if 15 or more of these people favor the new plan, then the majority (more than 50%) of the members of the union are in favor of the plan. Do you agree with this reasoning? Support your answer by finding the probability of obtaining 15 or more favorable responses if the percentage in favor is really only 50%.

72. A firm produces fiberglass tubing for use in protecting underground wiring. It is important that this tubing have no major cracks since this would allow water to seep into the wiring system. Suppose that past history indicates that the average number of major cracks is 4 per 10,000 feet of tubing. What is the probability that, at most, 1 major crack will be found in a 5000-foot length of tubing? If more than 6 such cracks were found, would you begin to suspect the accuracy of the reported average value of 4 per 10,000 feet? Explain based on the probability that $X \geq 6$ when $\lambda = 4$.

The Language of Statistical Inference

To be a wise user of statistical methods it is necessary to recognize a situation that calls for a statistical approach. It is then essential to pose questions of practical concern that can be translated mathematically and answered statistically on the basis of observed data.

Two basic areas of statistical inference are estimation and hypothesis testing. Although related both mathematically and in practice, they entail slightly different circumstances. We introduce here the vocabulary that underlies these two areas and present a general description of the two problem types. Some of the terms presented here were discussed in earlier chapters. We review their meaning for you now.

5.1 The General Statistical Problem

We begin by considering three problems, each of which calls for a statistical approach for its solution.

Example 1   A business chain runs a special sale in each of its stores. At the end of the day the management wishes to get an idea, as quickly as possible, of the sales average per store for the chain. How can this value be approximated quickly without taking time to gather data from each store? That is, how can we determine the mean sales statistically?

Example 2   A psychologist feels that room color may be a factor in the psychological well-being of hospital patients. The psychologist feels that subdued colors (light blue, pale yellow, and pale green) are more restful than the traditional white. She wishes to determine patient opinion. Since this innovation could be costly, the psychologist wishes to approximate the percentage of hospital patients who prefer color. Further, she intends not only to determine the numerical value but also report on the accuracy of the approximation.

Example 3

The IRS feels from past experience that tax forms filed during the week before the filing deadline are more prone to error than are those that were filed earlier. The proportion of returns with errors among the early filers has been 10% in past years. The tax service wants to verify for the current year its contention concerning late filers. How can it get the evidence needed without examining every tax form filed during the last filing week?

These problems are all statistical in nature and thus have common characteristics. In any statistical problem, there is a collection of objects either real or hypothetical under study. This collection is called the **population**. In Example 1, the population is real and consists of all stores that run the sale; in Example 2, the population is real and consists of all current hospital patients. However, social scientists often view the latter population as being hypothetical in the sense that current hospital patients are seen to be typical of all future hospital patients. Consequently, the population is thought of simply as being "hospital patients" present and future. In Example 3, the population is real, and consists of all tax forms filed during the week before the filing deadline.

The purpose of a statistical study is to draw conclusions about or to make inferences about the characteristics of the population. A population, mathematically speaking, is a set, and must be well defined. That is, given a specific object, it must be possible to determine whether or not that object is a member of the population. In conducting a statistical study, one must define the population explicitly.

There are many random variables associated with any population. However, in a particular study usually only a few are of interest.

Example 4

In a study in criminology the population of interest is all persons 16 years old or older, convicted of a first felony in a particular state during the current year. We are interested in answering two questions:

1. What is the average number of years of formal education for the population?

2. Is it true that a majority of the population had been arrested at least once prior to the arrest that resulted in a conviction?

We are asking two practical questions, each of which can be translated easily into mathematical terms. Question 1 asks us to determine the value of $E[X]$ or μ where X is the number of years of formal education of members of the population. Question 2 asks us if p, the proportion of the population that has had a prior arrest, is greater than 50%.

Note that there is a difference in the type of question being asked in the preceding example. In the first case, we are asked to draw a conclusion about the value of a specific parameter associated with the random variable X, namely the mean. We begin with no preconceived notion about its value. In the second case, we have proposed that the value of p is greater than .5, prior to gathering our data. That is, we have hypothesized that $p > .5$. We wish to gain evidence that supports this contention. This difference is exactly the property that distinguishes an estimation problem from a hypothesis-testing problem. We summarize both situations.

One has at hand a population and a random variable of interest. Associated with this variable are certain pertinent parameters, θ, that characterize the variable and therefore characterize the population.

1. *Estimation*: No preconceived notion concerning the numerical value of θ exists before the study is run. The study is conducted. On the basis of the data obtained we attempt to approximate or estimate θ.

2. *Hypothesis Testing*: A specific idea concerning the true value of θ exists before the study begins. The data either do or do not support the hypothesized value.

To draw inferences about a population using statistical methods, a *random sample* must be drawn from the population. To understand what is meant by this term, let us return to our last example. Here we have a large population that consists of all persons 16 years old or older who are convicted of a felony for the first time during the current year in a particular state. Associated with the population is the random variable X, the number of years of formal education of the individual. We want to select a subset of n individuals from the population *at random*. That is, we want to select n persons for inclusion in our study in such a way that the selection of one individual neither precludes nor insures the selection of any other. In this way, the selection of one individual is independent of the selection of any other. This is usually what comes to mind when one hears the term *random sample*.

Note that, prior to the actual selection of individuals to be included in the study, X_1, the number of years of formal education of the first person selected, is a random variable. Similarly X_2, X_3, ..., X_n, the numbers of years of formal education of the others selected, are also random variables. This collection of n-independent identically distributed random variables is called a random sample from the distribution of X.

Once the individuals have been selected for the study, their records can be reviewed and we can determine the numerical value of X for each. Thus we generate a set of observations on the random variables X_1, X_2, X_3, ..., X_n which we denote by x_1, x_2, x_3, ..., x_n. This collection of n numbers can be thought of as a random sample of size n.

As you can see, the term random sample is used in three different ways in applied statistics. It may refer to the *objects* being studied, the *random variables* associated with those objects, or the *numerical values* assumed by those variables. It will be clear from the context of the discussion which is intended.

One of the easiest ways to select a random sample from a finite population is by means of a table of random digits. This table is generated in such a way that each of the digits has the same probability of appearing in a given position in the table as every other digit. Table IV of Appendix A is such a table. Its use is illustrated in Example 5.

Example 5 In reference to our last example, suppose that during the years in question there were 300 persons convicted of felonies for the first time in the state. The names of these persons are obtained and numbered from 001 to 300. Let us draw a random sample of size 5 from the population using Table IV of Appendix A. A portion of this table is given in Table 5.1. To do so, we select 5 random three-digit numbers from the table. Individuals whose names match the numbers selected constitute our random sample. In this way, control of the choice of individuals to be studied has been taken out of our hands; each person in the population has the same chance of being selected for the study as every other person.

To begin, a random starting point is chosen. This can be done by closing your eyes and touching the table with a pencil. Suppose the pencil point lands on the digit 2 circled in Table 5.1. We begin reading the table at that point. The table may be read any way we wish: across the row, down the column, every other digit across the row, or by any other scheme desired. The easiest way is to read the first three digits across the row to obtain the random number 289. The individual whose identification number is 289 has been selected as the first member of our random sample. Reading down the column, we find that the next random three-digit number is 635. Since our identification numbers go only to 300, this number is too large and is discarded. The next individual selected is number 094. The third, fourth, and fifth individuals selected are those numbered 103, 071, and 069, respectively. If the same random number had been obtained more than once it would have been discarded after the first selection. At this point our random sample consists of the five people who have been chosen for the study. We now interview or obtain the records for these people and determine the number of years of formal education for each. Suppose that we observe these data:

$$x_1 = 10 \qquad x_2 = 9 \qquad x_3 = 12 \qquad x_4 = 14 \qquad x_5 = 12$$

Table 5.1

77921	06907	11008
99562	72905	56420
96301	91977	05463
89579	14342	63661
85475	36857	43342
②8918	69578	88231
63553	40961	48235
09429	93969	52636
10365	61129	87529
07119	97336	71048

These five numbers, observations on the random variables X_1, X_2, X_3, X_4, and X_5, are also referred to as a *random sample*.

The sample selected in the previous example is very small. This choice of a small sample was done for illustrative purposes only. We do not mean to imply that samples this small are sufficient for making inferences about large populations.

Sampling is a complex topic. Many schemes have been devised for sampling from a finite population. We have shown you how to draw what is called a *simple random sample*. A simple random sample is a sample drawn in such a way that each subset of size n has the same probability of selection. Since we sample without replacement, the random variables $X_1, X_2, \ldots, X_n$ are not independent. However, in this text we assume that the populations sampled are large enough so that the removal of the sample does not appreciably affect its composition. In this way the elements of our sample can, for all practical purposes, be considered to be a collection of n independent identically distributed random variables. For this reason, whenever we use the term *random sample* in a technical sense we will mean the following:

Definition 5.1

A **random sample** from the distribution of X is a collection $X_1, X_2, \ldots, X_n$ of n independent random variables each with exactly the same distribution as X.

To illustrate, suppose that X is normal with mean 10 and variance 8. A random sample of size five from the distribution of X is a collection X_1, X_2, X_3, X_4, and X_5 of independent random variables. Each of these random variables is assumed to be normal with mean 10 and variance 8.

Care must be taken whenever sampling is done. Any inference drawn from a sample refers only to the population from which the sample is selected. For example, to assess the outcome of a Democratic primary election, the target population is the set of all persons who intend to cast a ballot in the primary. A random telephone poll with no attempt made to screen respondents could lead to serious errors. The population actually polled, the general public, is not the same as the target population.

Exercises 5.1

In exercises 1–10 a problem is described. In each case, decide whether you think that a statistical study is called for. If so, identify the population under study and decide whether the problem entails estimation or hypothesis testing.

1. An insurance saleswoman has 15 clients. She wants to answer the question, What is the average age of my clients?

2. A large insurance firm handles thousands of accounts per year. Executives want to get an idea of the average age of the firm's clients.

3. A physician thinks that a majority of patients that she sees do not really need her services. She wants to verify that this is true.

4. Executives of a firm that produces microprocessor chips want to get an idea of the percentage of defective chips produced per day.

5. To help determine the size of the gambling industry in Nevada, government officials want to determine the average size of a bet placed in Las Vegas casinos during a given year.

6. An economist thinks that a family of four spends more than $6,000 per year on the average on housing. He wants to obtain evidence that supports this theory.

7. A baseball player asks, "What is my batting average?"

8. A baseball writer claims that a majority of major league pitchers over the last twenty years have batting averages under 150.

9. An accountant at a large bank is asked, "What is the average balance outstanding for all our loan customers?"

10. The Council on Alcohol and Drug Abuse wants to determine the percentage of young adults between the ages of 18 and 22 who have serious problems with alcohol.

11. It was recently announced that the majority of residents of Detroit are not happy about the newly adopted automobile population guidelines. Further investigation showed that this conclusion was drawn by interviewing people in a random telephone poll. Criticize the original statement in the light of the method used for sampling.

12. To study the distribution of weight among students enrolled at a small college, a researcher sampled every tenth student entering the student center snack bar. Criticize this procedure.

13. To ascertain public opinion concerning the Equal Rights Amendment, a student made a door-to-door survey in a randomly selected collection of residential blocks during the hours of 9 to 5 on Mondays. Criticize this procedure.

14. To study the amount of money being spent on groceries per customer, a businessperson took as a sample all the sales slips obtained on a randomly selected Friday in December. Criticize this procedure.

15. Let the population consist of the numbered pages in this text. Use Table IV of Appendix A to draw a simple random sample of size 25 from this population. Let X denote the number of figures or drawings found per page. Record the numerical values of the random variables $X_1, X_2, X_3, \ldots, X_{25}$.

16. Table V of Appendix A gives the sex and systolic and diastolic blood pressure of the 120 patients seen at a particular clinic. Use the random digit table

to draw a simple random sample of size 20 from this population. Record the sex and each of the blood pressures of the individuals selected.

*17. A finite population contains 100 elements. A sample of size 10 is to be drawn. How many such samples are possible? What is the probability that a particular sample of size 10 will be chosen? What is the probability that a particular member of this population will be selected into the sample?

18. Suppose that X is normal with mean 4 and variance 7. Let X_1, X_2, and X_3 be a random sample of size three from the distribution of X. What is the distribution of X_1; of X_2; of X_3?

5.2   The Terminology of Estimation

Recall that in the general estimation problem one seeks to use the information gained from a random sample to approximate or estimate the value of some parameter θ associated with the population of interest. Two approaches may be taken. We may derive either a *point estimate* or an *interval estimate* of θ. Each method is illustrated briefly here. We will consider detailed procedures for specific parameters later.

To utilize the information contained in a random sample, one must have available certain useful and, for the most part, logical statistics. Note that we have not yet defined the term *statistic*. We must do so now.

Definition 5.2

Statistic

A **statistic** is a random variable whose value can be determined from the observations in a random sample.

There are several points to notice about this definition.

1. We have already dealt with many statistics even though they were not labeled as such. For example, given a random sample X_1, X_2, ..., X_n of size n, the random variables

$$\Sigma X \qquad \Sigma X^2 \qquad \Sigma X/n$$

are all statistics. Once the observed values of X_1, X_2, ..., X_n are determined, we can determine the value of the above random variables. In Example 5 of Section 5.1, we drew a random sample X_1, X_2, X_3, X_4, and X_5 from the distribution of X, the number of years of formal education of an individual convicted of a first felony. These variables assumed the numerical values $x_1 = 10$, $x_2 = 9$, $x_3 = 12$, $x_4 = 14$, and $x_5 = 12$. For this sample, the statistics ΣX, ΣX^2, and $\Sigma X/5$ assume the numerical values 57, 665, and 11.4, respectively.

2. Since a statistic is also a random variable, it makes sense to ask for the probability distribution for a statistic, the expected value, and the variance of a statistic, and the probability that a statistic will assume certain specified values. In short, any question asked about the behavior of a random variable can be asked about the behavior of a statistic. In this sense, we began our study of statistics formally in Chapter 3.

To approximate the true value of a parameter θ from sample data, we use a statistic. Such a statistic is called a *point estimator* for θ because when it is evaluated for a given sample a single number or point results. The number obtained is called a *point estimate* for θ. These terms are defined formally in the next two definitions.

Definition 5.3

Point Estimate

A **point estimate** for a parameter θ is a real number used to approximate the true value of θ.

Definition 5.4

Point Estimator

A **point estimator** for a parameter θ is a statistic used to generate point estimates for θ.

These definitions at first appear to be a bit circular; however, the difference between them is relatively easy to see. A point estimate is a real number, while a point estimator is a statistic and, therefore, a random variable.

Example 1

Consider again the criminology study of Example 4 of Section 5.1. We wish to estimate μ, the average number of years of formal education for all 300 members of the population. Common sense indicates that, in order to estimate this population mean based on a sample drawn from that population, one should use the mean or average value of the sample. That is, our point estimator for μ based on X_1, X_2, X_3, X_4, and X_5, a random sample of size five is

$$\Sigma X/5$$

This statistic provides a logical formula to use to generate an estimate for μ. For the observed sample:

$$x_1 = 10 \qquad x_2 = 9 \qquad x_3 = 12 \qquad x_4 = 14 \qquad x_5 = 12$$

this estimator assumes the value

$$(10 + 9 + 12 + 14 + 12)/5 = 11.4 \text{ years.}$$

This number, used to approximate the population mean, is a point estimate for μ.

A major drawback to point estimation is that we have only a single number at the end of the experiment. In the above example, we estimate μ to be 11.4, and we must take it or leave it. Is the figure really correct? More than likely, it is not correct, but we hope it is close. How sure are we that it is close? Not sure at all. To overcome this uncertainty, we turn to the method of interval estimation.

Example 2   Consider Example 2 of Section 5.1 in which a psychologist estimates the proportion of hospital patients who prefer a colored room, rather than one with white walls. Suppose that, of 500 patients interviewed, 300 indicated that a colored room would be preferable to white. Common sense points to the value $300/500 = .6$ as the point estimate for the true population proportion, p, of patients favoring a colored room. How accurate is this estimate? We cannot really say now. All we can say is that a reasonable procedure was used to arrive at this value. Since the decision to change to colors is costly, it would be better if we could give not only an idea of the value of p, but also some assurance about how much confidence can be placed in the estimate. That is, we would like to derive a method by which the point estimate .6 can be expanded to an interval of values that will contain the true value of p with some surety. For instance, can we say that we are 90% confident that the actual proportion of patients favoring a room with color is between .58 and .62, or perhaps between .55 and .65? We need a systematic method for determining the endpoints of such intervals based on the observed data. That is, we need an interval estimator to generate interval estimates.

The terms *interval estimate* and *interval estimator* are defined in a manner similar to that used in defining the point estimate and point estimator, and are given below.

Definition 5.5

Interval Estimate

An **interval estimate** for a parameter θ is an interval of real numbers that should contain the true value of θ.

Definition 5.6

Interval Estimator

An **interval estimator** for a parameter is an interval of the form $[L_1, L_2]$, where L_1 and L_2 are statistics. L_1 and L_2 are used to determine the endpoints of the interval and hence $[L_1, L_2]$ can be used to generate interval estimates.

Exact methods for determining the form of the statistics L_1 and L_2 for estimating specific parameters with a given degree of accuracy will be discussed in later chapters.

Exercises 5.2

19. A social worker wants to estimate μ, the average number of calls that come into a help center per day. A random sample of 10 days yields these observations:

$$x_1 = 5 \qquad x_2 = 7 \qquad x_3 = 3 \qquad x_4 = 5 \qquad x_5 = 6$$
$$x_6 = 8 \qquad x_7 = 12 \qquad x_8 = 0 \qquad x_9 = 1 \qquad x_{10} = 4$$

(a) Evaluate the statistics $\Sigma X, \Sigma X^2$.
(b) Find a point estimate for μ based on these data.

20. A motorist wants to estimate the average cost of insuring his car. There are 30 insurance carriers in his locality. A random sample of eight of these carriers is selected and called. These data result:

$$x_1 = 205 \qquad x_2 = 210 \qquad x_3 = 195 \qquad x_4 = 202$$
$$x_5 = 200 \qquad x_6 = 225 \qquad x_7 = 203 \qquad x_8 = 200$$

Find a point estimate for the mean cost of insuring the car for these 30 carriers based on the given data.

21. Use the data of exercise 15 to estimate the mean number of figures or drawings found on the numbered pages of this text.

22. Use the data of exercise 16 to find point estimates for the mean systolic and diastolic blood pressures of the individuals seen at the given clinic.

23. Personnel in a hospital emergency room want to estimate the proportion, p, of cases encountered that are true emergencies. Of 44,000 patients seeking treatment in a year's time, it is found that 8,800 of these are classed by the staff as emergencies. Find a point estimate for p.

24. Use the data of exercise 16 to find a point estimate for the proportion of male patients at the given clinic. Estimate the proportion of female patients at the clinic.

5.3   The Terminology of Hypothesis Testing and the Use of P Values

Recall that in a hypothesis testing problem, the researcher has in mind a specific notion concerning the population characteristic under study *before* the data are gathered. This implies that there are two hypotheses of interest: the hypothesis being proposed by the researcher and the negation of this hypothesis. The former, denoted by H_1, is called the **alternative hypothesis** or **research hypothesis**; the latter, denoted by H_0, is called the **null hypothesis**. The purpose of the study is to decide whether the evidence tends to refute the null hypothesis. These three general statements help in deciding how to state H_0 and H_1.

Guidelines for Stating Statistical Hypotheses

1. The null hypothesis is the hypothesis of no difference. Practically speaking, this means that hypotheses involving a specific numerical value of a param-

eter θ must involve the statement of equality as a part of the null hypothesis. The specific value proposed for θ in the null hypothesis is called its **null value**.

2. One should state what one is trying to detect or support as the alternative hypothesis. The null hypothesis is the negation of the alternative.

3. Statistical hypotheses are framed in the hope of being able to reject the null hypothesis and therefore to accept the alternative hypothesis.

Example 1   Consider the experiment of bringing a defendant to trial. In our system of justice, we use the phrase *innocent until proven guilty beyond a reasonable doubt*. Thus, a trial is a special hypothesis test. The null hypothesis is that there is no difference between the defendant and the ordinary person on the street. That is, the defendant is innocent.

$$H_0 : \text{innocent}$$

The purpose of a trial is to detect guilt. Hence the statement that the defendant is guilty is the alternative hypothesis. That is,

$$H_1 : \text{guilty}$$

Since the State brings the defendant to trial, the purpose of the trial is to present evidence that will enable the judge or jury to reject the null hypothesis of innocence and find the defendant guilty. The defendant is actually either innocent or guilty. Furthermore, the jury can find the defendant either innocent or guilty, based on the evidence. Hence, at the end of the trial, one of four situations will have occurred:

An innocent person will have been found guilty.

A guilty person will have been found guilty.

An innocent person will have been found innocent.

A guilty person will have been found innocent.

As you can see, two of these situations result in correct decisions being made, and the other two result in errors. These facts are summarized in Table 5.2.

Table 5.2

	The Defendant Is	
	Innocent	*Guilty*
The Jury Finds the Person		
Guilty	Error	Correct decision
Innocent	Correct decision	Error

Table 5.3

	H_0 True	H_1 True
Reject H_0	Type I error	Correct decision
Fail to Reject H_0	Correct decision	Type II error

The situation in any hypothesis testing problem is similar to the above. In any hypothesis test H_0 is either true or false. At the end of the statistical experiment, we will either reject H_0 or we will not, as the data indicate. Thus, we will find ourselves in exactly one of four situations:

1. We will have rejected H_0 when in fact it was true; we will therefore have committed what is known as a **Type I error**.
2. We will have made the correct decision of rejecting H_0 when in fact H_0 was not true.
3. We will have made the correct decision of failing to reject H_0 when in fact H_0 was true.
4. We will have failed to reject H_0 when in fact H_0 was not true; we will therefore have committed what is known as a **Type II error**.

These ideas are summarized in Table 5.3.

Once a sample has been drawn and the data collected we must decide whether or not to reject the null hypothesis. The decision is made based on the observed value of some statistic whose distribution is known, under the assumption that the null value is the true value of θ.

Such a statistic is called a **test statistic**. If this statistic takes on an unusual value assuming that the null value is correct, then we reject H_0; if the observed value is a commonly occurring one when the null value is the true value for θ, then we do not reject H_0.

There are two ways to distinguish between H_0 and H_1. The first, used extensively in the past, involves determining the set of values of the test statistic that will cause us to reject H_0 *prior* to evaluating the test statistic. This set of values is called the **rejection** or **critical region** for the test. We will commit a Type I error if the observed value of the test statistic falls into the critical region by chance even though the null value is the true value for θ. The probability that this will occur is called **alpha** (α), the **size of the test**, or the **level of significance** of the test. This probability is known prior to evaluating the test statistic. That is, it is preset by the researcher. There are several reasons for wanting to do this. It gives a clear-cut way to make a decision. The researcher cannot be charged with manipulating the results to suit himself or herself. Furthermore, if the consequences of making a Type I error are very serious, by presetting α, we are able to specify, before the fact, exactly how large a risk we are willing to tolerate. This method of testing a hypothesis is illustrated in Example 2.

Example 2

A soft-drink firm wants to advertise that p, the proportion of persons who favor its new diet cola over the best seller, exceeds .7. We want to gather statistical evidence to support this claim. Since we state what we want to support as the alternative hypothesis, the alternative hypothesis is that $p > .7$. The null hypothesis is the negation of this statement. Therefore we are testing:

$$H_0 : p \leq .7$$

$$H_1 : p > .7 \qquad \text{(more than 70\% favor the new cola)}$$

Note that the null value for p is .7. We will run a blind taste test on 20 randomly selected cola drinkers. Our test statistic will be X, the number of individuals who select the new cola. If the null value of p is correct, X has a binomial distribution with $n = 20$, $p = .7$ and $E[X] = np = 20(.7) = 14$. If approximately 14 members of the sample select the new cola, then we should not reject H_0, as this is what is expected under the assumption that $p = .7$. However, if we see somewhat more than 14 select the new cola, then we should reject H_0 since this result is unusual if $p = .7$. Let us agree to reject H_0 in favor of H_1, if the observed value of our test statistic lies in the set $C = \{18, 19, 20\}$. This set constitutes the critical region for our test. Recall that

$$\alpha = P\left[\begin{array}{c} \text{test statistic falls} \\ \text{in the critical region} \end{array} \middle| \begin{array}{c} \text{null value is the true} \\ \text{value for } p \end{array}\right]$$

$$= P[X \geq 18 \mid p = .7]$$

From Table I of Appendix A, we see that

$$\alpha = P[X \geq 18 \mid p = .7]$$

$$\alpha = 1 - P[X < 18 \mid p = .7]$$

$$\alpha = 1 - P[X \leq 17 \mid p = .7]$$

$$\alpha = 1 - .9645 = .0355$$

By setting the critical region prior to conducting the taste test we are presetting α at .0355. If we are able to reject H_0, we know that the probability of committing a Type I error is only .0355. That is, there will be a very small probability that we will be making a false advertising claim.

It is possible that the test statistic will not fall into the critical region even though H_0 is false and should be rejected. The probability that this will occur is called **beta** (β). If this does occur, then we will commit a Type II error. For a specified α level, the value of β depends on the particular value assumed as the alternative. To illustrate, let us find β for the test designed in Example 2.

Example 3

The critical region for our test is $C = \{18, 19, 20\}$. Suppose that, unknown to the manufacturers, the new cola is preferred by 80% of the population. What is the probability of committing a Type II error in this case? That is, what is the probability that our test statistic will fall outside the critical region when the true value of p is .8? From Table I of Appendix A, we see that

$$\beta = P[X \leq 17 \mid p = .8] = .7939$$

Our test, as designed, does not give us a high probability of being able to distinguish between $p = .7$ and $p = .8$. Beta is a function of the alternative in that, if the alternative value of p is changed, β will change also. For example, if the alternative value of p changes from .8 to .9, then

$$\beta = P[X \leq 17 \mid p = .9] = .3231$$

Note that as the distance between the null value of p and the alternative value increases, β decreases. This is to be expected. Common sense says that the farther apart two proportions lie the easier it should become to tell them apart.

The second method of choosing between H_0 and H_1 is to evaluate the test statistic and then determine the probability of observing a value of the test statistic at least as extreme as that observed under the assumption that the null value is the true value for θ. This probability is called the **P value** of the test. The P value is the smallest level at which we could preset α and still have been able to reject H_0. H_0 is rejected for "small" P values. As you can see, this method leaves the decision as to whether or not to reject H_0 open for debate as the word *small* is rather vague. However, this method of testing hypotheses is becoming more and more popular. Research reports and computer printouts now routinely include P values for all statistical tests conducted. Example 4 demonstrates the use of P values.

Example 4

A random sample of 20 students is to be selected to determine student opinion concerning a unilateral U.S. freeze on nuclear weapons. It is thought that a majority of students favors the freeze. Here the alternative hypothesis is that $p > .5$. The null hypothesis is the negation of this statement, namely, that $p \leq .5$. We are testing

$$H_0 : p \leq .5$$

$$H_1 : p > .5 \qquad \text{(a majority of students favors the freeze)}$$

When the data are gathered, it is found that 12 of the 20 students questioned favor the freeze. If the null value for p, .5, is correct, then the expected value of our test statistic is $np = 20(.5) = 10$. Logic dictates that H_0 should be rejected if X, the observed number in favor of the freeze, is somewhat larger than 10. Is 12 enough larger than 10 to cause us to reject H_0? To decide, let us find the P value of the test. That is, let us find

$$P[X \geq 12 \mid p = .5] = 1 - P[X < 12 \mid p = .5]$$

$$= 1 - P[X \leq 11 \mid p = .5]$$

From Table I of Appendix A, this probability is .2517. We have observed a value that is fairly common under the assumption that the null value is correct. We do not have sufficient evidence based on these data to reject H_0.

We have attempted to introduce only the basic terminology of hypothesis testing here. These ideas will become clearer in chapters to follow as we develop the methods used to test various specific hypotheses. However, there are a few subtle points which must be understood.

1. Rejecting H_0 is a strong statistical result. It leads to the acceptance of our research hypothesis. Since we can always either preset α or find the *P* value of the test, we will always know the probability of committing an error by doing so. That is, the probability of committing a Type I error will be known.

2. Being unable to reject H_0 is usually a weak statistical result. In practice, the exact value of θ of interest in the alternative hypothesis is seldom specified. For this reason, it is seldom possible to compute β, the probability of committing a Type II error. It is a bit unwise to make bold claims when we do not know the probability that we are in error! For this reason, the inability to reject H_0 does not lead to its acceptance necessarily. All that we can say safely is that we do not have sufficient evidence to reject the null hypothesis based on the given data.

3. Alpha, beta, and sample size are interrelated. We cannot expect experiments based on small sample sizes to yield small values for both α and β. These interrelations are pointed out in exercises 18–31. For this reason most research problems require a substantial amount of time and effort to be effective.

Exercises 5.3

25. The director of athletics at a university wishes to expand the intercollegiate athletics program; this entails doubling the current athletic fee. Before taking such a drastic step, the director decides to try to gain statistical support for the contention that the majority of students favor the fee increase and the expansion.
    (a) Set up the appropriate statistical hypotheses.
    (b) Explain exactly what occurs if a Type I error is committed.
    (c) Explain exactly what occurs if a Type II error is committed.
    (d) Discuss the practical consequences of committing each of these errors.

26. The United States Public Health Department has set an average bacteria count of 70 as its maximum acceptable level for clam-digging waters. If the average number is greater than 70, then the waters are declared unsafe. Since eating clams taken from unsafe waters may cause hepatitis, these water are closed to fishing.
    (a) Assume you work for the government agency in charge of monitoring

these waters. Set up the appropriate null and alternative hypotheses to gain evidence that the bacteria count is above the safe limit.

(b) Discuss the practical consequences of making a Type I and a Type II error.

27. A firm uses a bookkeeping system requiring an average of one hour per week per account to keep the accounts current.

(a) Assume that you are the sales representative for a firm that handles materials for a new bookkeeping system. You wish to gain statistical support for the contention that your product allows accounts to be handled faster on the average than the old system. Let μ denote the average time required by the new system. Set up the appropriate hypotheses.

(b) Discuss the practical consequences of making a Type I and a Type II error.

28. In Example 3 of Section 5.1, the IRS wants to gain evidence that p, the proportion of returns with errors among late filers exceeds .10.

(a) Set up the hypotheses needed to support this claim.

(b) We will test this claim by selecting a random sample of size 20 from among the returns filed during the last week before the filing deadline. Let us agree to reject H_0 if X, the number of returns with errors, is more than 4. What is the critical region for this test?

(c) Find alpha (α).

(d) Find beta (β) if in fact $p = .2$; if $p = .3$; if $p = .4$.

(e) Suppose that when the sample is selected we observe 6 returns with errors. Will H_0 be rejected? To what type error are we now subject?

29. Manufacturing pollution-control devices for automobiles is a relatively new field. It has been found that 20% of all such devices must be adjusted, repaired, or replaced within one year of purchase. Suppose that a new method of production is devised to reduce this figure. A company considers the adoption of the new procedure and needs statistical evidence to support the contention that the new process is superior to the old.

(a) Set up the hypotheses needed to get the required evidence.

(b) We will test the null hypothesis by sampling 15 of the new devices. Let us agree to reject H_0 if none of the devices is found to be defective. What is the critical region for the test?

(c) Find α.

(d) Find β if in fact $p = .1$. Can this test, as designed, distinguish well between a failure rate of 10% and a failure rate of 20%?

(e) If a failure rate of 10% is considered to be a substantial improvement over that of 20%, what can we do to redesign the test to reduce the probability of committing a Type II error?

30. The **power** of a statistical test is the probability that H_0 will be rejected when it is, in fact, false.

(a) Do we want the power of a test to be large or small?

(b) Computationally, the power of a test is given by $1 - \beta$. Thus, power depends on the value specified for θ in the alternative. Find the power of the test described in exercise 28 if $p = .2$; $p = .3$; and $p = .4$.

(c) Find the power of the test described in exercise 29 if $p = .1$.

31. A new drug rehabilitation program is to be studied. The old program is effective with 50% of the subjects who complete it. We want to show that the new program is effective for a high percentage of subjects. Thus we want to test

$$H_0 : p \le .5$$

$$H_1 : p > .5$$

However, we want the test to be designed so that an effectiveness rate of 60% will be likely to be detected. The test statistic is X, the number of subjects who are helped under the new program.

(a) For a sample of size 10, let us reject H_0 if X is in the set $\{8, 9, 10\}$. Find α for this test. Find β if $p = .6$. Find the power if $p = .6$. With a sample this small, is there a good chance of being able to detect an effectiveness rate of .6 while maintaining an α level of approximately .05?

(b) For a sample of size 15, we use a critical region of $\{11, 12, 13, 14, 15\}$. Find α. Find β if $p = .6$. Find the power if $p = .6$. Is this sample large enough to detect a p of .6 while maintaining an α level of approximately .05?

(c) If n is increased to 20, what critical region should be chosen to obtain an α level of approximately .05? What is β if $p = .6$? What is the power if $p = .6$?

(d) It should be clear that to distinguish between $p = .5$ and $p = .6$ with $\alpha \approx .05$ we must use a sample larger than 20. Use the normal approximation to the binomial distribution to show that for a sample of size 100, $P[X \ge 59] = .05$. Show that when $p = .6$, the power for this test is over .50.

32. A new drug is being tested for effectiveness in fighting the common cold. The makers of the drug claim that over 60% of all cold sufferers can get relief from their drug.

(a) Assume that you are working for the drug company and want to gain statistical evidence to back their claim. Set up the appropriate null and alternative hypotheses to back this claim. Discuss the practical consequences of making each possible error.

(b) When the drug is tried on a sample of 15 cold sufferers, 11 claim to have obtained relief. Find the P value of the test. Do you think H_0 should be rejected? Explain.

33. A random telephone poll taken in December showed an approval rate of 60% for the performance of the President of the United States in domestic affairs. It is claimed that this percentage has declined.

(a) Set up the appropriate hypotheses to substantiate this.

(b) A sample of size 20 yields 10 persons who approve of the President's performance in this area. Find the P value for the test. Do you think that the claim has been substantiated? Explain.

Vocabulary List and Key Concepts

population	**null value**
random sample	**Type I error**
statistic	**Type II error**
point estimate	**test statistic**
point estimator	**rejection or critical region**
interval estimate	**alpha (α), size of the test,**
interval estimator	**or level of significance**
alternative hypothesis or	**beta (β)**
research hypothesis	***P* value**
null hypothesis	**power**

Review Exercises

34. What is the difference between an estimate for a parameter θ and an estimator for θ?

35. What is the main advantage of an interval estimate over a point estimate?

36. An automobile insurance adjustor thinks that the proportion of claims filed that involve damages of less than $1,000 is less than .6.
    (a) Set up the hypotheses needed to verify his contention.
    (b) We will test his claim by sampling 20 claims from the office files. If the stated null value is correct, what is the expected number of claims involving damages of less than $1,000?
    (c) Let us agree to reject H_0 if X, the number of claims that involve damages of less than $1,000, is at most 8. What is the value of α for this test?
    (d) Find β if $p = .5$; if $p = .3$; if $p = .1$.
    (e) Find the power of the test for $p = .5$; for $p = .3$; for $p = .1$.
    (f) Suppose that when the sample is drawn, nine claims are for amounts less than $1,000. Can H_0 be rejected? To what type error are you now subject?

37. It is thought that more than 10% of all new cars sold have defects that will cause them to be returned to the shop for major repairs within one year.
    (a) Set up the appropriate null and alternate hypotheses to verify this contention.
    (b) A particular dealer samples 15 records from her files. It is found that three were returned for major repairs during the first year of operation. Find the P value for the test. Do you think that H_0 should be rejected based on these data? Explain.

38. A self-teaching computer guide is being field tested. It is thought that more than 80% of all users will be able to run a program successfully after reading the manual.
   (a) Set up the appropriate hypotheses to verify this contention.
   (b) Twenty randomly selected beginners are used to test the manual. It is found that 19 were successful. Find the P value of the test. Do you think that H_0 should be rejected? Explain.

39. It is reported that the P value for a particular test lies between .01 and .05. Would H_0 have been rejected if α had been preset at .05? at .01?

40. Let X be a binomial random variable with parameters $n = 10$ and $p = .5$.
   (a) Find $E[X]$.
   (b) Find Var X.
   (c) Let X_1, X_2, X_3, and X_4 be a random sample of size 4 from the distribution of X. Each of these random variables has a binomial distribution. What are the numerical values of n, p, μ, and σ^2 for each?

Inferences on a Single Mean and a Single Variance

Chapter 5 introduced the terminology and general ideas of two basic problems of statistical inference, estimation and hypothesis testing. We extend these concepts here by developing the theory and methods needed to make inferences about three commonly encountered parameters—μ (the mean), σ^2 (the variance), and σ (the standard deviation). This chapter formalizes the methods introduced earlier on an intuitive level.

6.1 Point Estimation of the Mean

We consider first the problem of point estimation of the mean. Recall the general situation. We have a population and a random variable X associated with that population. We have available a random sample $X_1, X_2, \ldots, X_n$ of size n from the distribution of X. We wish to use information gained from this sample to approximate the value of μ. To do so, we must devise a logical statistic to be used to generate point estimates.

Example 1   A social worker is interested in the random variable X, the amount of money spent per month for prescription drugs by welfare recipients in the district. The purpose of the study is to obtain a point estimate for μ, the average cost in dollars of prescription drugs across the population. By interviewing a random sample of clients, these data are obtained:

$x_1 = 16.00$	$x_5 = 55.00$	$x_9 = 28.00$	$x_{13} = 21.90$
$x_2 = 10.50$	$x_6 = 12.50$	$x_{10} = 26.30$	$x_{14} = 25.60$
$x_3 = 0.00$	$x_7 = 20.00$	$x_{11} = 30.00$	$x_{15} = 27.95$
$x_4 = 32.50$	$x_8 = 23.50$	$x_{12} = 42.00$	$x_{16} = 33.50$

197

How can these data be used to approximate μ? Common sense says that to estimate the theoretical average value of X we should simply use as the estimate the arithmetic average of the sample values. That is, the estimate for μ should be

$$\frac{16.00 + 10.50 + 0.00 + \cdots + 33.50}{16} = 25.33$$

Notationally we write $\hat{\mu} = \$25.33$ using the "hat" to indicate that the given value is only the estimate for the mean. Remember that $\hat{\mu}$ is not usually exactly equal to μ. The latter, μ, is the average amount spent per month by *all* welfare recipients in the district. The former, $\hat{\mu}$, is the mean of our sample of welfare recipients. There is no reason to expect that this or any other particular sample mean will equal exactly the true population mean. The best that we can hope for is that we have used an estimation procedure that generates estimates that will be close in value to μ.

Example 1 suggests the following sample statistic as a logical estimator for the mean μ of a distribution.

Definition 6.1

Sample Mean

Let $X_1, X_2, \ldots, X_n$ be a random sample of size n from a distribution with mean μ. The statistic

$$\bar{X} = \frac{X_1 + X_2 + X_3 + \cdots + X_n}{n} = \frac{\sum X}{n}$$

is called the **sample mean**.

How good is $\bar{X}$ as an estimator for μ? We know that it is intuitively logical. Is it also mathematically justifiable?

Before considering the estimator $\bar{X}$ in particular, let us pause to consider what properties are desirable in an estimator in general. It is rather difficult to say exactly what constitutes a good estimator, but a good one should produce estimates usually close in value to the estimated parameter. There are two criteria that, when used in conjunction with each other, will produce the desired closeness with high probability. An estimator $\hat{\theta}$ for a parameter θ should

1. be unbiased for θ

2. have small variance for large sample sizes n

The term *unbiased* is a technical term. Recall that since $\hat{\theta}$ is a statistic, it is also a random variable. It has associated with it a probability distribution. It makes sense to ask for the mean of this distribution. To say that a statistic $\hat{\theta}$ is an unbiased estimator for θ means that the theoretical average, or mean value, of the estimator is exactly equal to the parameter that it estimates. That is, the distribution of the statistic $\hat{\theta}$ has as its center of location the parameter θ being estimated. This idea leads us to define the term *unbiased* as in Definition 6.2.

Definition 6.2	**Unbiased Estimator** An estimator $\hat{\theta}$ for a parameter θ is said to be an **unbiased estimator** for θ if and only if $E[\hat{\theta}] = \theta$.

Is the sample mean, $\bar{X}$, an unbiased estimator for the population mean, μ? The fact that the answer to this question is yes is the point of the next theorem. Its proof is outlined in exercise 5.

Theorem 6.1	Let $X_1, X_2, \ldots, X_n$ be a random sample of size n from a distribution that has mean μ. Then $\bar{X}$ is an unbiased estimator for μ. That is, $E[\bar{X}] = \mu$.

Example 2 Consider the random variable X, the amount of money spent per month for prescription drugs by welfare recipients. (See Example 1.) We are interested in estimating μ, the mean value of this random variable. The fact that $\bar{X}$ is unbiased for μ is illustrated as follows: Assume that a sample of size 16 is obtained, and the resulting estimate for μ based on this sample is $\bar{x}_1 = 29.50$. A second sample of size 16, drawn independently from the first, is obtained and yields a sample mean of $\bar{x}_2 = 31.00$. A third independent sample yields $\bar{x}_3 = 31.75$. Imagine that this process of obtaining independent samples of size 16 and computing $\bar{x}$ for each sample is repeated over and over again. The fact that $\bar{X}$ is unbiased for μ means that the *average* value of these estimates should be close in value to μ.

The fact that an estimator $\hat{\theta}$ is unbiased for a parameter implies that in repeated sampling, the average value of the estimates should be close to θ. Unfortunately, being unbiased does not guarantee that any single estimate will be close. This is indeed a problem, since in most statistical studies, experiments are run only once. They are not repeated over and over again, and hence we will have only a single estimate for θ. For instance, in estimating μ, the average amount spent per month for prescription drugs by welfare recipients, we would in practice draw only one sample and would base our estimates on the data from this sample only.

To obtain an estimator $\hat{\theta}$ for a parameter θ that can be expected to produce an estimate reasonably close in value to θ when used *only once*, we must consider not only the mean of $\hat{\theta}$, but also its variance. Recall that Var $\hat{\theta}$ is a measure of the variability, or fluctuation, of $\hat{\theta}$ about its mean. If $\hat{\theta}$ is unbiased then Var $\hat{\theta}$ measures the fluctuation of $\hat{\theta}$ about the parameter θ. If this variance is small, most of the time the observed value of $\hat{\theta}$ will lie close to its mean of θ. Hence, if $\hat{\theta}$ were used to estimate θ in a one-shot operation, we could reasonably assume that the value obtained is close to the actual value of the parameter θ.

To complete the argument that $\bar{X}$ is mathematically desirable as an estimator for μ, as well as being intuitively logical, we must show that for large n, $\bar{X}$ has small variance. This is done with the help of the next theorem. Its proof is outlined in exercise 6 of this chapter.

Theorem 6.2

Let $X_1, X_2, \ldots, X_n$ be a random sample from a distribution with mean μ and variance σ^2. Then

$$\text{Var } \bar{X} = \sigma^2/n$$

This theorem says that the variance of $\bar{X}$ is the original variance divided by the sample size. For example, if we draw a sample of size 10 from a distribution with variance 25, then the variance of the statistic $\bar{X}$ is $\sigma^2/n = 25/10$.

Note that since σ^2 remains constant, as n becomes larger, σ^2/n becomes smaller. Practically speaking, this means that as the sample size increases the variance of the estimator $\bar{X}$ decreases. Since $\bar{X}$ is centered at μ, for all but very small samples we can expect the observed value of $\bar{X}$ to lie fairly close to the population mean μ. Thus, $\bar{X}$ satisfies our criteria for being a good estimator for μ. It is unbiased and has small variance for large n.

Example 3

The manager of a fast food restaurant must decide how many hot dogs to order each week. If the manager orders too many, there will be a storage problem; if too few, the supply may run out. To get some idea of the number required, a sample of 16 randomly selected weeks is obtained, and the following observations on the random variable X, the number of hot dogs sold per week, result.

$x_1 = 905$	$x_5 = 975$	$x_9 = 783$	$x_{13} = 900$
$x_2 = 1000$	$x_6 = 950$	$x_{10} = 1003$	$x_{14} = 789$
$x_3 = 800$	$x_7 = 600$	$x_{11} = 850$	$x_{15} = 913$
$x_4 = 795$	$x_8 = 925$	$x_{12} = 875$	$x_{16} = 810$

What is the average number of hot dogs being sold per week? That is, what is μ? This question cannot be answered from a sample, but the answer can be estimated. Using the sample mean $\bar{X}$ as our estimator for μ, we obtain

$$\hat{\mu} = \bar{x} = 867.06$$

That is, we conclude that, on the average, approximately 867 hot dogs are sold per week. The manager can use this figure to guide the ordering.

Several questions arise from this problem. We now have a single point esti-mate for μ, namely 867.06. How good is this number? Is a sample of size 16 large enough to give us confidence that this value is indeed close to the true value of μ? To answer these questions partially, we must turn to the method of interval estimation. We must extend the unbiased point estimate of 867.06 to an interval estimate in such a way that we can report the accuracy of our estimate. This is done in the next section.

Exercises 6.1

1. A traffic controller studies the road usage of a stretch of interstate I-81. A roadside traffic count yields data on the number of trucks passing the check-point during ten randomly selected one-hour periods:

   5   8   3   9   12   7   6   15   20   3

   Use these data to find a point estimate for μ, the mean number of trucks passing the checkpoint per hour.

2. Psychological studies of memory in humans make use of random word lists. In one such study each subject was given five minutes to study a list of 15 words and was then asked to list as many of these words as can be recalled. These data were obtained:

10	11	14	8	11	9	8	9
10	6	12	6	12	3	7	5
10	5	15	12	8	11	9	11

   Use these data to estimate the mean number of words that can be recalled.

3. An accountant is interested in the random variable X, the amount of money owed per active account. To get an idea of the average amount owed, the accountant randomly samples 36 accounts and obtains the following data (negative values indicate a credit balance):

15.00	75.00	8.13	0.00	14.75	−20.00
25.00	−15.00	5.20	12.75	6.80	18.26
36.40	0.00	2.00	−5.00	3.95	7.98
−10.00	0.00	37.62	0.00	0.00	31.50
8.00	5.00	0.00	25.00	0.00	0.00
150.00	13.50	0.00	32.50	12.50	4.25

   Use these data to find a point estimate for μ, the theoretical average amount owed per account.

4. Let X be a random variable with mean 5 and variance 20. For random samples of the specified size, find $E[\bar{X}]$ and Var $\bar{X}$. Note that as n increases, Var $\bar{X}$ decreases as expected.
   (a) $n = 2$
   (b) $n = 4$

(c) $n = 8$

(d) $n = 20$

(e) $n = 25$

*5. Show that $\bar{X}$ is an unbiased estimator for μ. [*Hint:* Write $\bar{X}$ as $(1/n)(X_1 + X_2 + \cdots + X_n)$ and use the rules of expected value given in exercise 37 of Chapter 3.] Remember that each of the random variables $X_1, X_2, \ldots, X_n$ has expected value μ.

*6. Show that Var $\bar{X} = \sigma^2/n$. [*Hint:* Write $\bar{X} = (1/n)(X_1 + X_2 + \cdots + X_n)$ and apply the rules of variance given in exercise 38 of Chapter 3.] Remember that each of the random variables $X_1, X_2, \ldots, X_n$ has variance σ^2.

*7. Recall that Chebyshev's Inequality, exercise 40 of Chapter 3, states that a random variable will lie within k standard deviations of its mean with probability of at least $1 - 1/k^2$. For each of the sample sizes indicated in exercise 4, how close is $\bar{X}$ expected to lie to the true mean of 5 with probability at least 0.75? Note that as n increases $\bar{X}$ should lie closer and closer to its true mean. That is, as n increases $\bar{X}$ should provide a very good estimate for μ.

*8. **(Weighted Means)** Assume that one has k independent random samples of sizes $n_1, n_2, n_3, \ldots, n_k$ from the same population. These samples each generate an unbiased estimator for the mean, namely, $\bar{X}_1, \bar{X}_2, \ldots, \bar{X}_k$.

(a) Show that the arithmetic average of the estimators $(\bar{X}_1 + \bar{X}_2 + \bar{X}_3 + \cdots + \bar{X}_k)/k$ is also unbiased for μ.

(b) The manager of a branch bank is attempting to estimate the average size of the checks that will be cashed at the bank's drive-in window each Friday. These data are obtained on five randomly selected Fridays:

$\bar{x}_1 = \$50$	$\bar{x}_2 = \$75$	$\bar{x}_3 = \$200$	$\bar{x}_4 = \$25$	$\bar{x}_5 = \$30$
$n_1 = 20$	$n_2 = 30$	$n_3 = 5$	$n_4 = 100$	$n_5 = 50$

Use the results of part (a) to obtain an unbiased estimate for μ, the average size of each check being cashed on a Friday. By averaging the five values of \$50, \$75, \$200, \$25, and \$30, to obtain the estimate for μ, each sample is being given equal importance or weight. Does this seem reasonable in this problem? Explain.

(c) To take sample sizes into account a weighted mean is used. This estimator, $\hat{\mu}_{wt}$, is given by

$$\hat{\mu}_{wt} = \frac{n_1 \bar{X}_1 + n_2 \bar{X}_2 + \cdots + n_k \bar{X}_k}{n_1 + n_2 + \cdots + n_k}$$

Show that $\hat{\mu}_{wt}$ is an unbiased estimator for μ.

(d) Use the data of part (b) to find the weighted estimate for the mean size of a check being cashed on Friday at this bank. Compare your answer to the estimate in part (b). Which estimate do you think is preferable?

*9. The manager of a snack bar at a student center is interested in the amount of money spent per student per visit to the snack bar. The manager obtains the following three samples based on three different days:

Sample 1		Sample 2		Sample 3	
0.50	1.16	0.85	0.76	1.18	0.98
1.50	1.25	3.25	0.29	0.28	0.47
0.75	0.21	0.16	0.89	0.59	1.35
2.25	0.53	1.31			
0.35	0.16				

(a) Find $\bar{x}_1, \bar{x}_2, \bar{x}_3$.
(b) Estimate the average amount spent per visit using a weighted mean.

6.2   The Central Limit Theorem and Interval Estimation of the Mean with the Variance Known

An intuitive definition of the term *interval estimator* was given in Chapter 5. Recall that an interval estimator is an interval $[L_1, L_2]$ where L_1 and L_2 are statistics. To determine these statistics in any given situation, we must be more specific concerning the nature of L_1 and L_2. Definition 6.3 states exactly how L_1 and L_2 should behave.

Definition 6.3

$100(1-\alpha)$% Confidence Interval

A **$100(1 - \alpha)$% confidence interval** for a parameter θ is an interval of the form $[L_1, L_2]$, in which L_1 and L_2 are statistics such that

$$P[L_1 \le \theta \le L_2] \approx 1 - \alpha$$

regardless of the actual value of θ.

Don't be overly concerned with this definition at this point. You will understand its meaning after you have seen it in use a few times!

Note that since the endpoints of a confidence interval are random variables, the interval itself can be thought of as a random interval. That is, the actual location of the interval along the number line can and does vary from sample to sample. The definition can be used to generate intervals that contain θ with whatever degree of confidence is desired. For instance, if we wish to generate a 95% confidence interval on θ, we would allow α to assume the value .05 in the above definition. We would then be looking for two statistics, L_1 and L_2, such that the probability of trapping the true value of θ between them is approximately $1 - .05 = .95$.

There is one general statement that helps in constructing most confidence intervals. That is, *to construct a $100(1 - \alpha)$% confidence interval on a parameter θ, find a random variable whose expression contains θ and whose probability distribution is known at least approximately.*

The implication of this to the problem of finding a confidence interval on the population mean, μ, is that we must now consider the probability distribution of the statistic $\bar{X}$. Surely any sensible random variable leading to a confidence interval on μ will involve this unbiased estimator for μ. The next two theorems concern the distribution of the statistic $\bar{X}$ in two distinct situations. The first gives the distribution of $\bar{X}$ when the random variable X under study has a normal distribution. We have partially stated the theorem in Theorems 6.1 and 6.2.

Theorem 6.3

> **Distribution of $\bar{X}$: Normal Case**
>
> Let $X_1, X_2, \ldots, X_n$ be a random sample of size n from a distribution that is normal with mean μ and variance σ^2. $\bar{X}$ is distributed normally with mean μ and variance σ^2/n.

Note that our previous theorems state that $E[\bar{X}] = \mu$ and $\text{Var } \bar{X} = \sigma^2/n$. This theorem gives us an additional result, namely, that, if we are sampling from a *normal distribution*, $\bar{X}$ is also a *normal random variable*.

Since $\bar{X}$ is normal with mean μ and variance σ^2/n we may use our usual method for standardizing a normal random variable (namely, by subtracting the mean and dividing by the standard deviation) to conclude that when sampling from a normal distribution

$$\frac{\bar{X} - \mu}{\sigma/\sqrt{n}} = Z$$

is *standard* normal.

The second theorem, the **Central Limit Theorem**, is a well-known theorem in mathematics. It was originally formulated in the early nineteenth century by Laplace and Gauss. However, Lyapunov, in 1901, was the first to give a formal rigorous proof of the theorem.

Theorem 6.4

> **Central Limit Theorem**
>
> Let $X_1, X_2, \ldots, X_n$ be a random sample from a distribution with mean μ and variance σ^2. For large n, $\bar{X}$ is approximately normal with mean μ and variance σ^2/n.

The difference between this theorem and the preceding one is quite simple. The Central Limit Theorem does *not* require the sample to be drawn from a normal distribution. The remarkable result is that $\bar{X}$ is approximately normal, nonetheless, with the approximation becoming better as n becomes larger.

Standardization again allows us to conclude that for large n

$$\frac{\bar{X} - \mu}{\sigma/\sqrt{n}}$$

is *approximately* standard normal regardless of the type of distribution from which the sample is drawn. It has been found through empirical investigations that, in practice, the normal distribution gives an excellent approximation for the distribution of $\bar{X}$ for n as small as 25. Example 1 illustrates the Central Limit Theorem.

Example 1

Each day for 30 consecutive days, 25 different computer programs are run on a remote terminal. The probability is 1/5 that any given program will run error-free. Let X_1 denote the number of error-free programs run on the first day, X_2 the number of error-free programs run on the second day, and so on, to the thirtieth day. Note that each of the above variables is binomial, with the number of trials being 25 and the probability of success being 1/5. Hence each has a mean of 25(1/5) = 5 and a variance of 25(1/5)(4/5) = 4. Assuming that the programs are different, and therefore run independently of one another, $X_1, X_2, \ldots,$ X_n constitutes a random sample of size 30 from a binomial distribution with mean 5 and variance 4. The average number of programs running error-free per day is

$$\bar{X} = \frac{X_1 + X_2 + \cdots + X_{30}}{30}$$

By the Central Limit Theorem this variable is approximately normal even though the variables that constitute the sample are not only *not normal*, but not even *continuous*. The standardization procedure implies that the variable

$$\frac{\bar{X} - \mu}{\sigma/\sqrt{n}} = \frac{\bar{X} - 5}{2/\sqrt{30}}$$

is approximately standard normal.

These two theorems are used to solve our problem. They enable us to locate statistics L_1 and L_2 satisfying the definition of the term $100(1 - \alpha)\%$ confidence interval on μ. That is, they allow us to locate statistics L_1 and L_2 such that

$$P[L_1 \leq \mu \leq L_2] \approx 1 - \alpha$$

We illustrate how this is done for a 90% confidence interval. The result is then easily extended to obtain any desired degree of confidence.

Example 2

Consider the problem of obtaining a confidence interval on the mean number of hot dogs sold per week at a fast food restaurant. We may begin by considering the random variable

$$\frac{\bar{X} - \mu}{\sigma/\sqrt{n}}$$

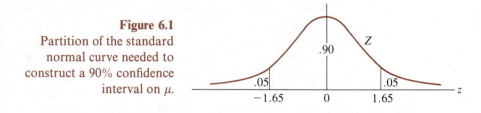

Figure 6.1
Partition of the standard
normal curve needed to
construct a 90% confidence
interval on μ.

This random variable satisfies our general criterion for the construction of a confidence interval. It involves the parameter μ being estimated, it is sensible (that is, it involves $\bar{X}$), and its distribution by the Central Limit Theorem is at least approximately normal. There is one obvious problem. What is σ? Let us assume for the sake of illustration that σ is known from past experience to be 100. Admittedly, this assumption is not realistic. In practice, if the mean is unknown and is being estimated, it is highly unlikely that the variance, and therefore the standard deviation, is known. The more realistic situation of constructing a confidence interval on μ when σ is unknown will be discussed in the next section. Let us construct a 90% confidence interval on μ. Recall that we are looking for two statistics L_1 and L_2 such that

$$P[L_1 \leq \mu \leq L_2] = .90$$

Since we are interested in developing a 90% confidence interval on μ, we begin by splitting the area under the standard normal curve into three regions as shown in Figure 6.1. To derive the 90% confidence interval, we need only write down a probability statement relative to this curve that can be set equal to .90. Such a statement is

$$P[-1.65 \leq Z \leq 1.65] = .90$$

In this case

$$Z = \frac{\bar{X} - \mu}{\sigma/\sqrt{n}}$$

Hence we obtain

$$P\left[-1.65 \leq \frac{\bar{X} - \mu}{\sigma/\sqrt{n}} \leq 1.65\right] = .90$$

Let us now algebraically simplify the above statement by isolating μ in the middle of the inequality.

$$P\left[-1.65 \leq \frac{\bar{X} - \mu}{\sigma/\sqrt{n}} \leq 1.65\right] = .90$$

$$P[-1.65(\sigma/\sqrt{n}) \leq \bar{X} - \mu \leq 1.65(\sigma/\sqrt{n})] = .90$$

$$P[-\bar{X} - 1.65(\sigma/\sqrt{n}) \leq -\mu \leq -\bar{X} + 1.65(\sigma/\sqrt{n})] = .90$$

$$P[\bar{X} - 1.65(\sigma/\sqrt{n}) \leq \mu \leq \bar{X} + 1.65(\sigma/\sqrt{n})] = .90$$

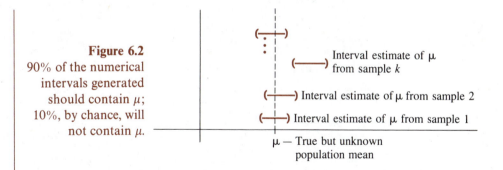

Figure 6.2
90% of the numerical intervals generated should contain μ; 10%, by chance, will not contain μ.

Interval estimate of μ from sample k

Interval estimate of μ from sample 2

Interval estimate of μ from sample 1

μ — True but unknown population mean

Since we are assuming that σ is known and we know the sample n, both expressions involved in the above inequality are statistics. Hence we can make the following identifications:

$$L_1 = \bar{X} - 1.65(\sigma/\sqrt{n}) \qquad L_2 = \bar{X} + 1.65(\sigma/\sqrt{n})$$

Thus, a 90% confidence interval for μ is

$$[L_1, L_2] = [\bar{X} - 1.65(\sigma/\sqrt{n}), \bar{X} + 1.65(\sigma/\sqrt{n})]$$

In our particular case, $\bar{x} = 867.06$, $\sigma = 100$, and $n = 16$. On the basis of our observed sample, we are 90% confident that the true average number of hot dogs being sold per week lies in the interval

$$[867.06 - 1.65(100/4), \ 867.06 + 1.65(100/4)]$$

$$= [867.06 - 41.25, \ 867.06 + 41.25] = [825.81, \ 908.31]$$

Note that the midpoint of this interval is 867.06, the unbiased point estimate for μ. What exactly does it mean to say that we are 90% confident that the true average number of hot dogs being sold per week is in the interval $[825.81, 908.31]$? We are simply saying that the random interval

$$[L_1, L_2] = [\bar{X} - 1.65(\sigma/\sqrt{n}), \bar{X} + 1.65(\sigma/\sqrt{n})]$$

$$= [\bar{X} - 1.65(100/4), \bar{X} + 1.65(100/4)]$$

used to generate this interval is such that if it were applied to many independently drawn samples, 90% of the numerical intervals so obtained would actually contain μ. We hope that our numerical interval is among these. The idea is illustrated in Figure 6.2.

The confidence interval developed in the last example applies specifically to 90% confidence. To obtain a general formula for a confidence interval on μ when σ^2 is known, we need only state one notational convention and then make the appropriate changes in the above formula.

Definition 6.4   Let Z be a standard normal random variable. By z_r, we mean that point such that $P[Z \geq z_r] = r$.

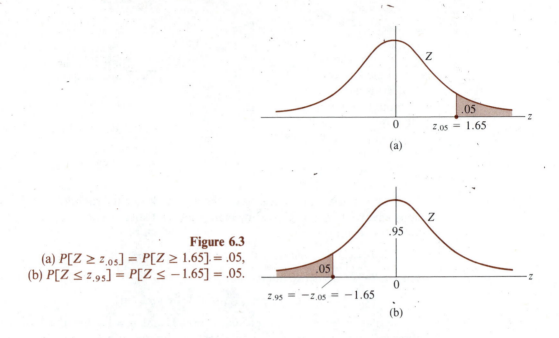

Figure 6.3
(a) $P[Z \geq z_{.05}] = P[Z \geq 1.65] = .05$,
(b) $P[Z \leq z_{.95}] = P[Z \leq -1.65] = .05$.

Note that z_r is the point such that the area under the standard normal curve to the *right* of z_r is r. For instance, $z_{.05}$ is the point on the standard normal curve such that the area to the *right* of $z_{.05}$ is .05. From the standard normal table (Appendix A, Table II) $z_{.05} = 1.65$. This idea is illustrated in Figure 6.3(a). The point with r to the left necessarily has $1 - r$ to the right and therefore is denoted by z_{1-r}. Due to the symmetry of the standard normal curve, $z_{1-r} = -z_r$. For instance, $z_{.95} = z_{1-.05} = -z_{.05} = -1.65$ as shown in Figure 6.3(b).

This notation allows us to generalize the discussion of Example 2. We need only realize that instead of dealing with the standard normal curve, split as shown in Figure 6.1, we are in fact dealing in general with the partition given in Figure 6.4.

The algebraic argument of Example 2 holds exactly as presented, with -1.65 replaced by the point $-z_{\alpha/2}$, and 1.65 replaced by the point $z_{\alpha/2}$. Thus, instead of the interval

$$[\bar{X} - 1.65(\sigma/\sqrt{n}), \bar{X} + 1.65(\sigma/\sqrt{n})]$$

Figure 6.4

Partition of the standard normal curve needed to construct a $100(1 - \alpha)\%$ confidence interval on μ.

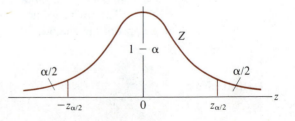

we obtain the general interval

$$[\bar{X} - z_{\alpha/2}(\sigma/\sqrt{n}),\ \bar{X} + z_{\alpha/2}(\sigma/\sqrt{n})]$$

This result is summarized in Theorem 6.5.

Theorem 6.5

Confidence Interval on the Mean: Variance Known

Let X_1, X_2, $\ldots$, X_n be a random sample of size n from a normal distribution with mean μ and *known* variance σ^2. Then a $100(1 - \alpha)\%$ confidence interval on μ is

$$[\bar{X} - z_{\alpha/2}(\sigma/\sqrt{n}),\ \bar{X} + z_{\alpha/2}(\sigma/\sqrt{n})]$$

Example 3

Consider the random variable X, the systolic blood pressure in adult males. Assume that X is normally distributed with unknown mean μ and known variance $\sigma^2 = 100$. A random sample of size 64 yields a sample mean of 118. Let us find a 95% confidence interval on the mean systolic blood pressure in adult males based on these data. To do so, we think of the standard normal curve as being split into three regions as shown in Figure 6.5. Substituting into the formula

$$[\bar{X} - z_{\alpha/2}(\sigma/\sqrt{n}),\ \bar{X} + z_{\alpha/2}(\sigma/\sqrt{n})]$$

we obtain the numerical interval

$$[118 - 1.96(10/8),\ 118 + 1.96(10/8)] \quad \text{or} \quad [115.55, 120.45]$$

We can be 95% confident that the true mean systolic blood pressure for adult males lies between 115.55 and 120.45.

We could just as easily have used the data to find a 99% confidence interval on μ. We need only split the normal curve as in Figure 6.6 and make the appropriate changes in the general formula.

The resulting 99% confidence interval is

$$[118 - 2.57(10/8),\ 118 + 2.57(10/8)] = [114.79, 121.21]$$

We can be 99% confident that the true mean systolic blood pressure lies between 114.79 and 121.21. Note that the higher the confidence required, the longer the

Figure 6.5
Partition of the standard normal curve needed to construct a 95% confidence interval on μ.

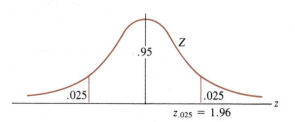

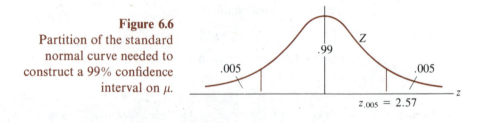

Figure 6.6
Partition of the standard
normal curve needed to
construct a 99% confidence
interval on μ.

interval becomes. This is to be expected since the more certain we want to be of our result, the more room we must leave for error.

Remember that to use the Z distribution to find a confidence interval for the mean, the population variance, σ^2, must be known from past experience. Note also that we are assuming either that X itself is normally distributed or that the sample size is large enough so that the Central Limit Theorem comes into play.

Exercises 6.2

10. Use the standard normal table (Table II, Appendix A) to find each of these points:

    (a) $z_{.025}$ (b) $z_{.975}$ (c) $z_{.01}$ (d) $z_{.99}$ (e) $-z_{.025}$
    (f) point z such that

    $$P[-z \leq Z \leq z] = .98$$

    (g) point z such that

    $$P[-z \leq Z \leq z] = .96$$

11. Find the indicated confidence intervals:

    (a) a 95% confidence interval on μ when $n = 36$, $\bar{x} = 10$, and $\sigma = 3$.
    (b) a 90% confidence interval on μ when $n = 25$, $\bar{x} = 8$, and $\sigma^2 = 4$.
    (c) a 99% confidence interval on μ when $n = 20$, $\bar{x} = -15$, $\sigma^2 = 16$.

12. Past experience indicates that the variance, σ_X^2, of the scores X obtained on the verbal portion of the Graduate Record Examination is 5625; similarly, the variance, σ_Y^2, for the scores Y on the mathematical portion is 2500.

    (a) A random sample of size 400 yields a sample mean of $\bar{x} = 600$. Find a 95% confidence interval on the true average score obtained on the verbal aptitude test.
    (b) A random sample of size 225 yields a sample mean of $\bar{y} = 500$. Find a 99% confidence interval on the true average score obtained on the mathematical aptitude test.
    (c) Find a 96% confidence interval on μ_Y based on the data from part (b). Compare the *lengths* of the intervals obtained. What is the implication of this result to the connection between interval length and degree of confidence?
    (d) Ten years prior to the year in which the above data were obtained, μ_X

was found to be 610. Can we safely conclude that the average score has declined over the past ten years? Explain briefly on the basis of the confidence interval constructed in part (a).

13. A sociologist makes a study of divorce patterns in a city. In the study, 500 divorced people are selected at random, and X is the number of years married before divorce. From this sample, it is found that $\bar{x} = 11.2$ years; assume that $\sigma = 4.2$.
    (a) Construct a 90% confidence interval on the average length of marriage before divorce.
    (b) Would you be surprised to hear a report that the average length of marriage before divorce is 15 years? Explain.

14. Consider the random variable X, the diastolic blood pressure in adult males. Assume that X is normally distributed with variance $\sigma^2 = 49$. Find a 90% confidence interval on μ based on these data:

80	72	87	76	82
79	90	88	72	70
84	89	92	73	84
75	85	84	89	88
72	70	80	82	86

Are these data consistent with a reported mean diastolic blood pressure of 80? Explain.

15. To justify staying open beyond 5:30 P.M., a businessman feels that he must average $50 per night in business between 5:30 and 8:00 P.M. A random sample of evening sales over a 36-day period yields a sample mean of $\bar{x} = \$45$. Assume that the variance of evening sales is 121, the same as during the day. Find a 95% confidence interval on μ, the mean evening sales. On the basis of this interval, do you feel that the store should remain open in the evening? Explain.

16. Past experience shows that the random variable X, the length of time one must wait to be served at the bank's drive-in window, is normally distributed with a mean of eight minutes and a standard deviation of two minutes. A new system is devised. The new system does not affect variability. A random sample of 225 cars reveals an average waiting time of 7.5 minutes. Construct a 99% confidence interval on μ, the mean waiting time under the new system. Does the new system appear to reduce the waiting time on the average? Explain.

17. A random sample of size 100 is drawn from a distribution with mean 25 and variance 36.
    (a) What is the expected value of $\bar{X}$?
    (b) What is the variance of $\bar{X}$?
    (c) Would you expect $\bar{X}$ to be approximately normally distributed? Explain.

Table 6.1

x	1	2	3	4
f(x)	1/4	1/4	1/4	1/4

Table 6.2

x	1	1.5	2.0	2.5	3.0	3.5	4.0
f(x)	1/16	2/16	3/16	4/16	3/16	2/16	1/16

18. A random sample of size 4 is drawn from a distribution with mean 10 and variance 9.
    (a) What is the expected value of $\bar{X}$?
    (b) What is the variance of $\bar{X}$?
    (c) Would you expect $\bar{X}$ to be approximately normally distributed? Explain.

*19. Consider a random variable X with probability function as shown in Table 6.1.
    (a) Is X normally distributed?
    (b) Find $E[X]$ and $E[X^2]$.
    (c) Find Var X.
    (d) List all 16 possible samples of size 2 that can be drawn from this distribution when sampling with replacement.
    (e) Find the numerical value of $\bar{X}$ for each sample.
    (f) Verify that the probability function for $\bar{X}$ is as shown in Table 6.2.
    (g) Sketch a relative frequency histogram for the distribution of $\bar{X}$. Does the histogram appear to be approximately bell shaped?
    (h) According to the Central Limit Theorem, what is the mean of $\bar{X}$? Use Table 6.2 to verify your answer.
    (i) According to the Central Limit Theorem, what is the variance of $\bar{X}$? Use Table 6.2 to verify your answer.

6.3   The *t* Distribution and Interval Estimation of the Mean with the Variance Unknown

Recall that in the previous section we were forced to make the unrealistic assumption that even though the mean of the variable X under study is unknown, we know its variance. This situation seldom arises in practice. Usually, in a statistical experiment, the investigator is the first to consider the problem. No one knows the value of σ^2. We consider here the more realistic problem of making inferences on the mean when the variance is assumed to be unknown.

When dealing with the problem of point estimation, the fact that the variance is known is irrelevant. None of the arguments concerning the desirability of $\bar{X}$ as a point estimator for μ require knowledge of the value of σ^2. It is not until the problem of confidence interval estimation arises that this knowledge becomes necessary. Hence, the sample mean is used as the point estimator for μ regardless of the situation.

To generate a $100(1 - \alpha)\%$ confidence interval on μ when σ is unknown, it is natural to look closely at the random variable used when σ is known. Perhaps this random variable can be modified to meet the present situation. Let us consider, then, the random variable

$$\frac{\bar{X} - \mu}{\sigma/\sqrt{n}}$$

There are two problems:

1. We do not know σ. Common sense would suggest that perhaps we can estimate σ from the data and replace σ by its estimated value in subsequent calculations.

2. When σ is known,

$$\frac{\bar{X} - \mu}{\sigma/\sqrt{n}}$$

   is approximately standard normal. If we replace σ by an estimator for σ, what will be the effect on the probability distribution? Is it still standard normal or does it change?

Once these problems are solved, the construction of a $100(1 - \alpha)\%$ confidence interval on μ is fairly routine.

Consider first the problem of estimating σ^2 and therefore σ from the data. To get an idea of how to begin, let us review the definition of σ^2. That is, let us recall exactly what we are estimating. By definition

$$\sigma^2 = \text{Var } X = E[(X - \mu)^2]$$

We are attempting to estimate the long-run theoretical average value of the square of the difference between the variable and its true mean. We have a random sample $X_1, X_2, \ldots, X_n$. We do not know the value of μ but we do have an unbiased estimator for μ, namely $\bar{X}$. We cannot observe $(X - \mu)^2$ for all X, but we can observe $(X - \bar{X})^2$ for each X in the sample. We cannot calculate the theoretical average squared difference $E[(X - \mu)^2]$ with only a sample available, but we can compute the arithmetic average of the squared differences between the sample values and the sample mean,

$$\frac{\Sigma(X - \bar{X})^2}{n}$$

Intuitively, this statistic should generate a fairly good estimate for σ^2, and it does. However, to be an *unbiased* estimator for σ^2, we must divide not by n, but by $n - 1$. We therefore follow the convention of defining the estimator for σ^2 in such a way that it is unbiased. You should be aware that this is not essential, and that others prefer division by n.

Definition 6.5

> **Sample Variance**
>
> Let $X_1, X_2, \ldots, X_n$ be a random sample from a distribution with mean μ and variance σ^2. The **sample variance**, denoted by S^2, is given by
>
> $$S^2 = \frac{\Sigma(X - \bar{X})^2}{n - 1}$$

Note that this is exactly the definition of the term *sample variance* used in Chapter 1. Note also that the fact that S^2 is unbiased for σ^2 implies that $E[S^2] = \sigma^2$. That is, the statistic S^2 has as its center of location σ^2.

Example 1 In evaluating faculty, each student is asked to rank instructors on a scale from 1 to 5, with rank 1 denoting excellent and rank 5 denoting unsatisfactory. A random sample of size 11 selected from student ratings given Doctor J. yields data on the random variable X, the rating received:

$$4, 4, 4, 5, 1, 3, 2, 1, 2, 5, 2$$

Estimate μ, the actual overall average rating, and σ^2, the variance of the ratings:

$$\hat{\mu} = \bar{x} = 3$$

$$\hat{\sigma}^2 = s^2 = \frac{\Sigma(x - 3)^2}{10}$$

$$= \frac{(4 - 3)^2 + (4 - 3)^2 + (4 - 3)^2 + \cdots + (2 - 3)^2}{10} = 2.2$$

How do we interpret this number? Unfortunately, by itself, it is not all that informative. However, if we know that the student ratings for Professor K. yield a sample mean of 3 and a sample variance of .5, we could use this to draw some rough comparisons between Doctor J. and Professor K. Even though they both have the same overall average rating, they are not alike in all respects. The higher variance for Doctor J. tends to indicate that student opinion is diverse. We could probably find a substantial number of students who think Doctor J. is excellent and an equally substantial number who think the opposite. Their average opinion is rather neutral. In the case of Professor K., we would probably find widespread agreement that he is average.

Recall that although the true variance of a random variable is

$$\sigma^2 = E[(X - \mu)^2]$$

this definition was rarely used in the actual computation of σ^2. In Chapter 3, we were able to use a computational formula for variance that was easier to work with than the above definition, namely

$$\sigma^2 = E[X^2] - (E[X])^2$$

The same holds for the sample variance, S^2. We rarely use the definition for sample variance in actual computations. We use instead a computational formula for sample variance that is easier to handle than the procedure required in the definition. The formula requires calculations of only ΣX and ΣX^2 from the data.

Theorem 6.6

Computational Formula for Sample Variance

Let $X_1, X_2, \ldots, X_n$ be a random sample of size n from a distribution with mean μ and variance σ^2.

$$S^2 = \frac{n\Sigma X^2 - (\Sigma X)^2}{n(n-1)}$$

Example 2

Consider the data on student evaluation of Doctor J. given in Example 1. To use the computational formula to compute s^2, we need to compute two quantities

$$\Sigma x^2 \quad \text{and} \quad \Sigma x$$

$$\Sigma x^2 = 4^2 + 4^2 + \cdots + 2^2 = 121$$

$$\Sigma x = 4 + 4 + \cdots + 2 = 33$$

Thus

$$s^2 = \frac{11(121) - (33)^2}{11(10)} = 2.2$$

as obtained previously. For small data sets of whole numbers, the advantage of the computational formula may not be evident. However, most electronic calculators, even inexpensive pocket calculators, are constructed so that Σx and Σx^2 can be obtained by simply entering the data.

Recall that our original problem was to find a point estimator for σ. It is natural to define this estimator as follows:

Definition 6.6

Sample Standard Deviation

Let $X_1, X_2, \ldots, X_n$ be a random sample of size n from a distribution with mean μ and variance σ^2. The **sample standard deviation**, denoted by S, is defined to be the nonnegative square root of the sample variance. That is,

$$S = \sqrt{S^2}$$

Note that even though S^2 is *unbiased for* σ^2, S *is not an unbiased estimator for* σ, that is, $E[S] \neq \sigma$. The proof of this statement is outlined in exercise 30. The sample standard deviation S is nevertheless the most commonly used estimator for σ.

Example 3

The sample standard deviation for the random variable X, the rating received by Doctor J. on student evaluations, is

$$s = \sqrt{s^2} = \sqrt{2.2} = 1.48$$

We have now accomplished our first goal—to develop a method for estimating σ from a sample. Now we need to investigate the effect of replacing σ by S in the random variable

$$\frac{\bar{X} - \mu}{\sigma/\sqrt{n}}$$

used previously to generate confidence intervals on μ. That is, we need to consider the distribution of the random variable

$$\frac{\bar{X} - \mu}{S/\sqrt{n}}$$

It can be shown that the distribution is affected when this change is made. That is,

$$\frac{\bar{X} - \mu}{S/\sqrt{n}}$$

is *not* standard normal. In fact, this random variable follows, at least approximately, the so-called **Student's t distribution**. This distribution was first studied in 1908 by W. S. Gosset; he published his result under the name "Student" because his employers did not approve of the publication of research by their employees. No confusion should result if we shorten the same for Student's t distribution to simply the t distribution. To be consistent with our previously adopted notational convention of denoting random variables by capital letters, we will denote a random variable that follows a t distribution by T. The graph of the probability function for a T random variable will be called a T curve.

Let us consider briefly the general properties of the family of T random variables:

Properties of T Random Variables

1. There are infinitely many T random variables, each identified completely by means of one parameter, v, called "degrees of freedom." Recall that in the case of a normal random variable, it is necessary to give both the mean and the variance to identify the normal curve completely. For the t distribution, we have to specify only one parameter, v, to identify the exact curve under study. We denote a T random variable with v degrees of freedom by T_v.

2. Each T random variable is continuous. Everything that we have said concerning continuous variables in general holds for T random variables. In particular, we will be dealing with smooth curves, and interpreting the area under these curves as probabilities. When we use the phrase T_{n-1} distribution, we mean that we are dealing with the T curve associated with a T random variable with $n - 1$ degrees of freedom.

3. The graph of the probability function for every T random variable is a symmetric, bell-shaped curve centered at 0; thus, the mean of every T random variable is 0. Recall that normal curves could be located or centered anywhere along the horizontal axis. This is not true of T curves. They are all centered at 0.

4. The parameter v is a shape parameter; as the number of degrees of freedom increases, the variance decreases. Recall that the variance in the case of continuous random variables determines the shape of the probability function. The higher the variance, the flatter the curve. For T random variables, as v increases, the variance decreases, implying that the higher the degrees of freedom, the more compact the bell becomes.

5. For large values of v, the T curve with v degrees of freedom, is approximated by the standard normal curve. This means that, as the degrees of freedom increase, the curves for T random variables essentially become superimposed on the curve for the standard normal random variable, Z. Hence, for large values of v, points on the T curve are very close in value to points on the Z curve, as shown in Figure 6.7.

Figure 6.7

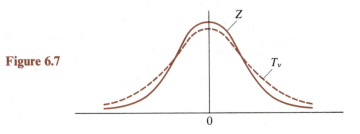

A partial summary of the cumulative distribution function for T random variables with selected degrees of freedom ranging from 1 to 120 is given in Table VI of Appendix A. Table VI is constructed so that the probabilities are listed as column headings, and the points associated with those probabilities are listed in the body of the table. We follow our previous convention of letting t_r denote the point such that the area under the T curve to the *right* of the point is r. That is, for a T random variable with v degrees of freedom, t_r is the point such that $P[T_v \geq t_r] = r$. We illustrate the use of Table VI in the next example.

Example 4   Consider a T random variable with $v = 15$ degrees of freedom. From Table VI of Appendix A, we conclude that

1. $P[T_{15} \leq .691] = .75$ [See Figure 6.8(a).]

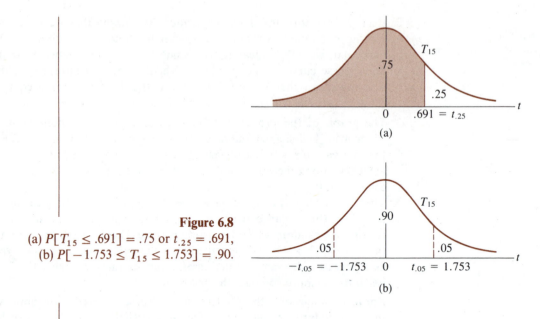

Figure 6.8
(a) $P[T_{15} \le .691] = .75$ or $t_{.25} = .691$,
(b) $P[-1.753 \le T_{15} \le 1.753] = .90$.

2. $t_{.25} = .691$

3. Due to the symmetry of the T curve, $t_{.75} = -t_{.25} = -.691$

4. The point t such that $P[-t \le T_{15} \le t] = .90$ is 1.753 [See Figure 6.8(b).]

The last row in Table VI is labeled ∞. The points listed in that row are actually points associated with the standard normal curve. The fact that as v increases, T curves approach the standard normal curve is seen by noting that, for each column in Table VI, as we descend the column, the numbers approach the number in the last row.

The next theorem provides the foundation for the construction of $100(1 - \alpha)\%$ confidence intervals on μ when σ^2 is unknown. It also provides the test statistic for testing hypotheses on the value of μ when the variance is not known.

Theorem 6.7

Let $X_1, X_2, \ldots, X_n$ be a random sample of size n from a normal distribution with mean μ and variance σ^2. Then

$$\frac{\bar{X} - \mu}{S/\sqrt{n}}$$

is a T random variable with $n - 1$ degrees of freedom.

There are several things to notice concerning this theorem. First, the value of v depends on the sample size. The number of degrees of freedom is the *sample size minus 1.* Second, there is an assumption of normality being made. However, it has been found that for distributions that are approximately bell-shaped, $(\bar{X} - \mu)/(S/\sqrt{n})$ is well approximated by the T_{n-1} distribution. Thus, we use the *t* distribution to make inferences on μ when σ^2 is unknown unless we have clear evidence that our sample is drawn from a distribution that is far from normal. Finally, the random variable $(\bar{X} - \mu)/(S/\sqrt{n})$ satisfies the criteria necessary for the construction of a $100(1 - \alpha)\%$ confidence interval on μ when σ^2 is unknown. It involves the parameter μ, and its distribution is at least approximately known.

To determine the formula for constructing a $100(1 - \alpha)\%$ confidence interval on the mean when the variance is unknown, we need only recall how this was done when σ^2 was assumed to be known. We started by considering the random variable

$$\frac{\bar{X} - \mu}{\sigma/\sqrt{n}}$$

that was known to have at least an approximate standard normal distribution. We now begin by considering the random variable

$$\frac{\bar{X} - \mu}{S/\sqrt{n}}$$

that has a *t* distribution with $n - 1$ degrees of freedom. Since these random variables are virtually identical in form, the algebraic argument used to generate the confidence interval

$$[\bar{X} - z_{\alpha/2}(\sigma/\sqrt{n}), \bar{X} + z_{\alpha/2}(\sigma/\sqrt{n})]$$

is exactly as before with σ being replaced by S and $z_{\alpha/2}$ being replaced by $t_{\alpha/2}$. Thus, the logical formula for the desired confidence interval is

$$[\bar{X} - t_{\alpha/2}(S/\sqrt{n}), \bar{X} + t_{\alpha/2}(S/\sqrt{n})]$$

This result is stated in Theorem 6.8.

Theorem 6.8

Confidence Interval on the Mean with Variance Unknown

Let $X_1, X_2, \ldots, X_n$ be a random sample of size n from a normal distribution with mean μ and unknown variance σ^2. A $100(1 - \alpha)\%$ confidence interval on μ is

$$[\bar{X} - t_{\alpha/2}(S/\sqrt{n}), \bar{X} + t_{\alpha/2}(S/\sqrt{n})]$$

Note that if the sample is drawn from a normal distribution, the above confidence interval is exact; otherwise, it is approximate, with the approximation becoming better for large n and for distributions that approach a bell shape.

Example 5   The Department of Health, Education, and Welfare is interested in the random variable X, the percentage increase in enrollment in state-supported colleges in the United States since 1974. Assume that the following observations are obtained on a random sample of size 25 (negative values indicate an actual decrease in enrollment):

5%	35%	−8%	.3%	5%
−1%	−30%	12%	0%	3%
−10%	16%	−5%	7%	7%
25%	−15%	2%	−17%	8%
0%	6%	9%	7%	3%

Is there statistical evidence that the enrollment in state-supported colleges is increasing, and if so, at what rate? To answer these questions, we compute a 95% confidence interval on μ, the average percentage increase across the country. Note that we do not know the value of σ^2; it must be estimated from the data. Hence, the confidence interval must be constructed by using the T distribution with 24 degrees of freedom. For this sample

$$\Sigma x = 64.30 \qquad \bar{x} = 2.57\% \qquad \Sigma x^2 = 4254.09$$

$$s^2 = [25(4254.09) - (64.3)^2]/(25)(24) = 170.36$$

$$s = \sqrt{170.36} = 13.05\%$$

The diagram of Figure 6.9 shows that the t point required in Theorem 6.8 is $t_{.025} = 2.064$ (see Table VI of Appendix A). Substituting these values into the expression given in the theorem we obtain the following 95% confidence interval on μ:

$$[.0257 - 2.064(.1305/5), .0257 + 2.064(.1305/5)] = [-.0282, .0796]$$

What do we conclude? We are 95% confident that the overall average increase in enrollment is somewhere between a 2.82% *decrease* and a 7.96% *increase*. Unfortunately, since 0 is included in this interval as are some negative numbers, we cannot conclude that the enrollment is actually increasing.

Figure 6.9
Partition of the T_{24} curve needed to construct a 95% confidence interval on μ.

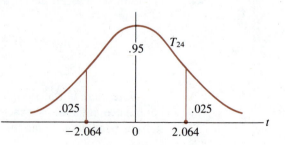

Example 6

In studying the efficiency of the federal court system we look at the random variable X, the time elapsed between the filing of a charge and the trial date. A random sample of 400 cases yields a sample mean of $\bar{x} = 10.1$ months with sample variance $s^2 = 4.3$. Let us find a 90% confidence interval on the average time between the filing of charges and the date of trial. Since we are estimating σ from the observed data, we should use the T-type confidence interval. However, since $n = 400$ calls for 399 degrees of freedom, which is not listed in Appendix A, Table VI, we must use the last line in the table (probability points for the standard normal curve) in the construction of the confidence interval. Substituting into the expression of Theorem 6.8, we obtain this 90% confidence interval on μ:

$$[10.1 - 1.645(2.07/20),\ 10.1 + 1.645(2.07/20)] = [9.93, 10.27]$$

We are 90% confident that the average wait for trial is between 9.93 and 10.27 months. Is this acceptable, or should the system be changed to reduce this time? That is a question for the public to answer.

One point should be emphasized here. We use the t distribution to find confidence intervals on the mean of a normal random variable whenever the population variance is unknown regardless of the size of the sample. If the sample size is so large that the required t points are not listed in the T table, then these points are approximated via the standard normal table. For convenience these Z points are listed in the last row of the T table.

Exercises 6.3

20. There has been some concern that young children are spending too much time watching television. In a study of television viewing these observations are obtained on the random variable X, the number of cartoon shows watched per child from 7:00 A.M. to 1:00 P.M. on a Saturday in Columbia, South Carolina:

    2, 1, 0, 8, 3, 6, 7, 0, 7, 6

    For these data,
    (a) Find Σx.
    (b) Find the point estimate for μ, the average number of cartoons watched by children on a Saturday morning in Columbia.
    (c) Find Σx^2.
    (d) Use the definition of sample variance to find the point estimate for σ^2.
    (e) Use the computational formula to find the point estimate for σ^2.
    (f) Find the point estimate for σ.

21. These data are obtained on the number of customers served at a fast food restaurant between the hours of 6:00 A.M. and 7:00 A.M. on 15 randomly selected days:

10	5	0	15	8
8	7	6	3	5
3	9	12	10	12

For these data,
(a) Find Σx.
(b) Find the point estimate for μ, the average number of customers served during this time period.
(c) Find Σx^2.
(d) Find point estimates for σ^2 and σ.

22. Use Table VI of Appendix A to find each of the following:
   (a) $t_{.05}$, $\nu = 15$
   (b) $t_{.25}$, $\nu = 29$
   (c) $t_{.75}$, $\nu = 17$
   (d) $t_{\alpha/2}$, $\nu = 19$, $\alpha = .10$
   (e) $-t_{\alpha/2}$, $\nu = 20$, $\alpha = .05$
   (f) point t such that $P[-t \leq T_{25} \leq t] = .95$
   (g) point t such that $P[-t \leq T_{28} \leq t] = .90$

23. Find the indicated confidence intervals:
   (a) a 95% confidence interval on μ when $n = 20$, $\bar{x} = 7$, and $s = 2$
   (b) a 90% confidence interval on μ when $n = 25$, $\bar{x} = 10$, and $s^2 = 4$
   (c) a 99% confidence interval on μ when $n = 30$, $\bar{x} = -5$, and $s^2 = 1$
   (d) a 95% confidence interval on μ when $n = 500$, $\bar{x} = 100$, and $s = 10$

24. The following data were obtained from a random sample of size 30 from the distribution of X, the percentage increase in blood-alcohol content after drinking four beers:

37.5	39.2	40.8	41.7	41.1	43.2
38.9	39.1	40.5	41.2	42.0	44.0
38.9	40.0	40.6	41.5	42.2	44.9
38.1	40.1	40.4	41.6	43.3	45.0
39.5	40.1	40.3	41.4	43.1	45.7

(a) Construct a stem-and-leaf diagram for these data. Does the normality assumption appear to be reasonable? Explain.
(b) Compute Σx, Σx^2.
(c) Find the point estimate for μ.
(d) Find the point estimate for σ^2.
(e) Find the point estimate for σ.
(f) Find a 90% confidence interval on μ.
(g) Authorities claim that four beers cause an average percentage increase in blood alcohol of 45%. On the basis of the above results, do you believe their claim?

25. To be sure to budget enough money to cover overtime expenses for employees, a pizza shop manager obtains the following observations on the random variable X, the number of hours worked overtime per week by an employee.

1.6	3.0	4.1	5.0	2.0	3.5	4.0	5.5
2.1	4.0	4.5	6.0	2.2	4.2	5.0	7.5

(a) Construct a stem-and-leaf diagram for these data. Does the normality assumption seem reasonable? Explain.
(b) Compute Σx, Σx^2.
(c) Find point estimates for μ, σ^2, σ.
(d) Find a 95% confidence interval on μ.
(e) The manager must pay \$5.00 per hour for each hour or fraction of an hour of overtime work. There are 10 employees. Approximately how much would you suggest be budgeted per week for overtime expenses to be 95% sure that these costs can be covered?

26. A psychological experiment was run to determine the effect that being praised has on the perceived likeability of the person giving the praise. The 200 subjects were allowed to work for a short time with a confederate of the experimenter on a mutual task. The subject was then asked some pertinent questions concerning the confederate and, from the answers received, a likeability score for the confederate (0 to 10) was obtained. The subject was then allowed to overhear an interview between the confederate and the experimenter in which the confederate was extremely complimentary toward the subject. The subject was then asked further questions concerning the confederate and a second likeability score for the confederate was obtained. The random variable of interest is X, the increase in likeability scores from the first to the second interview. For this experiment, a sample mean of $\bar{x} = 2.1$ and a sample variance of $s^2 = 169$ was obtained.

(a) Construct a 98% confidence interval on μ.
(b) On the basis of your answer to part (a), do you conclude that the likeability scores on the average show an increase from the first to the second interview?

27. A group of analysts from the New York Stock Exchange (NYSE) rated 100 randomly selected mutual-fund stocks on a performance scale from 1 to 5, where 1 is poor and 5 is excellent. A similar panel from the American Stock Exchange (AMX) rated the same group of stocks on the same scale. The random variable X is the *difference* in ratings for each stock (NYSE rating − AMX rating). A sample mean $\bar{x} = 0.2$ and a sample variance of $s^2 = 0.1$ was obtained.

(a) Construct a 95% confidence interval on μ.
(b) From part (a), do you conclude that the NYSE has a tendency to rate mutual-fund stocks higher? Explain.

28. Some economists feel that whenever the percentage of take-home pay devoted to meeting installment bills becomes greater than 20%, the consumer has taken on a dangerous credit overload. A random sample of 25 applications for a bank credit card produced the following observations on the random variable X, the percentage of take-home pay already being expended on installment bills:

19.3	17.1	21.1	18.5	22.3
20.9	13.6	17.1	25.6	24.4
14.7	23.3	24.5	22.0	17.7
24.1	21.9	18.9	18.1	21.7
21.9	16.0	19.1	17.6	19.8

(a) Find point estimates for the mean, variance, and standard deviation for X.

(b) Find a 95% confidence interval on the mean percentage of take-home pay devoted to meeting installment bills of applicants for this credit card.

*29. (**Sample Size for Estimating the Mean**) The length of a confidence interval depends on the confidence desired, the variability in the data, and the sample size. A confidence interval can be so long that it is almost useless. If σ is known or can be estimated from a small preliminary or pilot study, then it is possible to design an experiment in such a way that the resulting confidence interval will be short enough to give us a good idea of the value of μ. This is done by controlling the size of the sample. Let d denote the distance between the *center of the confidence interval* $\bar{X}$, *and the upper confidence bound,* $\bar{X} + z_{\alpha/2}\,\sigma/\sqrt{n}$ as shown in Figure 6.10. Note that $d = z_{\alpha/2}\,\sigma/\sqrt{n}$. Once the desired length of the confidence interval is specified, the sample size can be determined by solving this equation for n. In particular,

$$d = z_{\alpha/2}\,\sigma/\sqrt{n}$$
$$\sqrt{n} = z_{\alpha/2}\,\sigma/d$$

so

$$n = \frac{(z_{\alpha/2})^2\sigma^2}{d^2}$$

Figure 6.10
Confidence interval on μ of length $d = z_{\alpha/2}\,\sigma/\sqrt{n}.$

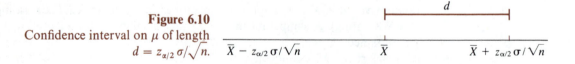

$\bar{X} - z_{\alpha/2}\sigma/\sqrt{n}$ $\bar{X}$ $\bar{X} + z_{\alpha/2}\sigma/\sqrt{n}$

Example | Suppose that we want to estimate the mean highway mileage obtained by a new model automobile to within two miles per gallon with 90% confidence.

How large a sample should be selected to accomplish this? Past experience indicates that the mileage obtained ranges from 15 to 35 mpg. The empirical rule suggests that the range of a data set covers approximately four standard deviations. Thus $\sigma \approx 20/4 = 5$ mpg. The point $z_{\alpha/2}$ is the point required to obtain a 90% confidence interval on μ. In this case, $z_{\alpha/2} = z_{.05} = 1.645$. Substituting,

$$n = \frac{(z_{\alpha/2})^2 \hat{\sigma}^2}{d^2} = \frac{(1.645)^2 (5)^2}{2^2} \approx 17$$

(a) It is thought that teenagers are spending between 0 and 8 hours per day playing video games. How large a sample is required to estimate the average time spent in this activity per day to within 1/2 hour with 95% confidence?

(b) In exercise 21, we estimate the variance in the number of customers served at a fast food restaurant between the hours of 6:00 A.M. and 7:00 A.M. daily to be 15.98. How large a sample is required to estimate the mean number served to within 2, with 99% confidence?

(c) A national survey organization wants to estimate the average number of years of formal education of adults over the age of 30. It is reasonable to assume that this random variable has a range of 25. How large a sample is required to estimate μ to within 6 months with 95% confidence?

*30. Show that S is not an unbiased estimator for σ. [*Hint:* Assume that S is unbiased. That is, assume that $E[S] = \sigma$. Note that Var $S > 0$. Use the fact that Var $S = E[S^2] - (E[S])^2$ to obtain a contradiction, thus proving the theorem indirectly.]

6.4   Testing a Hypothesis on the Mean

The theory already developed provides the foundation for testing hypotheses on the value of the mean of a random variable X. Recall briefly the general situation. We have a population and a random variable X associated with that population. We have available a random sample $X_1, X_2, \ldots, X_n$ of size n from the distribution of X. We have some preconceived notion concerning the actual value of μ. We wish to use information gained from this sample to confirm our hypothesized value.

Recall from Chapter 5 that there are three general rules to consider when setting up statistical hypotheses:

1. The **null hypothesis**, H_0, is the hypothesis of no difference.

2. Put whatever you are trying to detect or support as the **alternative hypothesis, H_1**.

3. Statistical hypotheses are always framed in hopes of being able to reject H_0 and therefore accept H_1.

We illustrate these ideas in Examples 1–3.

Example 1

Automatic pin setters for tenpin bowling should reset the pins about four seconds after the first ball is rolled. If the machine is too slow, bowlers get impatient, whereas if it is too fast, it hits the pins as they are falling. Either situation is undesirable. A machine is to be tested to see if it is operating correctly. The random variable of interest is X, the reset time for the machine. We wish to detect a situation in which the mean reset time, μ, is anything other than four seconds. Our alternative hypothesis is therefore

$$H_1: \mu \neq 4$$

The null hypothesis that the machine is operating correctly is

$$H_0: \mu = 4$$

Since shutting the machine down for repairs is troublesome and costly, we want to be sure that the test is run so that the chances of unnecessarily stopping the machine are small. That is, we want α, the probability of concluding that there is something wrong when in fact the machine is operating correctly, to be small.

Example 2

In computer operations time costs money. Hence any system or computing algorithm that reduces the time required to run a program is preferable to one that is slower. The system currently being used by an accounting firm requires a mean time of 45 seconds per program. A new system is being tested that, it is hoped, will reduce this mean time, and thereby save money. The random variable of interest is X, the time required to run a program under the new system. How should the null and alternative hypotheses be framed to get statistical evidence that the new system is superior to the old in this respect? Since we will not change systems unless we have evidence that the new system is superior to the old, we are trying to detect that fact. This statement then is taken as the alternative hypothesis. That is,

$$H_1: \mu < 45 \qquad \text{(mean running time of the new system is less than that of the old)}$$

The null hypothesis is the negation of H_1. Thus the null hypothesis is

$$H_0: \mu \geq 45$$

Obviously we would like to be able to reject H_0 in favor of H_1, since this would give us a way to improve our business operations. We will be able to run our test in such a way that α, the probability of rejecting H_0 when in fact $\mu = 45$ seconds, will be known. Thus, if H_0 is rejected and we decide to adopt the new system, we will be able to do so with confidence, since we will know the chance for error.

Example 3

Park officials are interested in building a camp store in Elkmont campground. However, based on the predicted number of customers, they think that the average amount of each sale must be greater than $2.00 to be profitable. They wish to design a statistical experiment to help them decide whether or not to build the store. How should the null and alternative hypotheses be framed to get the needed information? They want to build a store, so they want evidence that $\mu > \$2.00$—evidence that the store will be profitable. Since we put what we are trying to support as the alternative, our desired hypotheses are

$$H_0: \mu \leq \$2.00$$

$$H_1: \mu > \$2.00$$

We will be able to control α, the probability of erroneously rejecting H_0. Thus, if H_0 is rejected, we will be fairly sure that the store, if built, will be profitable.

As one can see from these examples, there are three general forms that hypothesis tests on μ can assume. These are given below with μ_0 representing the null value of the mean.

I $H_0: \mu = \mu_0$ II $H_0: \mu \geq \mu_0$ III $H_0: \mu \leq \mu_0$

$\quad\;$ $H_1: \mu \neq \mu_0$ $\quad\;$ $H_1: \mu < \mu_0$ $\quad\;\;$ $H_1: \mu > \mu_0$

$\quad$ Two-sided or $\quad$ Left-tailed test $\quad\;$ Right-tailed test

$\quad$ two-tailed test

Recall that α is the probability of rejecting H_0 given that the *null value is correct.* Recall also that the P value of a test is the probability of observing a value of the test statistic that is at least as extreme as that observed under the assumption that the *null value is correct.* To emphasize that both of these important probabilities are computed based on the stated null value, many statisticians prefer to express the null hypothesis as an equality. When this is done, the three forms become

I $H_0: \mu = \mu_0$ II $H_0: \mu = \mu_0$ III $H_0: \mu = \mu_0$

$\quad\;$ $H_1: \mu \neq \mu_0$ $\quad\;$ $H_1: \mu < \mu_0$ $\quad\;\;$ $H_1: \mu > \mu_0$

$\quad$ Two-sided or $\quad$ Left-tailed test $\quad\;$ Right-tailed test

$\quad$ two-tailed test

To see that there is no conflict here, consider our last example. Suppose we rewrite our hypotheses in the form

$$H_0: \mu = \$2.00$$

$$H_1: \mu > \$2.00$$

If our observed data allow us to reject the null value of $2.00 and conclude that the average sales at the camp store exceeds $2.00, then the same data surely allow us to reject even more soundly any value smaller than $2.00 as a possible value for μ! That is, by rejecting the null value, we are simultaneously rejecting any value smaller than the null value. A similar argument indicates that, in a left-tailed test by rejecting $H_0: \mu = \mu_0$ in favor of $H_1: \mu < \mu_0$, we are also rejecting $H_0: \mu \geq \mu_0$. You will probably see the null hypothesis in one-tailed tests expressed in both of these ways in your future work. They are both correct. You are free to choose whichever you like. However, to emphasize the importance of the null value in computing α and P values, *we will use the latter in the remainder of the text*. The important thing to remember, whichever way you choose, is to put whatever you are trying to detect or support as the alternative hypothesis.

Each of the three examples involves the general problem of testing a hypothesis on the mean value of a random variable X when its variance is unknown. We must devise a logical method for using data to help us run the test. That is, we must locate a statistic that allows us to test the hypothesis in such a way that α, the probability of committing a Type I error, is known. In this way, if we do reject H_0 and therefore accept the contention that we are trying to support, we will know, at least approximately, the chance that we are making an incorrect decision. What is the test statistic for such a test? There is one general statement to be kept in mind when attempting to devise any test statistic, namely, this:

> To test a hypothesis on the value of a parameter θ, we must be able to write down a statistic whose probability distribution is known (at least approximately) under the assumption that the null value of the parameter is correct.

Since we are testing a hypothesis about the true value of the mean of X, it is natural to begin by considering the unbiased estimator for this mean, namely, $\bar{X}$. It is also natural to compare this estimate for μ with the hypothesized value μ_0. We thus begin by looking at

$$\bar{X} - \mu_0$$

Note that for a left-tailed test, if H_1 is true, we would expect $\bar{X} - \mu_0$ to be negative. Large negative values of $\bar{X} - \mu_0$ should cause us to reject $H_0: \mu = \mu_0$ in favor of the alternative $H_1: \mu < \mu_0$. Similarly, in a right-tailed test, if H_1 is true, we would expect $\bar{X} - \mu_0$ to be positive. Hence, in this case, large positive values of $\bar{X} - \mu_0$ should cause us to reject $H_0: \mu = \mu_0$ in favor of the alternative $H_1: \mu > \mu_0$. For a two-tailed test, either large negative or large positive values of this difference should cause rejection of H_0.

We now have a logical beginning. However, we must have a test statistic whose distribution is known under the assumption that H_0 is true in the sense that the true mean is really equal to μ_0. Unfortunately, all that is known about the distribution of $\bar{X} - \mu_0$ is that under H_0 it is at least approximately normal with mean 0 and variance σ^2/n. Since σ^2 is not assumed known, we must turn again to the t distribution. Note that if H_0 is true, and μ is equal to the specified value, then

$$\frac{\bar{X} - \mu_0}{S/\sqrt{n}}$$

follows a t distribution with $n - 1$ degrees of freedom. We can denote this random variable by T_{n-1}. This statistic can be used then as the test statistic for testing hypotheses on the value of the mean. Our test is to reject H_0 in favor of H_1 if the observed value of this test statistic is too large, in either the positive or

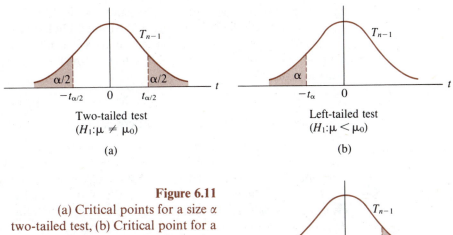

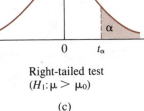

Figure 6.11
(a) Critical points for a size α two-tailed test, (b) Critical point for a size α left-tailed test, (c) Critical point for a size α right-tailed test.

negative sense, to have occurred by chance under H_0. The next obvious question is, How large is too large?

Recall that there are two methods that can be used to decide whether the observed value of the test statistic is large enough to cause us to reject H_0. We can preset α by determining a critical region prior to evaluating the test statistic. Alternatively, we can evaluate the test statistic, determine its P value, and then decide whether or not to reject H_0 based on the size of the P value. We illustrate the former method first.

The critical region chosen for a statistical test differs depending on the type of hypothesis being tested. In particular, the critical region used in a test depends on the form of the alternative hypothesis. To reject H_0 in favor of the alternative

$$H_1: \mu \neq \mu_0 \qquad \text{(two-tailed test)}$$

we would want to reject if the observed value of the test statistic

$$\frac{\bar{X} - \mu_0}{S/\sqrt{n}}$$

is either too large or too small. Thus, the critical region for the test should logically consist of both the upper and lower tail regions of the T_{n-1} distribution, since these values are the ones least likely to occur if H_0 is true. The critical region for such a test is shown in Figure 6.11(a). The critical region for a left-tailed test is the left-hand region of the T_{n-1} distribution; for a right-tailed test the critical region consists of the right-hand region of the T_{n-1} distribution. These critical regions are shown in Figure 6.11(b) and 6.11(c), respectively.

Notice that in the case of one-tailed tests, the inequality in the alternative actually points toward the end of the curve in which the critical region is located. Note also that since

$$\alpha = P[\text{reject } H_0 \,|\, \text{null value is the true value for } \mu]$$

$$\alpha = P[\text{test statistic falls in the critical region} \,|\, \mu = \mu_0]$$

the area under the curve within the critical region is α, the size of the test. The points $t_{\alpha/2}$, $-t_{\alpha/2}$, $-t_\alpha$, and t_α, shown in Figure 6.11, are called **critical points**. They are the points that define the critical region (or regions). H_0 is rejected in favor of H_1 if the observed value of the test statistic falls above or below the critical point, as the case may be. The exact numerical value of the critical point depends on the numerical value of α in the given problem and is determined from the T table.

Example 4

Let us test the hypothesis

$$H_0 \colon \mu = \$2.00 \qquad \text{(the store will not be profitable)}$$

$$H_1 \colon \mu > \$2.00 \qquad \text{(the store will be profitable)}$$

of Example 3. Since this is a right-tailed test, the critical region is located in the right tail of the appropriate t distribution. Since this is not a life-or-death situation, α can be moderate in size, say, $\alpha = .10$. That is, if we reject H_0 in favor of H_1, we will have only a 10% chance of having made the mistake of building the store when it is not likely to be profitable. We plan to test this hypothesis by observing a camp store in a similar situation and obtaining a sample of 25 observations on the random variable X, the amount of each sale. The distribution of the test statistic $(\bar{X} - 2)/(S/5)$ is T_{24} under H_0. The critical region and appropriate critical point is pictured in Figure 6.12. We reject H_0 if the observed value of the test statistic is greater than or equal to 1.318. Suppose that when the experiment is run these data are obtained:

2.75	6.25	3.50	3.01	5.10
5.06	4.50	4.17	2.57	3.15
3.98	2.37	2.03	1.02	5.38
1.57	1.00	1.16	1.07	3.12
0.75	0.10	0.25	3.09	4.10

For these data

$$\Sigma x = 71.05 \qquad \bar{x} = 2.84 \qquad \Sigma x^2 = 271.96$$

$$s^2 = \frac{n\Sigma x^2 - (\Sigma x)^2}{n(n-1)} = \frac{25(271.96) - (71.05)^2}{25(24)} = 2.92$$

$$s = \sqrt{2.92} = 1.71$$

Figure 6.12
Critical point for an
$\alpha = .10$ level right-tailed test
based on a sample of size 25.

Thus

$$\frac{\bar{x} - 2.00}{s/5} = \frac{2.84 - 2.00}{(1.71/5)} = 2.46$$

Since 2.46 is greater than the critical point 1.318, we reject H_0 and conclude that the mean sale exceeds \$2.00. If built, the store should be profitable.

The next example illustrates the use of a two-tailed critical region.

Example 5   The following data were obtained concerning the reset time of the automatic pin setter of Example 1.

4.1	2.5	3.5	3.8	3.2	4.6	4.1	3.0
3.5	4.1	4.3	3.6	4.0	3.7	4.5	3.9

Can we conclude from these data that the machine is not operating correctly? To see, let us test

$$H_0: \mu = 4 \quad \text{(the machine is operating correctly)}$$

$$H_1: \mu \neq 4 \quad \text{(the machine needs repair)}$$

Since we do not want to stop the machine for repairs unless we are fairly sure that something is actually wrong, let us select α to be .05. This means that if we do decide to shut down for repairs, there is only a 5% chance that we are doing so unnecessarily. The critical region for this two-tailed test is shown in Figure 6.13. The critical points, obtained from the T table with 15 degrees of freedom (sample size $- 1$) are -2.131 and 2.131. We reject H_0 if the observed value of the

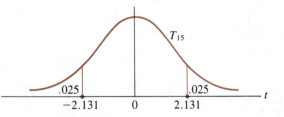

Figure 6.13
Critical points for an
$\alpha = .05$ level two-tailed test
based on a sample of size 16.

test statistic falls below -2.131 or above 2.131. From the data

$$\Sigma x = 60.40 \qquad \bar{x} = 3.78 \qquad \Sigma x^2 = 232.62$$

$$s^2 = \frac{16(232.62) - (60.40)^2}{16(15)} = .31$$

$$s = \sqrt{.31} = .55$$

The observed value of the test statistic is

$$\frac{\bar{x} - 4.0}{s/\sqrt{n}} = \frac{3.78 - 4.0}{(.55/4)} = -1.60$$

Since this value is not below the critical point of -2.131, we are unable to reject H_0. We do not have statistical evidence that the machine is out of order.

In Example 6, we illustrate the use of P values in testing a statistical hypothesis.

Example 6   In Example 2, the hypotheses

$$H_0: \mu = 45$$

$$H_1: \mu < 45 \qquad \text{(the mean running time of the new system is less than that of the old system)}$$

were framed to help us decide whether or not to change to a new computer system. A random sample of 30 programs is selected and run under the new system. The mean time required for these runs is 44.5 seconds, with a sample standard deviation of 2 seconds. Is 44.5 enough smaller than 45 to conclude that the mean running time is really smaller under the new system or could a value this small have been obtained by chance even though there is no difference between the two systems? To answer this question, we compute the P value for the test. The observed value of the test statistic is

$$\frac{\bar{x} - \mu_0}{s/\sqrt{n}} = \frac{44.5 - 45.0}{(2/\sqrt{30})} = -1.37$$

Since the sample is of size 30, the test statistic follows a T_{29} distribution. From Table VI of Appendix A, we see that $P[T_{29} \leq -1.699] = .05$ and $P[T_{29} \leq -1.311] = .10$. Since the observed value of our test statistic lies between -1.311 and -1.699, the P value of the test lies between .05 and .10. If you consider this probability to be small, then reject H_0; otherwise, do not reject H_0. The decision is yours!

There are several practical points to keep in mind:

1. Since you should have a clear idea of the purpose of your study, you should be able to set up the null and alternative hypotheses prior to conducting your research.

2. If you reject the null hypothesis $\mu = \mu_0$ in favor of the alternative $\mu \neq \mu_0$, your natural response would be to ask whether μ is greater or less than the null value. For this reason, one-sided tests are more informative than two-sided ones.

3. We have presented what are called *T tests*. Any time you are testing a hypothesis on the mean of a normal distribution when σ^2 is unknown a T test is appropriate. For large samples, the required t points can be approximated via the standard normal table. Remember that these Z points are listed in the last row of your T table.

4. If you are testing a hypothesis on a mean and you have evidence that the random variable you are studying has a distribution that is not normal, then you should not run a T test unless the sample size is large. The meaning of the term *large* is debatable. Most statisticians feel comfortable running a T test as long as the sample size is at least 25. When in doubt, you should use one of the nonparametric tests of location given in Chapter 12.

5. The probability α can be set by the experimenter. When dealing with small samples, the actual choice of α is limited by the available tables. In our case, we are limited to values of .40, .25, .1, .05, .025, .01, .005 and .0005. When dealing with large samples, we are not restricted in this manner, since we then use the standard normal curve.

6. The use of P values is becoming more and more widespread. Its main advantage over the practice of presetting α is that it gives the researcher a little more flexibility in interpreting the results of an experiment.

7. Finding the P value in a one-tailed test is straightforward. However, the two-tailed case is not so clear cut. If the distribution of the test statistic, under the assumption that H_0 is true, is symmetric, then it is reasonable to double the one-tailed P value. Since the t distribution is symmetric, this is the natural procedure to use in conducting a T test. If the distribution of the test statistic under H_0 is not symmetric, the issue is more complex. Various suggestions have been offered by statisticians, but no consensus has been reached yet. However, the most common procedure is to report a two-tailed P value that is twice the one-tailed value. This is the procedure that we will use.

We have not discussed hypothesis tests on the mean when σ^2 is assumed to be known because this situation rarely arises in practice. However, if you should be faced with this problem, the solution is easy. The test is performed in almost the same way as a T test. The only difference is that the test statistic used is

$$Z = \frac{\bar{X} - \mu_0}{\sigma/\sqrt{n}}$$

Critical points and P values are found using Table II of Appendix A.

Exercises 6.4

31. Find the critical point(s) for each of the tests on the mean described below:
    (a) right-tailed test with $\alpha = .05$, $n = 20$
    (b) right-tailed test with $\alpha = .10$, $n = 30$
    (c) right-tailed test with $\alpha = .01$, $n = 200$
    (d) left-tailed test with $\alpha = .05$, $n = 19$
    (e) left-tailed test with $\alpha = .01$, $n = 25$
    (f) two-tailed test with $\alpha = .05$, $n = 25$
    (g) two-tailed test with $\alpha = .10$, $n = 30$
    (h) two-tailed test with $\alpha = .01$, $n = 200$

32. Based on these data, can the null hypothesis be rejected in favor of the given alternative at the specified α level?
    (a) $H_0: \mu = 5$, $\alpha = .05$, $n = 20$, $\bar{x} = 5.3$, $s = 1.2$
        $H_1: \mu > 5$
    (b) $H_0: \mu = 5$, $\alpha = .10$, $n = 30$, $\bar{x} = 5.3$, $s = 1.2$
        $H_1: \mu > 5$
    (c) $H_0: \mu = 5$, $\alpha = .01$, $n = 200$, $\bar{x} = 5.3$, $s = 1.2$
        $H_1: \mu > 5$
    (d) $H_0: \mu = 10$, $\alpha = .05$, $n = 19$, $\bar{x} = 9.3$, $s = 2$
        $H_1: \mu < 10$
    (e) $H_0: \mu = 10$, $\alpha = .01$, $n = 25$, $\bar{x} = 9.3$, $s = 2$
        $H_1: \mu < 10$
    (f) $H_0: \mu = 6$, $\alpha = .05$, $n = 25$, $\bar{x} = 5.1$, $s = 3$
        $H_1: \mu \neq 6$
    (g) $H_0: \mu = 6$, $\alpha = .10$, $n = 30$, $\bar{x} = 5.1$, $s = 3$
        $H_1: \mu \neq 6$
    (h) $H_0: \mu = 6$, $\alpha = .01$, $n = 200$, $\bar{x} = 5.1$, $s = 3$
        $H_1: \mu \neq 6$

33. A pharmaceutical firm conducts research relative to the effectiveness of a measles vaccine. The random variable used is X, the antibody strength of an individual injected with the vaccine. It is felt that this variable is normally distributed. A vaccine produced by a competitor yields an average antibody strength of 1.9. The firm wishes to gain statistical support for their contention that their product produces a higher average antibody strength.
    (a) Set up the appropriate null and alternative hypotheses for gaining this support.
    (b) Discuss briefly the practical consequences of making a Type I and a Type II error.
    (c) What is the critical point for a size $\alpha = .05$ test of the hypothesis in part (a) based on a sample size 16?
    (d) The following sample size 16 is obtained. Use it to test the hypothesis of part (a) at the $\alpha = .05$ level

1.2	1.9	2.7	2.2	3.0	1.8	3.1	2.4
2.5	1.5	1.7	2.2	2.4	2.6	2.3	2.1

34. The accepted United States standard for exposure to microwave radiation is 10 microwatts per square centimeter. Citizens of a small residential area lying near a large television transmitting station feel that the station is polluting the air and pushing the microwave readings above the safe limit. To gain statistical support for their contention, they obtain the following readings on 20 randomly selected days (data are in microwatts per square centimeter):

10	11	12	10	11	9	10	10	11	8
12	13	9	11	10	8	9	7	11	10

(a) Set up the hypotheses necessary to gain statistical support for the contention that the mean emission is above the safe limit of 10.

(b) Test the hypothesis based on the above data at the $\alpha = .05$ level.

(c) To what type error are you now subject?

35. A psychologist claims that after taking a course in assertiveness training, a client's score on an assertiveness test increases by more than 10%. A sample size of 225 yields a mean increase of .11 and a sample standard deviation of .14. Do these data substantiate the claim at the $\alpha = .05$ level? To what type error is the psychologist now subject? Discuss the practical consequences of making this type of error.

36. The average number of days missed per month by workers in a large corporation is 2.5. It is claimed that the introduction of flex time will reduce this figure.

(a) Set up the appropriate null and alternative hypotheses for verifying the claim.

(b) Discuss the practical consequences of making a Type I and a Type II error.

(c) Find the critical point for an $\alpha = .01$ level test based on a sample size of 30.

(d) Flex time is tried for a period of time with a sample of 30 workers. The average number of days missed per month is 2.1 with a sample standard deviation of .70. Can H_0 be rejected at the $\alpha = .01$ level? To what type error are you now subject?

37. The average national score on the math portion of the SAT for a given year is 450. A school administrator is investigating the scores obtained by local students by testing

$$H_0: \mu = 450$$

$$H_1: \mu \neq 450 \qquad \text{(local students were not at the national}$$
$$\text{level in terms of SAT test scores)}$$

at the $\alpha = .05$ level.

(a) Find the critical points for the test based on a sample of size 25.

(b) When the data are analyzed it is found that $\bar{x} = 445$ and $s = 15$. Can H_0 be rejected at the $\alpha = .05$ level? Interpret you results in a practical sense. To what type error are you now subject?

38. Table V of Appendix A lists the sex and systolic and diastolic blood pressure of the 120 patients seen at a particular clinic. Use the random digit table (Table IV of Appendix A) to draw a random sample of size 20 from this population.

    (a) Use the data obtained on the systolic blood pressure to test

    $$H_0: \mu = 120$$

    $$H_1: \mu > 120$$

    at the $\alpha = .10$ level.

    (b) Use the data obtained on the diastolic blood pressure to test

    $$H_0: \mu = 78$$

    $$H_1: \mu < 78$$

    at the $\alpha = .10$ level.

39. In testing

    $$H_0: \mu = 10$$

    $$H_1: \mu > 10$$

    based on a random sample of size 20, the observed value of the T statistic is 3. What is the P value of the test? Do you think that H_0 should be rejected?

40. In testing

    $$H_0: \mu = 5$$

    $$H_1: \mu < 5$$

    based on a random sample of size 24, the observed value of the T statistic is -2.00. What is the P value for this test? Do you think that H_0 should be rejected?

41. In testing

    $$H_0: \mu = 2$$

    $$H_1: \mu \neq 2$$

    based on a random sample of size 16, the observed value of the T statistic is 1.5. What is the P value for this test? Do you think that H_0 should be rejected?

42. A manufacturer of lithium calculator batteries wants to advertise that the average lifespan of such a battery exceeds 1300 hours.

    (a) Set up the null and alternative hypotheses needed to substantiate this claim.

    (b) What are the practical consequences of making a Type I error?

    (c) A sample of 16 batteries is tested and it is found that $\bar{x} = 1325$ hours and $s = 30$ hours. Find the P value for the test. Do you think that H_0 should be rejected? Explain.

43. Television executives claim that commercial breaks average 30 seconds in length. A consumer advocate group thinks that the average break exceeds this value.
    (a) Set up the null and alternative hypotheses needed to obtain evidence to support the claim of the advocate group.
    (b) What has occurred if a Type II error is committed?
    (c) Data are gathered on 28 randomly selected commercial breaks. It is found that $\bar{x} = 32$ seconds and $s = 9$ seconds. Find the P value for the test. Do you think that H_0 should be rejected? Explain.

44. Past studies indicate that it takes five minutes on the average to memorize a list of 15 random words. A psychologist claims that a course in mnemonics tends to reduce the average time required to memorize such a list.
    (a) Set up the null and alternative hypotheses required to support the psychologist's claim.
    (b) A sample of 20 persons who completed the mnemonics course required an average of 4.0 minutes with a standard deviation of 2.3 minutes to memorize the random word list. Find the P value of the test. Do you think that H_0 should be rejected? Explain.

45. A social worker wants to gain statistical evidence that the average amount of money being spent per month for heating by welfare recipients in her region is more than the $80 per month allowed by the government.
    (a) Set up the appropriate null and alternative hypotheses for gaining this evidence.
    (b) The following observations were obtained on the random variable X, the amount spent per month for heating:

86.19	90.29	87.62	78.59	79.12
85.03	81.71	85.74	82.22	83.55
81.32	88.05	83.56	90.68	87.55
87.06	85.12	84.16	85.49	84.32

    Find $\bar{x}$ and s.
    (c) Find the P value for the test. Do you think that the social worker has the evidence that she wants? Explain.

46. The average time that a college basketball team takes to set up a shot is 30 seconds. Advocates of the 45-second shot clock claim that the clock has not affected this average. To test

    $H_0: \mu = 30$

    $H_1: \mu \neq 30$ (shot clock has affected the average)

    a random sample of games played using the clock is selected. The mean time taken to set up a shot based on 400 opportunities is found to be 28.8 with $s = 8$. What is the P value of the test? Be careful! Remember this is a two-tailed test. Do you think that H_0 should be rejected? Explain.

47. A study of the length of patient confinement, X, is conducted at a hospital. The purpose of the study is to refute the commonly accepted mean figure of five days or less. A sample of size 36 yields a sample mean of $\bar{x} = 6.2$ days and a sample variance of $s^2 = 5.2$. Do these data tend to refute the hypothesized figure? Support your answer on the basis of the P value of the test.

COMPUTING SUPPLEMENT B

The SAS procedure PROC MEANS can be used to test a hypothesis on μ when σ^2 is unknown. It can also be used to generate the terms $\bar{x}$ and $s/\sqrt{n}$ needed to construct confidence intervals on μ. We illustrate by testing the hypothesis $H_0: \mu = 2.00$ of Example 4.

Statement	*Purpose*
DATA STORE;	names the data set
INPUT PRICE @ @;	there is only one variable, PRICE, in the data set; @@ indicates that more than one data point will appear per line
HOMEAN = 2;	indicates that the null value for μ is 2
SPRICE = PRICE − HOMEAN;	creates a new variable, SPRICE, by subtracting the null value from each data point; E[SPRICE] = 0 and Var[SPRICE] = Var[PRICE]; testing E[PRICE] = 2 is equivalent to testing E[SPRICE] = 0
CARDS;	signals that data follow
2.75 6.25 3.50 3.10 5.10 5.06 4.50 4.17 2.57 3.15 . . . . 0.75 0.10 0.25 3.09 4.10	data; more than one observation per line with a space between each value
;	signals the end of data
PROC MEANS T PRT MEAN VAR STDERR;	asks for the observed value of the test statistic and its P value to be printed; asks for the statistics $\bar{x}$, s^2, $s/\sqrt{n}$ to be printed
TITLE TESTING HO:MEAN PRICE = 2;	titles the output

The output of this program follows.

```
                        TESTING HO: MEAN PRICE = 2

VARIABLE         T       PR>[T]        MEAN         VARIANCE       STD ERROR
                                                                   OF MEAN

PRICE          8.32      0.0001      2.84200000    2.91820000     0.34165480
HOMEAN           .         .         2.00000000    0.00000000     0.00000000
SPRICE         2.46①     0.0213②    0.84200000    2.91820000     0.34165480
```

Note that the P value given in ② is for a two-tailed test. Since we are running a one-tailed test, the actual P value is half of that listed on the printout. The observed value of the T statistic is given in ①.

6.5   The Chi-Square Distribution and Interval Estimation of the Variance

We discussed point estimation of σ^2 and σ so that we could develop confidence intervals on the mean and test hypotheses on μ when σ^2 is unknown. There are problems for which interest centers not on the mean but on the variance. We consider problems of this sort in this section. Typical problems are given in Examples 1–3.

Example 1   In the manufacturing of a drug, the potency of the drug varies. This variation must not be allowed to become too great, otherwise the drug will not be safe for human use. Some bottles of the drug may be too weak and therefore ineffective; others, too strong and therefore dangerous.

Example 2   A psychological experiment is to be run. It is thought that the age of the individual, although not under study, could influence the results. It is desirable that there be a diversity of ages within the population. That is, the variable *age* should have large variance so that the experiment will be free of age bias.

Example 3   A sociological experiment is to be run on a large group of people. For the experiment to be a success, there must be some diversity of opinion concerning the responsibility of government in righting historical social inequities. However, the diversity must not be so great that it causes internal conflicts within the population. That is, the variance of opinion within the population should be neither too large nor too small.

We extend here our study of variance to include the problems of confidence interval estimation of σ^2 and σ. To accomplish this, we must first introduce another useful family of continuous random variables called the **chi-square distribution** (X^2).

Let us consider the general properties of this family of variables.

Properties of Chi-Square Random Variables

1. There are infinitely many X^2 random variables, each identified completely by one parameter v called *degrees of freedom*. In this respect the X^2 random variable is similar to the T random variable discussed in the previous section. A chi-square random variable with v degrees of freedom is denoted by X_v^2.

2. Each X^2 random variable is continuous. Again we deal with smooth curves and identify areas with probabilities.

3. The X^2 random variable cannot assume negative values. All X^2 curves lie completely to the right of the vertical axis.

4. The graph of the probability function for every X^2 random variable is an asymmetric curve. A typical X^2 curve is shown in Figure 6.14.

5. The parameter v is a shape and a location parameter. It can be shown that

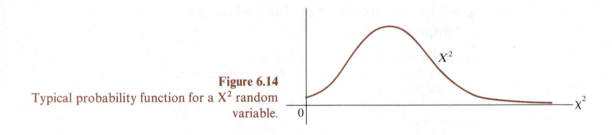

Figure 6.14

Typical probability function for a X^2 random variable.

$E[X_\nu^2] = \nu$ and $\text{Var}(X_\nu^2) = 2\nu$. Unlike the T family of random variables, as the number of degrees of freedom increases, the variance also increases. Thus, for X^2 random variables, the higher the degrees of freedom, the flatter the probability function curve becomes.

A partial summary of the cumulative distribution for X^2 random variables with degrees of freedom ranging from 1 to 30 is given in Table VII of Appendix A. Table VII is read in a manner similar to that of the T table. That is, probabilities are listed as column headings with the points associated with those probabilities listed within the body of the table. We again denote by χ_r^2 that point such that $P[X_\nu^2 \geq \chi_r^2] = r$.

Example 4

Consider a chi-square random variable with $\nu = 20$ degrees of freedom; from Table VII of Appendix A, we can conclude that

1. $P[X_{20}^2 \leq 12.4] = .10$

2. $\chi_{.05}^2 = 31.4$

3. $\chi_{.95}^2 = 10.9$

4. Points a and b such that $P[a \leq X_{20}^2 \leq b] = .95$ and $P[X_{20}^2 > b] = .025$ are

   $a = \chi_{.975}^2 = 9.59 \qquad b = \chi_{.025}^2 = 34.2$

We can also conclude that $E[X_{20}^2] = 20$ and $\text{Var}\, X_{20}^2 = 40$.

With this distribution available, it is possible to construct $100(1 - \alpha)\%$ confidence intervals on σ^2, provided we are sampling from a normal distribution. Again we look for a random variable whose expression involves the parameter being estimated, in this case σ^2, and whose distribution is known. Since S^2 is the unbiased estimator for σ^2, it is natural to assume that the desired random variable involves S^2 in some way. This can be seen in the following theorem.

Theorem 6.9

Let $X_1, X_2, \ldots, X_n$ be a random sample from a distribution that is normal with mean μ and variance σ^2. Then

$$\frac{(n - 1)S^2}{\sigma^2}$$

is a chi-square random variable with $n - 1$ degrees of freedom.

Note that once again the number of degrees of freedom depends on the sample size. Note also that a normality assumption is being made. If this assumption is not reasonably well met, then the above random variable may have a distribution that is far from chi-square. Hence, in any study, the normality assumption should be checked before applying the results of this section. Several methods for testing for normality are presented in Chapter 10 and may be studied at this time, if desired.

Let us illustrate the manner in which this theorem is used to generate $100(1 - \alpha)\%$ confidence intervals on σ^2.

Example 5 | When one-pound bags of potato chips are filled, there is some variation in weight. Not all bags contain exactly one pound. Company officials, from past experience, feel that a maximum variance of 1.5 is acceptable. If the variance becomes larger than this, two problems develop: Some bags do not have enough chips and draw consumer complaints, whereas others have too many chips and are unacceptable to the company. These observations are obtained on the random variable X, the number of ounces of potato chips in a one-pound bag. Assume that X is normal.

17.86	17.42	15.91	14.19
14.52	17.11	18.11	19.25
15.82	13.27	13.71	15.80
14.85	17.38	14.28	16.85

Let us construct a 90% confidence interval on σ^2. Recall that we are looking for two statistics, L_1 and L_2, such that $P[L_1 \leq \sigma^2 \leq L_2] = .90$. Since the sample is of size 16, we are dealing with a chi-square random variable with 15 degrees of freedom (sample size − 1). Since we want a 90% confidence interval, we divide the curve into three regions as shown in Figure 6.15.

We need only write down a probability statement relative to this curve that can be set equal to .90. Such a statement is

$$P[7.26 \leq 15S^2/\sigma^2 \leq 25.0] = .90$$

We now algebraically rewrite the above inequality isolating σ^2 in the middle as follows:

$$P[1/25.0 \leq \sigma^2/15S^2 \leq 1/7.26] = .90$$
$$P[15S^2/25.0 \leq \sigma^2 \leq 15S^2/7.26] = .90$$

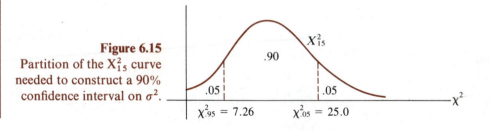

Figure 6.15
Partition of the X_{15}^2 curve needed to construct a 90% confidence interval on σ^2.

Thus, the desired 90% confidence interval is

$$[15S^2/25.0,\ 15S^2/7.26]$$

For these data

$$\Sigma x = 256.33 \qquad \Sigma x^2 = 4153.43$$

$$s^2 = \frac{n\Sigma x^2 - (\Sigma x)^2}{n(n-1)} = \frac{16(4153.43) - (256.33)^2}{(16)(15)}$$

$$= 3.12$$

Substituting this value of s^2 into the formula

$$[15S^2/25.0,\ 15S^2/7.26]$$

we obtain

$$[15(3.12)/25.0,\ 15(3.12)/7.26] = [1.87,\ 6.45]$$

as the desired 90% confidence interval on σ^2. Recall that this means that the procedure used to generate this interval is such that 90% of the time the intervals produced contain the true value of the population parameter σ^2.

What can we conclude? We are 90% confident that the true variance is at least as large as 1.87, which is considerably beyond the acceptable maximum of 1.5. Hence the company should make an effort to correct the situation.

To generalize the argument of the previous example to include any desired degree of confidence, we need only adjust the chi-square points appropriately. In particular, the point 25.0 is replaced by $\chi^2_{\alpha/2}$ and the point 7.26 becomes $\chi^2_{1-\alpha/2}$. The resulting general formula is stated in the next theorem.

Theorem 6.10

Confidence Interval on the Variance

Let $X_1, X_2, \ldots, X_n$ be a random sample from a distribution that is normal with mean μ and variance σ^2. A $100(1-\alpha)\%$ confidence interval on σ^2 is

$$\left[\frac{(n-1)S^2}{\chi^2_{\alpha/2}},\ \frac{(n-1)S^2}{\chi^2_{1-\alpha/2}}\right]$$

Example 6

The following data were obtained on the ages of 20 randomly selected individuals from the population of Example 2:

31	34	35	36	41	40	38	42	37	41
35	26	26	37	30	39	40	37	35	36

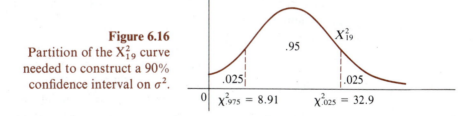

Figure 6.16
Partition of the X_{19}^2 curve
needed to construct a 90%
confidence interval on σ^2.

Let us construct a 95% confidence interval on σ^2 based on the above data. We first divide the X_{19}^2 curve as shown in Figure 6.16, locating the points $\chi_{.025}^2$ and $\chi_{.975}^2$ in Table VII of Appendix A.

For the above data

$$\Sigma x = 716 \qquad \Sigma x^2 = 26034$$

$$s^2 = \frac{20(26034) - (716)^2}{20(19)} = 21.12$$

Substituting these values into

$$[(n-1)S^2/\chi_{\alpha/2}^2, \ (n-1)S^2/\chi_{1-\alpha/2}^2]$$

we obtain

$$\left[\frac{19(21.12)}{32.9}, \frac{19(21.12)}{8.91}\right] = [12.20, 45.04]$$

We can be 95% confident that the actual variance in age in the population lies between 12.20 and 45.04.

To obtain a $100(1 - \alpha)\%$ confidence interval on σ when X is normal, it is natural to take the square root of the endpoints of the interval of Theorem 6.10.

Theorem 6.11

Confidence Interval on the Standard Deviation

Let $X_1, X_2, \ldots, X_n$ be a random sample from a distribution that is normal with mean μ and variance σ^2. A $100(1 - \alpha)\%$ confidence interval on σ is

$$\left[\sqrt{\frac{(n-1)S^2}{\chi_{\alpha/2}^2}}, \sqrt{\frac{(n-1)S^2}{\chi_{1-\alpha/2}^2}}\right]$$

Example 7

A 90% confidence interval on the standard deviation in the number of ounces of potato chips in a one-pound bag (Example 5) is

$$[\sqrt{1.87}, \sqrt{6.45}] = [1.37, 2.54]$$

A 95% confidence interval on the standard deviation of the age within the population of Example 6 is

$$[\sqrt{12.20}, \sqrt{45.04}] = [3.49, 6.71]$$

There is one further point to note here. The chi-square table available to us lists degrees of freedom from 1 to 30. What do we do if a confidence interval is desired and the sample size is greater than 31? Once again we turn to the normal curve. Exercise 54 of this chapter gives a transformation that can be used to approximate chi-square points from the standard normal curve for large samples.

Exercises 6.5

48. Use the chi-square table to find each of these points:
    (a) $\chi^2_{.025}$, $v = 15$
    (b) $\chi^2_{.975}$, $v = 15$
    (c) $\chi^2_{.01}$, $v = 20$
    (d) $\chi^2_{.99}$, $v = 20$
    (e) points a and b such that $P[a \le X^2_{10} \le b] = .90$ and $P[X^2_{10} > b] = .05$.
    (f) points a and b such that $P[a \le X^2_{29} \le b] = .95$ and $P[X^2_{29} > b] = .025$.

49. Find the indicated confidence intervals:
    (a) a 95% confidence interval on σ^2 when $n = 20$ and $s^2 = 8$
    (b) a 95% confidence interval on σ when $n = 20$ and $s^2 = 8$
    (c) a 90% confidence interval on σ^2 when $n = 25$ and $\sigma^2 = 12$
    (d) a 90% confidence interval on σ when $n = 25$ and $\sigma^2 = 12$

50. A job calls for mixing two types of sand for use in the production of cast-iron forms. The sand mixing is done partially by hand and calls for some intuitive judgments on the operator's part. Ideally, 1/4 of the mixture should be white sand, 3/4 red. However, some variability is inevitable. The following observations are obtained on the random variable X, the percentage of white sand in the mixture:

    26   22   20   24   25   28   30   21   26   23

    (a) Find point estimates for σ^2 and σ.
    (b) Find a 95% confidence interval on σ^2.
    (c) Find a 95% confidence interval for σ.

51. A school psychologist is interested in the program for learning-disabled children in the district. One variable under study is the IQ of children in the program. A sample of 25 children randomly selected from among those admitted to the program during the current school year yields a sample variance of 100. Find 90% confidence intervals on σ^2 and σ. Would you be surprised to hear someone claim that the standard deviation in IQ scores for the children entering the program this year exceeded 15? Explain.

52. The owner of a record store is attempting to assess the ages of his customers to assist him in ordering the proper mix of records. A random sample of 31 customers yields a sample mean of 23 years with a sample standard deviation of 5 years.

(a) Construct a 95% confidence interval on the average age of the customers.

(b) Construct a 95% confidence interval on the variance in ages of the customers.

(c) Construct a 95% confidence interval on the standard deviation in the ages of the customers.

53. Consider the random variable X, the time (in seconds) that it takes a kernel of popcorn to pop. In order not to burn large amounts of corn while also leaving large amounts unpopped, the variance of this variable should be small. Assume that X is normal and that a random sample of 28 kernels of popcorn produced a sample standard deviation of five seconds. Use this information to construct a 95% confidence interval on σ^2 and σ.

*54. (**Normal Approximation to Chi-Square**) The following result can be used to approximate chi-square points for $v > 30$. Let X_v^2 be a chi-square random variable with v degrees of freedom with $v > 30$. Then

$$\chi_r^2 \approx (1/2)[z_r + \sqrt{2v - 1}]^2$$

where z_r is that point such that $P[Z \geq z_r] = r$ and Z is the standard normal random variable.

Example

Suppose we want to find a 95% confidence interval on σ^2 based on a sample of size 60. Here $v = n - 1 = 59$. We need to approximate $\chi_{.025}^2$ and $\chi_{.975}^2$ using the normal curve. From the standard normal table, we see that $z_{.025} = 1.96$ and $z_{.975} = -z_{.025} = -1.96$. Therefore

$$\chi_{.025}^2 = (1/2)[z_{.025} + \sqrt{2v - 1}]^2$$

$$= (1/2)[1.96 + \sqrt{2(59) - 1}]^2$$

$$= 81.62$$

$$\chi_{.975}^2 = (1/2)[z_{.975} + \sqrt{2v - 1}]^2$$

$$= (1/2)[-1.96 + \sqrt{2(59) - 1}]^2$$

$$= 39.22$$

Approximate:

(a) $\chi_{.05}^2$, $v = 60$ (b) $\chi_{.975}^2$, $v = 50$ (c) $\chi_{.90}^2$, $v = 100$

*55. To set a standard for what is to be considered a "normal" cholesterol reading for individuals in the 20- to 29-year age group, a sample of 500 apparently healthy individuals is obtained. Blood tests are run and the resulting mean cholesterol level is 180 milligrams per deciliter with a standard deviation of 30 milligrams per deciliter.

(a) Find a 95% confidence interval on the mean cholesterol level for healthy individuals in this age group.

(b) Find 95% confidence intervals on σ^2 and σ.

*56. A survey is conducted to estimate the variability in the length of the active sentence given to individuals convicted of similar crimes. A random sample of size 200 yields a sample standard deviation of 8 years. Find 90% confidence intervals on σ^2 and σ.

*57. Let $X_1, X_2, \ldots, X_n$ be a random sample from a distribution that is normal with mean μ and variance σ^2. Use the fact that $(n-1)S^2/\sigma^2$ is X_{n-1}^2, properties of the X^2 distribution, and the rules of expectation and variance, to prove that S^2 is unbiased for σ^2 and that $\text{Var}(S^2) = 2\sigma^4/(n-1)$.

6.6 Testing a Hypothesis on the Variance (Optional)

In this section we consider the problem of testing hypotheses concerning the value of a population variance or standard deviation. Recall that this means that prior to conducting our research we will have a value for σ^2 or σ in mind. The purpose of the experiment is to confirm the proposed value.

To test a hypothesis concerning the value of σ^2, it is necessary to devise a logical test statistic. This statistic must be one whose distribution is known under the assumption that the null hypothesis is true in the sense that the null value σ_0^2 is the true value for σ^2. Again there are three forms that such a test may assume, namely the following:

I	H_0: $\sigma^2 = \sigma_0^2$	II	H_0: $\sigma^2 = \sigma_0^2$	III	H_0: $\sigma^2 = \sigma_0^2$
	H_1: $\sigma^2 \neq \sigma_0^2$		H_1: $\sigma^2 < \sigma_0^2$		H_1: $\sigma^2 > \sigma_0^2$
	Two-tailed test		Left-tailed test		Right-tailed test

Forms II and III are the ones usually of interest in applications. The natural choice for a test statistic in each case is $(n-1)S^2/\sigma_0^2$. This statistic is intuitively appealing since it utilizes the unbiased estimator for σ^2, namely S^2. Furthermore, if the data are drawn from a normal distribution and if H_0 is true, it follows a chi-square distribution with $(n-1)$ degrees of freedom. Thus, if H_0 is true, we would expect to observe the test statistic with a value close to the theoretical mean value of $(n-1)$. Note that if, in fact, $\sigma^2 < \sigma_0^2$ we would expect S^2 to be somewhat smaller in value than σ_0^2, implying that $(S^2/\sigma_0^2) < 1$, and hence that $[(n-1)S^2/\sigma_0^2] < (n-1)$. Thus, in the case of a left-tailed test, we should reject H_0 in favor of H_1 if the observed value of the test statistic is too small. Similarly, a right-tailed test calls for rejection of H_0 in favor of H_1 if the observed value of the test statistic appears too large. Again, remember that the terms *too small* and *too large* are used in the probabilistic sense. The critical regions and critical points for a size α test of each type of hypothesis are illustrated in Figure 6.17(a–c). We recall that in each case, the null hypothesis is rejected in favor of the alternative, if the observed value of the test statistic falls in the appropriate critical region.

Example 1 In the manufacturing of tennis balls a certain amount of variation in internal pressure is inevitable. Past experience indicates that the random variable X, the internal pressure of a tennis ball, is normally distributed with a mean of 28 pounds per square inch and a variance of .25. A new manufacturing process that

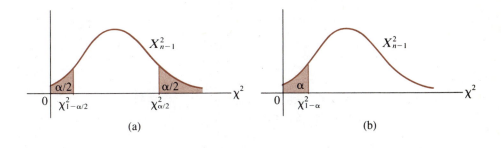

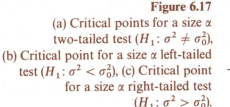

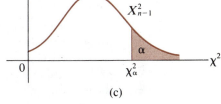

Figure 6.17
(a) Critical points for a size α two-tailed test ($H_1: \sigma^2 \neq \sigma_0^2$), (b) Critical point for a size α left-tailed test ($H_1: \sigma^2 < \sigma_0^2$), (c) Critical point for a size α right-tailed test ($H_1: \sigma^2 > \sigma_0^2$).

maintains the mean of 28 pounds is being tested. It is hoped that the new procedure will reduce the variance, thus producing balls with a more uniform bounce. A random sample of size 25 yields these observations:

28.20	27.31	28.68	27.98	27.99
28.04	27.47	28.57	28.12	28.75
28.36	27.96	28.30	28.29	28.40
27.46	27.99	27.94	27.76	27.91
27.59	27.71	28.60	27.91	27.82

The hypotheses to be tested are

$$H_0: \sigma^2 = .25$$

$$H_1: \sigma^2 < .25$$

Since changing manufacturing procedures is costly, let us agree to run a size $\alpha = .05$ test. In this way, we have only a small chance of recommending the change to the new process, when in fact it is no better than the old. Since the sample is of size 25, the test statistic, $24S^2/.25$, is distributed as a chi-square random variable with 24 degrees of freedom under H_0. The test is a left-tailed test with critical point as pictured in Figure 6.18.

We reject H_0 in favor of H_1 if the observed value of the test statistic is less than 13.8. For these data

$$\Sigma x = 701.11 \qquad \Sigma x^2 = 19665.80 \qquad s^2 = .15$$

Hence

$$24s^2/.25 = 24(.15)/.25 = 14.40$$

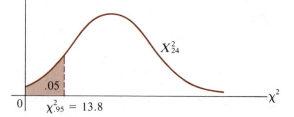

Figure 6.18
Critical point for an $\alpha = .05$ level left-tailed test
based on a sample of size 25.

Since $14.40 \not< 13.80$, we are unable to reject H_0. We do not have sufficient statistical evidence to justify changing the method of manufacture.

As when testing hypotheses on μ, we can test a hypothesis concerning σ^2 or σ using the P value approach. This idea is illustrated by testing a hypothesis on the standard deviation of a random variable X. This is done by first converting the problem to an equivalent one concerning the variance. We will illustrate this in the next example.

Example 2

In the manufacturing of a particular drug its potency is based on its percentage of codeine by weight. A standard deviation of one-tenth of one percent (0.1%) is acceptable, but deviations larger than that cannot be tolerated. A random sample of 25 bottles of the drug is to be tested for variability. Our hypothesis is therefore

$$H_0: \sigma = .10$$

$$H_1: \sigma > .10$$

Note that this hypothesis is equivalent to

$$H_0: \sigma^2 = .01$$

$$H_1: \sigma^2 > .01$$

Hence we can test the desired hypothesis by testing the latter one in the usual manner. The test is a right-tailed test based on the X_{24}^2 curve. The sample of size 25 yields a sample variance of $s^2 = .017$. Is this value too large to have occurred by chance? Is the drug exhibiting too much variability to be safely marketed? To decide, we compute the value of the test statistic $(n-1)S^2/\sigma_0^2$ and determine its P value.

In this case $(n-1)s^2/\sigma_0^2 = 24(.017)/.01 = 40.8$. From the chi-square table, we see that $P[X_{24}^2 \geq 39.4] = .025$ and $P[X_{24}^2 \geq 43.0] = .01$. Since the observed value of our test statistic, 40.8, lies between 39.4 and 43.0, the P value of the test lies between .01 and .025. Since this probability is very small, we reject H_0 and conclude that the drug exhibits too much variability to be marketed safely.

Remember that the methods presented in this chapter assume that sampling is from a *normal* distribution. We have checked this assumption only roughly via a stem-and-leaf diagram or assumed it to be true in the exercises and examples presented. In conducting a statistical study this assumption can and should be checked before analyzing data via the methods demonstrated here. Chapter 10 presents several ways to do so. If it is found that the data are from a distribution that is not at least approximately normal, these methods should not be used. Rather, some nonparametric procedure should be employed. These procedures are discussed in Chapter 12.

Exercises 6.6

58. Use the chi-square table to find the critical point(s), for each of the tests on σ^2 described below:
    (a) right-tailed test, $\alpha = .05$, $n = 20$
    (b) right-tailed test, $\alpha = .10$, $n = 30$
    (c) left-tailed test, $\alpha = .05$, $n = 25$
    (d) left-tailed test, $\alpha = .025$, $n = 24$
    (e) two-tailed test, $\alpha = .05$, $n = 28$
    (f) two-tailed test, $\alpha = .10$, $n = 16$

59. Based on these data, can the null hypothesis be rejected in favor of the given alternative at the specified α level?
    (a) $H_0: \sigma^2 = 4$, $\alpha = .05$, $n = 20$, $s^2 = 5.5$
        $H_1: \sigma^2 > 4$
    (b) $H_0: \sigma^2 = 4$, $\alpha = .10$, $n = 30$, $s^2 = 5.5$
        $H_1: \sigma^2 > 4$
    (c) $H_0: \sigma^2 = 2$, $\alpha = .05$, $n = 25$, $s^2 = 1.2$
        $H_1: \sigma^2 < 2$
    (d) $H_0: \sigma^2 = 2$, $\alpha = .025$, $n = 24$, $s^2 = 1.2$
        $H_1: \sigma^2 < 2$
    (e) $H_0: \sigma^2 = 9$, $\alpha = .05$, $n = 28$, $s^2 = 8.3$
        $H_1: \sigma^2 \neq 9$
    (f) $H_0: \sigma^2 = 9$, $\alpha = .10$, $n = 16$, $s^2 = 8.3$
        $H_1: \sigma^2 \neq 9$

60. A bowler currently bowls a hook. The variance of his scores has been 225. To become a more consistent bowler, he has switched to a straight ball.
    (a) Set up the null and alternative hypotheses needed to confirm that the variance in his scores has been reduced.
    (b) Find the critical point for an $\alpha = .10$ level test based on a sample of size 30.
    (c) Over a total of 30 games his average has not changed noticeably, but he has observed a sample variance of 169. Is there sufficient evidence at the $\alpha = .10$ level to conclude that the bowler has more control when bowling a straight ball? (Assume approximate normality.)

61. A study on the sound level of television commercials and programs is run. The variable of interest X, the sound level of the program in decibels. The variance of this variable should not be allowed to become too large. It is felt that a standard deviation of 2 or fewer decibels is acceptable.

(a) Set up the null and alternative hypotheses needed to detect a situation in which the variance in sound level has become too large.

(b) A random sample of 25 programs yields a sample standard deviation of 2.2 decibels. Use this information to test the hypotheses of part (a) at the $\alpha = .05$ level.

62. Psychologists are interested in the effect of crowding on human behavior. A 100-question timed test in basic arithmetic is devised. It is known from past experience that the scores obtained are approximately normally distributed with a variance of 16 when the test is administered with only the subject and the experimenter present. To study the effect of crowding, 20 subjects are placed in the same small room and given the test. It is felt that the variability of the scores will increase because some individuals will view the situation as a challenge and perform better than they normally would, whereas others will feel threatened by the presence of the other subjects and will tend to perform below their capabilities.

(a) Set up the null and alternative hypotheses needed to support the theory that crowding increases the variance in scores.

(b) Find the critical point for an $\alpha = .05$ level test.

(c) When the experiment is conducted, a sample variance of $s^2 = 20$ is found. Can H_0 be rejected at the $\alpha = .05$ level? To what type error are you now subject?

63. Approximate the P value when testing H_0 versus the given alternative based on these data. Decide whether or not you would reject H_0.

(a) $H_0: \sigma^2 = 10, \quad n = 20, \quad s^2 = 12.5$
$H_1: \sigma^2 > 10$

(b) $H_0: \sigma^2 = 10, \quad n = 30, \quad s^2 = 6$
$H_1: \sigma^2 < 10$

(c) $H_0: \sigma^2 = 10, \quad n = 25, \quad s^2 = 9$
$H_1: \sigma^2 \neq 10$

64. It is felt that the variability in the price charged for prescription glasses has become too large. A standard deviation of \$15 is thought to be acceptable to take into account local sources of supply, transportation costs, and operating expenses. The following observations were obtained on the cost of filling a particular prescription (assume normality):

39.47	38.57	49.66	36.77	31.64
24.11	15.91	35.78	45.08	29.21
12.76	70.99	23.02	25.42	44.83

Do these data support the contention that variability in prices has become too large? Explain by finding the P value of the test.

65. A foundry produces engine blocks for automobiles. To be acceptable they must possess a minimum hardness. The process currently in use produces a variance in the hardness reading of 0.09. This amount of variability tends to produce too many engine blocks that do not meet specifications. A sample of

16 blocks is produced using a slightly different chemical mix, and a sample variance of .04 is observed. Is this sufficient evidence to conclude that the new procedure has lowered variability? (Assume approximate normality.) Explain by finding the P value for the test.

66. In order to be satisfactory to both manufacturer and consumer, the amount of Tootsie Roll in a Tootsie Pop, X should be fairly uniform. Assume that X is normally distributed and that the standard deviation of X should not exceed 0.2 ounces. To maintain this standard, a periodic check is made of the Tootsie Pops produced. Set up and test the appropriate hypotheses based on the following random sample:

| 0.69 | 0.52 | 1.09 | 0.93 | 0.90 | 0.97 | 0.86 | 0.74 |
| 0.88 | 0.92 | 0.89 | 0.91 | 1.36 | 1.12 | 1.03 | 1.31 |

Can H_0 be rejected? Explain on the basis of the P value of the test. What type error may be involved? Discuss the practical consequences of making this error.

*67. The length of time that a particular model of a pocket calculator functions after being recharged is a normal random variable with a mean of 12 hours. It is thought that σ exceeds one-half hour.

(a) Set up the null and alternative hypotheses needed to verify that the standard deviation in operating times is more than one-half hour.

(b) Find the critical point for an $\alpha = .10$ level test based on a sample of size 50. (*Hint*: Use the normal approximation of exercise 54.)

(c) A quality control inspector tested 50 calculators and obtained a sample standard deviation of .6. Is this sufficient evidence at the $\alpha = .10$ level to conclude that the standard deviation is larger than one-half hour?

*68. A test is devised to measure an individual's opinion concerning the responsibility of government in righting historical social inequities. To have an acceptable population for experimentation, the sociologist feels that a desired variance on this test is $\sigma^2 = 5$. A variance significantly different from 5 will endanger the experiment. We thus want to test

$$H_0: \sigma^2 = 5$$

$$H_1: \sigma^2 \neq 5$$

(a) Find the critical points for an $\alpha = .10$ level test based on a sample of size 100. Use the normal approximation of exercise 54.

(b) When the experiment is run, a sample variance of $s^2 = 4.6$ is obtained. Can H_0 be rejected at the $\alpha = .10$ level?

Vocabulary List and Key Concepts

unbiased estimator	**null value**
100(1 − α)% confidence interval	**null hypothesis**
central limit theorem	**alternative hypothesis**

sample variance

sample standard deviation

critical points

sample mean

weighted mean

P value

chi-square distribution

Student's *t* distribution

alpha (α)

research hypothesis

Review Exercises

69. What is an unbiased point estimator for the mean of a random variable regardless of its distribution?

70. Let X_1, X_2, and X_3 be a random sample from a distribution with mean 8 and variance 36. What is $E[\bar{X}]$? What is Var $\bar{X}$? If the sample is drawn from a normal distribution, what type of random variable is $\bar{X}$? If the sample is drawn from a distribution that is not normal, can we use the Central Limit Theorem to conclude that $\bar{X}$ is approximately normal? Explain.

71. Under what circumstances is a T-type confidence interval appropriate?

72. A 95% confidence interval on a mean is constructed based on a given set of observations. Would a 99% confidence interval on μ based on the same data be longer or shorter than the 95% interval? Explain.

73. Is S^2 an unbiased estimator for σ^2?

74. Is S an unbiased estimator for σ?

75. When sample sizes are large, what distribution can be used to approximate t points?

76. What two properties are desirable in a point estimator?

77. A researcher reports that, based on his observations, he is 95% confident that the mean total blood protein in a healthy adult is 7.25 grams per deciliter. Explain what this means.

Use the following data to answer exercises 78–81.

A study of creditors' payment patterns was conducted by a retail firm. A sample of 16 randomly selected accounts yielded these observations on X, the time that elapses between the billing date and the date when payment is received (time is in days). Assume that X is normally distributed.

10	17	8	36	8	12	10	18
25	16	7	14	30	3	60	7

78. Find a point estimate for μ, the average elapsed time.

79. Find point estimates for σ^2 and σ.

80. Find a 95% confidence interval on the mean elapsed time. Based on this interval, would you be surprised to hear a claim that the mean elapsed time exceeds 28 days? Explain.

81. Find a 95% confidence interval on the standard deviation of X.

82. In testing a hypothesis on μ, what is the form of the alternative in a right-tailed test? a left-tailed test? a two-tailed test?

83. In testing a hypothesis on μ, when is a T test appropriate?

Use the following data to answer exercises 84–92:

Use of the telephone for personal reasons on company time costs businesses hundreds of thousands of dollars each year. An executive thinks that the average time spent per day on personal calls by her employees exceeds two hours. Calls are monitored on 25 randomly selected days and these observations are obtained on X, the total time in hours spent on noncompany business by the employees.

0.1	2.8	1.5	0.6	2.2
1.4	0.5	4.0	1.4	3.1
2.9	1.3	2.0	3.0	2.3
2.3	2.4	2.2	2.8	2.5
3.0	1.6	1.7	3.8	2.7

84. Construct a stem-and-leaf diagram for these data using stems of 0, 0, 1, 1, 2, 2, 3, and 3. Is there reason to suspect that X is not at least approximately normally distributed?

85. Find a point estimate for μ.

86. Find point estimates for σ^2 and σ.

87. Set up the null and alternative hypotheses needed to support the contention that the average time spent per day on personal calls by these employees exceeds two hours. Is the test a right-tailed or a left-tailed test?

88. Find the critical point for an $\alpha = .05$ level test of the hypothesis set up in exercise 87.

89. Evaluate the appropriate test statistic. Based on the observed value of this statistic, can H_0 be rejected? To what type error are you now subject? What are the practical consequences of making this error?

*90. Set up the null and alternative hypotheses needed to support the contention that the standard deviation of X is less than one hour.

*91. Find the critical point for an $\alpha = .05$ level test of the hypothesis set up in exercise 90.

*92. Evaluate the test statistic required to test the hypothesis of exercise 90. Can H_0 be rejected at the $\alpha = .05$ level?

Use the following data to answer exercises 93–96:

Test anxiety has been studied extensively by both educators and psychologists. The average score obtained by freshmen on a questionnaire designed to measure test anxiety has been 5 in past years. A new program has been put into effect to help students overcome test anxiety. These questionnaire scores are obtained on 16 randomly selected students who have participated in the program.

3.0	3.4	2.5	3.6	4.2	1.1	4.5	6.2
5.7	4.0	5.9	2.7	3.3	3.8	3.5	4.7

93. Construct a stem-and-leaf diagram for these data. Does the assumption that these scores are normally distributed appear to be reasonable?

94. Set up the null and alternative hypotheses needed to support the contention that the new program tends to reduce the average score of students on the test anxiety questionnaire.

95. Based on these data, can H_0 be rejected? Explain based on the P value of your test.

96. Find a 95% confidence interval on the standard deviation in scores based on these data.

7

Inferences on Proportions

We consider here two related topics. The first is the making of inferences on a single proportion p; the second is the comparison of two proportions p_1 and p_2. The statistical methods used are based on the binomial and normal random variables studied in detail in Chapter 4.

7.1 Estimating a Proportion

We begin by considering the problem of point estimation of a proportion. The general situation is as follows: there is a population under study, and each member is classed as either having, or failing to have, some specified trait. The **proportion**, p, of objects in the population with the trait is unknown and is to be estimated. The idea is visualized in Figure 7.1. Common sense indicates that to estimate p, we should obtain a random sample from the population, determine

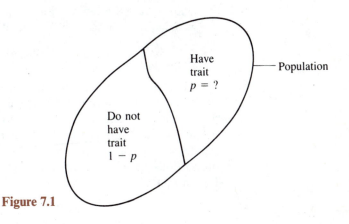

Figure 7.1

the proportion of the sample with the trait, and use that value as our estimate of p. That is, our estimator for p should be $\hat{p} = X/n$, where X is the number in the sample with the trait and n is the sample size. This estimator is referred to as the **sample proportion**. An example will demonstrate the idea.

Example 1

An "instapoll" in a newspaper reports that in a sample of size 500 obtained by a random telephone poll, 285 are opposed to the President's reversal of his tax rebate proposal. Estimate the proportion p of the population (area residents with telephones) that opposes the President. Since each member of the population either does or does not oppose the President, the population is divided as shown in Figure 7.2.

For this problem the observed sample proportion is $x/n = 285/500 = .57$. We estimate that 57% of the population is opposed to the President in this matter. Note that if the population consisted of 100,000 persons, we could further estimate that $.57(100,000) = 57,000$ persons opposed the plan.

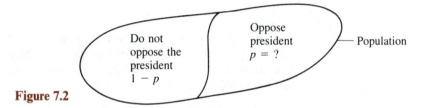

Figure 7.2

The idea of estimating a population proportion via a sample proportion is logical. Is it also mathematically sound? Does it possess the two desirable properties of a point estimator, of being unbiased and of having small variance for large sample sizes? The fact that it does is the point of the following theorem.

Theorem 7.1

The sample proportion

$$\frac{X}{n} = \frac{\text{number of objects in the sample with a specified trait}}{\text{sample size}}$$

is an unbiased estimator for the proportion p of objects in the sampled population with the trait. Furthermore,

$$\text{Var}(X/n) = p(1-p)/n$$

This theorem can be verified easily using properties of the binomial distribution. Its derivation is outlined in exercise 16 of this chapter. Note that since p is a population parameter its value does not change. Thus $p(1-p)$ is a constant. This means that the variance of the sample proportion, $p(1-p)/n$, decreases as the sample size increases. Practically speaking, this assures us that point estimates for proportions based on relatively large samples are usually fairly accurate. (See Figure 7.3.)

'Oh, oh, I just discovered 79 percent of my rats have cancer . . . and I didn't inject them with anything yet!'

Figure 7.3
Copyrighted by the Chicago Tribune. Used with permission.

As in the past, there are times when an interval estimate is preferable to a point estimate. We want to develop a means of constructing $100(1 - \alpha)\%$ confidence intervals on the proportion p of objects in a population with a given trait. To do so, we must find a random variable whose expression involves p and whose distribution is known at least approximately. This is done by recalling from Chapter 4 that when sample sizes are large, a binomial random variable can be approximated via a normal random variable. In particular, if X is binomial with parameters n and p, then X can be approximated by a normal random variable with mean np and variance $np(1 - p)$. That is, we can consider X to be approximately normal with mean np and variance $np(1 - p)$. If we then standardize X, we can conclude that the random variable

$$\frac{X - np}{\sqrt{np(1 - p)}}$$

is approximately *standard* normal. By dividing both the numerator and denominator of this random variable by n, we see that the random variable

$$\frac{(X/n) - p}{\sqrt{p(1 - p)/n}}$$

is approximately standard normal.

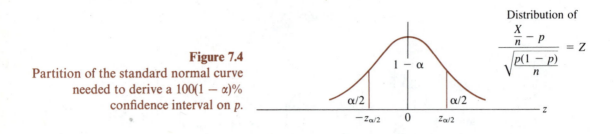

Figure 7.4
Partition of the standard normal curve
needed to derive a $100(1 - \alpha)\%$
confidence interval on p.

We thus have the random variable needed to construct a $100(1 - \alpha)\%$ confidence interval on p. We now look for two statistics L_1 and L_2 such that

$$P[L_1 \leq p \leq L_2] = 1 - \alpha$$

regardless of the actual value of p.

We begin the derivation of the confidence interval by considering the graph shown in Figure 7.4. Note that

$$P[- z_{\alpha/2} \leq [(X/n) - p]/\sqrt{p(1 - p)/n} \leq z_{\alpha/2}] = 1 - \alpha$$

$$P[- z_{\alpha/2}\sqrt{p(1 - p)/n} \leq (X/n) - p \leq z_{\alpha/2}\sqrt{p(1 - p)/n}] = 1 - \alpha$$

$$P[(X/n) - z_{\alpha/2}\sqrt{p(1 - p)/n} \leq p \leq (X/n) + z_{\alpha/2}\sqrt{p(1 - p)/n}] = 1 - \alpha$$

It appears that the desired confidence interval is

$$(X/n) \pm z_{\alpha/2}\sqrt{p(1 - p)/n}$$

However, there is a definite problem. The endpoints L_1 and L_2 for a random interval must be *statistics*. Since p is unknown, neither endpoint above is a statistic. There are two ways to overcome this difficulty.

Confidence Intervals on p

1. We can replace p by its unbiased estimator X/n to obtain

$$(X/n) \pm z_{\alpha/2}\sqrt{(X/n)[1 - (X/n)]/n}$$

2. Since a result from calculus states that $p(1 - p)$ will be at most $1/4$, regardless of the value of p, we can replace this term by $1/4$ to obtain

$$(X/n) \pm z_{\alpha/2}\sqrt{1/4n}$$

Either of the above methods produces acceptable confidence intervals on p, for large n. The second method produces numerical intervals that are slightly longer than those obtained by means of the first method unless the sample proportion assumes the value $1/2$. In this case the two methods coincide.

Although these formulas may look complicated, they are easy to use. To see how they work, let us consider an example.

Example 2 | Let us construct a 90% confidence interval on the proportion p of people in the instapoll survey population opposed to the President's plan to reverse the tax

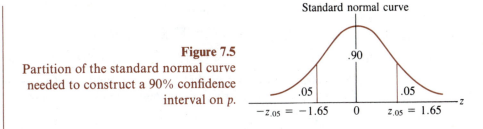

Figure 7.5

Partition of the standard normal curve
needed to construct a 90% confidence
interval on p.

rebate proposal. From Example 1, we know that our point estimate for p is
$\hat{p} = x/n = 285/500 = .57$. The partition of the standard normal curve needed to
construct a 90% confidence interval on p is shown in Figure 7.5.

Using method 1, we obtain

$$.57 \pm 1.65\sqrt{.57(1 - .57)/500} = [.5335, .6065]$$

Using method 2, we obtain

$$.57 \pm 1.65\sqrt{1/4(500)} = [.5331, .6069]$$

Notice that there is a slight difference between the two, with method 2 being
slightly longer. One would, of course, not apply both methods in a single
problem. This was done here only for purposes of illustration and comparison.
Pick the method that appeals to you and stick to it!

The next example points out the importance of planning an experiment prior
to data collection.

Example 3

A researcher wants to estimate the percentage p of white males in the 18- to
24-year age group suffering from hypertension. Let us construct a 95% confidence interval on this percentage based on a random sample of size 25 using
method 2. We are dealing with the standard normal curve shown in Figure 7.6.

Suppose that the sample yields 20 individuals with hypertension. The
unbiased point estimate for p is

$$\hat{p} = 20/25 = 4/5 = .80$$

Figure 7.6

Partition of the standard normal curve
needed to construct a 95% confidence
interval on p.

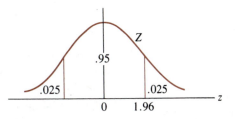

The 95% confidence interval is

$$(x/n) \pm z_{.025}\sqrt{1/4n} = .80 \pm 1.96\sqrt{1/4(25)}$$

$$= [.80 \pm .196] = [.604, .996]$$

We can be 95% confident that the true percentage of white males in the 18- to 24-year age group suffering from hypertension is between 60.4% and 99.6%.

Note that in the preceding example there is one obvious serious problem. Namely, the interval obtained, although constructed correctly, is so wide as to be practically worthless. To say that p lies between 60.4% and 99.6% is not very informative. How can this difficulty be overcome? How can an experiment be designed so that the confidence interval obtained is short enough to be truly meaningful? This can be done by noting that the endpoints of the interval, $X/n \pm z_{\alpha/2}\sqrt{1/4n}$, depend not only on the desired confidence but also on the sample size. The larger the value of n, the shorter and more precise the interval becomes. The crucial question to be asked *before the experiment is conducted* is, How large a sample should be taken to estimate p with a specified degree of accuracy? The next example shows how to answer this question.

Example 4

Suppose that in the hypertension study of Example 3, we want to estimate the proportion of white males in the 18- to 24-year age group with hypertension in such a way that we are 90% sure that the sample proportion X/n lies within 1% of the true proportion p. Since any confidence interval of p is centered at X/n, we need only choose n in such a way that the distance d from the center of the interval to either endpoint is .01. The situation is diagrammed in Figure 7.7.

To determine n, we solve the equation

$$d = 1.65\sqrt{1/4n} = .01$$

for n as follows:

$$1.65\sqrt{1/4n} = .01$$

$$(1.65)^2(1/4n) = (.01)^2$$

$$1/n = 4(.01)^2/(1.65)^2$$

$$n = (1.65)^2/4(.01)^2 = 6{,}806.25 \approx 6{,}807$$

Figure 7.7
Distance from the center of the interval to the endpoint is $d = 1.65\sqrt{(1/4n)} = .01$.

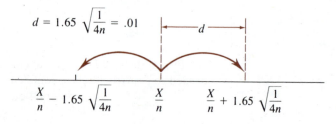

To estimate p with the desired degree of accuracy, we must sample at least 6807 individuals. Note that if we use the observed sample proportion as a point estimate for p based on this size sample, we could be 90% sure that it differs from the true value of p by at most .01.

The above argument can be generalized to give a formula for determining the sample size required to be $100(1 - \alpha)\%$ certain that X/n will differ from p by at most any specified amount d. We need only make two changes. First, we replace the point 1.65, which is specifically associated with a 90% confidence interval, with the general point $z_{\alpha/2}$. Second, we replace the number .01, the specific degree of accuracy required in our example, with d, the accuracy required in general. These substitutions lead to the following formula for determining sample size prior to experimentation:

Sample Size Required to Estimate p

The sample size required to estimate p to within a specified distance d with a specified confidence is

$$n = z_{\alpha/2}^2/4d^2$$

An interesting aspect of this formula is that, for large populations, the sample size required to estimate p is not dependent on the size of the population. Another interesting point is that, even if the sampled population is very large, we can usually estimate p accurately with a relatively small sample. The next example should convince you that this is true!

Example 5

To be an accredited college library, a library must contain at least 70% of the books listed as *essential*. The list contains 83,000 items. To estimate the proportion p of books on the list in a particular library, a random sample of size n is selected from the list. Each item in the sample can be classed as being in or not being in the library. How large a sample should be taken in order to be 95% confident that the sample proportion X/n lies within .02 of the true proportion p? In this problem, d is specified to be .02. Since we want to be 95% confident of being within .02 of the true value of p, $z_{\alpha/2} = z_{.025} = 1.96$. Hence

$$n = z_{.025}^2/4d^2 = (1.96)^2/4(0.02)^2 = 2{,}401$$

A sample of size 2,401 will give the desired accuracy with probability .95.

Exercises 7.1

1. A survey is conducted among high school seniors to estimate the proportion p that intends to enter the armed services upon graduation. In a random sample of size 10,000 it was found that 2,015 planned enlistment within a

year of graduation. Use these data to find a point estimate for p. Based on this estimate, how many seniors in a senior class of 300 do you estimate will enlist within a year?

2. In trying to assess the impact of a political campaign, the campaign workers for J. C. randomly select 110 individuals from the registered voter lists in the district. Each is asked to identify J. C.; only 15 do so correctly. Find a point estimate for the proportion p of registered voters in the district that can correctly identify the candidate. Based on this figure, estimate how many of the 10,000 registered voters in the district can identify the candidate.

3. The lead level in a child's body is considered to be dangerously high if it exceeds 30 micrograms per deciliter. Children come into contact with lead from a variety of sources, but are particularly susceptible to exposure from eating paint from toys, furniture, and other objects. A random sample of 1,000 children living in public housing projects in a particular city revealed that 200 of them had dangerous levels of lead in their bodies. Estimate the proportion of children with dangerously high lead levels in this population. If there are 20,000 children living in the projects, estimate the number of children with dangerously high lead levels.

4. A random sample of size n is drawn from a population consisting of various types of objects. The proportion of objects with trait A is denoted by p. The sample contains x objects with the trait.
   (a) If $n = 100$ and $x = 70$, construct a 95% confidence interval on p using method 1.
   (b) If $n = 200$ and $x = 100$, construct a 90% confidence interval on p using method 1.
   (c) If $n = 50$ and $x = 20$, construct a 95% confidence interval on p using method 2.
   (d) If $n = 75$ and $x = 15$, construct a 99% confidence interval on p using method 2.

5. Items produced on an assembly line are sampled each day. The proportion of defective items produced is denoted by p. Let n denote the sample size and let x denote the number of defective items found in the sample.
   (a) If $n = 70$ and $x = 10$, construct a 95% confidence interval on p using method 1.
   (b) Would a 95% method-2 confidence interval on p based on the data of part (a) be longer or shorter than that found in part (a)? Verify your answer by constructing such an interval.

6. Use method 1 to find a 90% confidence interval on the proportion of high school seniors that intends to enter the armed services upon graduation. Use the data of exercise 1.

7. Use method 2 to find a 95% confidence interval on the proportion of voters that can identify J. C. correctly. Use the data of exercise 2.

8. Use either method 1 or 2 to find a 90% confidence interval on the proportion

of children with dangerously high lead levels in the population described in exercise 3. Use the data given there.

9. A large department store chain is considering opening a new store in a town of 15,000 people. Before making the decision, a market survey is conducted. Of 200 persons interviewed, 165 indicated they would patronize the new store. Find a 95% confidence interval on the proportion p of people that will patronize the store. Use this to give a 95% confidence interval on the number of people who will patronize the store.

10. In a recent consumer survey, a random sample of 1000 American families produces 600 who consider not being in debt an important personal value. Construct a 96% confidence interval on the proportion p of American families that considers this important.

11. Determine how large a sample is required to estimate p in each of these cases:
   (a) to within .02 with 90% confidence
   (b) to within .05 with 95% confidence
   (c) to within .03 with 80% confidence

12. A sociologist is interested in the proportion p of juveniles appearing in juvenile court in New York City that comes from broken homes. How large a sample should the sociologist take to be 90% sure that the sample proportion lies within .01 of p?

13. A college administrator wishes to estimate the proportion p of students that favors the abolition of football as an inter collegiate sport on their campus. How large a sample should the administrator take to be 95% sure that the sample proportion lies within .03 of p?

14. Prior to a national election, a sample of voters is to be selected and interviewed. How large a sample is needed to estimate the proportion of voters that intends to vote for the Democratic candidate to within .02 with 95% confidence.

15. Use the table of random digits, Table IV of Appendix A, to select a random sample of size 10 from the population of patients listed on Table V of Appendix A.
   (a) Find a 95% confidence interval on the proportion of females.
   (b) Find a 95% confidence interval on the proportion of patients whose systolic blood pressure exceeds 120.
   (c) Find a 95% confidence interval on the proportion of patients whose diastolic blood pressure is less than 90.

*16. Show that the sample proportions is an unbiased estimator for p and that its variance is given by $p(1 - p)/n$. (*Hint:* When sampling from large populations, the random variable X, the number of objects in the sample with the trait, can be viewed as being at least approximately binomial with parameters n and p. Use this assumption and the rules for expectation and variance to show that $E[X/n] = p$ and that $\text{Var}(X/n) = p(1 - p)/n$.)

7.2   Testing a Hypothesis on a Proportion

In Chapter 5 we saw how to test a hypothesis on a proportion when the sample size is small. The test statistic used is X, the number of objects in the sample with the trait being studied. Here, we consider the problem of testing a hypothesis on p when the sample size is large.

Hypothesis testing on the value of p follows the usual pattern. The same general forms for the null and alternative hypotheses are possible. These are

I $H_0: p = p_0$ II $H_0: p = p_0$ III $H_0: p = p_0$

$H_1: p \neq p_0$ $H_1: p < p_0$ $H_1: p > p_0$

Two-tailed test   Left-tailed test   Right-tailed test

The test statistic for testing each of the above is

$$\frac{(X/n) - p_0}{\sqrt{p_0(1 - p_0)/n}}$$

This statistic is logical in that it compares the sample proportion to the hypothesized population proportion. If H_0 is true, then this statistic is approximately standard normal for large n. Note that if p is actually less than the hypothesized value p_0, we would expect the sample proportion X/n to assume a value less than p_0, forcing $(X/n) - p_0$, and thus the test statistic, to be negative. Hence we should reject H_0 in favor of the alternative $H_1: p < p_0$ for large negative values of the test statistic. A similar argument implies that H_0 should be rejected in favor of $H_1: p > p_0$ whenever the test statistic assumes large positive values.

Recall that there are two ways to decide whether to reject H_0. We can select α and then reject H_0 if the observed value of the test statistic lies in the critical region determined by α or we can evaluate the test statistic and reject H_0 if the observed P value is small. These methods are illustrated in the next two examples.

Example 1

A government study of hiring practices is to be run. It is felt that the percentage of minorities hired by heavy industry should reflect the percentage of minorities living within the area. The percentage should not be allowed to become too large (perhaps indicating favoritism) or too small (perhaps an indication of discrimination). The percentage of minorities living in a given locality is .2. A sample of size 100 is to be selected from those employed in heavy industry in the area, and the percentage of minorities is to be obtained. We thus want to test

$H_0: p = .2$

$H_1: p \neq .2$

based on a sample of size 100. Since rejecting H_0, when in fact H_0 is true, could lead to conflict between the government and the industries, let us run the test at the $\alpha = .05$ level. The critical region for the test is shown in Figure 7.8.

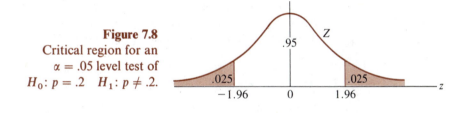

Figure 7.8
Critical region for an
$\alpha = .05$ level test of
$H_0: p = .2$ $H_1: p \neq .2$.

We will reject H_0 if the observed value of the test statistic

$$\frac{(X/n) - p_0}{\sqrt{p_0(1 - p_0)/n}}$$

lies above 1.96 or below -1.96. When the test is run, it is found that 17 of the 100 persons sampled are minorities. That is, 17% of the sampled group are minorities. Is this figure different enough from the desired figure of 20% to conclude that something is amiss, or could this value reasonably have occurred by chance, even though actually 20% of those employed overall are in fact minorities? To decide, we compute the value of the test statistic as follows:

$$\frac{(X/n) - p_0}{\sqrt{p_0(1 - p_0)/n}} = \frac{(17/100) - .2}{\sqrt{.2(.8)/100}}$$

$$= -.03/.04 = -.75$$

Since $-.75$ is neither below the critical point of -1.96 nor above that of 1.96, we are unable to reject H_0. We do not have statistical evidence of hiring irregularities in this locality.

Example 2   A manufacturer feels that it is competitive if at least 30% of the word processors used in governmental offices in each state are made by his company. The company periodically samples to detect situations in which this proportion has fallen below its target. In such cases, it embarks on an expensive advertising campaign. Whatever is to be detected should be placed in the alternative hypothesis; hence the company is interested in testing

$$H_0: p = .30$$

$$H_1: p < .30$$

In Virginia a random sample of size 500 reveals that 118 word processors were made by this manufacturer. Should an advertising campaign be undertaken in this state? For these data

$$\frac{(X/n) - p_0}{\sqrt{p_0(1 - p_0)/n}} = \frac{(118/500) - .3}{\sqrt{.3(.7)/500}} = -3.12$$

From the standard normal table we see that $P[Z \leq -3.12] = .0009$. That is, the P value, the probability of observing a value as small as -3.12 or smaller when the null value is the true value for p, is .0009. Since this probability is very small, we reject H_0 and conclude that $p < .3$. Based on this decision we recommend that an advertising campaign be launched in Virginia.

Exercises 7.2

17. Find the critical point(s) for testing $H_0: p = p_0$ in each of these cases:
    (a) a left-tailed test with $\alpha = .10$
    (b) a left-tailed test with $\alpha = .05$
    (c) a right-tailed test with $\alpha = .025$
    (d) a right-tailed test with $\alpha = .01$
    (e) a two-tailed test with $\alpha = .10$
    (f) a two-tailed test with $\alpha = .05$

18. A random sample of size n is drawn from a population. The proportion of objects in the population with a specified trait is p. The number of objects in the sample with the trait is x.
    (a) If $n = 100$ and $x = 70$, can we reject $H_0: p = .8$ in favor of $H_1: p < .8$ at the $\alpha = .10$ level? at the .05 level?
    (b) If $n = 25$ and $x = 15$, can we reject $H_0: p = .4$ in favor of $H_1: p > .4$ at the $\alpha = .025$ level? at the .01 level?
    (c) If $n = 50$ and $x = 31$, can we reject $H_0: p = .5$ in favor of $H_1: p \neq .5$ at the $\alpha = .10$ level? at the .05 level?

19. A woman claims to have ESP. We decide to test her claim by performing the following experiment: 25 cards are placed face down in front of her; each card is either black or red. In each case, she is asked to identify the color of the card. If she has ESP, the probability p of a correct answer should be greater than .5. Set up the appropriate null and alternative hypotheses for detecting the presence of ESP. What is the critical point for a size $\alpha = .1$ test of this hypothesis? When the experiment is run, she correctly identifies 18 cards. What conclusion can be drawn? What type error may be committed?

20. It is thought that more than 60% of the investment analysts in this country feel that the major issue affecting the solar energy industry is that of falling energy prices. Set up the null and alternative hypotheses needed to support this claim. Find the critical point for an $\alpha = .05$ level test. A random sample of 100 analysts is selected and interviewed. It is found that 65 of them hold the opinion that falling energy prices are the major issue. Based on these data, can H_0 be rejected?

21. The campaign manager for the incumbent candidate thinks that a majority of the voters in the district plan to vote for her candidate. Set up the null and alternative hypotheses needed to support her claim. Find the critical point for an $\alpha = .01$ level test. In a random survey of 1,000 voters, 515 indicated that they intend to vote for the incumbent. Is this sufficient evidence to claim that a majority of voters in the district plan to vote for this candidate? To what type error are you now subject? What are the practical consequences of making this error?

22. A random sample of size n is drawn from a population. The proportion of objects in the population with a specified trait is p. The number of objects in the sample with the trait is x.
    (a) If $n = 25$ and $x = 10$, what is the P value for testing $H_0: p = .5$ versus $H_1: p < .5$?
    (b) If $n = 30$ and $x = 25$, what is the P value for testing $H_0: p = .8$ versus $H_1: p > .8$?
    (c) If $n = 100$ and $x = 80$, what is the P value for testing $H_0: p = .75$ versus $H_1: p \neq .75$?

23. A medical researcher wants to show that a muscle relaxant produces drowsiness in a smaller proportion of patients suffering from a particular disorder than the 40% in which this occurs under the standard treatment. Set up the appropriate hypotheses for gaining this evidence. If 70 of 200 patients tested reported drowsiness, do you feel that the claim was substantiated? Justify your answer on the basis of the P value of the test.

24. It is felt that more than 22% of the nonfederal dams in this country are potentially hazardous. Set up the hypothesis appropriate to substantiate this claim. Of one hundred randomly selected nonfederal dams, 25 are potentially hazardous. If, on the basis to this result, you decide to reject H_0 in favor of H_1, what would be the P value for your test?

25. In past years, the suicide attempt rate among teenagers in the United States has been .01. Psychologists and physicians think that this rate has increased because of the increase in drug abuse and instability in home life. They wish to gain statistical support for their contention. Set up the appropriate hypothesis for gaining this support. In a sample of 2,000 teenagers selected and observed for a period of time, 25 attempt suicide. Do you think that the contention of the psychologists and physicians has been substantiated? Explain, based on the P value of your test.

7.3 Estimating the Difference in Proportions

We begin with a brief description of the general problem. There are two populations under study and a single trait of interest. Each member of each population can be classed as either having, or failing to have, the trait. The proportion of objects in each population with the trait is unknown. We wish to compare these proportions by means of independent random samples drawn from the respective populations. The idea is visualized in Figure 7.9. Again, comparisons can be done via estimation or hypothesis testing procedures. We consider estimation methods in this section.

To find point and interval estimates for the difference between two proportions, we need only extend the ideas presented for the one-sample case. In particular, we estimate p_1 by X_1/n_1, the sample proportion based on a random sample of size n_1 drawn from the first population. That is, we estimate p_1 by

$$\frac{X_1}{n_1} = \frac{\text{number in sample one with the trait}}{\text{size of sample one}}$$

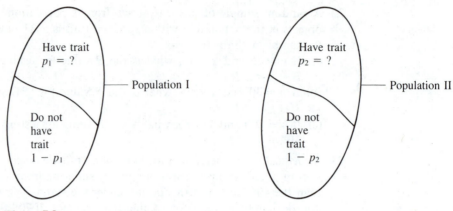

Figure 7.9

Similarly, we estimate p_2 by

$$\frac{X_2}{n_2} = \frac{\text{number in sample two with the trait}}{\text{size of sample two}}$$

the sample proportion based on an independent random sample of size n_2 drawn from the second population. Common sense suggests

$$\frac{X_1}{n_1} - \frac{X_2}{n_2}$$

as a point estimator for $p_1 - p_2$. That is,

$$\widehat{p_1 - p_2} = \frac{X_1}{n_1} - \frac{X_2}{n_2}$$

To illustrate the idea, let us consider a simple example from the field of social psychology.

Example 1   A study is to be run to ascertain the effect of prestige on changing attitudes. Two populations are of interest: persons whose opinion is contradicted by an individual of unknown background and persons whose opinion is contradicted by an individual of prestige. The problem is to compare the proportion of persons in each population that changes opinion after being contradicted. Psychologists conduct the study as follows: $n_1 = 50$ subjects are randomly selected and asked to evaluate a painting done by an obscure artist. Their opinions are noted. Each is then allowed to read an evaluation of the painting that differs markedly from the subject's and is credited to Pablo Picasso. Each subject is then asked to reevaluate the painting, and it is found that $x_1 = 40$ have changed their opinion. A similar procedure is followed with $n_2 = 60$ subjects, except that the critic is identified as J. Smith, an art major at a local university. In this case, it is found that $x_2 = 30$ have changed their opinion. From this experiment, we obtain the following estimates:

$$\hat{p}_1 = x_1/n_1 = 40/50 = .80 \qquad \hat{p}_2 = x_2/n_2 = 30/60 = .5$$

$$\widehat{p_1 - p_2} = (x_1/n_1) - (x_2/n_2) = .80 - .50 = .30$$

The next theorem shows that the procedure just used to estimate the difference between p_1 and p_2 is not only logical, but is also mathematically sound. The theorem states that the estimator $(X_1/n_1) - (X_2/n_2)$ is unbiased for $p_1 - p_2$. That is, the observed values of the estimator fluctuate about the true difference in population proportions. Furthermore, the theorem states that the variance of the estimator is $p_1(1 - p_1)/n_1 + p_2(1 - p_2)/n_2$. Since the numerator of each of the terms of this sum is constant, as the sample sizes increase, the variance of the estimator will decrease. This gives us the assurance that for reasonably large values of n_1 and n_2 we can expect the point estimate $\widehat{p_1 - p_2}$ to lie close to the actual value of $p_1 - p_2$. The theorem also is useful in that it will provide the basis for constructing confidence intervals on $p_1 - p_2$.

Theorem 7.2

Let X_1/n_1 and X_2/n_2 be the sample proportions based on independent random samples of sizes n_1 and n_2 drawn from populations with parameters p_1 and p_2, respectively. $(X_1/n_1) - (X_2/n_2)$ is an unbiased estimator for $p_1 - p_2$ and

$$\text{Var}[(X_1/n_1) - (X_2/n_2)] = p_1(1 - p_1)/n_1 + p_2(1 - p_2)/n_2$$

We are again faced with the problem of determining exactly how much confidence can be placed in a point estimate. In Example 1 we generated a single point estimate for the difference in the proportion of persons changing their opinion after being contradicted by a person of prestige and the proportion changing their opinion after being contradicted by a relatively unknown person, namely $\widehat{p_1 - p_2} = .30$. We know that this estimate is unbiased. Since the sample sizes used are fairly large, this estimate should be close to the true difference between p_1 and p_2. How close is this value really? We need to be able to extend the estimate to an interval estimate so that we will have some idea of its accuracy.

To be more precise concerning the value of $p_1 - p_2$, we need to construct a $100(1 - \alpha)\%$ confidence interval on this difference. We look for two statistics L_1 and L_2 such that

$$P[L_1 \le p_1 - p_2 \le L_2] = 1 - \alpha$$

regardless of the actual value of the difference $p_1 - p_2$. To find such statistics we must, as in the past, find a random variable whose expression involves the estimated parameter, in this case $p_1 - p_2$, and whose distribution is known at

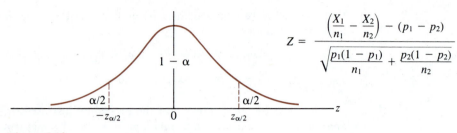

Figure 7.10
Partition of the standard normal curve needed to construct a $100(1 - \alpha)\%$ confidence interval on $p_1 - p_2$.

least approximately. Using techniques of mathematical statistics, we can show that the random variable $(X_1/n_1) - (X_2/n_2)$ is approximately normal for large values of n_1 and n_2. We may therefore use the results of Theorem 7.2 to standardize this variable by subtracting its mean of $p_1 - p_2$ and dividing by its standard deviation of

$$\sqrt{p_1(1 - p_1)/n_1 + p_2(1 - p_2)/n_2}$$

to conclude that the random variable

$$\frac{[(X_1/n_1) - (X_2/n_2)] - (p_1 - p_2)}{\sqrt{p_1(1 - p_1)/n_1 + p_2(1 - p_2)/n_2}}$$

is approximately standard normal. In constructing the $100(1 - \alpha)\%$ confidence interval on $p_1 - p_2$, we then deal with the standard normal curve partitioned as shown in Figure 7.10.

Using the usual procedure for generating $100(1 - \alpha)\%$ confidence intervals, we write down a probability statement that can be set equal to $1 - \alpha$. Such a statement is

$$P[-z_{\alpha/2} \le Z \le z_{\alpha/2}] = 1 - \alpha$$

or, in this case

$$P\left[-z_{\alpha/2} \le \frac{[(X_1/n_1) - (X_2/n_2)] - (p_1 - p_2)}{\sqrt{p_1(1 - p_1)/n_1 + p_2(1 - p_2)/n_2}} \le z_{\alpha/2}\right] = 1 - \alpha$$

We algebraically isolate $p_1 - p_2$ in the middle of the above inequality to obtain

$$P\left[\left(\frac{X_1}{n} - \frac{X_2}{n_2}\right) - z_{\alpha/2}\sqrt{\frac{p_1(1 - p_1)}{n_1} + \frac{p_2(1 - p_2)}{n_2}} \le p_1 - p_2\right.$$

$$\left. \le \left(\frac{X_1}{n_1} - \frac{X_2}{n_2}\right) + z_{\alpha/2}\sqrt{\frac{p_1(1 - p_1)}{n_1} + \frac{p_2(1 - p_2)}{n_2}}\right] = 1 - \alpha$$

As was the case when dealing with a single proportion, the endpoints above are not statistics because they involve the unknown parameters p_1 and p_2. This problem can be overcome by means of either suggestion given in Section 7.1. Both suggestions are recapitulated below:

Confidence Intervals on $p_1 - p_2$

1. Replace p_1 and p_2 by their unbiased estimators to obtain

$$\left(\frac{X_1}{n_1} - \frac{X_2}{n_2}\right) \pm z_{\alpha/2} \sqrt{\frac{\frac{X_1}{n_1}\left(1 - \frac{X_1}{n_1}\right)}{n_1} + \frac{\frac{X_2}{n_2}\left(1 - \frac{X_2}{n_2}\right)}{n_2}}$$

2. Replace both $p_1(1 - p_1)$ and $p_2(1 - p_2)$ by 1/4 to obtain

$$\left(\frac{X_1}{n_1} - \frac{X_2}{n_2}\right) \pm z_{\alpha/2} \sqrt{\frac{1}{4n_1} + \frac{1}{4n_2}}$$

Recall that, in general, the second interval will be longer than the first. Either interval can be used. The choice is yours!

Example 2

To illustrate the use of method 1, let us construct a 95% confidence interval on the difference in the proportion of persons changing their opinion after being contradicted by a person of prestige and the proportion changing their opinion after being contradicted by a relatively unknown person. We know from Example 1 that $\widehat{p_1 - p_2} = .8 - .5 = .3$. From the standard normal table, $z_{\alpha/2} = z_{.025} = 1.96$. The confidence interval, based on method 1, is given by

$$.3 \pm 1.96\sqrt{.8(.2)/50 + .5(.5)/60}$$

or [.1318, .4682]. Note that, since this interval does not contain 0 and is entirely positive, we can conclude that $p_1 > p_2$. The prestige of the critic does appear to have an effect on the subjects.

Exercises 7.3

26. Use method 1 to find these confidence intervals:
    (a) a 95% confidence interval on $p_1 - p_2$ when $x_1 = 10$, $n_1 = 50$, $x_2 = 12$, $n_2 = 48$
    (b) a 90% confidence interval on $p_1 - p_2$ when $x_1 = 20$, $n_1 = 40$, $x_2 = 10$, $n_2 = 25$

27. A national survey is conducted to compare the percentage of clergy favoring the ordination of women to the priesthood, p_1, to the percentage of laymen in favor of such a move, p_2. Samples of size 1,000 are selected from each population. It is found that 800 clergy and 650 laymen are in favor of the ordination of women. Construct a 96% confidence interval on $p_1 - p_2$ based on these data using method 1. Is there evidence that one of these proportions exceeds the other? Explain.

28. There has been some talk at the local university concerning a proposal to turn all dorms into coed dorms. The office of student affairs is interested in

determining the proportions of male and female students favoring this proposal, and in determining the difference, if any, in these proportions between
the two groups. A random sample of 50 males reveals 35 in favor of the
proposal, whereas a random sample of 75 females yields 45 in favor.
- (a) Find a point estimate for the proportion of males in favor; the proportion of females in favor.
- (b) Find a point estimate for the difference between the proportion of males
favoring the proposal and the proportion of females in favor.
- (c) Find a 95% confidence interval for the difference in part (b). Use
method 1.
- (d) Do you feel that it is safe, on the basis of your result in part (c), to claim
that the proportion of males favoring the proposal is higher than that of
females? Explain briefly.

29. Use method 2 to find the confidence intervals in parts (a) and (b):
- (a) a 95% confidence interval on $p_1 - p_2$ when $x_1 = 10$, $n_1 = 50$, $x_2 = 12$,
$n_2 = 48$
- (b) a 90% confidence interval on $p_1 - p_2$ when $x_1 = 20$, $n_1 = 40$, $x_2 = 10$,
$n_2 = 25$
- (c) Should the intervals found in parts (a) and (b) be longer or shorter than
those found in exercise 26? Compare the intervals numerically to verify
your answer.

30. In an experiment designed to study the effect of fear as a behavior modifier,
psychologists performed the following experiment: 400 randomly selected
students were randomly divided into two groups of 200 each. Each group
was urged to get a swine flu shot. Group I was shown slides and given a gory
verbal description of the effects of swine flu, the high degree of contagion of
the disease, and the danger of death resulting from contracting the disease.
The presentation was made in such a way that it was extremely frightening.
Group II was simply given a brochure describing the disease, and no attempt
was made to induce fear in the subjects. Of the 200 subjects in group I, 44
elected to receive the vaccine, whereas 38 of those in group II did so.
- (a) Find a point estimate for p_1, the proportion of students that would be
inoculated if the frightening approach were used as an inducement; find a
point estimate for p_2, the proportion inoculated using the low-key
approach as an inducement.
- (b) Find a point estimate for $p - p_2$.
- (c) Find a 94% confidence interval on $p_1 - p_2$. Use method 2.
- (d) On the basis of your answer to part (c), do you feel that there is really
much difference in the results obtained? Explain.
- (e) If it can be safely assumed that students react in much the same way as
the general public, what method of presentation would you suggest that
public health officials use in trying to persuade the general public to take
a flu shot? Why?

31. The produce manager of a large supermarket has two possible suppliers of fresh tomatoes. Both offer about the same financial deal. One of the biggest problems is that a large number of tomatoes are damaged in shipping. To decide which supplier to use, the manager wants to investigate the proportions p_1 and p_2 of damaged tomatoes received from each supplier. Over a period of several months the following data are obtained:

	Supplier I	Supplier II
Total shipped	2000	1600
Damaged	300	160

  (a) Find a 96% confidence interval on $p_1 - p_2$. Use method 2.
  (b) On the basis of this interval, do you feel that there is enough difference between the two to justify discontinuing business with one or the other supplier? Explain.

32. In a statewide poll of 2,000 men and 2,010 women, 980 men and 1,025 women report that they are opposed to the death penalty in all circumstances.
  (a) Find a 90% confidence interval for the difference in the proportion of men opposed to the death penalty and the corresponding proportion among women. Use method 2.
  (b) Do you feel that there is substantial evidence that a real difference exists? Explain.

33. Use the table of random digits, Table IV of Appendix A, to select a random sample of 10 males and 10 females from the patients listed in Table V of Appendix A.
  (a) Find point estimates for p_1 and p_2, the proportion of males and the proportion of females whose systolic blood pressure exceeds 120, respectively.
  (b) Find a 95% confidence interval on $p_1 - p_2$. Is there evidence of a difference in proportions for these groups?

7.4   Comparing Two Proportions: Hypothesis Testing

If there is a preconceived notion of the relationship between two proportions, and the purpose of the experiment is to gain evidence to support this notion, then a test of hypothesis is in order. The hypothesized difference between proportions can be any value whatsoever. Following our usual notational convention, we denote the null value by $(p_1 - p_2)_0$. We are then testing hypotheses of these forms:

I $H_0: p_1 - p_2 = (p_1 - p_2)_0$ II $H_0: p_1 - p_2 = (p_1 - p_2)_0$

   $H_1: p_1 - p_2 \neq (p_1 - p_2)_0$ $H_1: p_1 - p_2 < (p_1 - p_2)_0$

   Two-tailed test   Left-tailed test

III $H_0: p_1 - p_2 = (p_1 - p_2)_0$
     $H_1: p_1 - p_2 > (p_1 - p_2)_0$
        Right-tailed test

To develop a logical test statistic, consider the approximately standard normal random variable

$$\frac{[(X_1/n_1) - (X_2/n_2)] - (p_1 - p_2)}{\sqrt{p_1(1 - p_1)/n_1 + p_2(1 - p_2)/n_2}}$$

used to derive confidence intervals on $p_1 - p_2$. Since we are interested in the distribution of the test statistic under the assumption that the null value is correct, we replace $p_1 - p_2$ with the null value $(p_1 - p_2)_0$. Since we do not know the true values of p_1 and p_2, we replace these proportions with their estimators X_1/n_1 and X_2/n_2, respectively. In this way we obtain the approximately standard normal test statistic

$$\frac{[(X_1/n_1) - (X_2/n_2)] - (p_1 - p_2)_0}{\sqrt{(X_1/n_1)(1 - X_1/n_1)/n_1 + (X_2/n_2)(1 - X_2/n_2)/n_2}}$$

The use of this statistic is demonstrated in Example 1.

Example 1   A business firm is comparing two copiers. Copier I costs more than copier II. The makers of copier I argue that the proportion of acceptable copy produced by their machine exceeds that of their competitor by more than .10. If this can be substantiated statistically, then machine I will be more economical in the long run and will be purchased; otherwise, machine II will be purchased. Tests of the two machines yields these data:

$n_1 = 1,000$ copies run	$n_2 = 900$ copies run
$x_1 = 900$ acceptable copies	$x_2 = 711$ acceptable copies

Do these data support the contention of the makers of copier I?

Here, the null value is $(p_1 - p_2)_0 = .10$. To gain support for the contention of the makers of copier I, we test

$H_0: (p_1 - p_2)_0 = .10$

$H_1: (p_1 - p_2) > .10$

Let us preset α at .05. In this way, if we buy copier I we can be very sure that we are not making an error. The critical point for this right-tailed test is 1.65. The estimates for p_1, p_2 and $p_1 - p_2$ are

$\hat{p}_1 = x_1/n_1 = 900/1000 = .90$

$\hat{p}_2 = x_2/n_2 = 711/900 = .79$

$\widehat{p_1 - p_2} = .90 - .79 = .11$

Note that the estimated difference between p_1 and p_2 is .11. Is this enough larger than the null value of .10 to declare that $p_1 - p_2 > .10$, or could this difference have occurred reasonably by chance even though the true difference in proportions is only .10? To answer this question we test H_0. The observed value of the test statistic is

$$\frac{[(X_1/n_1) - (X_2/n_2)] - (p_1 - p_2)_0}{\sqrt{(X_1/n_1)(1 - X_1/n_1)/n_1 + (X_2/n_2)(1 - X_2/n_2)/n_2}}$$

$$= \frac{(.90 - .79) - (.10)}{\sqrt{.90(.10)/1000 + .79(.21)/900}}$$

$$= .604$$

Since this value does not exceed the critical point 1.65, we are unable to reject H_0. We do not have sufficient evidence to support the claim of the makers of copier I. Due to the differences in initial costs, we will purchase copier II.

Although the hypothesized difference between proportions can be any realistic value, the most commonly encountered null value is zero. In this case we will be concerned with testing hypotheses of these forms:

I $H_0: p_1 = p_2$	II $H_0: p_1 = p_2$	III $H_0: p_1 = p_2$
$H_1: p_1 \neq p_2$	$H_1: p_1 < p_2$	$H_1: p_1 > p_2$
Two-tailed test	Left-tailed test	Right-tailed test

To test these hypotheses, we can use the test procedure just developed with $(p_1 - p_2)_0$ assuming the value zero. Some statisticians do this. However, others prefer to use a **pooled** procedure. To see what this entails, recall that the variance of the random variable $(X_1/n_1) - (X_2/n_2)$ is

$$p_1(1 - p_1)/n_1 + p_2(1 - p_2)/n_2$$

If the null hypothesis is true, $p_1 = p_2$. We can denote this common population proportion by p. Thus, if the null hypothesis is true, we can substitute p for both p_1 and p_2 to obtain this expression for the variance of $(X_1/n_1) - (X_2/n_2)$:

$$p(1 - p)/n_1 + p(1 - p)/n_2 = p(1 - p)(1/n_1 + 1/n_2)$$

The Z random variable that we dealt with previously now assumes the form

$$\frac{(X_1/n_1) - (X_2/n_2) - 0}{\sqrt{p(1 - p)(1/n_1 + 1/n_2)}}$$

Our problem is obvious. We need a way to estimate the common population proportion p. This is easy to do. If p_1 and p_2 are really identical, then we have two samples drawn from the same population. Let us combine or pool these samples into one large sample and estimate p in the usual way. When we do this, we obtain this pooled estimator for p:

$$\hat{p} = \frac{X_1 + X_2}{n_1 + n_2} = \frac{\text{number in the combined sample with the trait}}{\text{size of the combined sample}}$$

Replacing the unknown common population proportion p by its estimator, we obtain this Z test statistic for testing $H_0: p_1 = p_2$:

$$\frac{(X_1/n_1) - (X_2/n_2)}{\sqrt{\hat{p}(1 - \hat{p})(1/n_1 + 1/n_2)}}$$

The test is conducted in the usual manner. This sounds complicated but it is not difficult. A few numerical examples should clarify things.

Example 2

There has been some concern recently that air traffic controllers (population I), because of their constant exposure to radar, have a higher incidence of eye cataracts than does the general public (population II). To gain statistical support for this theory, we will test

$$H_0: p_1 = p_2$$

$$H_1: p_1 > p_2$$

at the $\alpha = .05$ level. The critical point for such a test is $z_{.05} = 1.65$. To conduct the study, 100 former air traffic controllers are randomly selected and interviewed. It is found that $x_1 = 6$ of them have developed cataracts in one or both eyes. In a random sample of 200 persons not overly exposed to radar, it is found that $x_2 = 7$ have developed cataracts. For these data,

$$\hat{p} = \frac{x_1 + x_2}{n_1 + n_2} = \frac{6 + 7}{100 + 200} \approx .04$$

The observed value of the test statistic is

$$\frac{[(x_1/n_1) - (x_2/n_2)] - 0}{\sqrt{\hat{p}(1 - \hat{p})(1/n_1 + 1/n_2)}} = \frac{6/100 - 7/200}{\sqrt{.04(.96)(1/100 + 1/200)}} = 1.04$$

Since this value does not exceed the critical point of 1.65, we are unable to reject H_0. These data do not allow us to conclude that air traffic controllers tend to be more susceptible to cataracts than the general population.

Example 3

Opponents of the Equal Rights Amendment claim that the proportion of males (population I) opposing the amendment is smaller than the proportion of females (population II) in opposition. A study is conducted to test

$$H_0: p_1 = p_2$$

$$H_1: p_1 < p_2$$

In running the test, a random sample of 500 males reveals 275 opposed to the legislation; a random sample of 520 females reveals 316 opposed. What conclusion can be drawn? The pooled estimate for the common population proportion is

$$\hat{p} = \frac{x_1 + x_2}{n_1 + n_2} = \frac{275 + 316}{500 + 520} \approx .58$$

The observed value of the test statistic is

$$\frac{[(x_1/n_1) - (x_2/n_2)] - 0}{\sqrt{\hat{p}(1 - \hat{p})(1/n_1 + 1/n_2)}} = \frac{[(275/500) - (316/520)] - 0}{\sqrt{.58(.42)(1/500 + 1/520)}} = -1.87$$

From the standard normal table, we see that the P value, $P[Z \le -1.87]$, is .0307. Since this value is small, we can reject H_0. There is evidence that the proportion of males opposing the amendment is smaller than the proportion of females in opposition.

Remember that you can use either P values or preset α to test H_0. Also remember that if the null value is not zero you must use the first statistic discussed here. If the null value is zero, you may pool or not, as you choose. However, we recommend that you do pool.

Exercises 7.4

34. In each case, use the data given to test the stated hypothesis:
    (a) $H_0: p_1 - p_2 = \quad .05, \quad x_1 = 20, \quad x_2 = 32, \quad \alpha = .05$
    $\quad H_1: p_1 - p_2 > \quad .05, \quad n_1 = 50, \quad n_2 = 100,$
    (b) $H_0: p_1 - p_2 = \quad .04, \quad x_1 = 5, \quad x_2 = 5, \quad \alpha = .10$
    $\quad H_1: p_1 - p_2 < \quad .04, \quad n_1 = 25, \quad n_2 = 30,$
    (c) $H_0: p_1 - p_2 = -.10, \quad x_1 = 10, \quad x_2 = 29, \quad \alpha = .05$
    $\quad H_1: p_1 - p_2 > -.10, \quad n_1 = 20, \quad n_2 = 50,$
    (d) $H_0: p_1 - p_2 = \quad .07, \quad x_1 = 12, \quad x_2 = 15, \quad \alpha = .10$
    $\quad H_1: p_1 - p_2 \neq \quad .07, \quad n_1 = 30, \quad n_2 = 30$

35. An automobile manufacturer is investigating the possibility of installing robots to paint the trim on cars as they pass through the assembly line. To be worthwhile it is thought that the robots must reduce the defective rate by at least 2%. Let p_1 denote the proportion of defective paint jobs produced by human workers and let p_2 denote the proportion of defective paint jobs produced by robots in test runs.
    (a) Set up the null and alternative hypotheses needed to get statistical evidence to support the contention that purchasing robots is worthwhile.
    (b) Based on these data, can H_0 be rejected at the $\alpha = .05$ level?

$$x_1 = 25 \qquad x_2 = 2$$
$$n_1 = 500 \qquad n_2 = 100$$

36. Makers of a nonprescription drug for preventing motion sickness are testing a new formula for the drug. If it can be shown that the proportion of users for whom the new formula is effective (p_1) exceeds that for the old formula (p_2) by more than .05, the new formula will be marketed. Otherwise, it will not be marketed.
    (a) Set up the null and alternative hypotheses needed to gain statistical evidence that the new formula should be marketed.

(b) Based on these data and the P value of the test, do you recommend that the new drug be marketed?

$$x_1 = 80 \qquad x_2 = 360$$

$$n_1 = 100 \qquad n_2 = 500$$

37. In each case, use the data given to test the stated hypothesis:
   (a) $H_0: p_1 = p_2,$ $x_1 = 18,$ $x_2 = 15,$ $\alpha = .05$
       $H_1: p_1 > p_2,$ $n_1 = 30,$ $n_2 = 30,$
   (b) $H_0: p_1 = p_2,$ $x_1 = 4,$ $x_2 = 4,$ $\alpha = .10$
       $H_1: p_1 < p_2,$ $n_1 = 25,$ $n_2 = 20,$
   (c) $H_0: p_1 = p_2,$ $x_1 = 35,$ $x_2 = 55,$ $\alpha = .05$
       $H_1: p_1 \neq p_2,$ $n_1 = 70,$ $n_2 = 100$

38. Legislation is introduced allowing voters to register on the day of an election. Some observers feel that this is a political maneuver and that the proportion of Democrats in favor of the legislation is higher than the proportion of Republicans in favor.
   (a) Set up the null and alternative hypotheses appropriate for gaining evidence to support this contention.
   (b) What conclusion can be drawn at the $\alpha = .10$ level if, in a random sample of 1,000 Democrats, 450 favor the legislation, and, in a random sample of 1,000 Republicans, 440 are in favor?

39. The director of athletics at a university wants to gain statistical support for the contention that the failure rate, p_1, among varsity athletes is smaller than the rate p_2 for nonathletes.
   (a) Set up the appropriate hypotheses for gaining this support.
   (b) The following data are obtained:

	Athletes	Nonathletes
Total sampled	150	200
Flunked out	30	43

   At the $\alpha = .05$ level, what conclusion can be drawn?
   (c) What type of error may be committed?

40. Researchers feel that eating citrus fruits during the winter months reduces the rate of contracting colds. To test this theory, 200 subjects were randomly divided into a control group and an experimental group each of size 100. Each member of the experimental group ate three oranges a day during the months of November through March. The control group followed a normal diet. At the end of this period the following data were obtained.

	Control	Experimental
Sampled	100	100
Number contracting at least one cold	48	43

(a) Set up the appropriate hypotheses for verifying the theory.

(b) Test the theory at the $\alpha = .10$ level.

(c) Discuss the practical consequences of the error that may be involved due to the action taken on the basis of the conclusion drawn in part (b).

41. Preliminary studies indicate that high blood pressure is more prevalent among blacks than it is among whites. To gather evidence to support this contention, random samples of size 1,000 each were obtained from these two groups and examined. Among blacks, 220 cases of high blood pressure were discovered, whereas only 197 cases were found among whites. Do you feel that this is enough evidence to support the claim? Support your answer briefly on the basis of the P value of the test.

42. There is some disagreement concerning the treatment of tonsillitis. Some doctors favor treatment with antibiotics while others favor surgery. It is felt that opinion concerning this problem is probably regional and that the percentage of doctors favoring surgery is not the same in the South as it is in the Northeast.

(a) Let p_1 represent the percentage of doctors in the South favoring surgery and p_2 the percentage of doctors in the Northeast in favor of this type of treatment. Set up the null and alternative hypotheses needed to gain evidence that these percentages are not the same.

(b) Random samples of 100 doctors selected from each region yield 30 in the South favoring surgery and 40 in the Northeast in favor of surgery. Based on these data should H_0 be rejected? Explain, based on the P value of the test.

43. A political scientist is interested in obtaining statistical data to support the contention that there are basic differences between Republicans and Democrats concerning the role of the Federal Government in health care. In particular, interest centers on the percentage of Democrats, p_1, and the percentage of Republicans, p_2, in favor of socialized medicine.

(a) Set up the null and alternative hypotheses for this two-tailed test.

(b) A random sample of 600 registered Democrats reveals 420 in favor of socialized medicine, whereas a random sample of 400 Republicans reveals 240 in favor. Do these data allow us to reject H_0? Explain, based on the P value of the test. Remember that the P value for a two-tailed test is double that of the apparent one-tailed probability.

Vocabulary List and Key Concepts

population proportion

sample proportion

pooled sample proportion

Review Exercises

44. What is the point estimator for p, the proportion of objects in a population with a given trait? Is this estimator unbiased for p?

45. What are the bounds for the two types of confidence intervals on p? For the same data, which of these usually produces longer intervals? When will they be of the same length?

46. What formula is used to determine the sample size needed to estimate a proportion to within a specified distance with a specified degree of accuracy?

47. Before beginning a limited bus service in a small town, the city council wants to estimate p, the proportion of its adult citizens in favor of the proposal.
    (a) How large a sample is required to estimate p to within .05 with 90% confidence?
    (b) When the survey is conducted, it is found that 90 of the 300 persons surveyed favor the service. Based on these data, construct a 90% confidence interval on p using method 1.
    (c) Based on the confidence interval found in part (b), would you be surprised to hear a claim that the proportion of adults in favor of the service is less than a majority? Explain.

48. In response to a government inquiry, the manager of a supermarket wants to estimate p, the proportion of sales that entailed the use of food stamps.
    (a) How large a sample is required to estimate p to within .03 with 95% confidence?
    (b) Of 1,150 sales sampled, it is found that 460 entailed the use of food stamps. Use method 2 to find a 95% confidence interval on p.

49. What is the test statistic for testing $H_0: p = p_0$? What is the approximate distribution of this test statistic when H_0 is true?

50. During the month of January, the proportion of citizens that is satisfied with the manner in which the President of the United States is performing his duties is estimated to be .60. Six months later, a new poll is conducted. It is thought that the proportion of satisfied citizens has increased due to an upsurge in the economy.
    (a) Set up the null and alternative hypotheses needed to support the research hypothesis.
    (b) When the new poll is conducted, it is found that 1,600 of the 2,500 persons interviewed are satisfied with the performance of the President. Do these data support the research hypothesis? Explain based on the P value of your test.

51. What is the point estimator for the difference in population proportions, $p_1 - p_2$? Is this estimator unbiased for $p_1 - p_2$?

52. What are the bounds for the two types of confidence intervals on $p_1 - p_2$?

53. In a study of retention of college students, a comparison is to be made

between the proportion of male (p_1) and the proportion of female (p_2) students who return to school following their freshman year. In sampling records over a five-year period, these data result:

$$x_1 = 350 \qquad x_2 = 408$$

$$n_1 = 500 \qquad n_2 = 600$$

Construct a 95% confidence interval on $p_1 - p_2$ using method 2. Based on this interval, can you conclude that the proportion of men returning clearly exceeds the proportion of women that return? Explain.

54. In comparing two proportions via hypothesis testing, we introduced two test statistics. What are these statistics? Under what circumstances would each be used?

55. What is the pooled estimator for a common population proportion?

56. It is thought that men and women approach the act of buying a car differently. Researchers think that men are more likely than women to return to a dealership after the first contact is made. Do these data support the contention that the proportion of males who return for a second visit exceeds the proportion of women who do so? Explain, based on the P value of the appropriate test.

Men	Women
$x_1 = 75$	$x_2 = 28$
$n_1 = 100$	$n_2 = 40$

57. NCAA officers claim that college basketball officials are getting better each year. To substantiate this claim, randomly selected game films for the last two years are selected and reviewed. The random variable studied is the number of games in which five or more officiating errors are detected. These data result:

Year 1	Year 2
$x_1 = 6$	$x_2 = 4$
$n_1 = 50$	$n_2 = 50$

(a) Set up the null and alternative hypotheses needed to substantiate the claim.
(b) What is the critical point for an $\alpha = .05$ level test of H_0?
(c) Can H_0 be rejected?

58. In medicine, changing from one form of treatment to another on a massive scale is usually very expensive. Before such a change is suggested, researchers want to be fairly certain that the increase in successful treatments is substantial. Suppose that a particular change will be recommended only if the proportion of successes using the new treatment exceeds the old by more than

.10. Do these data support the contention that a change to the new treatment is in order? Explain, based on the P value of the appropriate test.

New	Old
$x_1 = 15$	$x_2 = 60$
$n_1 = 20$	$n_2 = 100$

8

Comparison of Two Means

In the previous chapter we considered the problem of comparing two proportions. We are now faced with a similar situation in which emphasis is focused on the means μ_1 and μ_2 of two given populations. We shall consider methods for doing such a comparison under three distinct situations.

1. When the population variances σ_1^2 and σ_2^2 are equal.
2. When the population variances σ_1^2 and σ_2^2 are unequal.
3. When the data are paired.

As can be seen, it is necessary first to compare variances so that the proper method of analysis can be employed. We thus begin this chapter by considering methods by which two population variances may be compared based on independent samples drawn from these populations.

8.1 Comparison of Two Variances and the F Distribution

Often it is of interest to compare the variance of a variable X to that of a variable Y. We have seen examples of this in Chapter 3. However, in that chapter we dealt with situations in which the probability functions for the variables are known, and we were actually computing σ_X^2 and σ_Y^2 to make the comparison. In this section, we consider methods by which σ_X^2 and σ_Y^2 can be compared when all that is available is a random sample drawn independently from each of the distributions. The ability to make such a comparison is especially useful since it helps determine the choice of the test statistic to be used in testing hypotheses concerning the relationship between the means of the two variables.

Example 1

1. Two bowlers, Marcie and Peg, try out for an opening on the Wayside team; each has a 160 average. Since consistency is important in bowling, the bowler with the smallest variability in scores is most useful to the team. Let X denote the score bowled by Marcie and Y that bowled by Peg. We are interested in comparing σ_X^2 with σ_Y^2.

2. A horticulturist is interested in developing a strain of yew that will produce hedges of fairly uniform height when they are allowed to grow without trimming. He is experimenting with two different cross-matches and wishes to select the one that exhibits the smaller variability in growth.

3. A psychologist is interested in studying the influence that TV commercials have on the buying habits of viewers. The study will be based on either commercials advertising patent medicines or commercials advertising food products. For the study, the psychologist wants to use commercials of varying lengths and therefore needs to determine which population of commercials has the higher variability in run time.

The general situation is as follows: we have available two populations. Associated with the first is a random variable X with unknown variance $\sigma_X^2 = \sigma_1^2$; associated with the second is a random variable Y with unknown variance $\sigma_Y^2 = \sigma_2^2$. We have available independent random samples of sizes n_1 and n_2 as illustrated in Figure 8.1. We assume throughout that X and Y are normal or at least approximately so. We wish to compare σ_1^2 and σ_2^2.

Since S_1^2 is unbiased for σ_1^2 and S_2^2 is unbiased for σ_2^2, these statistics must surely come into play in making comparisons. Let us consider the statistic

$$S_1^2/S_2^2$$

Note that if $\sigma_1^2 = \sigma_2^2$, we would expect S_1^2 and S_2^2 to be close in value, thus forcing the ratio S_1^2/S_2^2 to be close in value to 1. If $\sigma_1^2 > \sigma_2^2$, we would expect S_1^2 to be greater than S_2^2 resulting in a value somewhat larger than 1. Similarly a small

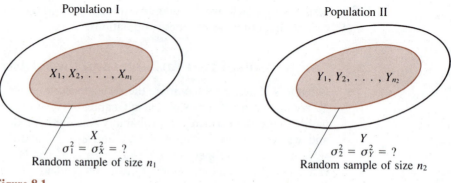

Population I

$X_1, X_2, \ldots, X_{n_1}$

X
$\sigma_1^2 = \sigma_X^2 = ?$
Random sample of size n_1

Population II

$Y_1, Y_2, \ldots, Y_{n_2}$

Y
$\sigma_2^2 = \sigma_Y^2 = ?$
Random sample of size n_2

Figure 8.1

value for S_1^2/S_2^2 is evidence that $\sigma_1^2 < \sigma_2^2$. Thus S_1^2/S_2^2 is a logical point estimator for σ_1^2/σ_2^2.

Example 2 | The following data were obtained on the length of time in minutes that it took to run 12 randomly selected commercials advertising food products and 10 randomly selected commercials advertising patent medicines:

Food (X)			Medicine (Y)		
1.7	.6	.9	1.3	1.1	2.0
1.5	1.0	2.6	1.1	.9	.5
2.0	.3	1.1	.3	.7	1.4
.8	2.1	1.2	1.3		

To compare σ_1^2, the variance in commercial length for commercials advertising food, to that of σ_2^2, the variance in commercial length for those of patent medicines, we compute the sample variances for the above data sets. In this case we obtain:

Food	Medicine
$\Sigma x = 15.8$	$\Sigma y = 10.6$
$\Sigma x^2 = 25.86$	$\Sigma y^2 = 13.4$
$s_1^2 = .46$	$s_2^2 = .24$

The point estimate for σ_1^2/σ_2^2 is

$$\widehat{\sigma_1^2/\sigma_2^2} = s_1^2/s_2^2 = .46/.24 = 1.92$$

This number is somewhat greater than 1, indicating perhaps that $\sigma_1^2 > \sigma_2^2$. There is an obvious question. Is there enough difference between the numbers 1 and 1.92 to conclude that, in fact $\sigma_1^2 > \sigma_2^2$? To answer this question, we must once again turn to the methods of interval estimation or hypothesis testing.

To construct $100(1 - \alpha)\%$ confidence intervals on σ_1^2/σ_2^2 or test hypotheses concerning the value of this ratio, it is necessary to pause briefly to introduce a fourth continuous probability distribution. This distribution, called *Snedecor's F*, or just the *F distribution*, plays a vital role in applied statistics.

The F random variable is a random variable that is in fact a quotient of two independent chi-square random variables divided by their respective degrees of freedom. Any F random variable is therefore identified by two parameters, each called *degrees of freedom*, one associated with the chi-square variable of the numerator, the other with the chi-square variable of the denominator. The exact definition of the F variable is given below.

Definition 8.1

F **Distribution**

Let $X^2_{v_1}$ and $X^2_{v_2}$ be independent chi-square random variables with v_1 and v_2 degrees of freedom, respectively. Then the ratio

$$\frac{X^2_{v_1}/v_1}{X^2_{v_2}/v_2}$$

follows an *F* **distribution** with v_1 and v_2 degrees of freedom.

We summarize briefly the properties of the *F* family of random variables:

Properties of *F* Random Variables

1. The *F* family is a family of continuous nonnegative random variables. Once again we deal with smooth curves and identify areas under these curves with probabilities.

2. Each *F* random variable is completely identified by specifying two parameters, v_1 and v_2, called degrees of freedom. v_1 is the number of degrees of freedom associated with the chi-square variable of the numerator and v_2 is the number of degrees of freedom associated with the denominator. v_1 and v_2 are always positive whole numbers; their values in an applied problem depend on the sample sizes in the experiment. We denote such a random variable by F_{v_1, v_2}.

3. The graph of the probability function for each *F* random variable is an asymmetric curve of the shape shown in Figure 8.2.

A partial table for the cumulative distribution function for selected *F* random variables is given in Appendix A, Table VIII. Table VIII gives for various values of v_1 and v_2, the point t such that

$$P[F_{v_1, v_2} \leq t] = .90 \quad P[F_{v_1, v_2} \leq t] = .975$$

$$P[F_{v_1, v_2} \leq t] = .95 \quad P[F_{v_1, v_2} \leq t] = .99$$

The notation $f_r(v_1, v_2)$ is used to denote the point associated with the F_{v_1, v_2} distribution with area r to the right. That is, $P[F_{v_1, v_2} \geq f_r(v_1, v_2)] = r$. The next example demonstrates the use of Table VIII.

Figure 8.2
Graph of the probability function for a typical *F* random variable.

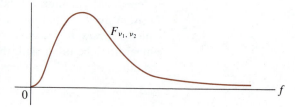

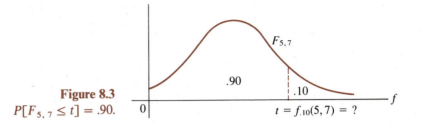

Figure 8.3
$P[F_{5,7} \leq t] = .90.$

Example 3

1. Let us find the point t such that

    $$P[F_{5,7} \leq t] = .90$$

    We are looking for the point t shown in Figure 8.3. To find this point, we go to the portion of Appendix A, Table VIII, labeled .90, look in the column headed $v_1 = 5$ and the row headed $v_2 = 7$ and read the desired number, 2.88. That is, $f_{.10}(5, 7) = 2.88$.

2. Find the point t, such that $P[F_{7,5} \leq t] = .90$. That is, find $f_{.10}(7, 5)$. This time we go to the portion of Table VIII, labeled .90, look in the column headed $v_1 = 7$ and the row headed $v_2 = 5$ and read the number, 3.37. Note that this is not the same number that was obtained before. The order in which the degrees of freedom are listed is important.

3. Find the point $f_{.90}(7, 5)$. This point is the left-tailed point shown in Figure 8.4. It cannot be read directly from Table VIII. However, it can be shown that any left-tailed point is the reciprocal of the corresponding right-tailed point with the degrees of freedom reversed. That is,

    $$f_{.90}(7, 5) = 1/f_{.10}(5, 7) = 1/2.88 = .347$$

4. Find the points a and b such that

    $$P[a \leq F_{6,9} \leq b] = .80$$

    and such that the area to the right of a is equal to the area to the left of b. These points are shown in Figure 8.5. From Table VIII, $f_{.10}(6,9) = 2.55$. Using the idea of item 3,

    $$f_{.90}(6, 9) = 1/f_{.10}(9, 6) = 1/2.96 = .34.$$

    The desired points are therefore $a = .34$ and $b = 2.55$.

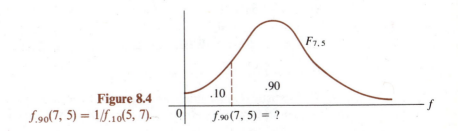

Figure 8.4
$f_{.90}(7, 5) = 1/f_{.10}(5, 7).$

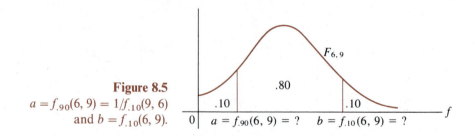

Figure 8.5
$a = f_{.90}(6, 9) = 1/f_{.10}(9, 6)$
and $b = f_{.10}(6, 9)$.

We now return to the problem that motivated our need for the F distribution, namely, that of comparing σ_1^2 with σ_2^2 via the statistic S_1^2/S_2^2. We will be concerned primarily with answering the question, Is $\sigma_1^2 = \sigma_2^2$? Therefore we will restrict our discussion to hypothesis testing. The derivation of confidence bounds for σ_1^2/σ_2^2 is outlined in exercise 10.

Hypothesis tests on the relationship between two variances take any of the usual three forms. These are

I	$H_0: \sigma_1^2 = \sigma_2^2$	II	$H_0: \sigma_1^2 = \sigma_2^2$	III	$H_0: \sigma_1^2 = \sigma_2^2$
	$H_1: \sigma_1^2 \neq \sigma_2^2$		$H_1: \sigma_1^2 < \sigma_2^2$		$H_1: \sigma_1^2 > \sigma_2^2$
	Two-tailed test		Left-tailed test		Right-tailed test

To test H_0 we need a test statistic. This statistic must be such that its distribution is known under the assumption that $\sigma_1^2 = \sigma_2^2$. That is, its distribution must be known when we assume that $\sigma_1^2/\sigma_2^2 = 1$. It is easy to derive this statistic. Recall from Theorem 6.9 that

$$(n_1 - 1)S_1^2/\sigma_1^2 \quad \text{is} \quad X_{n_1 - 1}^2$$

and

$$(n_2 - 1)S_2^2/\sigma_2^2 \quad \text{is} \quad X_{n_2 - 1}^2$$

If each of these chi-square variables is divided by its respective degrees of freedom and the ratio of the two is formed, we obtain the following random variable:

$$(S_1^2/\sigma_1^2)/(S_2^2/\sigma_2^2) = S_1^2 \sigma_2^2 / S_2^2 \sigma_1^2$$

By Definition 8.1, this random variable follows an F distribution with $v_1 = n_1 - 1$ and $v_2 = n_2 - 1$ degrees of freedom. If H_0 is true and $\sigma_1^2 = \sigma_2^2$, then $\sigma_2^2/\sigma_1^2 = 1$ and the statistic

$$S_1^2/S_2^2 \quad \text{is} \quad F_{n_1 - 1, \, n_2 - 1}$$

This statistic is appropriate as a test statistic for the above hypotheses.

To see how this statistic is used, consider Examples 4 and 5.

Example 4

A horticulturist experimenting with two cross-matches in yews wants to develop a strain with small variability in height. Suppose that he feels from rough observation that the first cross-match is superior to the second. He wishes to gain statistical evidence that his judgment is correct so that he can convince his employers to discontinue experimentation with the second match and begin to market the first. The following observations were obtained (height in feet):

Cross-match I			Cross-match II		
1.58	2.08	2.36	2.06	2.19	1.91
2.09	2.37	1.91	1.52	2.31	1.22
2.27	1.62	1.89	2.10	1.44	2.20
2.00	1.39	2.00	2.57		

Does the needed evidence exist? To decide, let us test

$$H_0: \sigma_1^2 = \sigma_2^2$$
$$H_1: \sigma_1^2 < \sigma_2^2$$

at the $\alpha = .05$ level. The test statistic S_1^2/S_2^2 is $F_{11, 9}$. The test is a left-tailed test with a critical point

$$f_{.95}(11, 9) = 1/f_{.05}(9, 11) = 1/2.90 = .34$$

We reject H_0 in favor of H_1 if the observed value of the test statistic is smaller than or equal to .34. For these observations

$$s_1^2 = .10 \quad \text{and} \quad s_2^2 = .18$$

The observed value of the test statistic is

$$s_1^2/s_2^2 = .10/.18 = .56$$

On the basis of this test, we are unable to reject H_0 at the $\alpha = .05$ level. The horticulturist does not yet have the evidence needed and should continue with the experimentation.

The P value can also be used to decide whether or not to reject a null hypothesis. The next example shows how this is done and also demonstrates a trick that makes computing a lower-tail F point unnecessary.

Example 5

These data constitute samples of bowling scores for two bowlers Peg (X) and Marcie (Y).

X				Y			
163	160	155	159	165	177	140	166
157	159	166	160	170	168	170	163
166	169	157	159	167	146	155	146
160	165			171	163	153	140

Do these data indicate that there is a difference in the variability of scores for these two bowlers? Since we have no preconceived notion as to the direction of any difference that we might discover, we want to test

$$H_0: \sigma_X^2 = \sigma_Y^2$$
$$H_1: \sigma_X^2 \neq \sigma_Y^2$$

Point estimates for these variances are $\hat{\sigma}_X^2 = 16.69$ and $\hat{\sigma}_Y^2 = 137.87$. In evaluating the test statistic S_1^2/S_2^2, either population can be considered to be population I. Since we can read right-tailed (large) F values directly from Table VIII, it is convenient to designate the larger of the two sample variances as S_1^2. In this case,

$$\hat{\sigma}_Y^2 = s_1^2 = 137.87 \quad \text{and} \quad \hat{\sigma}_X^2 = s_2^2 = 16.69$$

The observed value of the test statistic is

$$s_1^2/s_2^2 = 137.87/16.69 = 8.26$$

The P value is found by consulting Table VIII. We see that $f_{.01}(15, 13) = 3.82$. Our observed value does exceed this point. If we were conducting a right-tailed test we could reject H_0 with a P value less than .01. However, since our test is two-tailed, we take the P value to be *double* the one-tailed value. In this case, we can reject H_0 with $P < .02$.

Exercises 8.1

1. Use Table VIII of Appendix A to find:
   (a) $f_{.10}(7, 12)$
   (b) $f_{.10}(\infty, \infty)$
   (c) $f_{.05}(3, 8)$
   (d) $f_{.05}(\infty, 6)$
   (e) $f_{.95}(2, 6)$
   (f) $f_{.90}(3, 9)$
   (g) $f_{.90}(10, 10)$
   (h) Points a and b such that $P[a \leq F_{10, 9} \leq b] = .80$ and $P[F_{10, 9} \geq b] = .10$
   (i) Points a and b such that $P[a \leq F_{15, 8} \leq b] = .90$ and $P[F_{15, 8} \geq b] = .05$

2. In each case, test $H_0: \sigma_1^2 = \sigma_2^2$ based on the given information:

   (a) $H_1: \sigma_1^2 > \sigma_2^2$ $\quad s_1^2 = 16$ $\quad s_2^2 = 7.3$ $\quad \alpha = .10$
   $\quad\quad n_1 = 8$ $\quad n_2 = 13$

   (b) $H_1: \sigma_1^2 > \sigma_2^2$ $\quad s_1^2 = 20$ $\quad s_2^2 = 4.4$ $\quad \alpha = .05$
   $\quad\quad n_1 = 4$ $\quad n_2 = 9$

   (c) $H_1: \sigma_1^2 < \sigma_2^2$ $\quad s_1^2 = 8$ $\quad s_2^2 = 200$ $\quad \alpha = .05$
   $\quad\quad n_1 = 3$ $\quad n_2 = 7$

   (d) $H_1: \sigma_1^2 < \sigma_2^2$ $\quad s_1^2 = 5$ $\quad s_2^2 = 12.7$ $\quad \alpha = .10$
   $\quad\quad n_1 = 11$ $\quad n_2 = 11$

   (e) $H_1: \sigma_1^2 \neq \sigma_2^2$ $\quad s_1^2 = 18$ $\quad s_2^2 = 6.9$ $\quad \alpha = .05$
   $\quad\quad n_1 = 11$ $\quad n_2 = 10$

   (f) $H_1: \sigma_1^2 \neq \sigma_2^2$ $\quad s_1^2 = 3.6$ $\quad s_2^2 = 10$ $\quad \alpha = .05$
   $\quad\quad n_1 = 16$ $\quad n_2 = 9$

3. An investor is interested in speculating in the stock market. Two different stocks, each of which has shown a fair amount of fluctuation, are of interest. The investor wants to invest in the one that has the greater variability since it is felt this will increase the chances of a quick profit. To decide which stock to buy, a random sample of size 10 is obtained from the closing prices of these stocks over the last few months. The following observations are recorded:

Stock I		Stock II	
21.125	9.875	26.875	23.75
10.75	37.25	30.75	16.25
7.75	17.125	22.50	23.875
43.50	26.25	19.625	17.50
29.125	46.50	21.875	31.625

Test at the $\alpha = .05$ level, the hypothesis that the variance of the price of stock I is greater than that of stock II. What type of error may be involved in this situation?

4. An instructor claims that, during the spring quarter grade, patterns change and that in fact there is much more variability in grades during this quarter than there is during the rest of the year. Random samples of size 25 each are obtained from among this year's spring quarter and combined fall and winter quarter grades. Observed sample variances are $s_1^2 = .09$ and $s_2^2 = .04$, respectively. Set up and test at the $\alpha = .05$ level the appropriate hypothesis.

5. Science writers are thought to write more concisely than writers in the humanities. One measure of this difference is word length. As an experiment to compare these groups, randomly select three sentences from the first page of text in your history, literature, psychology, sociology, or philosophy book. Count and record the length of each word in each sentence. Then randomly select three sentences from the first page of text in this book and do the same. Use these data to test at the $\alpha = .10$ level the hypothesis that science writers exhibit less variability in word length than do writers in the humanities.

6. Approximate the P value for each of these tests:
   (a) $H_0: \sigma_1^2 = \sigma_2^2$ $s_1^2 = 17$ $s_2^2 = 4.25$
        $H_1: \sigma_1^2 > \sigma_2^2$ $n_1 = 10$ $n_2 = 8$
   (b) $H_0: \sigma_1^2 = \sigma_2^2$ $s_1^2 = 7$ $s_2^2 = 18.9$
        $H_1: \sigma_1^2 < \sigma_2^2$ $n_1 = 25$ $n_2 = 16$
   (c) $H_0: \sigma_1^2 = \sigma_2^2$ $s_1^2 = 4.5$ $s_2^2 = 1.5$
        $H_1: \sigma_1^2 \neq \sigma_2^2$ $n_1 = 16$ $n_2 = 16$

7. A sociologist wants to experiment with two different groups of people. The sociologist wants the age distribution within the sampled population to be as nearly alike as possible at the beginning of the experiment, both in terms of mean age and in terms of age dispersion.

(a) Set up the appropriate null and alternative hypotheses needed to detect a situation in which the variability in ages is not the same within the two populations.

(b) Random samples of size 121 are selected from each population. The sample variances for the two are found to be 225 and 289 respectively. Do these data indicate that there is a difference in the variance in ages between these groups? Explain based on the P value of your test.

8. In comparing the cost of repairs to small electrical appliances two groups of repair facilities are studied: large retail stores (population I), and smaller private repair shops (population II). It is felt that even though the mean repair costs may be the same for these two populations, there is a difference in variability.

(a) Set up the null and alternative hypotheses needed to gain statistical evidence that the population variances are not the same.

(b) Random samples of 30 appliances repaired at each of the above type of facilities are obtained. The respective sample variances were $s_1^2 = 25$ and $s_2^2 = 16$. What is the P value for the test? Do you think that H_0 should be rejected?

9. The EPA mileage rating on the highway for cars A and B is given as 30 miles to the gallon for both cars. The makers of car A claim that their auto exhibits less variability in mileage actually obtained than does car B. To support this claim, random samples of 25 cars of each type are tested. It is found that the sample variances for the mileage obtained are $s_A^2 = 1.21$ and $s_B^2 = 2.56$. Do they have the evidence they need to include the claim of less variability in their advertisements? Explain based on the P value of the test.

*10. **(Confidence Interval on σ_1^2/σ_2^2)** Show that the confidence bounds for σ_1^2/σ_2^2 are given by

$$L_1 = [1/f_{\alpha/2}(n_1 - 1, n_2 - 1)](S_1^2/S_2^2)$$
$$L_2 = [f_{\alpha/2}(n_2 - 1, n_1 - 1)](S_1^2/S_2^2)$$

(*Hint*: Consider the diagram given in Figure 8.6.) Note that $P[f_{1-\alpha/2}(n_1 - 1, n_2 - 1) \le S_1^2\sigma_2^2/S_2^2\sigma_1^2 \le f_{\alpha/2}(n_1 - 1, n_2 - 1)] = 1 - \alpha$. Isolate σ_1^2/σ_2^2 in the middle of this inequality.

*11. Find a 95% confidence interval on σ_1^2/σ_2^2 based on the data of Example 2. The experimenter wants to use commercials drawn from the population exhibiting the largest variability in run time. Which type of commercial would you suggest he use in the study? Your answer should be based on your confidence interval. (*Hint*: You can conclude that $\sigma_1^2 \ne \sigma_2^2$ only when the confidence interval constructed does *not* contain the number 1.)

*12. Two different chemicals are being tested for use on the nylon mantles used in camping lanterns. Evidence shows that each chemical produces about the

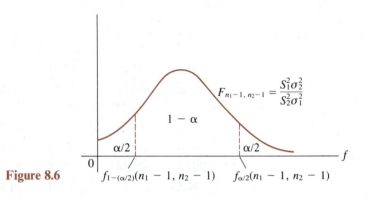

Figure 8.6

same average burn time. The one with the smaller variance is the more desirable since it produces a more dependable mantle. The following observations are obtained on the length of life of mantles treated with the respective chemicals (time in hours):

Chemical I			Chemical II		
17.12	15.23	18.06	16.87	26.91	14.74
16.02	25.45	17.78	18.40	21.66	15.59
15.28	30.04	23.13	17.24	19.83	20.77
20.77			16.06	20.10	20.61

(a) Find a point estimate for σ_1^2/σ_2^2, where σ_1^2 is the variance of the burn time for mantles produced using chemical I.
(b) Find a 90% confidence interval on σ_1^2/σ_2^2.
(c) Which chemical would you suggest using? The answer should be based on your answer to part (b). Explain briefly.

*13. There has been some speculation that college students today do not read as well as in the past because of the influence of television in their earlier school years. To obtain some comparative information, 50 college freshmen were randomly selected and randomly divided into two groups of 25 each. Each member of group I was asked to read a short illustrated story; group II viewed a film of the same story. A written test on the story was given to each subject. The sample variances obtained were $s_1^2 = 26.21$ and $s_2^2 = 27.35$. Construct a 90% confidence interval on σ_1^2/σ_2^2. Do you think that there is sufficient evidence, on the basis of this interval, to claim that $\sigma_1^2 < \sigma_2^2$? Explain.

*14. Two different resort areas use in their advertising, the statement that the average yearly temperature is 80. Since this statement can be misleading, random samples of daily temperature readings were obtained and the results were:

Area I		Area II	
83.24	84.19	65.11	66.64
85.66	75.03	84.98	66.49
81.37	92.08	82.87	69.55
78.20	79.84	96.29	84.82
73.40	68.04	81.43	85.78
81.18	79.82	77.68	83.86
77.88	81.54	75.29	81.19
80.52		85.83	84.98

Construct a 95% confidence interval on σ_1^2/σ_2^2. Can you make a statement, based on this confidence interval, concerning which area has the greater temperature variability? Explain.

*15. Two consumer research groups are vying for a large government contract. Since most subjective evaluations of consumer products will be ratings made by judges, government officials prefer to award the contract to a company that utilizes judges with consistent ratings. One measure of consistency is the variability of judges' scores on the same item. Before issuing the contract, a test is conducted in which ten judges from each company are asked to rate a single item. The sample variances are given below:

$$s_1^2 = .50 \qquad s_2^2 = .15$$

Find a 95% confidence interval on σ_1^2/σ_2^2. Is there evidence that a difference in variability exists?

8.2   Comparing Two Means: Pooled Estimation

We are concerned here with the problem of comparing the means of two populations based on information obtained from independent random samples drawn from those populations. We make the comparison by utilizing the t distribution. In this section we limit the discussion to estimation techniques.

Example 1

Two well-known pain relievers, one containing aspirin and the other containing no aspirin, advertise that they "get to the pain" in record time. To decide which brand to buy, we wish to compare the mean time that it takes to get relief from a headache when using one compound with the time required when using the other.

The general situation is as follows: we have two populations under study. Associated with the first is a random variable X with mean $\mu_X = \mu_1$ and variance $\sigma_X^2 = \sigma_1^2$; associated with the second is a random variable Y with mean $\mu_Y = \mu_2$ and variance $\sigma_Y^2 = \sigma_2^2$. The random variables X and Y are assumed to be normal or at least approximately so. We wish to compare μ_1 with μ_2 by estimating the

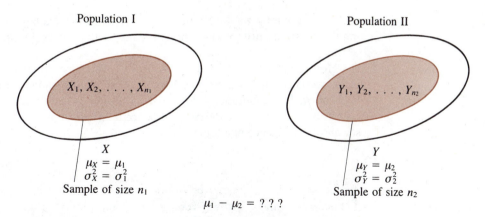

Figure 8.7

difference between these two means based on independent random samples drawn from the two populations. This idea is pictured in Figure 8.7.

The logical point estimator for the difference between the actual means μ_1 and μ_2 is the difference between the corresponding sample means denoted by $\bar{X}_1$ and $\bar{X}_2$, respectively. That is,

$$\widehat{\mu_1 - \mu_2} = \bar{X}_1 - \bar{X}_2$$

Example 2   To compare the speed with which an aspirin compound relieves a headache to that of a nonaspirin compound, 22 individuals are selected and randomly split into two groups. The first group is treated with the aspirin compound; the other receives the nonaspirin drug. These data on the time, in minutes, required to obtain relief result:

Population I *contains aspirin*			*Population II* *contains no aspirin*		
9.9	13.5	10.3	8.2	10.5	11.6
12.2	9.5	5.9	9.1	15.2	10.2
12.5	9.5	11.5	9.0	10.1	
8.0	9.6	11.9	17.3	9.7	

For these data $\bar{x}_1 = 10.36$, $\bar{x}_2 = 11.09$.

Thus, the estimated difference between the mean times to relief using the two products is

$$\widehat{\mu_1 - \mu_2} = \bar{x}_1 - \bar{x}_2 = 10.36 - 11.09 = -.73 \text{ minutes}$$

The fact that this difference is negative implies that perhaps $\mu_1 < \mu_2$. It appears that the product containing aspirin works slightly faster on the average than the product containing no aspirin. Is this really true? We cannot say with certainty with only a point estimate for the difference in means available. We defer answering the question until we know how to construct an interval estimate for $\mu_1 - \mu_2$.

How good is the estimator $\bar{X}_1 - \bar{X}_2$? The next theorem shows that this estimator is an unbiased estimator for $\mu_1 - \mu_2$. It also shows that when sampling from normal distributions, the statistic $\bar{X}_1 - \bar{X}_2$ is itself normal. This is important as it will allow us to derive confidence intervals on $\mu_1 - \mu_2$ and test hypotheses on this difference in a logical way.

Theorem 8.1

Let $\bar{X}_1$ and $\bar{X}_2$ denote the sample means based on independent samples of sizes n_1 and n_2 drawn from normal distributions means μ_1 and μ_2 and variances σ_1^2 and σ_2^2, respectively. Then $\bar{X}_1 - \bar{X}_2$ is normally distributed with mean $\mu_1 - \mu_2$ and variance $(\sigma_1^2/n_1) + (\sigma_2^2/n_2)$.

To generate $100(1 - \alpha)\%$ confidence intervals on $\mu_1 - \mu_2$ or to test hypotheses on the value of this difference, we will consider two distinct cases:

1. The population variances σ_1^2 and σ_2^2, although unknown, are assumed to be equal in value;

2. The population variances σ_1^2 and σ_2^2 are unknown and assumed to be unequal in value.

The first situation is handled by what is termed a *pooling procedure*; the second by the *Satterthwaite* procedure. This means that when making comparisons between two means, the first job of the experimenter is to consider the population variances. There may be some reason a priori for thinking that the population variances are equal. For example, past data might indicate that the variances in the scores of men on the law school entrance exams is the same as that for women; biological theory might predict that the variance in birth weights for boys is the same as that for girls; or psychological theory might indicate that there is no difference in the variance in IQ scores between whites and nonwhites. In cases such as these, the researcher might feel safe in pooling immediately. If no such evidence exists, then pooling without some careful thought is unwise. In either case, we suggests that a preliminary test for equality of population variances be conducted. That is, we suggest that you first test

$$H_0 : \sigma_1^2 = \sigma_2^2$$
$$H_1 : \sigma_1^2 \neq \sigma_2^2$$

using the F test introduced in the last section. We also suggest one small change in the F procedure when it is used prior to conducting a test to compare means: namely, that the α level set or the P value tolerated be relatively large. By this we mean that α or P can be as large as .2 or even .3 and we will still reject H_0 in this preliminary test. The reason for this is that if we are unable to reject H_0, we will pool. However, as is usually the case, when we are unable to reject H_0, we will not know β, the probability of error. Since α and β are interrelated, tolerating a relatively large value for α should insure that β is reasonably small. Thus, if we are unable to reject H_0 and elect to pool, the probability that we are in error in doing so should be rather small. In summary, before comparing two means we first compare variances. If we reject $H_0 : \sigma_1^2 = \sigma_2^2$, in favor of $H_1 : \sigma_1^2 \neq \sigma_2^2$ we use the Satterthwaite procedure to compare means; otherwise we pool. You should be aware to the fact that this topic is controversial. Some statisticians agree with our point of view, others do not. Your instructor can discuss this with you further.

Example 3

We have generated a point estimate for the difference in the mean time to relief using a remedy containing aspirin (μ_1) and the mean time to relief using a remedy containing no aspirin (μ_2). Namely,

$$\widehat{\mu_1 - \mu_2} = -.73 \text{ minutes}$$

This estimate is unbiased. Suppose we wish to extend this point estimate to an interval estimate to get an idea of the accuracy of the figure. We first pause to consider the population variances so that we can choose between the pooling procedure and the Satterthwaite procedure. That is, we test

$$H_0 : \sigma_1^2 = \sigma_2^2$$

$$H_1 : \sigma_1^2 \neq \sigma_2^2$$

Let us use an $\alpha = .10$ test. The critical region for the test is shown in Figure 8.8. For the data of Example 2, $s_1^2 = 4.48$ and $s_2^2 = 8.49$. The observed value of the test statistic S_1^2/S_2^2 is $s_1^2/s_2^2 = 4.48/8.49 = .53$. Since .53 is neither below the critical point of .44 nor above that of 2.42, we are unable to reject H_0. We do not have statistical evidence that the population variances are unequal. We therefore use the pooling method for obtaining a confidence interval on $\mu_1 - \mu_2$.

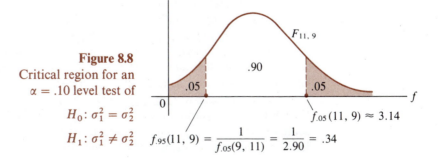

Figure 8.8
Critical region for an
$\alpha = .10$ level test of

$H_0 : \sigma_1^2 = \sigma_2^2$
$H_1 : \sigma_1^2 \neq \sigma_2^2$ $f_{.95}(11, 9) = \dfrac{1}{f_{.05}(9, 11)} = \dfrac{1}{2.90} = .34$

$F_{11, 9}$

.90

.05   .05

$f_{.05}(11, 9) \approx 3.14$

What is the purpose of pooling and how does one go about it? These questions can be answered quite easily. Recall that in the one-sample case of Chapter 6 we began by considering

$$\frac{\bar{X} - \mu}{\sigma/\sqrt{n}}$$

This random variable had a standard normal distribution. When σ^2 was unknown and had to be estimated by S, we became involved with the t distribution. This happens again.

Note that by Theorem 8.1, $\bar{X}_1 - \bar{X}_2$ is approximately normal with mean $\mu_1 - \mu_2$ and variance $(\sigma_1^2/n_1) + (\sigma_2^2/n_2)$. We may therefore standardize this variable by subtracting its mean and dividing by its standard deviation to conclude that

$$\frac{(\bar{X}_1 - \bar{X}_2) - (\mu_1 - \mu_2)}{\sqrt{(\sigma_1^2/n_1) + (\sigma_2^2/n_2)}}$$

is standard normal. Since we are assuming that $\sigma_1^2 = \sigma_2^2$ we may denote this common variance by σ^2 and conclude that

$$\frac{(\bar{X}_1 - \bar{X}_2) - (\mu_1 - \mu_2)}{\sigma\sqrt{(1/n_1) + (1/n_2)}}$$

is standard normal. Since σ is unknown, we must estimate it and hence will again use a t distribution.

The statistic used to estimate the common population variance σ^2 is called a *pooled estimator*. This name is appropriate since the procedure developed involves combining, or pooling, the information obtained from both available samples. Thus the purpose of pooling is to utilize all available data in estimating the common variance σ^2 and thus also the common standard deviation σ. How is this done? To answer this question, note that the unbiased estimators for the actual variances σ_1^2 and σ_2^2 are the corresponding sample variances S_1^2 and S_2^2. Since $\sigma_1^2 = \sigma_2^2$, each of these sample variances is in fact estimating the same thing. It is logical then to try to find a way to utilize both S_1^2 and S_2^2 in estimating σ^2.

We could simply average the two. However, we choose to use instead what is called a *weighted average*. This simply means that we use a procedure that takes into account the sizes of the samples upon which the sample variances are based. Intuition leads one to believe that the larger the sample size involved, the closer the sample variance will be to the true variance. Hence, the estimator based on the larger sample should receive more weight or importance than the other. If the sample sizes are equal, each should receive the same consideration. Keeping these ideas in mind, we can define what is called the *pooled variance* for σ^2.

Definition 8.2

Pooled Variance

Let S_1^2 and S_2^2 be sample variances based on samples of size n_1 and n_2, respectively, drawn from populations with a common variance σ^2. The **pooled variance**, denoted S_p^2, is given by

$$S_p^2 = \frac{(n_1 - 1)S_1^2 + (n_2 - 1)S_2^2}{n_1 + n_2 - 2}$$

Note that by multiplying the respective sample variances by $n_1 - 1$ and $n_2 - 1$, we accomplish exactly what we set out to do, namely weight S_1^2 and S_2^2 to reflect sample size. The divisor $n_1 + n_2 - 2$ is chosen so that S_p^2 is unbiased for σ^2.

Example 4

We have no reason to believe there is a difference in variance between the times to relief using the aspirin compound and the compound containing no aspirin. (See Example 3.) Hence we pool the estimates $s_1^2 = 4.48$ and $s_2^2 = 8.49$ to obtain a single estimate of the common population variance σ^2.

$$\hat{\sigma}^2 = s_p^2 = \frac{(n_1 - 1)s_1^2 + (n_2 - 1)s_2^2}{n_1 + n_2 - 2}$$

$$= \frac{11(4.48) + 9(8.49)}{12 + 10 - 2} = 6.28$$

Note that by multiplying s_1^2 by 11 and s_2^2 by 9 in the numerator, we give more weight to the estimate of σ^2 that is based on the larger-sized sample. Note also that the pooled estimate of the common population standard deviation σ is

$$\hat{\sigma} = s_p = \sqrt{s_p^2} = \sqrt{6.28} = 2.51$$

If we replace the unknown common population standard deviation σ by its estimator S_p in the expression

$$\frac{(\bar{X}_1 - \bar{X}_2) - (\mu_1 - \mu_2)}{\sigma\sqrt{(1/n_1) + (1/n_2)}}$$

we obtain the random variable

$$\frac{(\bar{X}_1 - \bar{X}_2) - (\mu_1 - \mu_2)}{S_p\sqrt{(1/n_1) + (1/n_2)}}$$

It can be shown that this random variable follows a t distribution with $n_1 + n_2 - 2$ degrees of freedom. This is a logical random variable to use in the construction of a $100(1 - \alpha)\%$ confidence interval on $\mu_1 - \mu_2$ when the population variances are assumed to be equal.

Rather than repeat an algebraic argument that you have seen several times already, let's look for a pattern in our previous results. In Chapter 6, we developed two confidence intervals. In particular we began with the Z random variable $(\bar{X} - \mu)/(\sigma/\sqrt{n})$ to derive the bounds $\bar{X} \pm z_{\alpha/2} \sigma/\sqrt{n}$ for a confidence interval on μ when σ^2 is known; we began with the T_{n-1} random variable $(\bar{X} - \mu)/(S/\sqrt{n})$ to derive the bounds $\bar{X} \pm t_{\alpha/2} S/\sqrt{n}$ for a confidence interval on μ when σ^2 is unknown. Both of these random variables are of the form

$$\frac{\text{estimator} - \text{parameter}}{\text{denominator}}$$

In each case, an algebraic argument resulted in bounds of the form

estimator $\pm$ (an appropriate probability point) $\times$ denominator

The $T_{n_1 + n_2 - 2}$ random variable

$$\frac{(\bar{X}_1 - \bar{X}_2) - (\mu_1 - \mu_2)}{S_p \sqrt{(1/n_1) + (1/n_2)}}$$

is also of the form:

$$\frac{\text{estimator } (\bar{X}_1 - \bar{X}_2) - \text{parameter } (\mu_1 - \mu_2)}{\text{denominator } (S_p \sqrt{(1/n_1) + (1/n_2)})}$$

For this reason, we know, without actually doing the algebraic derivation, that the confidence bounds for a $100(1 - \alpha)\%$ confidence interval on $\mu_1 - \mu_2$ will take the form

estimator $\pm$ (an appropriate probability point) $\times$ denominator

That is, the bounds are given by

$$(\bar{X}_1 - \bar{X}_2) \pm t_{\alpha/2} S_p \sqrt{(1/n_1) + (1/n_2)}$$

This result is summarized in Theorem 8.2 and illustrated in Example 5.

Theorem 8.2

Confidence Interval on $\mu_1 - \mu_2$: Variances Equal

A $100(1 - \alpha)\%$ confidence interval on $\mu_1 - \mu_2$ is

$$(\bar{X}_1 - \bar{X}_2) \pm t_{\alpha/2} S_p \sqrt{(1/n_1) + (1/n_2)}$$

where the point $t_{\alpha/2}$ is found relative to the t distribution with $n_1 + n_2 - 2$ degrees of freedom.

Example 5

To construct a 90% confidence interval on the difference between the mean times to relief using the aspirin and nonaspirin compounds based on the data of Example 2, we must utilize the t distribution with $n_1 + n_2 - 2 = 12 + 10 - 2 =$

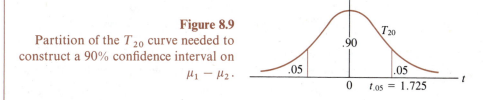

Figure 8.9

Partition of the T_{20} curve needed to construct a 90% confidence interval on $\mu_1 - \mu_2$.

20 degrees of freedom. This T curve with the appropriate probability point labeled is shown in Figure 8.9.

Since we already know that $\bar{x}_1 - \bar{x}_2 = -.73$ and that $s_p = 2.51$, the desired confidence interval is

$$(\bar{x}_1 - \bar{x}_2) \pm t_{\alpha/2}\, s_p \sqrt{(1/n_1) + (1/n_2)}$$

or

$$-.73 \pm 1.725(2.51)\sqrt{(1/12 + 1/10)}$$

This becomes $[-2.58, 1.12]$. We are 90% sure that the true difference in mean time to relief is between -2.58 and 1.12 minutes. Unfortunately, 0 lies in this interval, leaving open the distinct possibility that there is in fact no difference at all. We really do not have evidence that one remedy works any faster on the average than the other does.

In pooling variances, we are assuming normality and equal variances. Some thought should be given to the validity of each of these assumptions. The normality assumption can be checked visually via a stem-and-leaf diagram or a histogram. Several analytic tests are available for testing for normality. One of these is given in Chapter 10. The F test as described here can be used to detect situations in which the population variances are unequal. Our point of view is that we will pool unless our data indicates that we should not do so. It is true that there is some risk involved. To minimize this risk we run the F test at an unusually large α level. In this way we reduce the probability that we are pooling when in fact pooling is inappropriate. Furthermore, studies have shown that the effects of slightly unequal variances are minimized if the sample sizes are the same. For this reason, we suggest that experiments be designed with equal sample sizes whenever possible.

Exercises 8.2

16. Let $s_1^2 = 25$, $n_1 = 25$, $s_2^2 = 15$, $n_2 = 30$.
   (a) Test $H_0: \sigma_1^2 = \sigma_2^2$ at the $\alpha = .20$ level
        $H_1: \sigma_1^2 \neq \sigma_2^2$
   (b) If H_0 is not rejected, find s_p^2 and s_p.
   (c) Why is s_p^2 closer to s_2^2 than it is to s_1^2?

17. Let $s_1^2 = 30$, $n_1 = 16$, $s_2^2 = 28$, $n_2 = 16$.
    (a) Test $H_0: \sigma_1^2 = \sigma_2^2$ at the $\alpha = .20$ level
          $H_1: \sigma_1^2 \neq \sigma_2^2$
    (b) Find s_p^2. Don't use your calculator!
    (c) Why is s_p^2 equal to the average of s_1^2 and s_2^2 in this case?

18. Find the indicated confidence interval based on the given data. Assume that $\sigma_1^2 = \sigma_2^2$.
    (a) Find a 95% confidence interval on $\mu_1 - \mu_2$ when

    $$\bar{x}_1 = 8, \qquad \bar{x}_2 = 4$$
    $$s_1^2 = 3, \qquad s_2^2 = 2$$
    $$n_1 = 15, \qquad n_2 = 15$$

    (b) Find a 90% confidence interval on $\mu_1 - \mu_2$ when

    $$\bar{x}_1 = 7, \qquad \bar{x}_2 = 9$$
    $$s_1^2 = 6, \qquad s_2^2 = 5.5$$
    $$n_1 = 10, \qquad n_2 = 8$$

    (c) Find a 99% confidence interval on $\mu_1 - \mu_2$ when

    $$\bar{x}_1 = 25, \qquad \bar{x}_2 = 23$$
    $$s_1^2 = 8, \qquad s_2^2 = 9$$
    $$n_1 = 15, \qquad n_2 = 27$$

19. The cost of books for college students is an important random variable. A study is run to compare the mean cost per year for students of the liberal arts to that for students of the natural sciences. The following observations are obtained.

Population I liberal arts		Population II natural sciences	
220.47	189.34	180.59	198.66
190.52	202.81	186.84	196.59
197.23	208.82	195.34	177.15
197.37	212.03	194.26	206.81
215.12	203.34	196.46	208.40

    (a) Find the point estimate for $\mu_1 - \mu_2$. Which group appears to spend more on the average for books?
    (b) Find s_1^2 and s_2^2. Use these values to test

    $$H_0: \sigma_1^2 = \sigma_2^2 \qquad \text{at the } \alpha = .20 \text{ level}$$
    $$H_1: \sigma_1^2 \neq \sigma_2^2$$

(c) Do you think that pooling is appropriate? If so, find s_p^2 and s_p.

(d) Find a 90% confidence interval on $\mu_1 - \mu_2$. Is there concrete evidence that a real difference exists? Explain.

20. There has been some discussion among amateur gardeners about the virtues of black plastic versus newspapers as weed inhibitors for growing tomatoes. To compare the two, several rows of tomatoes are planted. Black plastic is used around nine randomly selected plants and newspaper around the remaining ten. All plants start at virtually the same height and receive the same care. The variable of interest is the height after a month's growth. The following observations are obtained (height, in feet):

Population I black plastic		Population II newspaper	
1.80	1.25	2.57	1.34
1.29	2.61	1.59	1.43
1.13	1.60	1.78	1.06
2.92	2.06	1.37	1.44
2.20		1.22	1.12

(a) Find the point estimate for $\mu_1 - \mu_2$. Which method appears to produce the best growth?

(b) Find s_1^2 and s_2^2. Use these to test for equality of variances at the $\alpha = .20$ level. Can H_0 be rejected?

(c) Find s_p^2 and s_p. Construct a 95% confidence interval on $\mu_1 - \mu_2$. Does there appear to be a real difference in the mean heights of plants grown using these two methods for controlling weeds? Explain.

21. A study is run in a large school district to compare the achievement scores of fifth-graders before and after busing is ordered to achieve integration. A random sample of 121 scores is selected from among the scores made by fifth-graders in 1960. A sample mean of 5.25 (fifth grade, third-month level) is obtained with a sample variance of .37. A sample of 121 scores is selected from among the scores made by fifth-graders in 1970. A sample mean of 5.50 (fifth grade, sixth-month level) is obtained with a sample variance of .44.

(a) Test $H_0: \sigma_1^2 = \sigma_2^2$.

(b) If H_0 is not rejected, find s_p^2 and s_p.

(c) Find a 95% confidence interval on $\mu_1 - \mu_2$. Does there appear to be a real difference in mean scores for these years? Explain

22. A guidance counselor wants to compare the mean length of time it takes to complete two different vocational preference checklists. Ten people were randomly selected and asked to complete checklist I while ten others completed checklist II. The following information was obtained:

$$\bar{x}_1 = 54.3, \ s_1^2 = 16.00 \qquad \bar{x}_2 = 48.1, \ s_2^2 = 12.2$$

Find a 90% confidence interval on $\mu_1 - \mu_2$. Does there appear to be any real difference in the average times involved? If so, which list should be used in order to save time?

*23. Use the rules for expectation to show that $\bar{X}_1 - \bar{X}_2$ is an unbiased estimator for $\mu_1 - \mu_2$.

*24. Use the rules for expectation to show that S_p^2 is an unbiased estimator for the common population variance σ^2.

8.3 Comparing Two Means: Pooled-T Tests

In this section we discuss what is called a *pooled-T test*. This test is used to test hypotheses on the relationship between the means of two normal populations when their variances are assumed to be equal. Once again we are basing our conclusions on data obtained by drawing independent random samples from two populations.

Although the null value of $\mu_1 - \mu_2$ can be any real number whatsoever, the most commonly encountered value is 0. In this case, the hypotheses take one of these three forms:

I $H_0: \mu_1 - \mu_2 = 0$ II $H_0: \mu_1 - \mu_2 = 0$ III $H_0: \mu_1 - \mu_2 = 0$

$H_1: \mu_1 - \mu_2 \neq 0$ $H_1: \mu_1 - \mu_2 < 0$ $H_1: \mu_1 - \mu_2 > 0$

Two-tailed test   Left-tailed test   Right-tailed test

These forms are usually expressed in the following equivalent way:

I $H_0: \mu_1 = \mu_2$ II $H_0: \mu_1 = \mu_2$ III $H_0: \mu_1 = \mu_2$

$H_1: \mu_1 \neq \mu_2$ $H_1: \mu_1 < \mu_2$ $H_1: \mu_1 > \mu_2$

Two-tailed test   Left-tailed test   Right-tailed test

The test statistic for testing any of these hypotheses is

$$\frac{(\bar{X}_1 - \bar{X}_2) - 0}{S_p\sqrt{(1/n_1) + (1/n_2)}}$$

If the null value of 0 is correct, that is, if $\mu_1 = \mu_2$, then this statistic follows a t distribution with $n_1 + n_2 - 2$ degrees of freedom. Tests are performed as you would expect. In the case of a left-tailed test, we reject H_0 in favor of H_1 if the observed value of the test statistic is a large negative number; in a right-tailed test, H_0 is rejected for large positive values of the test statistic; large negative or large positive values result in the rejection of H_0 in the case of a two-tailed test. As you might guess, this procedure is also somewhat controversial. We are faced with the same problem as that encountered earlier. Namely, when should we pool? We shall take the same approach here as we did when constructing a confidence interval on $\mu_1 - \mu_2$ under these circumstances. Namely, we suggest that a preliminary F test for equality of variances be conducted at a relatively large α level or P value. If $H_0: \sigma_1^2 = \sigma_2^2$ is not rejected, pool the data. To minimize the effects of erroneously pooling, we suggest that experiments be designed with equal sample sizes whenever possible. We demonstrate the use of the pooled-T test in the next two examples.

Example 1 | Makers of two small cars, A and B, each claim that their auto gets superior gas mileage on the highway. To decide the matter once and for all the makers of car A decide to run a carefully controlled experiment designed to get statistical evidence that the average number of miles per gallon on the highway obtained by car A is greater than that of car B. We wish to test

$$H_0: \mu_1 = \mu_2$$

$$H_1: \mu_1 > \mu_2$$

where μ_1 represents the mean gas mileage obtained by car A. The experiment is designed with $n_1 = n_2 = 25$. However, one car of type B is found to be defective. For this reason, tests are run on 25 cars of type A and 24 of type B. The variable of interest in each case is the number of miles per gallon obtained over a 100-mile test track. The observed sample statistics are:

$$\bar{x}_1 = 37.5, \quad s_1^2 = 0.36, \qquad \bar{x}_2 = 35.2, \quad s_2^2 = .25$$

Since $s_1^2/s_2^2 = .36/.25 = 1.44$ does not cause rejection of $H_0: \sigma_1^2 = \sigma_2^2$ even at the $\alpha = .20$ level, ($f_{.10} = 1.72$), we may pool s_1^2 and s_2^2 to obtain

$$s_p^2 = \frac{24(.36) + 23(.25)}{\sqrt{25 + 24 - 2}} = .31$$

$$s_p = \sqrt{.31} = .56$$

The critical point for a size $\alpha = .05$ test based on the t distribution with $n_1 + n_2 - 2 = 47$ degrees of freedom is approximately 1.684. Note that this point is actually the point for a test based on 40 degrees of freedom. It is used as a conservative approximation to $t_{.05}$ (47 degrees of freedom), since the latter is not listed in Table VI of Appendix A. We reject H_0 if the observed value of the test statistic falls on or above this point. By conservative, we mean that the critical point of 1.684 is actually a bit larger than it need be. Thus, if we are able to reject H_0, the actual probability of a Type I error is somewhat less than the apparent level of .05. For the data given, the test statistic

$$\frac{(\bar{X}_1 - \bar{X}_2) - 0}{S_p\sqrt{(1/n_1) + (1/n_2)}}$$

assumes the value

$$\frac{37.5 - 35.2}{.56\sqrt{(1/25) + (1/24)}} = 14.37$$

Obviously, we can reject H_0 and conclude that the average gas mileage of a car of type A is greater than that of type B. Keep in mind that we are subject to making a Type I error. We may be rejecting H_0 when in fact it is true. We may be in a position as promoters of car A of making false advertising claims; however, the probability of this occurring is small.

Example 2

A study is conducted to compare the effectiveness of two compounds for use in undercoating cars. Each of the compounds is thought to act as a rust inhibitor. That is, the compound helps to slow the development of rust on exposed portions of the automobile. The random variable studied is X, the time in months that elapses before the first signs of rust appear. Since the compounds are both experimental, there are no preconceived notions as to their behavior. The primary concern of the researchers is to compare the mean elapsed times for the two compounds. Thus, the primary hypothesis to be tested is $H_0: \mu_1 = \mu_2$ versus $H_1: \mu_1 \neq \mu_2$. These data are obtained.

Compound I	*Compound II*
$n_1 = 9$	$n_2 = 9$
$\bar{x}_1 = 16$ months	$\bar{x}_2 = 15$ months
$s_1 = 10.1$ months	$s_2 = 10$ months

Before testing our primary null hypothesis, we first compare variances. For these data,

$$s_1^2/s_2^2 = (10.1)^2/(10)^2 = 1.02$$

the P value for this test based on the F distribution with 8 and 8 degrees of freedom exceeds .20 ($f_{.10} = 2.59$). Since this P value is large, we do not reject H_0. We elect to pool variances. For these data,

$$s_p^2 = [8(10.1)^2 + 8(10)^2]/16 = 101.005$$

and

$$s_p = \sqrt{101.005} \approx 10.05$$

The observed value of the pooled-T statistic is

$$\frac{(\bar{x}_1 - \bar{x}_2)}{s_p\sqrt{(1/n_1 + 1/n_2)}} = \frac{16 - 15}{10.05\sqrt{(1/8 + 1/8)}} = .199$$

The P value for a one-tailed test, based on the t distribution with 16 degrees of freedom, exceeds .25 ($t_{.25} = .69$). Thus the P value for our two-tailed test exceeds .50. Since this probability is large, we are unable to reject H_0. We do not have sufficient statistical evidence to claim that there is a difference in the mean elapsed times for these two compounds.

In summary, we have introduced a widely used procedure for comparing the means of two normal populations when we assume that their variances are equal. This pooled-T test is based on the t distribution with $n_1 + n_2 - 2$ degrees of freedom. Remember that when the number of degrees of freedom involved exceeds that listed in your T table you may approximate critical points and P values via the standard normal curve. These Z points are listed in the last row of your T table.

If you encounter a situation in which it is assumed that $\sigma_1^2 = \sigma_2^2$ but in which the null value is not zero, then H_0: $\mu_1 - \mu_2 = (\mu_1 - \mu_2)_0$ can be tested via the statistic

$$\frac{(\bar{X}_1 - \bar{X}_2) - (\mu_1 - \mu_2)_0}{S_p \sqrt{(1/n_1 + 1/n_2)}}$$

This statistic follows a t distribution with $n_1 + n_2 - 2$ degrees of freedom. Tests are conducted exactly as described in this section.

Exercises 8.3

25. Assuming that $\sigma_1^2 = \sigma_2^2$, test each of these hypotheses:
    (a) H_0: $\mu_1 = \mu_2$ $\bar{x}_1 = 8$ $\bar{x}_2 = 6$ $\alpha = .05$
        H_1: $\mu_1 > \mu_2$ $s_1^2 = 12$ $s_2^2 = 10$
                                 $n_1 = 16$ $n_2 = 16$
    (b) H_0: $\mu_1 = \mu_2$ $\bar{x}_1 = 7$ $\bar{x}_2 = 9$ $\alpha = .025$
        H_1: $\mu_1 < \mu_2$ $s_1^2 = 4$ $s_2^2 = 2.6$
                                 $n_1 = 25$ $n_2 = 24$
    (c) H_0: $\mu_1 = \mu_2$ $\bar{x}_1 = 16$ $\bar{x}_2 = 13$ $\alpha = .10$
        H_1: $\mu_1 \neq \mu_2$ $s_1^2 = 10$ $s_2^2 = 14$
                                 $n_1 = 41$ $n_2 = 30$

26. The admissions office at a local university claims that there is not a significant difference in the SAT scores of its entering male students and those of its entering female students. A group of skeptics thinks that the females' scores are on the average higher than the males' scores. To gain support for their theory, they obtain the following observations on the scores for both groups.

Population I SAT scores, Male		Population II SAT scores, Female	
500	800	616	1100
550	810	650	900
700	300	750	875
1200	650	400	570
1500	900	1000	1400

(a) Set up the appropriate null and alternative hypotheses for gaining the evidence desired.
(b) Test for equality of variances at the $\alpha = .20$ level.
(c) Test the hypotheses of part (a) at the $\alpha = .10$ level.
(d) What type of error may we be inadvertently committing?

27. A cigarette manufacturer is experimenting with two new filters for its filter-tip cigarettes. Filter I is cheaper to make than filter II and will be used unless there is statistical evidence that filter II is more effective than filter I. In a test to see if this is the case, 25 cigarettes are made using each type of filter. The cigarettes are mechanically smoked and the amount (in milligrams) of trapped tar and nicotine recorded. The following information is obtained.

$$\bar{x}_1 = 1.10, \quad s_1^2 = .05 \qquad \bar{x}_2 = 1.13, \quad s_2^2 = .07$$

(a) Set up the null and alternative hypotheses needed to support the claim that filter II tends to trap more tar and nicotine than filter I.
(b) Test for equality of variances at the $\alpha = .20$ level.
(c) Test the hypotheses of part (a) at the $\alpha = .05$ level. What action would you recommend that the company take?

28. Approximate the P value for each of these tests. Assume $\sigma_1^2 = \sigma_2^2$.

(a) $\quad H_0: \mu_1 = \mu_2 \qquad \bar{x}_1 = 12 \qquad \bar{x}_2 = 10$
$\qquad\quad H_1: \mu_1 > \mu_2 \qquad s_1^2 = 9 \qquad\; s_2^2 = 8$
$\qquad\qquad\qquad\qquad\qquad\quad n_1 = 11 \qquad n_2 = 11$

(b) $\quad H_0: \mu_1 = \mu_2 \qquad \bar{x}_1 = 7 \qquad\; \bar{x}_2 = 10$
$\qquad\quad H_1: \mu_1 < \mu_2 \qquad s_1^2 = 12 \qquad s_2^2 = 12.5$
$\qquad\qquad\qquad\qquad\qquad\quad n_1 = 25 \qquad n_2 = 25$

(c) $\quad H_0: \mu_1 = \mu_2 \qquad \bar{x}_1 = 5 \qquad\; \bar{x}_2 = 6$
$\qquad\quad H_1: \mu_1 \neq \mu_2 \qquad s_1^2 = 4 \qquad\; s_2^2 = 4.1$
$\qquad\qquad\qquad\qquad\qquad\quad n_1 = 121 \qquad n_2 = 61$

29. A personnel psychologist is concerned with staffing a department that is involved in assembling a small electronic control panel. This job requires a great deal of manual dexterity. For this reason, the psychologist thinks that women, because of their smaller hands, are better suited for the job than men. To test this theory the psychologist selects 16 male and 16 female applicants for the job and gives them a test designed to measure the type of manual dexterity needed for the job. The following results are obtained:

$$\bar{x}_{male} = 22.5, \; s_{male}^2 = 24.01 \qquad \bar{x}_{female} = 25.2, \; s_{female}^2 = 16.0$$

If higher scores indicate a higher degree of manual dexterity, does the psychologist have statistical evidence that the theory is correct? Explain based on the P values of the tests that you conduct.

30. Students in an introductory psychology course were asked to rate a prospective faculty member from a written description. The same description was given to each student, except that one-half (selected at random) of the descriptions were labeled Doctor Joe Brown and the other half were labeled Doctor Lois Brown. The ratings based on a ten-point scale (with ten the highest) were as follows:

Joe: 10, 9.5, 9.3, 9.0, 8.8, 8.5, 8.5, 7.6, 7.5, 9.2

Lois: 9.4, 9.1, 8.3, 7.7, 7.4, 7.3, 7.3, 7.2, 7.0, 6.9

Do these data support the contention that the mean rating assigned to the faculty member when known as Lois is less than that received when known as Joe? Explain based on the P values of the tests that you conduct.

31. A random sample of 40 fifth-graders was randomly divided into two groups of 20 each. Both groups were given the same lesson to learn, but group I read the lesson from a standard textbook, while group II studied the lesson from a computer. At the end of the hour, all the students were given a test to determine how well they had learned the lesson. Group I had a mean grade of $\bar{x}_1 = 71.3$ with $s_1 = 9.2$, while group II had a mean grade $\bar{x}_2 = 79.7$ with $s_2 = 8.7$. Do these data support the contention that students tend to score better when taught via a computer? Explain based on the P values of the tests that you conduct.

8.4   Comparing Two Means: Variances Unequal

In the last two sections we considered methods for comparing means when we are unable to detect any difference in the population variances. We now turn our attention to the problem of comparing means when our F test does lead us to conclude that σ_1^2 and σ_2^2 are not equal.

Recall that we have been considering the random variable

$$\frac{(\bar{X}_1 - \bar{X}_2) - (\mu_1 - \mu_2)}{\sqrt{(\sigma_1^2/n_1) + (\sigma_2^2/n_2)}}$$

If we have convincing evidence that $\sigma_1^2 \neq \sigma_2^2$, it makes no sense to pool. There is no common population variance to be estimated. Instead, we simply substitute for these unknown variances their unbiased estimators S_1^2 and S_2^2 to obtain

$$\frac{(\bar{X}_1 - \bar{X}_2) - (\mu_1 - \mu_2)}{\sqrt{(S_1^2/n_1) + (S_2^2/n_2)}}$$

It has been found that this random variable follows approximately the t distribution with

$$v = \frac{[(S_1^2/n_1) + (S_2^2/n_2)]^2}{\dfrac{(S_1^2/n_1)^2}{n_1 - 1} + \dfrac{(S_2^2/n_2)^2}{n_2 - 1}}$$

degrees of freedom.

Confidence intervals and tests of hypotheses on $\mu_1 - \mu_2$ can be made in the usual manner by means of the above T random variable. In particular, $100(1 - \alpha)\%$ confidence intervals on $\mu_1 - \mu_2$ are given by

Confidence Interval on $\mu_1 - \mu_2$ When $\sigma_1^2 \neq \sigma_2^2$

$$(\bar{X}_1 - \bar{X}_2) \pm t_{\alpha/2}\sqrt{(S_1^2/n_1) + (S_2^2/n_2)}$$

where the point $t_{\alpha/2}$ is found in the T table with v degrees of freedom.

Hypothesis tests are run using the following as the test statistic:

Test Statistic for Testing H_0: $\mu_1 = \mu_2$ When $\sigma_1^2 \neq \sigma_2^2$

$$\frac{(\bar{X}_1 - \bar{X}_2) - 0}{\sqrt{(S_1^2/n_1) + (S_2^2/n_2)}}$$

which follows a t distribution with v degrees of freedom.

The formula given for computing the degrees of freedom was proposed by F. E. Satterthwaite in 1946. This **Satterthwaite procedure** has been widely accepted. However, you should be aware of the fact that other formulas do exist. The issue of how best to estimate the number of degrees of freedom associated with the random variable

$$\frac{\bar{X}_1 - \bar{X}_2}{\sqrt{(S_1^2/n_1 + S_2^2/n_2)}}$$

has not yet been resolved. You should also realize that it is unusual for v as computed via the Satterthwaite procedure to be an integer. Since the number of degrees of freedom associated with a T random variable must be a positive integer, we suggest that you round the calculated value of v down to the nearest positive integer. Examples 1 and 2 should make this point clear.

Example 1 A businessman is interested in determining the store hours that will be most profitable to him. He is considering two different schedules: schedule 1 (9:30–5:30), and schedule 2 (10:00–6:00). He has no idea which schedule will be better and hence wishes to estimate the difference in mean sales for the two schedules via a 95% confidence interval. Over a period of time, the following data are obtained on sales using these two schedules:

Schedule 1	Schedule 2
$\bar{x}_1 = 570.00$	$\bar{x}_2 = 600.00$
$n_1 = 25$	$n_2 = 25$
$s_1^2 = 1600$	$s_2^2 = 625$

We first compare population variances by testing

$$H_0: \sigma_1^2 = \sigma_2^2$$

$$H_1: \sigma_1^2 \neq \sigma_2^2$$

Since $s_1^2/s_2^2 = 1600/625 = 2.56$ causes rejection of H_0 at the $\alpha = .05$ level $(f_{.025}(24, 24) = 2.27)$, we do have evidence that the population variances are different. We use the Satterthwaite procedure to compute the number of degrees of freedom as shown

$$v = \frac{[(s_1^2/n_1) + (s_2^2/n_2)]^2}{\dfrac{(s_1^2/n_1)^2}{n_1 - 1} + \dfrac{(s_2^2/n_2)^2}{n_2 - 1}} = \frac{[(1600/25) + (625/25)]^2}{\dfrac{(1600/25)^2}{24} + \dfrac{(625/25)^2}{24}} = 40.27 \approx 40$$

We enter the T table, Table VI of Appendix A, with $v = 40$ and see that $t_{.025} = 2.021$. A 95% confidence interval on $\mu_1 - \mu_2$ is given by

$$(\bar{x}_1 - \bar{x}_2) \pm t_{\alpha/2}\sqrt{(s_1^2/n_1) + (s_2^2/n_2)}$$

or

$$(570 - 600) \pm 2.021\sqrt{(1600/25) + (625/25)}$$

After completing this computation, we see that the 95% confidence interval on the difference in mean sales is $[-49.07, -10.93]$. Since 0 does not lie in this interval, we can conclude that schedule 2 produces a higher average sale than schedule 1. However, whether the difference is enough to make schedule 2 worth the effort is not a statistical decision. It is a business decision that must be made by the businessperson.

Example 2   Sociologists believe that some characteristics once associated with college-age students are now found among high school students. One such characteristic is the age at which an individual acquires his or her first automobile. Twenty-five persons between the ages of 30 and 40 are randomly selected and the age at which they obtained their first automobile is determined. It is found that $\bar{x} = 22.3$ and $s_1^2 = 4.52$. Similar statistics for 25 individuals between the ages of 16 and 30 reveal $\bar{x}_2 = 18.7$ and $s_2^2 = 2.00$. Let us test

$$H_0: \mu_1 = \mu_2$$

$$H_1: \mu_1 > \mu_2$$

To do so, we first consider the ratio $s_1^2/s_2^2 = 4.52/2.00 = 2.26$. Since $f_{.05}(24, 24) = 1.98$, we can reject $H_0: \sigma_1^2 = \sigma_2^2$ at the $\alpha = .10$ level and conclude that $\sigma_1^2 \neq \sigma_2^2$. We will not pool variances. The number of degrees of freedom associated with our T test is

$$v = \frac{[(s_1^2/n_1) + (s_2^2/n_2)]^2}{\dfrac{(s_1^2/n_1)^2}{n_1 - 1} + \dfrac{(s_2^2/n_2)^2}{n_2 - 1}} = \frac{[(4.52/25) + (2.00/25)]^2}{\dfrac{(4.52/25)^2}{24} + \dfrac{(2.00/25)^2}{24}} = 42.5 \approx 42$$

The observed value of the test statistic is

$$\frac{(\bar{x}_1 - \bar{x}_2) - 0}{\sqrt{(s_1^2/n_1) + (s_2^2/n_2)}} = \frac{(22.3 - 18.7)}{\sqrt{(4.52/25) + (2/25)}} = 7.05$$

Based on the T_{42} distribution, the probability of observing a value this large or larger is less than .0005. That is, $P < .0005$. We can reject H_0 and conclude that individuals today do tend to obtain their first car at an earlier age than in the past.

Let us remind you that the procedure that we have suggested for comparing population means is somewhat controversial. Statistics is an art as well as a science. Differences of opinion do exist. We have presented what we believe is a reasonable approach to the problem. Our opinion is shared by some statisticians but not by others. Unfortunately, there are problems involved regardless of whether you pool or do not pool. The only advice that we can give is to think carefully about the problem at hand. Try to take into account all of the information at your disposal when making your choice concerning pooling. Then make the best decision that you can based on your data.

Exercises 8.4

32. Estimate the number of degrees of freedom associated with the statistic

$$\frac{\bar{X}_1 - \bar{X}_2}{\sqrt{S_1^2/n_1 + S_2^2/n_2}}$$

based on these data:
(a) $n_1 = 20$ $\quad n_2 = 25$
$\quad s_1^2 = 30$ $\quad s_2^2 = 80$
(b) $n_1 = 10$ $\quad n_2 = 9$
$\quad s_1^2 = 25$ $\quad s_2^2 = 5$
(c) $n_1 = 18$ $\quad n_2 = 18$
$\quad s_1^2 = 8$ $\quad s_2^2 = 2$

33. Find the indicated confidence intervals. Assume that $\sigma_1^2 \neq \sigma_2^2$.
(a) Find a 95% confidence interval on $\mu_1 - \mu_2$ when

$$n_1 = 20 \quad n_2 = 25$$
$$s_1^2 = 30 \quad s_2^2 = 80$$
$$\bar{x}_1 = 100 \quad \bar{x}_2 = 95$$

(b) Find a 90% confidence interval on $\mu_1 - \mu_2$ when

$$n_1 = 10 \quad n_2 = 9$$
$$s_1^2 = 25 \quad s_2^2 = 5$$
$$\bar{x}_1 = 30 \quad \bar{x}_2 = 26$$

(c) Find a 99% confidence interval on $\mu_1 - \mu_2$ when

$$n_1 = 18 \qquad n_2 = 18$$
$$s_1^2 = 8 \qquad s_2^2 = 2$$
$$\bar{x}_1 = 25 \qquad \bar{x}_2 = 27$$

34. To estimate the difference in the mean starting salaries for male and female faculty within the university system of a particular state, random samples of size 31 are obtained from each group. These data are obtained on the salaries (in thousands of dollars):

$$\bar{x}_{male} = 16, \quad s_{male}^2 = 40 \qquad \bar{x}_{female} = 14.5, \quad s_{female}^2 = 250$$

(a) Test

$$H_0: \sigma_{male}^2 = \sigma_{female}^2$$

$$H_1: \sigma_{male}^2 \neq \sigma_{female}^2$$

Can H_0 be rejected? What is the approximate P value of your test?

(b) Find a 95% confidence interval on $\mu_1 - \mu_2$. Can you conclude from this confidence interval that there is a difference in the mean starting salaries for these two groups? Explain.

35. A new chemical treatment that does not involve the use of pressure (called NT1) has been devised to improve the life of fence posts. For a random sample of 122 similar cedar posts, 61 are subjected to NT1 and the others are treated with the creosote pressure process (which we will call CPP). The posts are then tagged and placed in the same type of soil in a field. The life span (in years) is noted for each post. From these data the following calculations were made:

mean of post treated with NT1: $\bar{x}_1 = 25.9, \quad s_1^2 = 31.36$

mean of post treated with CPP: $\bar{x}_2 = 22.3, \quad s_2^2 = 17.64$

(a) Can we conclude at the $\alpha = .20$ level that $\sigma_1^2 \neq \sigma_2^2$?

(b) Find a 95% confidence interval on $\mu_1 - \mu_2$. Based on this interval can we conclude that posts treated with NT1 tend to last longer than those treated with CPP? Explain.

36. Test each of these hypotheses. Assume that $\sigma_1^2 \neq \sigma_2^2$. If no α level is given, find the P value of the test.

(a) $H_0: \mu_1 = \mu_2$ $n_1 = 12$ $n_2 = 10$ $\alpha = .05$
     $H_1: \mu_1 > \mu_2$ $s_1^2 = 15$ $s_2^2 = 3$
                   $\bar{x}_1 = 10$ $\bar{x}_2 = 7$

(b) $H_0: \mu_1 = \mu_2$ $n_1 = 16$ $n_2 = 16$ $\alpha = .01$
     $H_1: \mu_1 < \mu_2$ $s_1^2 = 18$ $s_2^2 = 6$
                   $\bar{x}_1 = 9$ $\bar{x}_2 = 11$

(c) $H_0: \mu_1 = \mu_2$ $n_1 = 8$ $n_2 = 8$ $\alpha = .10$
     $H_1: \mu_1 \neq \mu_2$ $s_1^2 = 2$ $s_2^2 = 7$
                   $\bar{x}_1 = 5$ $\bar{x}_2 = 5.5$

(d) $H_0: \mu_1 = \mu_2$ $n_1 = 10$ $n_2 = 10$
    $H_1: \mu_1 > \mu_2$ $s_1^2 = 8$ $s_2^2 = 2$
                           $\bar{x}_1 = 12$ $\bar{x}_2 = 11$

(e) $H_0: \mu_1 = \mu_2$ $n_1 = 100$ $n_2 = 100$
    $H_1: \mu_1 < \mu_2$ $s_1^2 = 36$ $s_2^2 = 9$
                           $\bar{x}_1 = 25$ $\bar{x}_2 = 26$

(f) $H_0: \mu_1 = \mu_2$ $n_1 = 22$ $n_2 = 12$
    $H_1: \mu_1 \neq \mu_2$ $s_1^2 = 16$ $s_2^2 = 4$
                           $\bar{x}_1 = 27$ $\bar{x}_2 = 25$

37. There is some concern that TV commercial breaks are becoming longer. The following observations are obtained on the length of commercial breaks for the 1984 viewing season (January '84–December '84) and the current season (time in minutes):

Population I 1984			Population II current		
2.42	2.16	2.04	2.28	2.39	2.04
2.00	2.35	2.23	2.36	2.63	2.25
1.17	2.40	1.95	2.05	2.29	2.31
1.18	1.47	1.38	2.45	2.39	2.44
2.32	2.82	2.42	2.64	2.11	2.57
1.84			2.62		

(a) Find the point estimate for $\mu_1 - \mu_2$. Does the concern seem to have any basis in fact?

(b) Test

$$H_0: \sigma_1^2 = \sigma_2^2 \quad \text{at the } \alpha = .20 \text{ level}$$

$$H_1: \sigma_1^2 \neq \sigma_2^2$$

Can H_0 be rejected?

(c) Set up and test the appropriate hypotheses for gaining evidence to support the contention that commercial breaks are becoming longer today than in the past.

38. In an experiment to measure free chlorine in water, samples composed of dry powder and ampoules of solution to be mixed according to exact instructions to produce a sample of known chlorine concentration are sent to selected reputable laboratories. Seventeen labs are asked to analyze the solution using the DPD method (method 1) and 26 are asked to use the methyl orange method (method 2). It is believed that method 1 yields results that are on the average higher than method 2. The following information is obtained.

$$\bar{x}_1 = 72.1765, \quad s_1^2 = 156.1544 \qquad \bar{x}_2 = 64.7308, \quad s_2^2 = 4190.0725$$

(a) Is the Satterthwaite procedure an appropriate method for comparing means? Explain.

(b) Set up and test the appropriate hypotheses for gaining evidence that method 1 tends to produce higher chlorine readings than method 2.

39. Students have expressed the opinion that they perform better on handwritten tests than they do on typed tests. To test this theory, an instructor gives identical tests simultaneously to two statistics classes that, from past experience, generally perform comparably on examinations. One test is typed, the other handwritten. The following results are obtained.

Handwritten	Typed
$\bar{x}_1 = 72.1$	$\bar{x}_2 = 68.5$
$s_1^2 = 64.1$	$s_2^2 = 33.74$
$n_1 = 25$	$n_2 = 21$

Does there appear to be statistical evidence that the students are right? Use whichever T procedure is appropriate for these data.

40. Two drug treatments are being compared for potential use in heart transplants. The drugs act as immunosuppresants—they repress the body's natural tendency to reject the transplant. Male ACI rats serve as donors; male Lewis-Brown Norway rats serve as recipients. These rats are known to be poor matches. It is thought that sodium salicylate alone will tend to produce a longer survival time than sodium salicylate used with the drug azathioprine. Do these data support this contention? Explain by describing the statistical tests that you conduct and reporting the P values of these tests.

Sodium salicylate alone (I)			Sodium salicylate and azathioprine (II)		
14.1	17.5	16.6	15.1	14.5	16.8
15.1	15.9	16.4	15.7	16.0	14.9
18.2	16.0	16.5	13.9	17.2	15.3

41. Use the table of random digits, Table IV of Appendix A to select a random sample of ten males and ten females from the patients listed in Table V of Appendix A.
   (a) Find the point estimates for the mean systolic and mean diastolic blood pressures for each group.
   (b) Find point estimates for the variance in the diastolic and systolic blood pressures for each group.
   (c) Test $H_0: \sigma_{male}^2 = \sigma_{female}^2$ where these variances refer to the systolic blood pressure for each group.
   (d) Test $H_0: \mu_{male} = \mu_{female}$ where these means refer to the systolic blood pressure for each group. Use whichever T test is appropriate.
   (e) Test $H_0: \sigma_{male}^2 = \sigma_{female}^2$ where these variances refer to the diastolic blood pressure for each group.
   (f) Test $H_0: \mu_{male} = \mu_{female}$ where these means refer to the diastolic blood pressure for each group. Use whichever T test is appropriate.

COMPUTING SUPPLEMENT C

The procedure PROC TTEST can be used to test for equality of both variances and means by utilizing either the pooled-T test or the Satterthwaite procedure. This SAS procedure is illustrated by using the data of exercise 40 of this chapter to test $H_0: \mu_1 = \mu_2$.

Statement	*Purpose*
DATA TRANSPLT;	names the data set
INPUT TRT TIME @ @;	indicates two variables named *TRT* (treatment) and *TIME* (survival time) are to be entered; @ @ indicates that there will be more than one observation per line
CARDS; 1 14.1 1 17.5 1 16.6 1 15.1 1 15.9 1 16.4 1 18.2 1 16.0 1 16.5 2 15.1 2 14.5 2 16.8 2 15.7 2 16.0 2 14.9 2 13.9 2 17.2 2 15.3 ;	indicates that data follows data values
PROC TTEST;	signals end of data calls for a T test to be run
CLASS TRT;	names the variable that identifies the two populations being compared
TITLE1 TESTING FOR EQUALITY; TITLE2 OF; TITLE3 MEANS AND VARIANCES;	titles the output

The output of this program is shown below:

```
                            TESTING FOR EQUALITY
                                    OF
                            MEANS AND VARIANCES
                              TTEST PROCEDURE
VARIABLE: TIME
    TRT     N        MEAN          STD DEV        STD ERROR       MINIMUM         MAXIMUM
     1      9    16.25555556     1.20945350      0.40315117     14.10000000     18.20000000
     2      9    15.48888889     1.06000524      0.35333508     13.90000000     17.20000000

VARIANCES            T         DF       PROB > |T|

UNEQUAL          1.4301 ⑥    15.7 ⑤     0.1723        ⑦

EQUAL            1.4301 ⑧    16.0       0.1719        ④

FOR HO: VARIANCES ARE EQUAL, F' = 1.30 WITH 8 AND 8 DF  ①  PROB > F' = 0.7180  ②
```

Note that ① gives the observed value of the test statistic for testing H_0: $\sigma_1^2 = \sigma_2^2$ and ② gives the P value for the two-tailed test. The observed value of the test statistic for testing H_0: $\mu_1 = \mu_2$ by using the pooled-T test is given by ② with the P value for the two-tailed test shown in ④. Since we are conducting a one-tailed test our P value is *half* that shown in ④. The approximate number of degrees of freedom for the Satterthwaite procedure is shown in ⑤. The observed value of the test statistic and the two-tailed P value using the Satterthwaite procedure are shown in ⑥ and ⑦ respectively. One should first consider the results of the F test for equality of variances and then use whichever T test is appropriate to test for equality of means.

   The program can be adjusted by making changes in the names of the input variables and the data set. Also, the output of the program can be used to compute confidence intervals on $\mu_1 - \mu_2$ since all required summary statistics are given on the printout.

8.5   Comparison of Two Means: Paired-*T* Test

In this section we discuss the problem of comparing two means for samples that are not independent. This situation arises quite naturally when observations occur in pairs. Sometimes pairs are rather natural; at other times they are created by the researcher. Regardless of how pairing occurs, the purpose of pairing is the same. Namely, it is used to control the effect of some *extraneous* variable. An extraneous variable is a variable that is not of direct interest but which, if not controlled, could interfere with our ability to detect actual differences in population means. The test statistic used here is again a *T* statistic. Hence the name **paired-*T* procedure**. In psychological literature the term *dependent* or *correlated T* is used. Example 1 should clarify the idea.

Example 1

1. There are two supermarkets in the neighborhood. A shopper is interested in determining which offers the best prices on the average. We could simply select a random sample of items from market I and another independent sample of items from market II and then use the techniques developed earlier to compare μ_1 with μ_2. However, in this process there is always a chance that our samples will not really reflect the average prices within the market accurately. For example, by chance, our sample of items from market I might contain a number of very expensive items while our sample from market II might contain only small, inexpensive items. It should be evident that we need to control the extraneous variable *item type* in order to make a true comparison of average prices. To do so, we do comparative shopping. We price a randomly selected item from market I; we then price the same item or one that is comparable at market II. In this way, we create a sample that consists of a collection of paired items. It is obvious that the sample of items from market II is not independent from that of market I. In fact, the items selected from market II are completely determined by those selected from the first market. The pairs here are not natural pairs; they are intentionally created by the researcher for the express purpose of controlling the effects of differences that might occur in item types if samples were drawn independently.

2. A company is preparing to market a new sunscreen. It will be competing with the current favorite. Company officials wish to gain statistical evidence that their brand is more effective than their competitor's product, so that they may safely include this claim in their advertising. To gain this evidence we obtain a group of subjects willing to participate in the study. We could randomly divide this group into two subgroups and use the company brand with one subgroup and the competing brand with the other. However, this procedure ignores a rather serious problem. Namely, the effectiveness of sunscreens varies widely with different skin types. We need to control this extraneous variable in our experiment. To do so, we apply the company brand to one arm and leg of each subject and the competitor's product to the other arm and leg. At the end of three hours' exposure to the sun, the degree

of burn for each individual for each product is determined by making use of skin temperature and color. Hence, each individual generates a natural pair of observations. These observations are obviously not independent, since they are each related to the same individual.

The general situation is as follows: we have two populations under study. Associated with the first is a random variable X with mean $\mu_X = \mu_1$ and variance $\sigma_X^2 = \sigma_1^2$; associated with the second is a random variable Y with mean $\mu_Y = \mu_2$ and variance $\sigma_Y^2 = \sigma_2^2$. We have at hand n observations on the random variable X, and, paired with these, n observations on Y. We make no assumptions concerning σ_1^2 and σ_2^2. We wish to compare μ_1 with μ_2 by considering the parameter $\mu_1 = \mu_2$. The idea is visualized in Figure 8.10.

Theoretically speaking, we will not do anything particularly unusual or new to handle this problem. In fact, several of the problems of Chapter 6 were actually paired-T problems in disguise.

We approach the problem by introducing a new random variable D defined by

$$D = X - Y$$

We assume that D is normal. The rules for expectation can be used to show that the mean value of D is the difference between the mean value of X and the mean value of Y. That is,

$$\mu_D = \mu_X - \mu_Y = \mu_1 - \mu_2$$

Note that, the mean of D is exactly equal to the parameter $\mu_1 - \mu_2$ in which we are interested. The variance of D is denoted by σ_D^2. Note that since we have

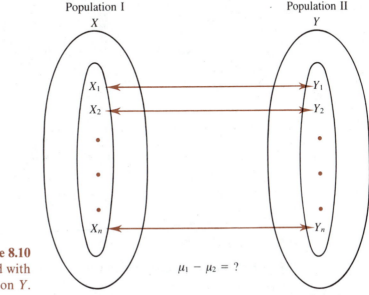

Population I
X

Population II
Y

X_1 Y_1
X_2 Y_2

X_n Y_n

$\mu_1 - \mu_2 = ?$

Figure 8.10
Each observation on X is paired with an observation on Y.

available a collection of n ordered pairs $(X_1, Y_1), (X_2, Y_2), \ldots, (X_n, Y_n)$, we can use these pairs to generate a random sample of n observations on D as follows:

$$D_1 = X_1 - Y_1, D_2 = X_2 - Y_2, \ldots, D_n = X_n - Y_n$$

It should be obvious at this point that we are now back in the context of Chapter 6. We are interested in making inferences on the value of the mean of a single normal random variable D based on information available from a random sample of size n drawn from that distribution. We use exactly the same techniques used earlier.

The unbiased point estimator for $\mu_1 - \mu_2$ is simply the sample mean $\bar{D}$; the unbiased estimator for σ_D^2 is the sample variance S_D^2. Confidence intervals and tests of hypotheses are performed by making appropriate use of the fact that

$$\frac{\bar{D} - (\mu_1 - \mu_2)}{(S_D/\sqrt{n})}$$

is distributed as a t random variable with $n - 1$ degrees of freedom. Since this random variable also assumes the familiar form

$$\frac{\text{estimator} - \text{parameter}}{\text{denominator}}$$

the formula for a $100(1 - \alpha)\%$ confidence interval on $\mu_1 - \mu_2$ based on paired data is

Confidence Interval on $\mu_1 - \mu_2$ When Data Are Paired

$$\bar{D} \pm t_{\alpha/2} S_D/\sqrt{n}$$

where the point $t_{\alpha/2}$ is found relative to the t distribution with $n - 1$ degrees of freedom.

The test statistic used to test any of the three hypotheses:

I $H_0: \mu_1 = \mu_2$	II $H_0: \mu_1 = \mu_2$	III $H_0: \mu_1 = \mu_2$
$H_1: \mu_1 \neq \mu_2$	$H_1: \mu_1 < \mu_2$	$H_1: \mu_1 > \mu_2$
Two-tailed test	Left-tailed test	Right-tailed test

is

$$(\bar{D} - 0)/(S_D/\sqrt{n})$$

which follows a t distribution with $n - 1$ degrees of freedom.

Example 2   The shopper described in Example 1, part 2 compares the prices on 20 common household items and obtains these paired observations and differences:

Price at market I, x	Price at market II, y	d = x − y
$1.50	$1.34	$.16
1.85	1.75	.10
.79	.68	.11
4.59	4.57	.02
1.79	1.83	−.04
1.83	1.71	.12
1.32	1.28	.04
.35	.47	−.12
.33	.46	−.13
3.25	3.15	.10
.37	.36	.01
.59	.54	.05
2.09	2.01	.08
1.90	1.83	.07
1.98	1.87	.11
2.07	1.95	.12
6.00	6.06	−.06
3.49	3.29	.20
1.15	1.12	.03
1.60	1.68	−.08

The observed sample statistics are

$$\bar{d} = .0445$$

$$s_D^2 = .0084$$

$$s_D = .0914$$

The unbiased point estimate for the difference in the mean price at the two markets is

$$\widehat{\mu_1 - \mu_2} = \bar{d} = .0445$$

That is, the average price at market I appears to be about 4 cents higher than at market II.

The general formula for a $100(1 - \alpha)\%$ confidence interval on $\mu_1 - \mu_2$ is

$$\bar{D} \pm t_{\alpha/2}(S_D/\sqrt{n})$$

where the t point involved is associated with the t distribution with $n - 1$ degrees of freedom. Suppose we wish to find a 95% confidence interval on $\mu_1 - \mu_2$. Reading from Table VI of Appendix A with $n - 1 = 20 - 1 = 19$ degrees of freedom, we see that the point $t_{.025}$ in this situation is 2.093. Thus, the desired 95% confidence interval is

$$.0445 \pm 2.093(.0914/\sqrt{20}) \quad \text{or} \quad [.0017, .0873]$$

Since this interval does not contain 0 and consists entirely of positive numbers, we can be 95% sure that on the average, prices at market I are in fact slightly higher than those at market II.

Example 3   To get the statistical evidence needed to support the claim that a particular sunscreen, X is more effective than its nearest competitor, Y, officials select 25 subjects to participate in the experiment described in Example 1, part 2. Burns are rated on a continuous scale from 0 to 20, with higher scores representing a more severe burn. These data are obtained:

Experimental brand, x	Competitor, y	Difference x − y
1.3	7.1	− 5.8
6.0	7.5	− 1.5
4.3	2.0	2.3
19.1	19.3	− .2
7.5	4.3	3.2
2.0	7.5	− 5.5
5.0	6.0	− 1.0
7.9	8.3	− .4
8.9	8.7	.2
9.2	11.3	− 2.1
6.2	7.5	− 1.3
3.0	2.5	.5
6.9	7.1	− .2
7.6	8.3	− .7
8.2	6.9	1.3
15.3	15.7	− .4
14.9	13.8	1.1
6.1	7.3	− 1.2
7.9	8.3	− .4

Experimental brand, x	Competitor, y	Difference x − y
17.5	17.9	−.4
6.1	7.3	−1.2
5.1	4.9	.2
13.7	13.5	.2
14.2	17.1	−2.9
18.1	19.2	−1.1

The makers of brand X are attempting to get evidence that their product is superior. That is, they wish to show that the mean rating with X is smaller than that of the competitor. This contention then should be included as the alternative hypothesis.

Let us test at the $\alpha = .05$ level the hypotheses

$$H_0: \mu_X = \mu_Y$$

$$H_1: \mu_X < \mu_Y$$

The test statistic for such a test is

$$\frac{\bar{D} - 0}{(S_D/\sqrt{n})}$$

which, under H_0, follows a t distribution with $n - 1$ degrees of freedom. In this particular problem, the test statistic follows the t distribution with 24 degrees of freedom. The critical point for a left-tailed test with $\alpha = .05$ is $t_{.95} = -1.711$. The observed sample statistics are

$$\bar{d} = -.69 \qquad s_D^2 = 3.91 \qquad s_D = 1.98.$$

The observed value of the test statistic is

$$\frac{\bar{d} - 0}{(s_D/\sqrt{n})} = \frac{-.69}{(1.98/5)} = -1.74$$

Since $-1.74 < -1.711$, we reject H_0 at the $\alpha = .05$ level and conclude that brand X, on the average, does a better job of preventing sunburn than the current favorite. We can be fairly confident in making this claim in our advertising, since the probability of our being wrong is only .05.

To summarize, a paired-T procedure is used when we want to compare the means of two populations based on paired data. Pairing is done to control the effects of an extraneous variable. Since we actually work with a single population, the population of differences, no comparison between the original population variances is needed. We are assuming the population of differences is normally distributed.

Exercises 8.5

42. A manufacturer is interested in showing that a newly developed gasoline additive increases the gasoline mileage obtained by most makes of cars. To do so, the manufacturer obtains five new cars. Each car is tested using premium unleaded gasoline and then using premium unleaded gasoline with the additive. The variable of interest each time is the number of miles per gallon obtained. The data from the tests are:

x (no additive)	y (with additive)	d = x − y (difference)
15	16	
17	15	
12	15	
25	27	
35	35	

(a) Complete the above chart by finding the five difference scores generated by these paired observations.

(b) Find an unbiased estimate for the mean difference in mileage with and without the additive.

(c) Find an unbiased estimate of the variance in the difference in mileage obtained with and without the additive.

(d) Find s_D.

43. A physical education instructor is trying to decide whether the three-step or four-step approach is preferable for beginning bowlers. The instructor allows each of 18 beginners to bowl three games, and, using as a basis the averages obtained on these games, pairs them so that the members of each pair have approximately the same average score. The first member of each pair is taught the three-step approach and the second, the four-step approach. At the end of the quarter, the change in average for each bowler is recorded. The following data were obtained:

x	y	x	y
15	8	14	5
6	0	12	2
8	1	29	18
22	14	35	26
7	−3		

where x = three-step approach; y = four-step approach.

(a) Find the point estimate for the difference in the mean scores obtained using these two methods.

(b) Find a 90% confidence interval on $\mu_X - \mu_Y$. On the basis of this interval, do you think that one method is superior to the other? Explain.

44. In the manufacture of golf balls two procedures are used. Method I utilizes a liquid center and method II, a solid center. To compare the distance obtained using both types of balls, 12 golfers are allowed to drive a ball of each type, and the length of the drive (in yards) is measured. The following results are obtained:

I	II	I	II
180.0	172.7	195.2	188.9
215.8	202.5	117.6	108.8
140.6	128.1	199.0	186.5
182.7	173.9	179.5	175.9
193.8	180.7	122.3	112.7
100.2	88.7	106.7	99.8

(a) Find the point estimate for the difference in the mean scores obtained.

(b) Find a 90% confidence interval on $\mu_1 - \mu_2$. Is there reason to believe that one ball tends on the average to yield greater distances on the drive?

45. A new barbecue sauce called BBQ is supposed to make chicken more flavorful. Twenty chickens are selected and BBQ is applied to one leg (at random) while the traditional sauce is applied to the other. A panel of experts rates the legs (after cooking) on a scale of 1 through 5 with 5 as the highest flavor rating. Let X be the BBQ rating and Y be the traditional sauce rating. For this sample of 20 pairs,

$$\bar{d} = .124 \quad \text{where} \quad d = x - y \quad \text{and} \quad s_D^2 = .013$$

(a) Construct a 90% confidence interval for the difference in the mean ratings $\mu_X - \mu_Y$.

(b) What can you conclude about the flavor rating of BBQ as compared with the traditional sauce?

46. A micro method has been developed for measuring the amount of carbon monoxide in the air. Researchers want to compare this method with the standard iodine pentoxide method. In order to do so, nine samples were analyzed by each of the two methods, and the following results were obtained (results are in parts per million):

micro	95	184	40	261	215	26	56	128	155
standard	90.5	184.6	44.8	320	244.7	25.8	66.2	137.8	137.8

(a) Construct a 95% confidence interval on $\mu_{\text{micro}} - \mu_{\text{standard}}$.

(b) Is there evidence that these two methods do not yield identical readings on the average?

47. A physician suspects that the office balance beam scale may be giving consistently higher readings for the weights of patients than those obtained on the small bathroom scales that the patients use in their homes. To test this

hypothesis, the physician asks 10 patients to record their weights when they are fully clothed before leaving home. They are weighed fully clothed as soon as they get to the office. The following observations are obtained:

x	y	x	y
150	153	110	112
147	146	179	181
203	207	132	132
171	173	138	139
129	127	210	213

where x = home weight; y = office weight.

Test the appropriate hypotheses at the $\alpha = .10$ level.

48. There is some feeling among psychologists that the IQ of a first-born child, X, tends to be slightly higher than that of the second born, Y. To test the theory, observations are obtained on 200 siblings. The following data are obtained, where $D = X - Y$.

$$\Sigma d = 860 \qquad \Sigma d^2 = 4494$$

Test the appropriate hypotheses at the $\alpha = .10$ level.

49. A paint firm wants to advertise that its brand X lasts longer without peeling than that of the nearest competitor. To gain statistical support for this theory, the firm has 40 houses painted in such a way that half of the front of each house is painted with its brand and the other half with that of the competitor. In this way, each brand of paint is exposed as nearly as possible to the same conditions. The houses are observed, and the number of months until peeling occurs with the firm's brand, X, and the competitor's brand, Y, is recorded. The difference $X - Y$ is found for each house. The following data are obtained:

$$\Sigma d = 22.0 \qquad \Sigma d^2 = 278.87$$

Test the appropriate hypotheses at the $\alpha = .10$ level. Do you think that the firm should go ahead with the planned advertising? Explain on the basis of the consequences involved in making a Type I error.

50. There is some concern among basketball fans that officials are intimidated by the home crowd and tend to call more fouls on the visitors than on the home team. To get support for this point of view, 20 games are randomly selected and the number of fouls called on the home team and the number called on the visitors are recorded. The following data are obtained:

home	12	15	19	8	23	20	30	32	15	21
visitor	14	15	24	17	20	22	27	35	18	24
home	7	18	16	21	23	19	15	6	10	9
visitor	9	17	16	22	26	18	23	10	9	13

Do these data support the contention that officials tend to call more fouls on the visitors than on the home team? Explain based on the P value of your test.

51. An experiment was conducted to compare aluminum baseball bats with wooden bats. Twenty-five major league players were selected and asked to use the two types of bats for a number of games in random order. The number of hits obtained with each type was recorded for each player. It was found that $\bar{d} = 3.2$ (d = the number of hits with the wooden bat minus the number of hits with the aluminum bat) and that $s_D^2 = .36$. Do these data support the contention that, on the average, hitters do better with wooden bats? Explain based on the P value of your test.

52. Fifteen subjects were chosen randomly from a class of freshmen to respond to a questionnaire relating to their attitudes toward life on a campus. Two years later, when they are juniors, the same subjects are asked to fill out the questionnaire again. Their scores on both occasions are shown below, with the higher scores indicating a generally more favorable attitude.

| *freshman:* | 22 | 32 | 18 | 29 | 36 | 25 | 25 | 23 | 30 | 27 | 28 | 26 | 27 | 30 | 31 |
| *junior:* | | 29 | 27 | 22 | 46 | 45 | 35 | 38 | 31 | 41 | 25 | 30 | 24 | 35 | 32 | 37 |

Do these data support the conention that the attitude of juniors tends to be more favorable than that of freshmen? What is the P value of your test?

COMPUTING SUPPLEMENT D

The SAS procedure PROC MEANS can be used to run a paired-T test. The method is illustrated by using the data of Example 3.

Statement	*Purpose*
DATA SUN;	names data set
INPUT X Y;	indicates that two variables, X and Y, will be used; each line will contain an X value followed by its corresponding Y value with a blank between
DIFF = X − Y;	forms a new variable named DIFF that is the difference between X and the corresponding Y
CARDS; 1.3 7.1 6.0 7.5 4.3 2.0	signals that data follows data
18.1 19.2	
; PROC MEANS MEAN STDERR T PRT;	signals end of data asks for the statistics $\bar{D}$ and $S_D/\sqrt{n}$ to be printed; asks for the value of the T statistic $\bar{D}/(S_D\sqrt{n})$ and its P value to be printed
TITLE PAIRED T TEST;	titles output

The output of this program follows:

PAIRED T TEST

VARIABLE	MEAN	STD ERROR OF MEAN	T	PR>|T|
X	8.88000000	1.01924809	8.71	0.0001
Y	9.57200000	1.00424964	9.53	0.0001
DIFF	−0.69200000 ①	0.39534458 ②	−1.75 ③	0.0928 ④

The observed value of $\bar{d}$ is given by ①. The value of $s_D/\sqrt{n}$ used in the construction of a confidence interval on $\mu_1 - \mu_2$ is given by ②. This is also the observed value of the denominator of our test statistic. The value of the test statistic $\bar{d}/(s_D/\sqrt{n})$ is given by ③. It agrees with that found in Example 3, apart from roundoff error. The P value of the two-tailed test is given by ④. Note that, since we are running a one-tailed test, the actual P value is half that reported on the printout.

Vocabulary List and Key Concepts

F distribution

pooled variance

pooled-*T* test

Satterthwaite procedure

paired-*T* test

Review Exercises

53. Before comparing two means based on independent random samples, what test should be conducted? What precautions do we take to minimize the probability of committing a Type II error? What precautions do we take to minimize the consequences of committing a Type II error?

54. If variances seem to be unequal, what procedure is used to compare means when samples are independent? In this case, how are the degrees of freedom determined? What do you do if the estimated degrees of freedom is not an integer?

55. If we cannot show that variances are unequal, what procedure is used to compare means? In this case, how is the common population variance estimated?

56. How are means compared when data are paired? Why is it not necessary to compare population variances in this case?

57. When sample sizes are so large that the number of degrees of freedom associated with these *T* tests are not found in your *T* table, how do you approximate the required *T* points?

Use the following data to answer exercises 58–60:

It is thought that the two sexes respond differently to heat stress. A group of ten men and ten women agree to participate in the study. These subjects were put through a rigorous exercise program. The environment was hot and a minimal amount of water was available. The random variable observed was the percentage of body weight lost. During the course of the experiment two of the women withdrew. These data are observed:

Men (I)		Women (II)	
2.9	3.7	3.0	3.8
3.5	3.8	2.5	4.1
3.9	4.0	3.7	3.6
3.8	3.6	3.3	4.0
3.6	3.7		

58. Find an unbiased estimate for the difference between the mean percentage of body weight lost for men and women. Subtract in the order of men minus women.

59. Test H_0: $\sigma_1^2 = \sigma_2^2$ at the $\alpha = .20$ level.

60. Find a 95% confidence interval on $\mu_1 - \mu_2$. Use whichever interval you think is appropriate taking into consideration the results of exercise 59. Based on this interval, would you be willing to claim that men tend to lose a higher percentage of their body weight than do women when exercising under conditions of extreme heat stress? Explain.

Use the following data to answer exercises 61–65:

A study is conducted to compare the mean occupational exposure to radioactivity among utility workers in 1973 to that during the current year. It is thought that new safety measures implemented in recent years will result in a lower current exposure rate. These data are obtained: (data are in rem)

1973			Current year		
.90	.91	.92	.40	.74	.63
.91	.85	1.10	.57	.59	.69
.94	.76	.93	.60	.85	.65
1.00	.97	1.30	.70	.61	.76
.81	.91	.94	.72	.89	.62
.96	.80		.81	.60	

61. Find an unbiased point estimate for $\mu_1 - \mu_2$ where μ_1 denotes the mean exposure to radioactivity among utility workers in 1973.

62. Test H_0: $\sigma_1^2 = \sigma_2^2$ at the $\alpha = .20$ level.

63. Set up the null and alternative hypotheses needed to gain statistical evidence to support the contention that the mean level of exposure to radioactivity among utility workers has been reduced.

64. Do you think that pooling is appropriate?

65. Based on your answer to exercise 64, test the hypotheses of exercise 63 at the $\alpha = .05$ level.

Use the following data to answer exercises 66–68:

A study is conducted to investigate the effect of physical exercise on the serum-cholesterol level of the individual. It is thought that regular exercise will reduce the level of cholesterol in the blood. Blood samples are taken on 11 individuals before and after training. These data result:

Subject	Pretraining cholesterol level in mg/DL	Posttraining cholesterol level in mg/DL
1	182	198
2	232	210
3	191	194
4	200	220
5	148	138
6	249	220
7	276	219
8	213	161
9	241	210
10	380	313
11	262	226

66. Set up the null and alternative hypotheses required to support the contention that this exercise program tends to reduce the serum-cholesterol level.

67. Find a point estimate for $\mu_1 - \mu_2$ where μ_1 is the cholesterol level prior to beginning the exercise program.

68. What type of T test is appropriate for testing the hypotheses in exercise 66? Test the hypotheses. Do you think that H_0 should be rejected? Explain based on the P value of your test.

Use the following data to answer exercises 69–71:

Manufacturers of a new quick-start heater think that their heater can raise the temperature of a room much faster than can that of their nearest competitor. They think that the difference between mean heating times is more than ten seconds. To gain statistical support for this claim, a small test room is devised. The random variable studied is the time in seconds required to raise the temperature of the test room from 60°F to 70°F. These data are gathered:

Competitor (I)		Quick start (II)	
69.3	52.6	28.6	30.6
56.0	34.4	25.1	31.8
22.1	60.2	26.4	41.6
47.6	43.8	34.9	21.1
53.2	48.1	29.8	36.0
23.2	13.8	28.4	37.9
		38.5	13.9
		30.2	

69. Set up the null and alternative hypotheses needed to establish the manufacturer's claim.

70. Find a point estimate for $\mu_1 - \mu_2$. Does this estimate tend to support the research hypothesis?

71. Test the null hypothesis of exercise 69. Do you think that the research hypothesis has been substantiated? Explain based on the P values of the tests that you have conducted.

Use the following data to answer exercises 72–73:

A firm has two possible sources for its computer hardware. A study is conducted to compare the prices offered. These data result:

Item	Price I	Price II
1	6,000	5,900
2	575	580
3	15,000	15,000
4	150,000	145,000
5	76,000	75,000
6	5,650	5,600
7	10,000	9,975
8	850	870
9	900	890
10	3,000	2,900

72. Find point estimates for μ_1, μ_2 and $\mu_1 - \mu_2$. Based on the point estimate for $\mu_1 - \mu_2$, do you think that there is probably a difference in the average prices charged by these two companies?

73. Find a 95% confidence interval on $\mu_1 - \mu_2$. Does the confidence interval indicate that $\mu_1 \neq \mu_2$? Explain.

74. The cost of repairing an electronic circuit may depend on the stage of production at which the circuit is found to be defective. It is claimed that the variance in the cost of repairing a circuit is smaller when it fails as it is being installed in the system, rather than when it has already been put in use in the field. These data are obtained:

Field failure	System failure
$n_1 = 25$	$n_2 = 21$
$\bar{x}_1 = \$120$	$\bar{x}_2 = \$65$
$s_1^2 = 100$	$s_2^2 = 30$

(a) Set up the null and alternative hypotheses needed to support the claim of the researchers.

(b) Test H_0 at the $\alpha = .01$ level.

Regression and Correlation

In this chapter, we take a brief look at two very important statistical topics. These are correlation and regression. Correlation studies are designed to measure the strength of the linear relationship between two random variables X and Y; regression studies are designed to describe the relationship between the mean of a random variable Y and one or more other variables. Usually, the primary purpose of a regression study is prediction. We want to develop an equation by which the value of Y can be predicted reasonably well, based on knowledge of the values assumed by the other variables in the equation. We begin by considering the problem of simple linear regression.

9.1 Introduction to Simple Linear Regression

In **simple linear regression** two variables are involved. Usually the role of each is clearly defined. The first, denoted by Y, is the variable of primary concern. Its behavior is uncertain and uncontrolled by the experimenter; that is, Y is a random variable. We want to make inferences about the mean of Y and about Y itself. We think that both the mean of Y and the value assumed by Y depends somewhat on the value assumed by some other variable x. Furthermore, we think that the relationship between the mean of Y and x is linear. Since we think that the behavior of Y depends on x, Y is called the **dependent variable**. Our job is to use observed data to develop a linear equation, an equation whose graph is a straight line, that expresses the mean of Y in terms of x. The values of x used in developing this equation are often controlled by the experimenter; that is, x is not considered to be a random variable. Rather, we select the values of x to be used in our experiment and then observe the value or values assumed by the random variable Y at these points. Since the values chosen for x do not depend on Y, x is called the **independent variable**.

337

A few examples should make these ideas clear.

Example 1

1. A physician wants to predict the concentration of a particular drug in the bloodstream (Y) based on knowledge of the length of time (x) since the drug was administered to the patient.

2. A physical therapist wants to study the relationship between the number of hours of aerobic exercises performed per week (x) and the number of pounds lost by individuals participating in a weight-reducing program (Y).

3. An economist wants to develop an equation by which the price of wheat futures in Chicago (Y) can be predicted from knowledge of the amount of precipitation received during the growing season in the Midwest farm belt (x).

4. A psychiatrist wants to predict the length of time that a patient will be kept under treatment (Y) based on knowledge of the severity of the original diagnosis (x).

Note that there is a subtle difference in the variable x in the above examples. In the first two cases, the numerical values of x actually used in an experiment to generate a prediction equation can be selected specifically by the experimenter. The experimenter can allow as much time as needed to elapse between the administration of the drug and the measurement of the concentration of drug in the bloodstream; he or she can design therapeutic programs to include preset numbers of hours of aerobic exercises. A study such as this in which the value of x can be preselected is called a **designed regression study**. In the last two cases, the values of x used to develop the prediction equation are not controlled by the researcher. The economist does not control rainfall in the midwest; he observes it. The psychiatrist does not control the severity of the initial diagnosis of a mental patient; she observes it. Studies such as these in which the values of x used to develop the prediction equation are not specifically selected by the researcher are called *observational studies*. Regardless of the situation, the problem of regression is the same—to find a reasonable prediction equation. The mathematical procedures used in both cases are identical.

If we know that the independent variable has assumed a specific value x, what is our best choice for the predicted value of Y? For example, if we know that a drug has been in the system for five minutes, what is the predicted concentration of the drug in the bloodstream? That is, What is Y, given that $x = 5$? Using a notation similar to that used for conditional probabilities in Chapter 2, we write, What is $Y \mid x = 5$? Since different people react differently to the same drug, even though it is given in the same concentration at the same time, we know that $Y \mid x = 5$ is a random variable. It has a theoretical mean $\mu_{Y\mid x=5}$. Common sense indicates that the most logical prediction for the value of Y when $x = 5$ is the average value of Y for that particular value of x. The choice of the value 5 is arbitrary. For any specified value x, intuition points to $\mu_{Y\mid x}$ as the best predicted value for $Y \mid x$. This idea provides the motivation for the definition of the term *curve of regression of Y on x*.

Definition 9.1

Regression of Y on x

Let x be a mathematical variable and let Y be a random variable. The **curve of regression of Y on x** is the graph of the mean value of Y for various values of x. That is, it is the graph of $\mu_{Y|x}$.

Typical regression curves are shown in Figure 9.1. Note that these curves are theoretical in nature. They are the graphs of the theoretical mean for the random variable Y as a function of the variable x. They serve as the ideal curve for predicting the value of Y based on knowledge of x. As in all statistical problems, we will not usually know the exact equations for these curves. Our problem is to estimate these equations from observed data.

We are concerned here with simple linear regression. The phrase *simple linear regression* implies two things. The word *simple* refers to the fact that the predicted value of the dependent variable Y is based on knowledge of the value of exactly one independent variable x. If the value of Y is based on knowledge of more than one independent variable, then we are dealing with what is called *multiple regression*. The word *linear* refers to the fact that the curve of regression is a straight line. In Figure 9.1(a) the regression is nonlinear, whereas it is linear in Figure 9.1(b). The formal definition of a linear regression curve is given in Definition 9.2.

Definition 9.2

Linear Regression

A curve of regression of Y on x is said to be linear if and only if

$$\mu_{Y|x} = \alpha + \beta x$$

for α and β real numbers.

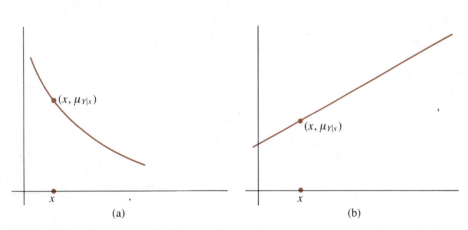

(a)   (b)

Figure 9.1
(a) Nonlinear regression curve, (b) linear regression curve.

Since we shall be attempting to estimate the numbers α and β from a set of observations on x and Y, let us review the geometric meaning of these numbers. Beta, the coefficient of x, is called the **slope** of the line. It tells us the direction and steepness of the line. Positive values of β indicate that the line rises from left to right; negative values indicate a line that falls from left to right. Large values of β, in either the positive or negative sense, indicate very steep curves. These ideas are demonstrated in Figure 9.2(a), (b), and (c). A value of 0 for β indicates that the line is horizontal. A horizontal regression line implies that knowledge of the value of x does not help in predicting Y. See Figure 9.2(d). The number α is called the **Y intercept** of the line. It gives the place at which the line intersects the Y axis. In the regression context, it gives the predicted value of Y when $x = 0$. See Figure 9.2(e).

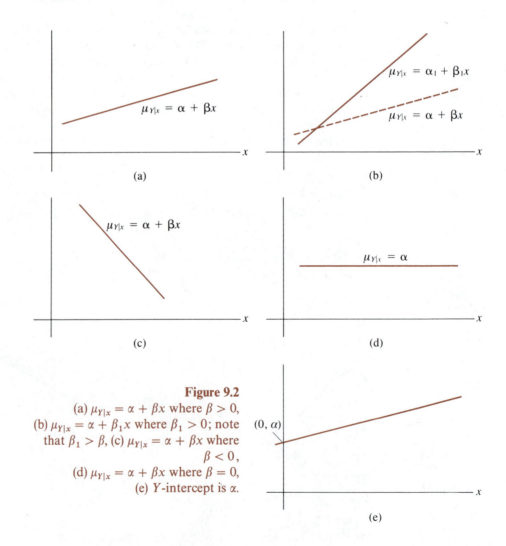

Figure 9.2
(a) $\mu_{Y|x} = \alpha + \beta x$ where $\beta > 0$,
(b) $\mu_{Y|x} = \alpha + \beta_1 x$ where $\beta_1 > 0$; note that $\beta_1 > \beta$, (c) $\mu_{Y|x} = \alpha + \beta x$ where $\beta < 0$,
(d) $\mu_{Y|x} = \alpha + \beta x$ where $\beta = 0$,
(e) Y-intercept is α.

The first job in attempting to develop a regression equation is to decide whether or not a linear regression curve appears to be appropriate. To do so we obtain a random sample. In simple linear regression, a random sample is properly viewed as consisting of a collection of pairs of the form

$$\{(x_1,\ Y\,|\,x_1), (x_2,\ Y\,|\,x_2), \ldots, (x_n,\ Y\,|\,x_n)\}$$

Here the first member of each pair is a numerical value of the independent variable x; the second member of each pair is the random variable Y considered at the point $x = x_i$. When no confusion could result by doing so, we shorten the notation by letting $Y_i = Y\,|\,x_i$. Thus, we usually think of our sample as being a collection of pairs of the form

$$\{(x_1,\ Y_1), (x_2,\ Y_2), \ldots, (x_n,\ Y_n)\}$$

Once the experiment is conducted and we observe the values assumed by the random variables $Y_1, Y_2, \ldots, Y_n$, we have available a collection of pairs of real numbers of the form

$$\{(x_1,\ y_1), (x_2,\ y_2), \ldots, (x_n,\ y_n)\}$$

One approach to the problem of determining whether or not linear regression is appropriate is to plot what is called a *scattergram* of the data points. This is a picture of the data in which the pairs $(x,\ y)$ are plotted as points in a coordinate plane. If the points tend to form a linear trend, then we use a simple linear regression method to obtain a prediction equation; otherwise, we do not. This idea is illustrated in Examples 2 and 3.

Example 2

Suppose that the psychiatrist mentioned in Example 1 devises a rating scale that ranges in value from 1, mild disorder, to 10, extremely severe. Each patient at the time of the original diagnosis can be rated at some point x along this scale. The patient is then observed and the length of time of treatment y is noted. We thus generate a pair of numbers for each patient. Suppose for a sample of 25 patients we obtain the following data (time is in months):

(1.1, 3.0)	(2.0, 4.0)	(3.5, 5.2)	(5.2, 7.0)	(7.2, 12.0)
(1.1, 5.0)	(2.0, 4.2)	(3.5, 5.4)	(5.2, 6.0)	(7.2, 12.5)
(1.3, 4.0)	(2.7, 4.6)	(4.0, 6.0)	(6.0, 9.0)	(8.0, 16.0)
(1.3, 3.5)	(3.5, 5.0)	(4.0, 6.3)	(6.1, 10.0)	(8.0, 15.5)
(1.3, 2.0)	(3.5, 4.9)	(4.0, 6.5)	(6.1, 11.0)	(8.0, 13.0)

These observations can be plotted on a scattergram as shown in Figure 9.3. The points in the scattergram do not lie on a straight line. However, there is a linear trend to the data. This trend is what we are seeking. It identifies the problem as being one of simple linear regression. In this case, we may assume that the graph of $\mu_{Y|x}$ is a straight line and we may visualize the theoretical line of regression of Y on x as shown in Figure 9.4.

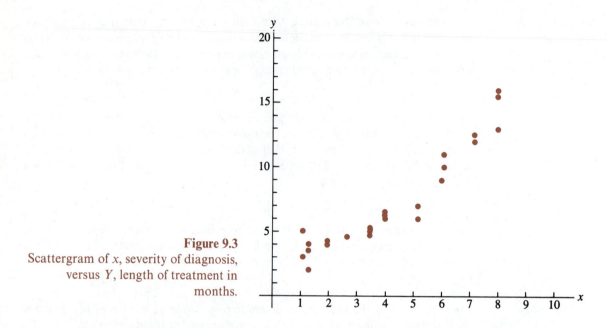

Figure 9.3
Scattergram of *x*, severity of diagnosis, versus *Y*, length of treatment in months.

Our problem in the previous example is to estimate the equation for the straight line by means of the observed data. Note that several straight lines could be drawn to include some of the data points. Also, many lines could be drawn so that the points are close to the lines. Which of the lines should we choose as our estimate of $\mu_{Y|x}$? Which of the lines best fits the data points? These questions will be answered in the next section.

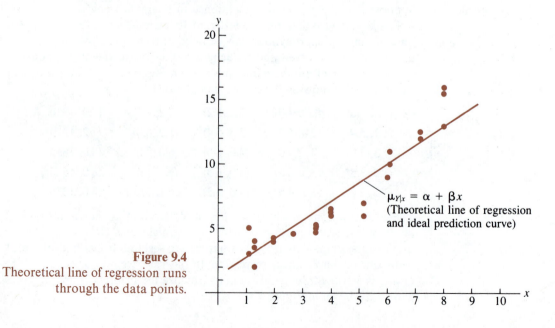

$\mu_{Y|x} = \alpha + \beta x$
(Theoretical line of regression and ideal prediction curve)

Figure 9.4
Theoretical line of regression runs through the data points.

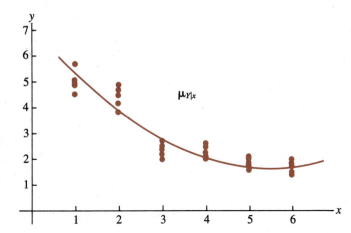

Figure 9.5

Scattergram of x, time elapsed since administration of the drug, versus Y, the concentration of drug in the bloodstream.

Example 3

To help predict the concentration of a drug in the bloodstream based on knowledge of the length of time since the drug was administered, a physician obtains these data: (time is in minutes and concentration is in percentage per cubic centimeter)

(1.0, 5.0)	(2.0, 4.9)	(3.0, 2.5)	(4.0, 2.6)	(5.0, 2.1)
(6.0, 1.5)	(1.0, 5.1)	(2.0, 4.7)	(3.0, 2.7)	(4.0, 2.3)
(5.0, 2.0)	(6.0, 1.9)	(1.0, 5.7)	(2.0, 3.8)	(3.0, 2.0)
(4.0, 2.5)	(5.0, 1.9)	(6.0, 1.7)	(1.0, 4.9)	(2.0, 4.2)
(3.0, 2.4)	(4.0, 2.1)	(5.0, 1.8)	(6.0, 2.0)	(1.0, 4.5)
(2.0, 4.5)	(3.0, 2.2)	(4.0, 2.0)	(5.0, 1.6)	(6.0, 1.4)

Is a linear regression approach to this problem feasible? To help determine this, consider the scattergram shown in Figure 9.5. It appears that a curve of the form shown in the figure fits the data better than any straight line. We would not use the methods of linear regression to estimate $\mu_{Y|x}$.

Exercises 9.1

In each of these exercises, plot a scattergram for the given data and decide whether or not linear regression seems applicable.

1. Let x denote the number of hours that a student studied for a statistics test, and Y the grade obtained on that test. The following observations are obtained:

(.25, 60)	(.75, 60)	(1.0, 65)	(1.5, 70)	(1.75, 70)	(2.5, 75)
(3.0, 85)	(.5, 62)	(.75, 69)	(1.0, 70)	(1.5, 75)	(2.0, 73)
(2.5, 82)	(3.0, 82)	(.5, 65)	(.75, 71)	(1.25, 68)	(1.5, 65)
(2.0, 79)	(2.75, 85)	(3.0, 88)	(.5, 80)	(1.0, 58)	(1.25, 71)
(1.75, 78)	(2.0, 60)	(2.75, 79)	(3.5, 90)		

2. Let x denote the amount of money spent per week on advertising by a small fast-food shop, and Y the total sales per week. These observations are obtained:

(10, 824)	(20, 976)	(30, 180)	(40, 239)	(50, 467)
(10, 218)	(20, 329)	(30, 836)	(40, 991)	(50, 323)
(10, 603)	(20, 796)	(30, 795)	(40, 509)	(50, 865)
(10, 439)	(20, 856)	(30, 926)	(40, 424)	(50, 385)

3. Let x denote the IQ of a small child, and Y the child's attention span when exposed to a new task involving the manipulation of building blocks. An experiment produces these observations (time is in minutes):

(70, .5)	(80, 1.5)	(95. 2.1)	(110, 3.5)	(120, 2.0)	(130, 1.3)
(70, .8)	(85, 2.0)	(95, 2.5)	(110, 3.7)	(120, 1.8)	(135, 1.1)
(75, .9)	(85, 1.8)	(98, 2.6)	(110, 4.0)	(120, 1.5)	(140, 1.0)
(77, 1.0)	(90, 2.1)	(100, 2.8)	(112, 2.1)	(129, 1.2)	(140, .75)
(77, 1.1)	(90, 2.3)	(100, 3.0)	(115, 2.0)	(130, 1.2)	(150, .5)
(77, 1.3)	(92, 2.5)	(105, 3.3)	(115, 2.1)	(130, 1.0)	(150, .3)

4. Let x denote the average speed of a specific make and model of auto over a 100-mile test trip, and Y the gas mileage obtained. The following results are observed:

(35, 27.1)	(40, 27.2)	(45, 23.2)	(50, 19.5)	(55, 17.2)	(60, 17.2)
(35, 28.2)	(40, 25.0)	(45, 24.0)	(50, 20.1)	(55, 19.1)	(60, 15.0)
(35, 30.0)	(40, 22.0)	(45, 21.1)	(50, 22.3)	(55, 18.5)	(60, 16.3)
(35, 26.7)	(40, 24.1)	(45, 20.0)	(50, 17.8)	(55, 18.7)	(60, 14.9)

5. Let x denote the number of years of formal education, and Y the income of an individual at age 30. The following observations are obtained on these variables (income is in thousands of dollars):

(8, 8)	(12, 10)	(14, 15)	(16, 21)	(20, 17)
(8, 10)	(12, 12)	(14, 14)	(16, 18)	(20, 20)
(10, 7)	(12, 13)	(16, 12)	(16, 40)	(20, 15)
(12, 5)	(12, 14)	(16, 14)	(16, 10)	(22, 25)
(12, 15)	(14, 11)	(16, 16)	(16, 17)	(24, 35)

COMPUTING SUPPLEMENT E

The SAS procedure PROC PLOT produces scattergrams. To illustrate, we ask SAS to construct a scattergram of the data of Example 2.

Statement	*Purpose*

```
DATA PSYC;                          names the data set

INPUT X Y;                          names the variables

LABEL Y=TIME;                       labels variable Y
LABEL X=RATING;                     labels variable X

CARDS;                              data follows

1.1  3.0
1.1  5.0                            data; one x and its corresponding
1.3  4.0                            y per line
 .    .
 .    .
 .    .
8.0 13.0                            signals end of data
;                                   asks for a scattergram
PROC PLOT;

PLOT Y*X;                           first named variable, Y, appears
                                    along the vertical axis

TITLE1    IS REGRESSION;            titles the output
TITLE2    LINEAR?;
```

The scattergram produced is shown below.

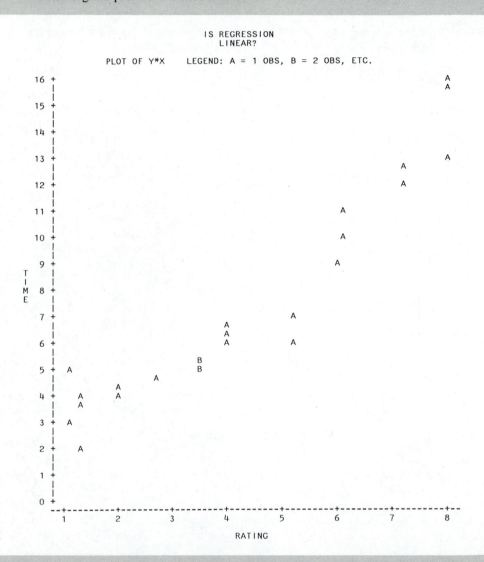

9.2   The Method of Least Squares

We now consider the problem of estimating the line of regression of Y on x. The procedure used is called the **method of least squares**. We illustrate this technique through the following example.

Example 1

Let x denote the number of hours of aerobic exercises performed per week, and Y the number of pounds lost by an individual participating in a weight-reduction program. These observations are obtained on the pair of variables (x, Y):

(1, .5)	(2, .7)	(3, 1.1)	(4, 1.3)	(5, 1.6)
(1, .8)	(2, .65)	(3, 1.2)	(4, 1.29)	(5, 1.62)
(1, .6)	(2, .71)	(3, 1.0)	(4, 1.32)	(5, 1.64)
(1.5, .7)	(2.5, 1.0)	(3.5, 1.0)	(4.5, 1.2)	(5.5, 1.7)

The scattergram for these data, together with the imagined theoretical line of regression, is shown in Figure 9.6. Our job is to estimate the line $\mu_{Y|x} = \alpha + \beta x$ based on the observed data. This, in fact, means that we must estimate α, the y intercept of the theoretical line of regression, and β, the slope of this line. The estimates for these parameters are denoted by a and b. The estimated regression line and thus the estimated value of y for a given value x is

$$\hat{y} = \hat{\mu}_{Y|x} = a + bx$$

The reasoning behind the method of least squares is quite simple. From among the many straight lines drawn through the scattergram, we pick out the one that is closest to, or best fits, the observations. The fit involved is "best" in the sense that it minimizes the sum of the squares of the differences between the observed values of Y and the values estimated for Y via the fitted regression line. We denote the difference between the ith observed value of Y, y_i, and the estimated value of Y at this point, $a + bx_i$, by e_i. That is,

$$e_i = y_i - (a + bx_i)$$

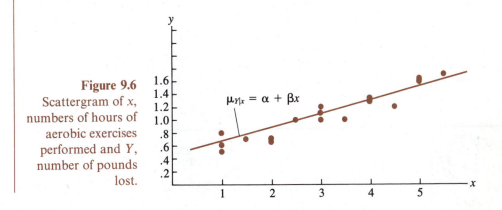

Figure 9.6
Scattergram of x, numbers of hours of aerobic exercises performed and Y, number of pounds lost.

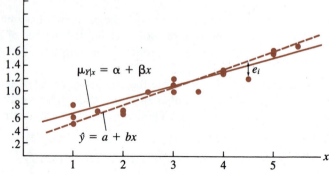

Figure 9.7
$\mu_{Y|x} = \alpha + \beta x$ is the theoretical line of
regression and ideal prediction curve;
$\hat{y} = a + bx$ is the estimated line of
regression and the equation used to
make predictions; e_i is the difference
between the estimated value of Y at x_i
and the observed value of Y at x_i.

These differences are called **residuals**. This idea is illustrated in Figure 9.7.

Note that

$$e_1 = y_1 - (a + bx_1)$$

$$e_2 = y_2 - (a + bx_2)$$

$$\vdots$$

$$e_{20} = y_{20} - (a + bx_{20})$$

The sum of the squares of these differences or residuals is denoted by **SSE** and is given by

$$\text{SSE} = \Sigma e^2 = \Sigma[y - (a + bx)]^2$$

To minimize this sum, methods of differential calculus must be used. However, it can be shown that the values of a and b that minimize SSE are

$$b = \frac{20\Sigma xy - \Sigma x \Sigma y}{20\Sigma x^2 - (\Sigma x)^2}$$

$$a = \bar{y} - b\bar{x}$$

Thus, to fit a regression line to this particular set of observations, we need to compute four quantities:

$$\Sigma y \quad \Sigma x \quad \Sigma xy \quad \Sigma x^2$$

These quantities are found to be:

$$\Sigma y = 21.63 \qquad \Sigma xy = 77.66$$

$$\Sigma x = 62.50 \qquad \Sigma x^2 = 236.25$$

Thus,

$$b = \frac{20(77.66) - (21.63)(62.5)}{20(236.25) - (62.50)^2} = .25$$

$$a = (21.63)/(20) - .25(62.5/20) = .30$$

The estimated regression line is

$$\hat{\mu}_{Y|x} = .30 + .25x$$

Suppose we are now asked to predict the average amount of weight that would be lost by individuals who exercise for $x = 2.1$ hours per week. Our best estimate for this average weight loss is

$$\hat{\mu}_{Y|x} = .30 + .25(2.1) = .83 \text{ pounds}$$

This is also our best estimate for the amount of weight that would be lost by a particular individual who exercises 2.1 hours per week.

The procedure used and the results obtained in Example 1 can be generalized easily. To do so, we first note that we are estimating the parameters α and β. We have called the estimates for these parameters a and b. We shall denote the estimators for α and β by A and B respectively. This notation is consistent with most past usage. The parameter being estimated is denoted by a Greek letter, the statistic used to estimate the parameter is denoted by a capital letter, and the observed value of the statistic is denoted by a lowercase letter. We now note that the only change needed to make the formulas given in Example 1 for a and b applicable to any problem is to replace the number 20, the sample size in our example, by n, the sample size in general. The resulting estimators for α and β are given in Theorem 9.1.

Theorem 9.1

The least squares estimators for α and β in a problem in simple linear regression are

$$B = \frac{n\Sigma xY - \Sigma x\Sigma Y}{n\Sigma x^2 - (\Sigma x)^2}$$

$$A = \bar{Y} - B\bar{x}$$

Example 2

The scattergram of Figure 9.4 on page 342 suggests that linear regression is applicable in predicting the length of time that a patient will be kept under treatment based on the severity of the original diagnosis. Let us fit a regression line to these data:

(1.1, 3.0)	(2.0, 4.0)	(3.5, 5.2)	(5.2, 7.0)	(7.2, 12.0)
(1.1, 5.0)	(2.0, 4.2)	(3.5, 5.4)	(5.2, 6.0)	(7.2, 12.5)
(1.3, 4.0)	(2.7, 4.6)	(4.0, 6.0)	(6.0, 9.0)	(8.0, 16.0)
(1.3, 3.5)	(3.5, 5.0)	(4.0, 6.3)	(6.1, 10.0)	(8.0, 15.5)
(1.3, 2.0)	(3.5, 4.9)	(4.0, 6.5)	(6.1, 11.0)	(8.0, 13.0)

From the data

$$\Sigma xy = 979.02 \qquad \Sigma y = 181.6$$

$$\Sigma x = 105.8 \qquad \Sigma x^2 = 579.96$$

$$b = \frac{25(979.02) - (105.8)(181.6)}{25(579.96) - (105.8)^2} = 1.59$$

$$a = (181.6/25) - 1.59(105.8/25) = .53$$

The estimated line of regression is $\hat{\mu}_{Y|x} = .53 + 1.59x$. If an individual's original diagnosis indicates that $x = 5$, what is the predicted length of treatment? Based on our estimated line of regression,

$$\hat{y} = .53 + 1.59(5) = 8.48 \text{ months}$$

A word of caution needs to be added. One should not extend or project a regression line beyond the given data points. For instance, in Example 1, our observations on x extend from $x = 1$ through $x = 5.5$. For values in this range we can use the regression line $\hat{y} = .30 + .25x$ to predict the amount of weight loss, because for these values we have evidence that the regression is linear. However, outside of this range we have no indication at all of linearity. In fact, we have absolutely no information concerning the behavior of Y for these values of x. Hence, for these values, we have no basis for making inferences.

Exercises 9.2

Use these observations on x and Y to answer exercises 6–11.

$$(1, 10) \qquad (2, 12) \qquad (3, 14) \qquad (4, 20) \qquad (5, 21)$$

6. What are the observed values of Σx, Σx^2, Σy, Σxy?

7. What are the least squares estimates for α and β?

8. What is the equation for the estimated regression line? Does this line rise or fall from left to right?

9. What is the predicted value of Y when $x = 2.5$?

10. What is the predicted value of the average value of Y when $x = 2.5$?

11. Would it be wise to try to use the equation of exercise 8 to predict the value of Y when $x = 10$? Explain.

Use these observations on x and Y to answer exercises 12–16:

$$(1, -2) \qquad (2, -4) \qquad (3, -9) \qquad (4, -15) \qquad (5, -18) \qquad (6, -21)$$

12. What are the observed values of Σx, Σx^2, Σy, Σxy?

13. What are the least squares estimates for α and β?

14. What is the equation for the estimated regression line? Does this line rise or fall from left to right?

15. What is the predicted value of Y when $x = 3.5$?

16. What is the predicted value of the average value of Y when $x = 3.5$?

17. It is known that humidity influences evaporation. A study is conducted to examine the relationship between relative humidity (x), and the percentage of evaporation of the solvent in paint while spray painting (Y). These data are obtained:

$$n = 25 \qquad \Sigma x = 1314.90 \qquad 28\% \le x \le 75\%$$

$$\Sigma x^2 = 76{,}308.53 \qquad \Sigma xy = 11{,}824.44 \qquad \Sigma y = 235.7$$

(a) Find the estimated line of regression.
(b) As x increases, would you expect the value of Y to increase or decrease? Explain.
(c) Estimate the percentage evaporation of the paint solvent on a day in which the relative humidity is 50%.

18. A study is conducted to examine the relationship between x, an individual's score on an obesity index, and Y, the individual's resting metabolic rate. This rate is measured in milliliters of oxygen consumed per minute. Higher values of x denote a larger degree of obesity. These data result:

$$n = 43 \qquad \Sigma x^2 = 53{,}515.25 \qquad 0 \le x \le 50$$

$$\Sigma x = 1482.5 \qquad \Sigma xy = 379{,}207.5 \qquad \Sigma y = 10{,}719$$

(a) Find the estimated line of regression.
(b) As x increases would you expect Y to increase or decrease? Explain.
(c) Estimate the average resting metabolic rate of individuals who score 40 on the obesity scale.

19. Let x denote the number of pounds of salt applied per 100 yards of highway during winter snows, and Y the number of potholes observed for the same 100-yard stretch of street in the spring. Assume that the following observations have been obtained.

$$(10, 5) \qquad (15, 10) \qquad (15, 9) \qquad (8, 3) \qquad (30, 16) \qquad (10, 6)$$

(a) Plot a scattergram for these data.
(b) Find the value of each of the following quantities:

$$\Sigma x \qquad \Sigma x^2 \qquad \Sigma y \qquad \Sigma xy$$

(c) Find the estimated values for α and β.
(d) Find the equation for the estimated regression line.
(e) How many potholes would you expect to see if 18 pounds of salt were used per 100-yard stretch of highway?

20. Let x denote the number of weeks that a subject has been on a special diet, and Y the number of pounds lost during that period. The following observations are obtained:

    (3, 9)   (7, 12)   (2, 2)   (4, 13)   (8, 12)   (3, 8)

    (a) Plot a scattergram for these data.
    (b) Find the value of each of the following quantities:

    $$\Sigma x \quad \Sigma x^2 \quad \Sigma y \quad \Sigma xy$$

    (c) Find the equation for the estimated regression line of Y on x.
    (d) Use this line to predict the average weight loss for a 5-week period.

21. Let x denote the temperature in degrees Celsius between 3:00 and 4:00 in the afternoon during the summer, and Y the demand for electrical power on a scale of 1 to 10, with 10 representing peak demand. The following data are obtained over a 10-day period:

    (21, 4)   (31, 7)   (25, 5)   (34, 8)   (30, 7)
    (23, 4.5)   (32, 6)   (30, 6)   (38, 9)   (38, 9.5)

    (a) Plot a scattergram for the above data to ascertain whether or not linear regression looks feasible.
    (b) Find the value of each of the following quantities:

    $$\Sigma x \quad \Sigma x^2 \quad \Sigma y \quad \Sigma xy$$

    (c) Find the equation for the estimated regression line.
    (d) Use the estimated regression line to predict the demand for electricity when the temperature hits 35 degrees Celsius (95 degrees Fahrenheit).

22. Let x denote the height of a father in feet, and Y the adult height of his eldest son. Assume that the following observations have been obtained:

    (5.5, 5.8)   (5.7, 6.0)   (5.8, 5.8)   (6.2, 6.4)
    (5.6, 5.7)   (5.7, 5.6)   (6.0, 6.1)   (6.4, 6.5)
    (5.6, 6.0)   (5.7, 6.1)   (6.0, 6.2)   (6.4, 6.2)
    (5.65, 6.1)   (5.75, 6.0)   (6.1, 6.0)   (6.6, 6.7)

    (a) Plot a scattergram for the above data to ascertain whether or not linear regression seems applicable.
    (b) Compute

    $$\Sigma x \quad \Sigma x^2 \quad \Sigma y \quad \Sigma xy$$

    (c) Find estimates for α and β.
    (d) Find the equation for the estimated regression line.
    (e) Use the estimated line of regression to predict the adult height of a son born to a man who is 6.3 feet tall.

23. These data are obtained on x, the length of time in weeks that a promotional project has been in progress at a small business, and Y, the percentage increase in weekly sales over the period just prior to the beginning of the campaign.

    (1, 10)   (2, 10)   (3, 18)   (4, 20)   (1, 11)   (3, 15)
    (1, 12)   (2, 15)   (3, 17)   (4, 19)   (2, 13)   (4, 16)

    (a) Plot a scattergram for these data.
    (b) Find the equation for the estimated line of regression.
    (c) Use the estimated regression line to predict the percentage increase in sales if the campaign has been in progress for 1.5 weeks.

24. Consider the data of exercise 1. Find the estimated line of regression of Y on x. Use this to predict the grade of a student who studies for 1.7 hours.

25. Consider the data of exercise 4. Find the estimated line of regression of Y on x. Use this line to predict the average gas mileage obtained by drivers averaging 42 miles per hour.

26. Find the estimated line of regression of Y on x for the data of exercise 5. Use this to predict the income at age 30 of an individual with 15 years of formal education.

27. Each of the following pairs represent mile run times (in seconds) by world class runners during the given year: (54, 239.4) represents the record set by Bannister in 1954 of 3:59.4, while Ryun in 1966 ran the mile in 3:51.3.

    (54, 239.4)   (58, 236.2)   (64, 234.1)   (68, 231.8)
    (54, 238.0)   (60, 235.3)   (64, 234.9)   (70, 232.0)
    (56, 238.1)   (60, 234.8)   (66, 231.3)   (70, 231.9)
    (56, 238.5)   (62, 235.1)   (66, 232.7)   (72, 231.4)
    (58, 234.5)   (62, 234.4)   (68, 231.4)   (72, 231.5)

    (a) Use these data to find the estimated line of regression.
    (b) Use the estimated line from part (a) to predict a time for the year 2000.
    (c) Does this example suggest limits for reasonable predictions from the regression line?

COMPUTING SUPPLEMENT F

The procedure PROC REG is used to estimate the line of regression of Y on x. The data of Example 2 is used to illustrate how this is done.

Statement	*Purpose*
`DATA PATIENT;`	names the data set
`INPUT X Y @@;`	two variables x and Y are used; @@ indicates that each line may contain more than one pair (x, y)
`CARDS;`	indicates that data follows

`1.1  3.0  2.0  4.0`
`3.5  5.2  7.2  12.0`
`.    .    .    .`
`.    .    .    .`
`6.1 11.0  8.0 13.0`
`5`
	data
	
	last real data points
	. indicates that the Y value for an x value of 5 is "missing"; this allows PROC REG to estimate Y when $x = 5$
	
	signals end of data

`PROC REG;`	calls for regression procedure
`MODEL Y = X/P;`	variable to the left of " $=$ " is identified as the dependent variable; variable to the right of " $=$ " is identified as being independent; P asks for predicted values for all values of x in data set to be computed.

```
                                           SAS
DEP VARIABLE: Y
                           SUM OF              MEAN
SOURCE         DF          SQUARES            SQUARE       F VALUE       PROB>F

MODEL          1           335.104            335.104    181.547.0.0001
ERROR          23          42.453898          1.845822
C TOTAL        24          377.558

               ROOT MSE        1.358610       R-SQUARE         0.8876
               DEP MEAN        7.264000       ADJ R-SQ         0.8827
               C.V.            18.70333

                           PARAMETER          STANDARD      T FOR H0:
VARIABLE       DF          ESTIMATE           ERROR      PARAMETER=0      PROB> [T]

INTERCEP       1           0.526545①          0.569095        0.925        0.3645
X              1           1.592026②          0.118156       13.474        0.0001

                           PREDICT
OBS        ACTUAL          VALUE         RESIDUAL

  1         3.000          2.278         0.722226
  2         5.000          2.278         2.722
  3         4.000          2.596         1.404
  4         3.500          2.596         0.903821
  5         2.000          2.596        -.596179
  6         4.000          3.711         0.289403
  7         4.200          3.711         0.489403
  8         4.600          4.825        -.225016
  9         5.000          6.099        -1.099
 10         4.900          6.099        -1.199
 11         5.200          6.099        -.898637
 12         5.400          6.099        -.698637
 13         6.000          6.895        -.894650
 14         6.300          6.895        -.594650
 15         6.500          6.895        -.394650
 16         7.000          8.805        -1.805
 17         6.000          8.805        -2.805
 18         9.000         10.079        -1.079
 19        10.000         10.238        -.237905
 20        11.000         10.238         0.762095
 21        12.000         11.989         0.010866
 22        12.500         11.989         0.510866
 23        16.000         13.263         2.737
 24        15.500         13.263         2.237
 25        13.000         13.263        -.262755
 26           .            8.487③

SUM OF RESIDUALS                         -7.99361E-15
SUM OF SQUARED RESIDUALS                  42.4539
```

The estimate for α is given by ① ; that for β is shown in ② . The predicted value of y when x is 5 is given by ③ .

9.3   The Simple Linear Regression Model (Optional)

As you have seen, the method of least squares allows us to find point estimates for α and β. With these available we can find point estimates for $\mu_{Y|x}$ and $Y|x$. In order to extend these point estimates to interval estimates, we must make some additional assumptions concerning the random variables $Y|x_1, Y|x_2, \ldots, Y|x_n$.

In simple linear regression we are assuming that the mean of $Y|x_i$ is $\alpha + \beta x_i$ for $i = 1, 2, \ldots, n$. That is, we are assuming that the mean values for these random variables lie on a straight line. Random variables usually deviate from their theoretical mean value. These deviations are assumed to be due to random or chance influences that cannot be controlled. Let us denote the difference between $Y|x_i$ and its mean, $\alpha + \beta x_i$, by E_i. Hence we are defining the random variables E_i by

$$E_i = Y|x_i - (\alpha + \beta x_i) \qquad i = 1, 2, \ldots, n$$

We assume that $E_1, E_2, \ldots, E_n$ is a random sample from a normal distribution with mean 0 and variance σ^2. We can rewrite the above equation in the form

$$Y|x_i = (\alpha + \beta x_i) + E_i \qquad i = 1, 2, \ldots, n$$

It can be shown that the assumptions made on $E_1, E_2, \ldots, E_n$ guarantee that the random variables $Y|x_1, Y|x_2, \ldots, Y|x_n$ are independent, normally distributed random variables with means $\alpha + \beta x_i$ and common variance σ^2. These ideas are illustrated in Figure 9.8. Note that the mean values of $Y|x_1, Y|x_2, \ldots, Y|x_n$ may differ but their variances are the same. Thus, the associated normal curves may differ in location but they all have the same shape.

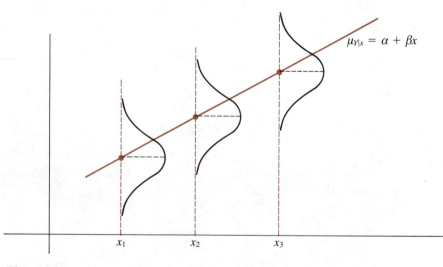

Figure 9.8
At each value x_i, $Y|x_i$ is normally distributed with mean $\alpha + \beta x_i$ and variance σ^2. Thus each curve has the same shape but they may differ in location.

In summary, to make inferences on α, β, $\mu_{Y|x}$, and $Y|x$ other than point estimates we adopt this model for simple linear regression.

Simple Linear Regression Model

$$Y \,|\, x_i = (\alpha + \beta x_i) + E_i$$

where E_i are independent, normally distributed random variables each with mean 0 and variance σ^2.

The model states that the observed value of Y for a given value x_i depends on two things: $\alpha + \beta x_i$, the mean value of Y for this value of x, and E_i, a random or unexplained deviation from this mean value.

Our first job is to estimate the common variance σ^2. Recall that variance is the average value of the square of the difference between a random variable and its theoretical mean value. Hence, for a given value x of the independent variable, the variance of the random variable $Y|x$ is

$$E[(Y \,|\, x - (\alpha + \beta x))^2]$$

Since we do not know the true values for α and β, we must estimate σ^2 based on our sample

$$\{(x_1, \, Y \,|\, x_1), (x_2, \, Y \,|\, x_2), \ldots, (x_n, \, Y \,|\, x_n)\}$$

To do so, we replace $Y|x$ by our sample values, and replace α and β by their estimators A and B in the above expression. We then sum over the sample values and "average" the results. As in the past, the "average" that we use is not a true arithmetic average. Rather, we divide not by n, but by $n-2$. In this way the estimator that we obtain for σ^2 can be shown to be unbiased. The estimator is:

$$\hat{\sigma}^2 = \Sigma[Y \,|\, x - (A + Bx)]^2/(n-2)$$

We can simplify things notationally by letting $Y \,|\, x_i = Y_i$. Thus

$$\hat{\sigma}^2 = \Sigma[Y - (A + Bx)]^2/(n-2)$$

The numerator of this expression should be familiar to you. It is the sum of the squares of the deviations of the random variables Y about the fitted regression line. That is, it is SSE written in estimator form. In summary, we shall use this statistic to estimate σ^2, the common population variance of the random variables $Y_1, Y_2, \ldots, Y_n$.

unbiased estimator for σ^2: $\hat{\sigma}^2 = \text{SSE}/(n-2)$

A simple example will illustrate the idea. We will show you a computational shortcut to use with larger data sets later.

Example 1

Consider these data:

x	y
0	1
1	3
2	2
5	6

For these data

$$\Sigma x = 8 \qquad \Sigma y = 12 \qquad \Sigma xy = 37 \qquad \bar{x} = 2$$

$$\Sigma x^2 = 30 \qquad \Sigma y^2 = 50 \qquad n = 4 \qquad \bar{y} = 3$$

The point estimates for the slope and intercept of the estimated line of regression are

$$b = \frac{n\Sigma xy - \Sigma x\Sigma y}{n\Sigma x^2 - (\Sigma x)^2} = \frac{4(37) - 8(12)}{4(30) - 8(8)} = \frac{13}{14}$$

$$a = \bar{y} - b\bar{x} = 3 - (13/14)(2) = 16/14$$

The estimated line of regression is

$$\hat{\mu}_{Y|x} = 16/14 + (13/14)x$$

The estimated values of Y at each of our observed x values are

$$\hat{y}_1 = a + bx_1 = 16/14 + (13/14)(0) = 16/14$$

$$\hat{y}_2 = a + bx_2 = 16/14 + (13/14)(1) = 29/14$$

$$\hat{y}_3 = a + bx_3 = 16/14 + (13/14)(2) = 42/14$$

$$\hat{y}_4 = a + bx_4 = 16/14 + (13/14)(5) = 81/14$$

The residuals, the differences between the observed values of Y and the estimated values of Y at each of our observed x values, are

$$e_1 = y_1 - \hat{y}_1 = 1 - 16/14 = -2/14$$

$$e_2 = y_2 - \hat{y}_2 = 3 - 29/14 = 13/14$$

$$e_3 = y_3 - \hat{y}_3 = 2 - 42/14 = -14/14$$

$$e_4 = y_4 - \hat{y}_4 = 6 - 81/14 = 3/14$$

These residuals are pictured in Figure 9.9. SSE, the sum of the squares of the residuals, is given by

$$\text{SSE} = \Sigma e^2 = (-2/14)^2 + (13/14)^2 + (-14/14)^2 + (3/14)^2 = 27/14$$

An unbiased estimate for σ^2 based on these data is

$$\hat{\sigma}^2 = \text{SSE}/(n-2) = (27/14)/(4-2) = 27/28$$

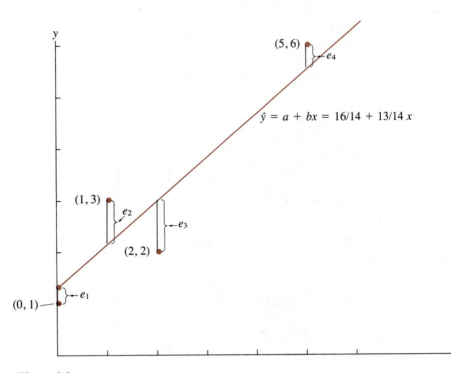

Figure 9.9
The residuals e_1, e_2, e_3, and e_4 give the difference between the data points and the estimated regression line.

It should be clear that calculating SSE directly from its definition is time consuming when sample sizes are large. To alleviate this problem, we introduce a notational convention and some computational shortcuts that will allow us to express and compute SSE more easily. The derivations of these shortcuts are not hard but they are not particularly instructional. For this reason, we shall simply state them here and use them in future computations. They are particularly helpful because most calculators are programmed to allow you to find the individual terms involved by simply entering the data points.

Computational Shortcuts

1. $S_{yy} = \Sigma(y - \bar{y})^2 = [n\Sigma y^2 - (\Sigma y)^2]/n$

2. $S_{xx} = \Sigma(x - \bar{x})^2 = [n\Sigma x^2 - (\Sigma x)^2]/n$

3. $S_{xy} = \Sigma(x - \bar{x})(y - \bar{y}) = [n\Sigma xy - \Sigma x\Sigma y]/n$

The formulas for S_{xy} and S_{yy} are expressed assuming that we are working with the observed values of Y. To avoid complicating the notation, we shall use S_{xy} and S_{yy} even when Y is still in random variable form. This should cause no problem, since you should be able to tell whether you are working with Y or observed values of Y from the context of the problem.

The first thing to notice is that we can now express B, the estimator for the slope of the line of regression, in compact form. In particular,

Computational Formula for the Slope of the Regression Line

$$B = \frac{n\Sigma xY - \Sigma x \Sigma Y}{n\Sigma x^2 - (\Sigma x)^2} = \frac{S_{xy}}{S_{xx}}$$

It can be shown that SSE is given by

Computational Formula for SSE

$$\text{SSE} = S_{yy} - BS_{xy}$$

To illustrate, we reconsider the data of Example 1.

Example 2

In Example 1, we found that

$$\Sigma x = 8 \qquad \Sigma y = 12 \qquad \Sigma xy = 37 \qquad \bar{x} = 2 \qquad b = 13/14$$

$$\Sigma x^2 = 30 \qquad \Sigma y^2 = 50 \qquad n = 4 \qquad \bar{y} = 3 \qquad \text{SSE} = 27/14$$

For these data,

$$S_{yy} = [n\Sigma y^2 - (\Sigma y)^2]/n = [4(50) - (12)^2]/4 = 56/4$$

$$S_{xx} = [n\Sigma x^2 - (\Sigma x)^2]/n = [4(30) - (8)^2]/4 = 56/4$$

$$S_{xy} = [n\Sigma xy - \Sigma x \Sigma y]/n = [4(37) - 8(12)]/4 = 52/4$$

Using the shortcut formulas,

$$b = S_{xy}/S_{xx} = (52/4)/(56/4) = 52/56 = 13/14$$

$$\text{SSE} = S_{yy} - bS_{xy} = 56/4 - (13/14)(52/4) = 108/56 = 27/14$$

As you can see, these results agree with those obtained earlier.

The computational formulas presented here will be used extensively in the next section. The exercises that follow should provide practice in their use.

Exercises 9.3

Use these data to answer exercises 28–37.

x	y
1	6
2	7
3	10
4	13

28. Find Σx, Σx^2, and S_{xx}.
29. Find Σxy, Σy, and S_{xy}.
30. Find the estimated slope of the regression line.
31. Find the estimated intercept of the regression line.
32. Find the equation for the estimated regression line.
33. Find the predicted value of Y at each of the observed x values.
34. Find the four residuals and use these to compute SSE directly.
35. Find Σy^2 and S_{yy}.
36. Use the computational formula for SSE to verify your answer in exercise 34.
37. Estimate σ^2.

Use the following data to answer exercises 38–50:

Let x denote the number of baggers on duty at a supermarket, and let Y denote the length of time, in minutes, that it takes for a customer to get through the check-out line.

x	y	x	y
3	9	5	15
4	11	7	17
5	14	8	17
10	26		

38. Plot a scattergram.
39. Find Σx, Σx^2, and S_{xx}.
40. Find Σxy, Σy, and S_{xy}.
41. Find the estimated slope of the regression line.
42. Find the estimated intercept of the regression line.
43. Graph the estimated regression line on the scattergram.
44. Find the predicted value of Y at each of the observed x values.
45. Find the seven residuals and indicate them graphically on your scattergram.
46. Find SSE directly from the residuals.

47. Find Σy^2 and S_{yy}.

48. Use the computational formula for SSE to verify your answer to exercise 46.

49. Estimate σ^2.

50. Predict the time required to get through the check-out line when 8 baggers are at work.

9.4   Confidence Interval Estimation (Optional)

Confidence intervals can be found on four different entities associated with simple linear regression. These are α, the intercept of the regression line; β, the slope of the regression line; $\mu_{Y|x_0}$, the mean value of Y when x assumes a specific value x_0; and $Y|x_0$, the value of Y itself when x assumes the value x_0. The two intervals of primary importance are the last two mentioned. We restrict our discussion to these intervals.

We begin by considering the problem of constructing a confidence interval on the mean value of Y when $x = x_0$. Recall that the point estimator for this parameter is

$$\hat{\mu}_{Y|x_0} = A + Bx_0$$

As in the past, to extend this point estimator to a confidence interval we must consider its probability distribution. To do this, we must consider the distribution of the random variables A and B. Although the derivations are beyond the scope of this text, it can be shown that the model assumptions made earlier guarantee the following things:

Distribution of A and B

1. A is normally distributed with mean α and variance $\sigma^2 \Sigma x^2 / n S_{xx}$.

2. B is normally distributed with mean β and variance σ^2 / S_{xx}.

Note that these facts say, among other things, that A and B are unbiased estimators for α and β respectively. Since $\hat{\mu}_{Y|x_0}$ is expressed in terms of A and B, its distribution depends on the distribution of A and B. It can be shown that the distribution of $\hat{\mu}_{Y|x_0}$ is as follows:

Distribution of $\hat{\mu}_{Y|x_0}$

$\hat{\mu}_{Y|x_0}$ is normally distributed with mean $\alpha + \beta x_0 = \mu_{Y|x_0}$ and variance $[1/n + (x_0 - \bar{x})^2 / S_{xx}]\sigma^2$.

If we now standardize our estimator, we see that the random variable

$$\frac{\hat{\mu}_{Y|x_0} - \mu_{Y|x_0}}{\sigma\sqrt{(1/n) + (x_0 - \bar{x})^2 / S_{xx}}}$$

is standard normal. Since we do not know σ, we must estimate it from our data. As in the past, when we replace the true value of σ by its estimator $\hat{\sigma}$, the resulting random variable no longer follows the standard normal distribution. Rather, it follows a t distribution. This time, the number of degrees of freedom associated with the T random variable is $n - 2$. That is,

$$\frac{\hat{\mu}_{Y|x_0} - \mu_{Y|x_0}}{\hat{\sigma}\sqrt{(1/n) + (x_0 - \bar{x})^2/S_{xx}}} = T_{n-2}$$

This random variable assumes the familiar form

(estimator − parameter)/denominator

Without doing the derivation, we know that the confidence interval will be of the form

estimator $\pm$ probability point $\cdot$ denominator

In this case, we obtain the following bounds for a $100(1 - \alpha)\%$ confidence interval on the mean value of Y for a specified value of x.

Confidence Interval on $\mu_{Y|x_0}$

$$\hat{\mu}_{Y|x_0} \pm t_{\alpha/2}\hat{\sigma}\sqrt{(1/n) + (x_0 - \bar{x})^2/S_{xx}}$$

where $\hat{\sigma} = \sqrt{SSE/(n - 2)}$ and the point $t_{\alpha/2}$ is found relative to the t distribution with $n - 2$ degrees of freedom.

This looks complicated. However, the calculations required are not hard. An example should make things clear!

Example 1

In Example 1 in Section 2, we considered the relationship between x, the number of hours of aerobic exercises performed per week, and Y, the number of pounds lost by an individual participating in a weight-reduction program. For the data given,

$$\Sigma x = 62.50 \qquad \Sigma y = 21.63 \qquad \Sigma xy = 77.66 \qquad \bar{x} = 3.13$$

$$\Sigma x^2 = 236.25 \qquad \Sigma y^2 = 26.11 \qquad \bar{y} = 1.08 \qquad n = 20$$

$$S_{xy} = [n\Sigma xy - \Sigma x\Sigma y]/n$$

$$= [20(77.66) - 62.50(21.63)]/20 = 10.07$$

$$S_{xx} = [n\Sigma x^2 - (\Sigma x)^2]/n$$

$$= [20(236.25) - (62.50)^2]/20 = 40.94$$

$$b = S_{xy}/S_{xx} = .25$$

$$a = \bar{y} - b\bar{x} = 1.08 - (.25)(3.13) = .30$$

As you can see these point estimates agree with those found earlier. Suppose we want to estimate the average amount of weight lost by individuals who exercise for $x_0 = 2.1$ hours per week. We know that the point estimate for this mean is

$$\hat{\mu}_{Y|x_0} = a + bx_0$$

$$= .30 + .25(2.1) = .83 \text{ pounds}$$

To extend this point estimate to a 95% confidence interval, we must estimate σ. For these data,

$$S_{yy} = [n\Sigma y^2 - (\Sigma y)^2]/n$$

$$= [20(26.11) - (21.63)^2]/20 = 2.72$$

and

$$\text{SSE} = S_{yy} - bS_{xy}$$

$$= 2.72 - .25(10.07) = .20$$

The estimated value of σ^2 is

$$\hat{\sigma}^2 = \text{SSE}/(n - 2) = .20/18 \approx .01$$

Taking the square root, we see that

$$\hat{\sigma} = \sqrt{\hat{\sigma}^2} = \sqrt{.01} = .10$$

A 95% confidence interval on $\mu_{Y|x_0}$ is

$$\hat{\mu}_{Y|x_0} \pm t_{\alpha/2}\hat{\sigma}\sqrt{(1/n) + (x_0 - \bar{x})^2 S_{xx}}$$

The point $t_{\alpha/2}$ is found relative to the t distribution with $n - 2 = 18$ degrees of freedom. Its value is $t_{.025} = 2.101$. Substituting, the 95% confidence interval is given by

$$.83 \pm (2.101)(.10)\sqrt{(1/20) + (2.1 - 3.13)^2/40.94}$$

or

$$.83 \pm .057$$

We are 95% confident that the average amount of weight lost by individuals exercising 2.1 hours per week lies between .773 and .887 pounds.

We can form what is called a **confidence band** on the true regression line. To do so, we choose several values of x and construct confidence intervals at these selected points. We then join the endpoints of these intervals with smooth curves. In this way, we obtain a band or strip which should contain the true regression line. Example 2 demonstrates the idea.

Example 2

Let us use previous data to construct a 95% confidence band on the true line of regression of Y, the number of pounds lost by an individual who participates in a weight-reduction program, on x, the number of hours of aerobic exercises performed per week. To do so, we select several x values within the range of our data. We want these values to be spread throughout our possible x values. Furthermore, it is convenient to let $\bar{x}$ be one of the points selected. For this reason, let us use the points 1, 2.1, 3.13, 4, and 5 in constructing our confidence band. From example 1,

$$\hat{\sigma} = .10 \qquad \bar{x} = 3.13 \qquad t_{.025} = 2.101 \qquad b = .25$$

$$n = 20 \qquad S_{xx} = 40.94 \qquad a = .30$$

Substituting into the formula

$$\hat{\mu}_{Y|x_0} \pm t_{\alpha/2}\hat{\sigma}\sqrt{(1/n) + (x_0 - \bar{x})^2/S_{xx}}$$

we obtain these confidence intervals:

Value of x	Confidence bands
1	$.55 \pm .084$ or .466 to .634
2.1	$.83 \pm .057$ or .773 to .887
3.13	$1.08 \pm .047$ or 1.033 to 1.127
4	$1.30 \pm .055$ or 1.245 to 1.355
5	$1.55 \pm .077$ or 1.473 to 1.627

The endpoints of these confidence intervals are plotted and joined by smooth curves to form the confidence band shown in Figure 9.10. We are 95% confident that the true line of regression lies within the band shown. Note that the band is narrowest at the point $x_0 = \bar{x} = 3.13$. This is not coincidental. At this point, $x_0 - \bar{x} = 0$ and the term $(x_0 - \bar{x})^2/S_{xx}$ drops out of our calculations making the term $\sqrt{(1/n) + (x_0 - \bar{x})^2/S_{xx}}$ a minimum.

The second problem that we consider is that of estimating the value of Y itself when x assumes the value x_0. We already know that the point estimator for $Y|x_0$ is the same as the point estimator for $\mu_{Y|x_0}$. That is,

$$\hat{Y}|x_0 = A + Bx_0$$

However, to extend this point estimator to a confidence interval requires some thought. We begin by considering the random variable $\hat{Y}|x_0 - Y|x_0$, the difference between the estimator for Y when $x = x_0$ and the value of Y itself when $x = x_0$. It can be shown that this random variable has the following distribution:

Distribution of $\hat{Y}|x_0 - Y|x_0$

The random variable $\hat{Y}|x_0 - Y|x_0$, is normally distributed with mean 0 and variance $[1 + (1/n) + (x_0 - \bar{x})^2/S_{xx}]\sigma^2$.

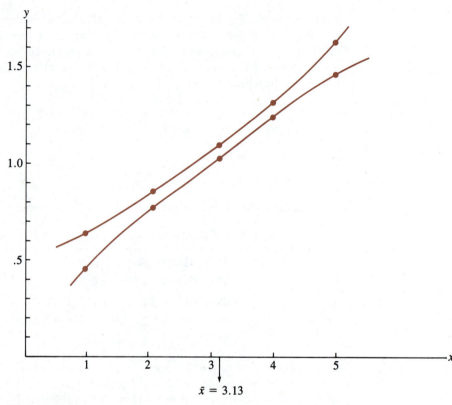

Figure 9.10
A 95% confidence band on the true regression line.

We can standardize this random variable to conclude that the random variable

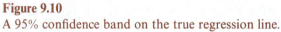

$$\frac{\hat{Y}\,|\,x_0 - Y\,|\,x_0}{\sigma\sqrt{1 + (1/n) + (x_0 - \bar{x})^2/S_{xx}}}$$

is standard normal. When we replace σ by its estimator $\hat{\sigma}$, we again obtain a T random variable with $n - 2$ degrees of freedom. This random variable,

$$T_{n-2} = \frac{\hat{Y}\,|\,x_0 - Y\,|\,x_0}{\hat{\sigma}\sqrt{1 + (1/n) + (x_0 - \bar{x})^2/S_{xx}}}$$

is similar in structure to those seen earlier. For this reason, we can conclude that the bounds for a $100(1 - \alpha)\%$ confidence interval on the value of Y when $x = x_0$ are

---

Confidence Interval on $Y \mid x_0$

$$\hat{Y} \mid x_0 \pm t_{\alpha/2}\,\hat{\sigma}\sqrt{1 + (1/n) + (x_0 - \bar{x})^2/S_{xx}}$$

where $\hat{\sigma} = \sqrt{\text{SSE}/(n-2)}$ and the point $t_{\alpha/2}$ is found relative to the t distribution with $n-2$ degrees of freedom.

Note that these bounds look very much like those used to determine a confidence interval on the mean value of Y when $x = x_0$. The difference in the two is that the former entails the term $\sqrt{(1/n) + (x_0 - \bar{x})^2/S_{xx}}$, whereas the corresponding term in the latter is a little larger, namely, $\sqrt{1 + (1/n) + (x_0 - \bar{x})^2/S_{xx}}$. This is to be expected since we should be able to predict the average response of a group with more precision than we can predict the response of an individual member of that group.

Example 3 illustrates this type of confidence interval.

Example 3

Let us construct a 95% confidence interval on the amount of weight that will be lost by an individual who exercises 2.1 hours per week. From previous work we know that

$$\hat{\sigma} = \ .10 \qquad \bar{x} = \ 3.13 \qquad t_{.025} = 2.101 \qquad b = .25$$

$$n = 20 \qquad S_{xx} = 40.94 \qquad a = \ .30$$

Substituting into the formula

$$\hat{\mu}_{Y \mid x_0} \pm t_{\alpha/2}\,\hat{\sigma}\sqrt{(1/n) + (x_0 - \bar{x})^2/S_{xx}}$$

we obtain the confidence bands

$$.83 \pm 2.101(.10)\sqrt{1 + (1/20) + (2.1 - 3.13)^2/40.94}$$

or

$$.83 \pm .218$$

We are 95% confident that an individual who exercises 2.1 hours per week will lose between .612 and 1.048 pounds. Note that, as expected, this interval is a little longer than that obtained for the average weight loss for all individuals who exercise 2.1 hours per week ($.83 \pm .057$).

Confidence bands can be constructed on the value of Y for given values of x. These bands are found in the same way as those found earlier. They will, of course, be a little wider than the corresponding band for the mean value of Y for given values of x. Remember that these bands are constructed in such a way that, a priori, there is a $100(1 - \alpha)\%$ probability that the observed value of Y for a given value of x will lie within the band. For example, a 95% confidence band is constructed in such a way that 95% of the time the observed value of Y should lie within the band. Values outside the band are considered to be unusual.

Exercises 9.4

Use the following data to answer exercises 51–54:

A study is conducted of the relationship between x, the weight of a car in tons; and Y, the gasoline mileage obtained. Ten properly tuned cars are driven for 1000 miles and the average number of miles per gallon is recorded. Summary statistics are

$$n = 10 \qquad \Sigma x^2 = 28.64 \qquad \Sigma y^2 = 2900.46$$

$$\Sigma x = 16.75 \qquad \Sigma y = 170 \qquad \Sigma xy = 282.41$$

51. Find $S_{xx}, S_{xy}, S_{yy}, \bar{x}$.
52. Estimate σ^2 and σ.
53. Find a 95% confidence interval on the average gasoline mileage obtained by all cars weighing two tons.
54. Find a 95% confidence interval on the gasoline mileage of an individual car weighing two tons.

Use the following data to answer excercises 55–61:

A study of the relationship between energy consumption and household incomes is conducted. These data are obtained on x, the household income in thousands of dollars, and Y, the energy consumption. Actual consumption is 10 times the reported figure in Btu.

x	y	x	y
21.0	1.9	60.0	6.5
30.0	3.0	75.0	7.0
40.0	4.5	88.0	9.0
55.0	5.0	95.0	9.5

55. Find $\Sigma x, \Sigma x^2, S_{xx}, \bar{x}$.
56. Find $\Sigma y, \Sigma y^2, S_{yy}$.
57. Find $\Sigma xy, S_{xy}$.
58. Estimate σ^2 and σ.
59. Find a 90% confidence interval on the mean energy consumption of all families with incomes of $50,000 per year.
60. Find a 90% confidence interval on the energy consumption of a particular family with an income of $50,000 per year.

61. Find a 90% confidence band on the energy consumption of individual families. Use x values of 25, 45, 58, and 90. Sketch the confidence band. Would it be unusual for a family with an income of $70,000 to have an energy consumption in excess of 8.0×10 Btu? Explain based on the confidence band.

62. Use the data of exercises 39–50 to construct a 95% confidence interval on the average time required to get through a check-out line when 8 baggers are at work.

63. Use the data of exercises 39–50 to construct a 95% confidence band on the time required for an individual to get through a check-out line when 8 baggers are at work. Use x values of 4, 6, 8, and 10. Would it be unusual for it to take less than ten minutes for an individual to get through the line under these conditions? Explain.

64. Use the data of exercise 23 to construct a 90% confidence interval on the percentage increase in sales for a particular campaign that has been in progress for 1.5 weeks.

65. Use the data of exercise 21 to construct a 90% confidence interval on the average demand for electricity on days when the temperature hits 35 degrees Celsius.

66. Use the data of exercise 22 to construct a 95% confidence interval on the adult height of the eldest son of a man who is 6.3 feet tall.

*67. Use the rules for expectation to show that $\hat{\mu}_{Y|x_0}$ is an unbiased estimator for $\mu_{Y|x_0}$.

*68. Use the rules of variance to show that

$$\text{Var}(\hat{\mu}_{Y|x_0}) = [(1/n) + (x_0 - \bar{x})^2/S_{xx}]\sigma^2$$

[Hint: Write $\hat{\mu}_{Y|x_0}$ in the form

$$\hat{\mu}_{Y|x_0} = A + Bx_0 = \bar{Y} - B\bar{x} + Bx_0 = \bar{Y} - B(\bar{x} - x_0)$$

and use the fact that $\bar{Y}$ and B are independent to complete the derivation.]

*69. Use the rules for variance to show that

$$\text{Var}(\hat{Y}|x_0 - Y|x_0) = [1 + (1/n) + (x_0 - \bar{x})^2/S_{xx}]\sigma^2$$

[Hint: $\text{Var}(\hat{Y}|x_0 - Y|x_0) = \text{Var}(\hat{Y}|x_0 + \text{Var}(Y|x_0).$]

9.5   Inferences on the Slope (Optional)

Recall that the first step in a regression problem is to look at the scattergram of the data. If linear regression is applicable, then a linear trend should be suggested by the plot. If the data form a pattern that is very obviously not linear, then no attempt should be made to fit a straight line prediction equation to the data. However, occasionally a data set arises in which the issue is not clear cut. A weak linear trend might be visible but it might not be pronounced enough to be sure that a straight line prediction equation will be useful. When this occurs, we need to settle the matter in a manner that is not subjective in nature. That is, we need a test that will allow us to distinguish between points that are randomly scattered and those that form at least a weak linear trend. In this section, we develop such a test.

Recall that our simple linear regression model is

$$Y \mid x_i = (\alpha + \beta x_i) + E_i$$

This model states that the value of Y when $x = x_i$ depends upon two things: the mean value of Y for this value of x, $\alpha + \beta x_i$; and a random or unexplained deviation from this mean value, E_i. Note that β is the coefficient of x. If $\beta = 0$, the x_i term drops from the model. This means that the variation in Y does not depend on x; rather, it is due to random or unexplained factors. If $\beta \neq 0$, the x_i term remains in the model. This means that a linear trend is present and that a linear regression equation should be useful in estimating the values of $\mu_{Y \mid x_0}$ and $Y \mid x_0$. Thus, to detect the presence of a linear trend in data that might be randomly scattered, we test

$H_0: \beta = 0$ (scatter is random)

$H_1: \beta \neq 0$ (simple linear regression is applicable)

The test statistic used to test

$H_0: \beta = 0$

$H_1: \beta \neq 0$

is easy to derive. We know that under our model assumptions, B, the estimator for β, is normally distributed with mean β and variance σ^2 / S_{xx}. Standardizing, we can conclude that the random variable

$$\frac{B - \beta}{\sigma / \sqrt{S_{xx}}}$$

is standard normal. We now replace β by its hypothesized value of 0 and σ by its estimator $\hat{\sigma}$. By now, you should guess that when we replace the true value of σ by its estimator $\hat{\sigma}$, the resulting random variable no longer follows a normal distribution. Rather, it follows a t distribution. The number of degrees of freedom associated with this T random variable is $n - 2$. That is, our test statistic for testing $H_0: \beta = 0$ is

Test Statistic for Testing $H_0: \beta = 0$

$$T_{n-2} = \frac{B}{\hat{\sigma} / \sqrt{S_{xx}}} \qquad \text{where } \hat{\sigma} = \sqrt{\text{SSE}/(n-2)}$$

If H_0 is rejected, we will use the estimated regression equation

$$\hat{Y}\,|\,x = A + Bx$$

to predict values of Y when $x = x_0$. Otherwise, we will not do so.
Example 1 illustrates the use of this statistic.

Example 1   Earlier, we considered the relationship between x, the number of hours of aerobic exercises performed by an individual, and Y, his or her weight loss. The scattergram for the data is given in Figure 9.6. Note that a linear trend appears to be present. To test to see if this linear trend is strong enough to be useful in helping to predict the value of Y for given values of x, we test

$$H_0: \beta = 0$$

$$H_1: \beta \neq 0$$

These summary statistics are available from our previous work:

$$S_{xx} = 40.94 \qquad b = .25 \qquad \hat{\sigma} = .10 \qquad n = 20$$

The observed value of the test statistic is

$$\frac{b}{\hat{\sigma}/\sqrt{S_{xx}}} = \frac{.25}{(.10)/\sqrt{40.94}} = 15.996$$

Based on the t distribution with $n - 2 = 18$ degrees of freedom, we can reject H_0. The P value for the two-tailed test is less than .001 ($t_{.0005} = 3.922$). We have strong statistical evidence that $\beta \neq 0$. We have good reason to suspect that a linear regression equation is useful in predicting the weight loss of an individual based on the number of hours of aerobic exercises performed each week. The estimates that we obtained earlier should be fairly accurate.

If your scattergram suggests a pattern that is nonlinear, then a prediction equation can still be developed. However, it will not be a straight line equation. There are many texts on the market that discuss this problem. The following references may be helpful to you: [12], [14], [25].

Exercises 9.5

Use these data to answer exercises 70–76.

x	y	x	y
1	2	5	9
2	3	6	11
3	6	6	12
4	7	7	14
5	10	8	15

70. Plot a scattergram for these data. Based on this plot, do you think that a straight line prediction equation is reasonable?

71. Find Σx, Σx^2, and S_{xx}.

72. Find Σy, Σy^2, S_{yy}, and $\bar{y}$.

73. Find Σxy and S_{xy}.

74. Estimate the true regression line.

75. Estimate σ^2 and σ.

76. Test

$$H_0: \beta = 0 \qquad \text{at the } \alpha = .05 \text{ level}$$

$$H_1: \beta \neq 0$$

Can H_0 be rejected? Based on this result, would you feel comfortable using x to help predict the value of Y via the regression line found in exercise 74?

77. Plot a scattergram for these data:

x	y	x	y
-3	9	0	.5
-3	8	1	1.5
-2	4	2	4
-1	1	3	9
0	0	3	10

Based on this plot, do you think that a straight line prediction equation is reasonable? Explain.

Use the following data to answer exercises 78–83:

x	y	x	y
1	5	5	5
2	2	6	1
3	1	6	3
4	3	7	2
5	1	8	4

78. Plot a scattergram for these data. Based on this plot, do you think that a straight line prediction equation is reasonable?

79. Find Σx, Σx^2, and S_{xx}.

80. Find Σy, Σy^2, S_{yy}, and $\bar{y}$.

81. Find Σxy and S_{xy}.

82. Estimate σ^2 and σ.

83. Test

$$H_0: \beta = 0$$
$$H_1: \beta \neq 0$$

Do you think that H_0 should be rejected? Explain, based on the P value of your test.

Would you feel comfortable using x to predict the value of Y via a straight line prediction equation?

Use the following data to answer exercises 84–87:

The headwaiter of a large restaurant wants to develop an equation by which he can estimate the length of time in minutes that a patron must wait to be served based on the number of tables currently occupied. These data are obtained:

x	y	x	y
5	10	20	19
5	12	20	21
10	14	25	23
12	13	27	30
15	18	30	35

84. Plot a scattergram for these data. Do you think that a straight line prediction equation is feasible?

85. For these data,

$$\Sigma x = 169 \qquad \Sigma y = 195 \qquad \Sigma xy = 3911$$
$$\Sigma x^2 = 3573 \qquad \Sigma y^2 = 4389 \qquad \bar{y} = 19.5$$

Use this information to find S_{xx}, S_{xy}, S_{yy}, SSE, $\hat{\sigma}^2$, and $\hat{\sigma}$.

86. Test

$$H_0: \beta = 0 \qquad \text{at the } \alpha = .05 \text{ level}$$
$$H_1: \beta \neq 0$$

If H_0 is rejected, estimate α and β, and find the estimated regression line.

87. If appropriate, use the estimated regression line found in exercise 86 to estimate the time that a customer must wait to be served if there are 28 tables occupied when he arrives at the restaurant.

Use the following data to answer exercises 88–90:

When a program is submitted to a computer for processing, it is placed in a queue. The response time is the time that elapses before the output of the program can be read by the programmer. A study is conducted by executives of a large business firm to investigate the relationship between x, the time of day, and Y, the response time in minutes. Time is measured from 0 (9:00 A.M.) to 7 (4:00 P.M.). These data result

x	y	x	y
0	.5	3.7	4.3
.5	.4	4.0	5.7
.5	.7	4.5	6.0
.5	.5	5.0	8.0
1.0	1.0	6.0	10.0
1.5	5.3	6.5	15.0
2.0	9.0	6.8	17.0
2.5	10.5	7.0	18.0
3.0	1.0	7.0	16.0
3.5	1.5	7.0	19.0

88. Plot a scattergram for these data. Does the scattergram suggest that a single straight line prediction equation might not be appropriate?

89. Let us break the data into two subsets. Let the first subset consist of the observations obtained from 9:00 A.M. up to but not including 12:00 noon. That is, let us consider these observations:

x	y	x	y
0	.5	1.0	1.0
.5	.4	1.5	5.3
.5	.7	2.0	9.0
.5	.5	2.5	10.5

For these data,

$$S_{xx} = 5.22 \qquad S_{yy} = 124.19$$

$$S_{xy} = 24.36 \qquad b = 4.67$$

Test

$$H_0: \beta = 0 \qquad \text{at the } \alpha = .05 \text{ level}$$

$$H_1: \beta \neq 0$$

If H_0 is rejected, find the estimated line of regression and use it to predict the response time for a program submitted at 10:45 A.M.

90. Consider the data gathered from 12:00 noon to 4:00 P.M. These values are

x	y	x	y
3.0	1.0	6.0	10.0
3.5	1.5	6.5	15.0
3.7	4.3	6.8	17.0
4.0	5.7	7.0	18.0
4.5	6.0	7.0	16.0
5.0	8.0	7.0	19.0

For these data,

$$S_{xx} = 26.35 \quad S_{yy} = 479.04$$

$$S_{xy} = 110.06 \quad b = 4.18$$

Test

$$H_0: \beta = 0 \quad \text{at the } \alpha = .01 \text{ level}$$

$$H_1: \beta \neq 0$$

If H_0 is rejected, find the estimated line of regression and use it to predict the response time for a program submitted at 3:45 P.M.

9.6   Multiple Regression (Optional)

In simple linear regression, we assume that the behavior of the dependent random variable Y is influenced by the value of a single independent variable x. In practice, things are not usually this simple! Here we consider **multiple linear regression.** Multiple linear regression comes into play in situations in which Y is assumed to be dependent upon more than one independent variable. For example, a physician might think that a man's blood pressure depends on both his weight and age; an economist might believe that the prime lending rate is related to the price of gold, the national trade deficit, and the rate of inflation; a realtor might suspect that the price obtained for a house depends on the age of the house and its livable square footage, the square footage not counting such things as basements, garages, and porches. In each of these cases, interest centers on a single random variable Y. We want to describe its behavior in terms of several independent variables.

In multiple regression, the independent variables are denoted by $x_1, x_2, \ldots, x_k$. We are assuming that the mean of Y is a function of these variables. If we let $\mu_{Y|x_1, x_2, \ldots, x_k}$ denote this mean value, then we are assuming that

$$\mu_{Y|x_1, x_2, \ldots, x_k} = \beta_0 + \beta_1 x_1 + \beta_2 x_2 + \cdots + \beta_k x_k$$

Our job is to use values of $Y, x_1, x_2, \ldots, x_k$ to estimate $\beta_0, \beta_1, \beta_2, \ldots, \beta_k$. The estimators for these parameters are denoted by $B_0, B_1, B_2, \ldots, B_k$ and their observed values by b_0, b_1, b_2, ..., b_k. Although the mathematics required to derive formulas for B_0, B_1, B_2, ..., B_k is beyond the scope of this text, the idea involved is the same as that involved in simple linear regression. Namely, we want to choose values for b_0, b_1, b_2, ..., b_k in such a way that the sum of the squares of the differences between the observed values of Y and the values estimated for Y by the regression equation are as small as possible. That is, we want to minimize

$$\text{SSE} = \Sigma[y - (b_0 + b_1 x_1 + b_2 x_2 + \cdots + b_k x_k)]^2$$

To analyze a multiple regression problem by hand is time consuming. It requires the computation of a large number of sums, sums of squares, and sums of cross-products. For this reason, most regression analysis is done by computer. However, to give you an idea of what is done, let us solve one relatively small problem by hand.

Example 1   Suppose that we want to develop an equation by which the price obtained for a house can be predicted based on its age and its livable square footage. Assume that these data are available:

x_1 (age in years)	x_2 (square footage in thousands)	y (price in thousands of $)
1	1	50
5	1	40
5	2	52
10	2	47
20	3	65

Remember that we are assuming that the average price of a house is a function of its age and its square footage. That is, we are assuming that

$$\mu_{Y|x_1, x_2} = \beta_0 + \beta_1 x_1 + \beta_2 x_2$$

We want to use the observed data to estimate β_0, β_1, and β_2. To do so, we must compute the same sorts of sums, sums of squares, and sums of cross-products that were necessary in simple linear regression. In particular, we need these quantities:

$$\Sigma x_1 = 1 + 5 + 5 + 10 + 20 = 41$$

$$\Sigma x_2 = 1 + 1 + 2 + 2 + 3 = 9$$

$$\Sigma x_1^2 = 1^2 + 5^2 + 5^2 + 10^2 + 20^2 = 551$$

$$\Sigma x_2^2 = 1^2 + 1^2 + 2^2 + 2^2 + 3^2 = 19$$

$$\Sigma y = 50 + 40 + 52 + 47 + 65 = 254$$

$$\Sigma x_1 y = 1(50) + 5(40) + 5(52) + 10(47) + 20(65) = 2280$$

$$\Sigma x_2 y = 1(50) + 1(40) + 2(52) + 2(47) + 3(65) = 483$$

In addition, we need the new cross-product $\Sigma x_1 x_2$. For these data,

$$\Sigma x_1 x_2 = 1(1) + 5(1) + 5(2) + 10(2) + 20(3) = 96$$

We want to find the values for b_0, b_1, and b_2 that minimize

$$\text{SSE} = \Sigma[y - (b_0 + b_1 x_1 + b_2 x_2)]^2$$

Calculus techniques can be used to show that the values that do this are solutions to these equations:

$$b_0 n + b_1 \Sigma x_1 + b_2 \Sigma x_2 = \Sigma y$$

$$b_0 \Sigma x_1 + b_1 \Sigma x_1^2 + b_2 \Sigma x_1 x_2 = \Sigma x_1 y$$

$$b_0 \Sigma x_2 + b_1 \Sigma x_1 x_2 + b_2 \Sigma x_2^2 = \Sigma x_2 y$$

For these data, we must solve these equations simultaneously:

$$5b_0 + 41b_1 + 9b_2 = 254$$

$$41b_0 + 551b_1 + 96b_2 = 2280$$

$$9b_0 + 96b_1 + 19b_2 = 483$$

There are a number of ways to solve such a system. However, there is no quick, easy method. Regardless of how you choose to attack the problem, the calculations required are time consuming and somewhat messy. After some work, we find that

$$b_0 = 17{,}953/543 \qquad b_1 = -103/543 \qquad b_2 = 5{,}820/543$$

You can check these solutions for yourself by substituting them back into the three equations given. They do work! Our estimated regression equation is

$$\hat{\mu}_{Y|x_1, x_2} = b_0 + b_1 x_1 + b_2 x_2$$

$$= (17{,}953/543) + (-103/543)x_1 + (5{,}820/543)x_2$$

Based on this equation, the estimated average selling price of 15-year-old houses with 1,800 square feet of living space is

$$\hat{\mu}_{Y|x_1, x_2} = (17{,}953/543) + (-103/543)(15) + (5{,}820/543)(1.8)$$

$$\approx 49.51$$

Don't forget that this average selling price is also the estimated selling price for a single fifteen-year-old house with 1,800 square feet of living space.

As you can see, solving even a small multiple regression problem by hand is not very practical. In practice, statisticians use one of the many computer packages available to analyze such data. In the next example, we show you how to write an SAS (statistical analysis system) program to analyze the data of Example 1.

Example 2

If you have been reading the computer supplements all along, then much of what we do will be familiar. If you have not read the previous supplements, then you will see for the first time how simple it is to use the SAS package. To begin, we pick a name for our data set. The name should remind us of the nature of our data set and should be a single word of at most 8 letters. Let us name the data set *house*. We inform the computer of our choice by entering

```
DATA HOUSE;
```

as the first line of our program. The semicolon signals the end of the first command. We next choose names for our variables. These names must also be at most 8 letters. Let us name our three variables *age*, *footage*, and *price*. We use an input statement to inform the computer of our variable names and the order in which they will appear in the data set. The input statement takes this form:

```
INPUT    AGE    FOOTAGE    PRICE;
```

We next inform the computer that the data follow. To do so we use this statement:

```
CARDS;
```

The data set is now inserted into the program. So that we can use the regression equation to estimate the price of a fifteen-year-old house with 1,800 square feet of living space, we insert the line

```
15    1800    .
```

as the last data line. The dot in the y position indicates that this y value is unknown and should be estimated. The end of the data is indicated via a line containing a single semicolon. Our program now looks like this:

```
DATA HOUSE;
INPUT    AGE    FOOTAGE    PRICE;
CARDS;

 1    1     50
 5    1     40
 5    2     52
10    2     47
20    3     65
15    1.8    .
 ;
```

We must now tell the computer what to do with the data. This is done via a procedure or "proc" command. The procedure that we want is the regression procedure. We write

```
PROC REG;
```

We must now tell the computer what model we are assuming. The "model" statement is used for this purpose. To get predicted Y values printed, this statement takes the form

```
MODEL    PRICE = AGE    FOOTAGE/P;
```

Note that the variable appearing on the left side of this equation is the dependent variable; those on the right are the independent variables. The statement "P" asks for predicted Y values to be printed. The entire program is

```
DATA HOUSE;
INPUT    AGE    FOOTAGE    PRICE;
CARDS;
   1    1    50
   5    1    40
   5    2    52
  10    2    47
  20    3    65
  15    1.8    .
;
PROC REG;
MODEL    PRICE = AGE    FOOTAGE/P;
```

The output of this program is shown in Figure 9.11 on page 380.

The portions of the output that are of interest to us now are these:

① This gives us the point estimate for β_0. Recall that we obtained a value of 17,953/543 for this parameter when we did the problem by hand. The SAS program gives us this value in decimal form.

② This is the point estimate for β_1, in decimal form. To verify this, convert our hand result, $-103/543$, to decimals.

③ This is the point estimate for β_2, in decimal form. Recall that our previous estimate was 5,820/543.

④ This is the point estimate for the price of a 15-year-old house with 1,800 square feet of living space. Note that this value agrees with the result that we obtained by hand.

You can see that it is very easy to get the computer to do the computations needed to estimate a regression equation. However, it will not think for you! We programmed the computer to estimate β_0, β_1, and β_2 under the assumption that x_1, the age of the house, and x_2, its square footage, are good predictors of price when used together. Is this really true? Is our model reasonable? If so, the predicted price of $49,510 for a 15-year-old house with 1,800 square feet of living space is probably fairly accurate; if not, this number is meaningless.

As pointed out in the last example, the real art of regression analysis lies in model selection. If we have k potential predictor variables, should we use them all, or will some subset of these variables do the job just as well or better? For example, should we use both the age and the square footage of a house in predicting the price of the house, or would one or the other of these variables alone do just as well or better than the two together? The output of the regres-

SAS

DEP VARIABLE: PRICE

SOURCE	DF	SUM OF SQUARES	MEAN SQUARE	F VALUE	PROB>F
MODEL	2	239.124	119.562	2.499	0.2858
ERROR	2	95.675875	47.837937		
C TOTAL	4	334.800			

ROOT MSE	6.916497	R-SQUARE	0.7142
DEP MEAN	50.800000	ADJ R-SQ	0.4285
C.V.	13.61515		

VARIABLE	DF	PARAMETER ESTIMATE	STANDARD ERROR	T FOR H0: PARAMETER=0	PROB>|T|
INTERCEP	1	33.062615 ①	10.506591	3.147 ⑤	0.0879 ⑧
FOOTAGE	1	10.718232 ③	9.727214	1.102 ⑦	0.3854 ⑩
AGE	1	-0.189687 ②	1.110581	-0.171 ⑥	0.8801 ⑨

OBS	ACTUAL	PREDICT VALUE	RESIDUAL
1	50.000	43.591	6.409
2	40.000	42.832	-2.832
3	52.000	53.551	-1.551
4	47.000	52.602	-5.602
5	65.000	61.424	3.576
6	.	49.510 ④	.

SUM OF RESIDUALS	7.10543E-15
SUM OF SQUARED RESIDUALS	95.67587

Figure 9.11
Multiple regression model for estimating price based on both age and square footage.

sion procedure gives us a clue to the answer to this question. In particular, the regression procedure automatically tests

$H_0: \beta_0 = 0$ (the constant term can be dropped from the model)

$H_0: \beta_1 = 0$ (the term involving variable x_1 can be dropped from the model)

$H_0: \beta_2 = 0$ (the term involving variable x_2 can be dropped from the model)

$\vdots$

$H_0: \beta_k = 0$ (the term involving variable x_k can be dropped from the model)

If all of these null hypotheses are rejected, then our stated model is a reasonable one. In this case, estimates made using the fitted regression equation should be quite good. If we fail to reject even one of these null hypotheses then there is evidence that our stated model is not the best model possible. In this case, we must continue to experiment with the data in hopes of discovering a better model. In the next example, we continue the analysis of our house data with this idea in mind.

Example 3 | Our proposed model states that

$$\mu_{Y|x_1, x_2} = \beta_0 + \beta_1 x_1 + \beta_2 x_2$$

This model suggests that x_1, the age of the house, and x_2, its square footage, should both be used to estimate the mean price of a house. To decide whether or not this model is reasonable, we test

$H_0: \beta_0 = 0$ (constant term can be dropped from the model)

$H_0: \beta_1 = 0$ (age can be dropped from the model)

$H_0: \beta_2 = 0$ (footage can be dropped from the model)

Each of these can be tested via a T statistic analogous to that used in the last section to test the null hypothesis that the slope of the regression line is zero. The observed values of these T statistics are given by ⑤, ⑥, and ⑦, respectively, in Figure 9.11. The corresponding P values for two-tailed tests are given by ⑧, ⑨, and ⑩. Note that two of these P values are very large (.8801 and .3854). Since we do not reject H_0 for large P values, we conclude that we do not have the best model at this point. We should look for a better one. Furthermore, we should not put much faith in our predicted price of \$49,510 for a 15-year-old house with 1,800 square feet of living space since we now have reason to suspect that our model is not very good.

There are a number of ways to decide which variables to include in a model. Some statisticians like to begin by placing all of the variables in the model. One by one they then eliminate the ones that don't appear to be useful. This method is called *backward elimination*. Others prefer to build a model from the ground up. They add variables one at a time until the addition of a new one does not appear to be worthwhile. This method of model determination is called *forward selection*. Still others like to try all possible combinations of variables and then pick the set which appears to do the best job. We cannot give the details of these and other methods in this text. However, the following references should be helpful: [12], [14], [25]. Let us complete our study of the house data by considering two other possible models.

Example 4

We know from Example 2 that the two-variable model originally proposed does not look good. Let us try two other possible models. The first is the single variable model that uses only age as a predictor. This model assumes the form

$$\mu_{Y|x_1} = \beta_0 + \beta_1 x_1$$

Note that this is a simple linear regression model. We can analyze the data by hand using the techniques of previous sections, or we can let SAS do it for us. To let SAS do the analysis, we need only make one change in our original program. In particular, we change the model statement to read

```
MODEL PRICE = AGE/P;
```

The output for this program is shown in Figure 9.12.

On the printout ① and ② give the observed values of the T statistics used to test

$$H_0: \beta_0 = 0 \qquad \text{(intercept is zero)}$$

$$H_0: \beta_1 = 0 \qquad \text{(age can be dropped from the model)}$$

respectively. The two-tailed P values are given by ③ and ④. Note that since the P value associated with the latter test is large (.1568), we do not reject H_0. We do have some reason to believe that the age variable can be dropped from the model. This means that age alone does not appear to be a good predictor of price.

Example 5

We now know that age and footage together do not seem to do a good job of predicting price, and that age alone does not work well either. The only possibility left is that footage can be used to predict price. The new model is

$$\mu_{Y|x_2} = \beta_0 + \beta_2 x_2$$

This is also a simple linear regression model. To let SAS analyze the new model, we change the model statement to read

```
MODEL PRICE = FOOTAGE/P;
```

The output of this program is shown in Figure 9.13 on page 384.

DEP VARIABLE: PRICE

SAS

SOURCE	DF	SUM OF SQUARES	MEAN SQUARE	F VALUE	PROB>F
MODEL	1	181.042	181.042	3.532	0.1568
ERROR	3	153.758	51.252638		
C TOTAL	4	334.800			

ROOT MSE	7.159095	R-SQUARE	0.5407
DEP MEAN	50.800000	ADJ R-SQ	0.3877
C.V.	14.09271		

VARIABLE	DF	PARAMETER ESTIMATE	STANDARD ERROR	T FOR H0: PARAMETER=0	PROB> [T]
INTERCEP	1	43.271881	5.127808	8.439 ①	0.0035 ③
AGE	1	0.918063	-0.488473	1.879 ②	0.1568 ④

OBS	ACTUAL	PREDICT VALUE	RESIDUAL
1	50.000	44.190	5.810
2	40.000	47.862	-7.862
3	52.000	47.862	4.138
4	47.000	52.453	-5.453
5	65.000	61.633	3.367
6	.	57.043	.

SUM OF RESIDUALS   3.55271E-14
SUM OF SQUARED RESIDUALS   153.7579

Figure 9.12
Simple linear regression model for estimating price based on age alone.

SAS

DEP VARIABLE: PRICE

SOURCE	DF	SUM OF SQUARES	MEAN SQUARE	F VALUE	PROB>F
MODEL	1	237.729	237.729	7.347	0.0731
ERROR	3	97.071429	32.357143		
C TOTAL	4	334.800			

ROOT MSE	5.688334	R-SQUARE	0.7101
DEP MEAN	50.800000	ADJ R-SQ	0.6134
C.V.	11.19751		

VARIABLE	DF	PARAMETER ESTIMATE	STANDARD ERROR	T FOR H0: PARAMETER=0	PROB> [T]
INTERCEP	1	34.214286 ⑤	6.626708	5.163 ①	0.0141 ③
FOOTAGE	1	9.214286 ⑥	3.399430	2.711 ②	0.0731 ④

OBS	ACTUAL	PREDICT VALUE	RESIDUAL
1	50.000	43.429	6.571
2	40.000	43.429	-3.429
3	52.000	52.643	-.642857
4	47.000	52.643	-5.643
5	65.000	61.857	3.143
6	.	50.800 ⑦	.

SUM OF RESIDUALS 0
SUM OF SQUARED RESIDUALS 97.07143

Figure 9.13
Simple linear regression model for estimating price based on square footage alone.

On the printout ① and ② give the observed values of the T statistics used to test

$$H_0: \beta_0 = 0 \qquad \text{(intercept is zero)}$$

$$H_0: \beta_2 = 0 \qquad \text{(footage can be dropped from the model)}$$

respectively. The two-tailed P values are given by ③ and ④. This time both P values are fairly small. We have evidence that neither β_0 nor β_2 is zero. This simple linear regression model appears adequate. The intercept and slope of the regression line are given by ⑤ and ⑥, respectively. Thus our estimated line of regression is

$$\hat{\mu}_{Y|x_2} = 34.214286 + 9.214286x_2$$

The estimated average price of houses with 1,800 square feet of living space is found by substituting 1.8 into this equation for x_2. Its value is shown in ⑦. It appears that our best estimate for the price of a house that is 15 years old and has 1,800 square feet of living space is $50,800.

As you can see, regression analysis is tricky. You must think carefully about what you are doing. If you encounter a research problem involving multiple regression, we suggest that you go to a professional statistician for help.

Exercises 9.6

Use the data of Example 1 to answer exercises 91–101.

91. Find Σy, $\bar{y}$ and S_{yy}.
92. Let $x_1 = x$ and find S_{xx} and S_{xy}.
93. Consider the model $\mu_{Y|x_1} = \beta_0 + \beta_1 x_1$. Estimate β_1.
94. Estimate σ^2 and σ for the model of exercise 93. Compare these answers to the numbers labeled *Mean Square Error* and *Root MSE* on the printout of Figure 9.12.
95. Test $H_0: \beta_1 = 0$ and verify that the observed value of the test statistic matches that given by ② on the SAS printout of Figure 9.12.
96. Let $x_2 = x$ and find S_{xx} and S_{xy}.
97. Consider the model $\mu_{Y|x_2} = \beta_0 + \beta_2 x_2$. Estimate β_2.
98. Estimate σ^2 and σ for the model of exercise 97. Compare these answers to the numbers labeled *Mean Square* and *Root MSE* on the printout of Figure 9.13.
99. Test $H_0: \beta_2 = 0$ and verify that the observed value of the test statistic matches that given by ② on the SAS printout of Figure 9.13.
100. Estimate β_0 and write the estimated regression equation.
101. Use the estimated regression equation to approximate the selling price of a 15-year-old house with 1,800 square feet of living space. Verify that your answer matches that given by ⑦ on the SAS printout of Figure 9.13.

SAS

DEP VARIABLE: DEMAND

SOURCE	DF	SUM OF SQUARES	MEAN SQUARE	F VALUE	PROB>F
MODEL	3	10.891591	3.630530	26.932	0.0002
ERROR	8	1.078409	0.134801		
C TOTAL	11	11.970000			

ROOT MSE	0.367153	R-SQUARE	0.9099
DEP MEAN	5.550000	ADJ R-SQ	0.8761
C.V.	6.615365		

VARIABLE	DF	PARAMETER ESTIMATE	STANDARD ERROR	T FOR H0: PARAMETER=0	PROB>|T|
INTERCEP	1	9.834300	0.827931	11.878	0.0001
COST	1	0.170931	0.086633	1.973	0.0839
EMPLOY	1	-0.504401	0.080378	-6.275	0.0002
INTEREST	1	-0.265237	0.107045	-2.478	0.0382

Figure 9.14
Multiple regression model for estimating demand based on cost, unemployment rate, and interest rate.

Use Figure 9.14 and the following data to answer exercises 102–104:

An experiment is conducted to develop an equation for predicting the demand for automobiles based on the cost of the car, the current unemployment rate, and the current interest rate. These data are obtained:

x_1 (cost in thousands of dollars)	x_2 (unemployment rate %)	x_3 (interest rate %)	Y (units sold in thousands)
6.5	8.0	10.0	4.0
6.0	5.0	11.0	5.0
5.9	3.5	9.0	7.0
8.2	5.1	10.5	6.0
8.0	6.0	12.0	5.0
8.5	6.5	14.5	4.5
9.0	5.8	11.0	6.2
9.5	4.1	11.2	6.4
10.0	4.5	12.0	5.8
8.5	3.0	10.8	6.7
10.0	4.5	11.6	6.1
11.5	8.0	14.0	3.9

The SAS printout for analyzing the model

$$\mu_{Y|x_1, x_2, x_3} = \beta_0 + \beta_1 x_1 + \beta_2 x_2 + \beta_3 x_3$$

is shown in Figure 9.14.

102. Suppose that we will reject $H_0: \beta_i = 0$, where $i = 0, 1, 2,$ and 3 if $P < .10$. Does the given model seem appropriate?

103. If the answer to exercise 102 is yes, find the estimated regression equation.

104. If the answer to exercise 102 is yes, use the estimated regression equation to estimate the demand for cars that cost $8,200 when the unemployment rate is 6.1% and the interest rate is 10%.

Use the following data to answer exercises 105–107:

An experiment is designed to develop an equation for predicting the diastolic blood pressure (Y) of an individual from knowledge of his or her age (x_1) and weight (x_2).

105. Figure 9.15 gives the SAS printout for analyzing the model

$$\mu_{Y|x_1, x_2} = \beta_0 + \beta_1 x_1 + \beta_2 x_2$$

Do you think that the two-variable model is appropriate? If so, find the estimated regression equation and use it to estimate the diastolic blood pressure of a 40-year-old man who weighs 175 pounds.

SAS

DEP VARIABLE: PRESSURE

SOURCE	DF	SUM OF SQUARES	MEAN SQUARE	F VALUE	PROB>F
MODEL	2	238.426	119.213	3.453	0.0905
ERROR	7	241.674	34.524825		
C TOTAL	9	480.100			

ROOT MSE	5.875783	R-SQUARE	0.4966
DEP MEAN	90.300000	ADJ R-SQ	0.3528
C.V.	6.506958		

VARIABLE	DF	PARAMETER ESTIMATE	STANDARD ERROR	T FOR H0: PARAMETER=0	PROB> |T|
INTERCEP	1	62.610997	21.872591	2.863	0.0242
WEIGHT	1	0.046235	0.117181	0.395	0.7049
AGE	1	0.442897	0.169796	2.608	0.0350

Figure 9.15
Multiple regression model for estimating diastolic blood pressure based on age and weight.

SAS

DEP VARIABLE: PRESSURE

SOURCE	DF	SUM OF SQUARES	MEAN SQUARE	F VALUE	PROB>F
MODEL	1	3.526323	3.526323	0.059	0.8139
ERROR	8	476.574	59.571710		
C TOTAL	9	480.100			

ROOT MSE	7.718271	R-SQUARE	0.0073	
DEP MEAN	90.300000	ADJ R-SQ	-0.1167	
C.V.	8.547366			

VARIABLE	DF	PARAMETER ESTIMATE	STANDARD ERROR	T FOR H0: PARAMETER=0	PROB>|T|
INTERCEP	1	83.838817	26.668433	3.144	0.0137
WEIGHT	1	0.037434	0.153862	0.243	0.8139

Figure 9.16
Simple linear regression model for estimating diastolic blood pressure based only on weight.

SAS

DEP VARIABLE: PRESSURE

SOURCE	DF	SUM OF SQUARES	MEAN SQUARE	F VALUE	PROB>F
MODEL	1	233.052	233.052	7.547	0.0252
ERROR	8	247.048	30.881060		
C TOTAL	9	480.100			

ROOT MSE	5.557073	R-SQUARE	0.4854
DEP MEAN	90.300000	ADJ R-SQ	0.4211
C.V.	6.154012		

VARIABLE	DF	PARAMETER ESTIMATE	STANDARD ERROR	T FOR H0: PARAMETER=0	PROB>|T|
INTERCEP	1	70.676929	7.356091	9.608	0.0001
AGE	1	0.440968	0.160519	2.747	0.0252

Figure 9.17
Simple linear regression model for estimating diastolic blood pressure based only on age.

106. Figure 9.16 on page 389 gives the SAS printout for analyzing the one-variable model

$$\mu_{Y|x_2} = \beta_0 + \beta_2 x_2$$

that includes only weight as a predictor of diastolic blood pressure. Do you think that this simple linear regression model is appropriate? If so, find the estimated regression line and use it to estimate the diastolic blood pressure of a 40-year-old man who weighs 175 pounds.

107. Figure 9.17 on page 390 gives the SAS printout for analyzing the simple linear regression model

$$\mu_{Y|x_1} = \beta_0 + \beta_1 x_1$$

that includes only age as a predictor of diastolic blood pressure. Do you think that this model is appropriate? If so, find the estimated regression line and use it to estimate the diastolic blood pressure of a 40-year-old man who weighs 175 pounds.

9.7   Introduction to Correlation

As we have seen in the previous sections, statistical regression deals with the relationship between an independent variable x and the mean, $\mu_{Y|x}$, of a dependent random variable Y. We have been particularly concerned with situations in which this relationship is linear in nature. In regression analysis, the variable x is not considered to be a random variable. Its values can be predetermined by the researcher.

In correlation studies, X is a random variable; its values are always determined by chance. We are interested in the relationship existing between the random variables X and Y themselves. We want to know if this relationship is linear. That is, we want to know if there exist real numbers α and β, $\beta \neq 0$, such that $Y = \alpha + \beta X$.

Probably the most often used measure of correlation is the Pearson coefficient of correlation, ρ. To have some understanding of exactly what this parameter measures and how to estimate it from observed data, we must pause for a moment to consider what is called the *covariance* between X and Y.

Definition 9.3   **Covariance**

Let X and Y be random variables with means μ_X and μ_Y, respectively. The **covariance** between X and Y, denoted by $\text{Cov}(X, Y)$, is

$$\text{Cov}(X, Y) = E[(X - \mu_X)(Y - \mu_Y)]$$

Roughly speaking, Cov(X, Y) is a measure of the manner in which X and Y vary together. If large values of X tend to be associated with large values of Y and conversely, then $X - \mu_X$ and $Y - \mu_Y$ tend to have the same algebraic sign. Thus, in this situation Cov(X, Y) assumes a positive value. If the reverse is true, and large values of X tend to be associated with small values of Y, then $X - \mu_X$ and $Y - \mu_Y$ tend to differ in sign, thus producing a negative covariance. If large values of X are just as likely to be associated with small values of Y as they are with large values, then Cov(X, Y) assumes a value of 0.

Keep in mind the fact that the covariance between two random variables is a theoretical parameter. It can be calculated exactly only if the probability distribution of the pair of random variables (X, Y) is known completely. Here we need only the ability to estimate this parameter from a collection of n observations on (X, Y). It is not hard to see how this can be done. Since

$$\text{Cov}(X, Y) = E[(X - \mu_X)(Y - \mu_Y)]$$

is a theoretical average value, it is reasonable to estimate this parameter via an analogous sample average. This leads us to propose

$$\Sigma(X - \bar{X})(Y - \bar{Y})/n$$

as an estimator for Cov(X, Y). This estimator is acceptable. However, if we want an unbiased estimator we must make one slight change. Namely, we must use $n - 1$ as our divisor rather than n. This should not surprise you. This is exactly the same change that we had to make earlier in order to obtain an unbiased estimator for the variance of X. This leads us to define the sample covariance as follows:

Definition 9.4

Sample Covariance

Let X and Y be random variables. The sample covariance and unbiased estimator for Cov(X, Y) is defined by

$$\widehat{\text{Cov}(X, Y)} = \Sigma(X - \bar{X})(Y - \bar{Y})/(n - 1) = S_{xy}/(n - 1)$$

Example 1

Researchers are investigating the relationship between obesity and high blood pressure in middle-aged males. The following data are obtained on the random variables X, the number of pounds over ideal weight of the subject, and Y, the systolic blood pressure:

(5, 115)   (20, 128)   (15, 120)   (10, 118)   (25, 130)   (28, 135)

Let us estimate the covariance between these two random variables. For these data,

$$\Sigma x = 103 \qquad \Sigma y = 746 \qquad \Sigma xy = 13145$$

Using the computational formula for S_{xy}, we see that

$$S_{xy} = [n\Sigma xy - \Sigma x \Sigma y]/n = [6(13145) - (103)(746)]/5 = 406.4$$

The estimated covariance between obesity and high blood pressure is

$$\widehat{\text{Cov}(X, Y)} = S_{xy}/(n - 1) = 406.4/5 = 81.28$$

Since this value is positive, we can conclude that a high degree of obesity tends to be associated with high blood pressure. The magnitude of this estimate has no meaning. We are interested only in the algebraic sign.

Even though the algebraic sign of the covariance gives us a clue to the nature of the relationship between X and Y, it is not easy to interpret covariance precisely. It is a parameter that can assume any real value but its magnitude has no real meaning. To overcome this difficulty, we consider the Pearson correlation coefficient, ρ. This parameter uses the information gained from the covariance. It has the added advantage that it is designed in such a way that its value always lies between -1 and 1 inclusive and its magnitude as well as its algebraic sign is meaningful. The algebraic sign is interpreted exactly as that of the covariance; its magnitude measures the strength of the linear relationship between X and Y. We now state the definition of ρ.

Definition 9.5

Pearson Coefficient of Correlation

Let X and Y be random variables with means μ_X and μ_Y and variances σ_X^2 and σ_Y^2, respectively. The **correlation**, ρ, between X and Y is

$$\rho = \frac{\text{Cov}(X, Y)}{\sqrt{(\text{Var } X)(\text{Var } Y)}}$$

It is not at all obvious that ρ measures the strength of the linear relationship between X and Y! However, the next theorem shows that this is in fact the case. Since this theorem is very important we include at least a partial proof. If you are interested in why things work and are familiar with the rules for expectation, try to follow the proof! Otherwise, concentrate on understanding the meaning of the theorem.

Theorem 9.2

Let α and β be real numbers with $\beta \neq 0$. Then a linear relationship exists between X and Y, that is, $Y = \alpha + \beta X$, if and only if $\rho = 1$ or $\rho = -1$.

Proof. Assume that $Y = \alpha + \beta X$ for $\beta \neq 0$. By definition

$$\rho = \frac{\text{Cov}(X, Y)}{\sqrt{(\text{Var } X)(\text{Var } Y)}}$$

Note that $\mu_Y = E[Y] = E[\alpha + \beta X] = \alpha + \beta E[X] = \alpha + \mu_X$.

$$\begin{aligned}
\text{Cov}(X, Y) &= E[(X - \mu_X)(Y - \mu_Y)] \\
&= E\{[X - \mu_X][(\alpha + \beta X) - (\alpha + \beta \mu_X)]\} \\
&= E\{[X - \mu_X][\beta(X - \mu_X)]\} \\
&= \beta E[(X - \mu_X)^2] \\
&= \beta \text{ Var } X
\end{aligned}$$

Using the rules for variance, $\text{Var } Y = \text{Var}(\alpha + \beta X) = \beta^2 \text{ Var } X$. Thus, we can conclude that

$$\rho = \frac{\text{Cov}(X, Y)}{\sqrt{(\text{Var } X)(\text{Var } Y)}} = \frac{\beta \text{ Var } X}{\sqrt{\text{Var } X \beta^2 \text{ Var } X}} = \frac{\beta \text{ Var } X}{|\beta| \text{ Var } X} = \frac{\beta}{|\beta|}$$

If $\beta > 0$, then $\rho = 1$ and if $\beta < 0$, $\rho = -1$. The proof of the converse is beyond the scope of this text.

Even if you are not interested in the proof of this theorem, don't miss its practical implications! The theorem says, in short, that if there is a linear relationship between X and Y, then this will be reflected in ρ assuming a value of either $+1$, **perfect positive correlation**, or -1, **perfect negative correlation**. The converse is also true: a correlation coefficient of $+1$ or -1 implies the existence of a perfect linear relationship between X and Y. Perhaps it is even more important to note what is not being said. If $\rho = 0$, in which case, we say that X and Y are **uncorrelated**, we are not saying that there is no relationship between X and Y. We are saying that if there is a relationship, it is *not linear*. These ideas are illustrated graphically in Figure 9.18.

Let us now turn to the problem of estimating ρ from a collection of n ordered pairs of observations on the random variables (X, Y). This is really no problem at all. We know that the unbiased estimator for $\text{Cov}(X, Y)$ is

$$\widehat{\text{Cov}(X, Y)} = S_{xy}/(n - 1)$$

We also know that the unbiased estimators for $\text{Var } X$ and $\text{Var } Y$ are

$$\widehat{\text{Var } X} = \Sigma(X - \bar{X})^2/(n - 1) = S_{xx}/(n - 1)$$

$$\widehat{\text{Var } Y} = \Sigma(Y - \bar{Y})^2/(n - 1) = S_{yy}/(n - 1)$$

When we replace the parameters by their estimators, we obtain

$$\hat{\rho} = \frac{S_{xy}/(n - 1)}{\sqrt{[S_{xx}/(n - 1)][S_{yy}/(n - 1)]}} = \frac{S_{xy}}{\sqrt{S_{xx} S_{yy}}}$$

In summary, this statistic is used to estimate the Pearson correlation coefficient.

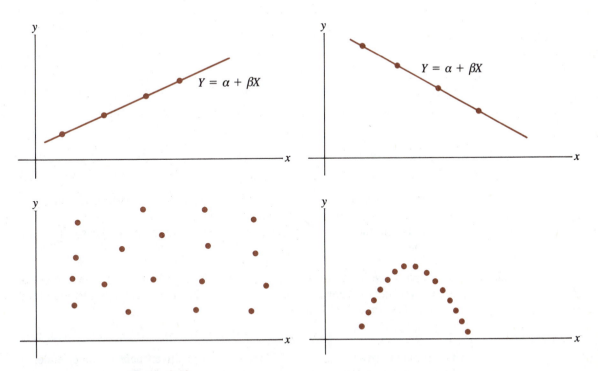

Figure 9.18
(a) Perfect positive correlation, $\rho = 1$. All points lie on the straight line $Y = \alpha + \beta X$ where $\beta > 0$. (b) Perfect negative correlation, $\rho = -1$. All points lie on the straight line $Y = \alpha + \beta X$ where $\beta < 0$. (c) Uncorrelated $\rho = 0$. Points are randomly scattered. (d) Uncorrelated $\rho = 0$. Points show relationship between X and Y, but the relationship is not linear.

Definition 9.6

Sample Correlation Coefficient

The sample correlation coefficient is denoted by $\hat\rho$ or R and is given by

$$R = \hat\rho = \frac{S_{xy}}{\sqrt{S_{xx}S_{yy}}} = \frac{n\Sigma XY - \Sigma X \Sigma Y}{\sqrt{[n\Sigma X^2 - (\Sigma X)^2][n\Sigma Y^2 - (\Sigma Y)^2]}}$$

When working with sample data, it is rare for R to assume one of the easily interpretable values of 1, -1, or 0. This means that some value judgments are usually in order. Unfortunately, there are no rigid rules for interpreting R that apply to all disciplines. We will use the rough scale given in Figure 9.19. Please realize that the interpretation of R depends somewhat on the problem at hand. A social scientist might consider an R value of .6 to be an indication of a strong linear relationship whereas an engineer might consider the same R value to be

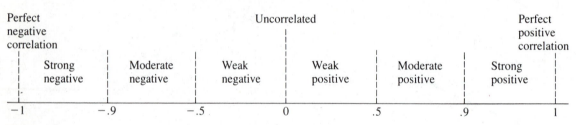

Figure 9.19
Guidelines for interpreting R.

too weak to be useful. To interpret a correlation coefficient with intelligence you must be familiar with the subject matter at hand.

To illustrate the use of this estimator, let us continue our analysis of the data of Example 1.

Example 2

In Example 1, we began to investigate the relationship between X, the number of pounds over ideal weight of an individual, and Y, the individual's systolic blood pressure. These data are available:

(5, 115)   (20, 128)   (15, 120)   (10, 118)   (25, 130)   (28, 135)

The estimated covariance is 81.28. Since this is positive, there is some tendency for large values of X to be paired with large values of Y. To investigate further, we consider the strength of this relationship. The scattergram for the above data is shown in Figure 9.20. There does appear to be a fairly strong linear association between X and Y, since these points tend to lie along a straight line sloping upward. We would therefore expect r, the value of the sample correlation coefficient, to be positive and close in value to 1. To compute the value of r, thus estimating the true value of ρ, we need the following quantities:

$$\Sigma x = 103 \quad \Sigma x^2 = 2,159 \quad \Sigma xy = 13,145 \quad \Sigma y = 746 \quad \Sigma y^2 = 93,058$$

Thus

$$r = \frac{n\Sigma xy - \Sigma x \Sigma y}{\sqrt{[n\Sigma x^2 - (\Sigma x)^2][n\Sigma y^2 - (\Sigma y)^2]}}$$

$$= \frac{6(13145) - (103)(746)}{\sqrt{[6(2159) - (103)^2][6(93058) - (746)^2]}}$$

$$= \frac{78870 - 76838}{\sqrt{2345(1832)}}$$

$$= .98$$

As expected, r is positive and close in value to 1, thus substantiating analytically what has already been observed graphically, namely a strong positive linear relationship between the number of pounds overweight of the subject and his

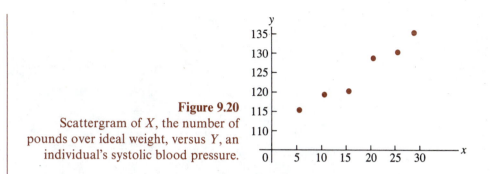

Figure 9.20
Scattergram of X, the number of pounds over ideal weight, versus Y, an individual's systolic blood pressure.

systolic blood pressure. Practically speaking, we are saying that, as a person becomes more and more overweight, his blood pressure in most cases tends to increase.

Example 3 Let X denote the percentage of A's given by an instructor and Y the overall average rating obtained by the instructor on student evaluations on a scale from 0 (poor) to 4 (excellent). The following data are obtained:

(20, 2.2)   (8, 3.4)   (50, 3.1)   (5, 3.8)   (60, 3.7)
(22, 1.9)   (7, 3.6)   (60, 3.5)   (15, 2.5)   (65, 2.0)
(25, 1.8)   (15, 2.8)   (55, 3.6)   (45, 3.3)   (90, 3.9)
(10, 3.1)   (35, 2.6)   (45, 3.2)   (40, 2.8)

Let us first examine the scattergram in Figure 9.21 to see if there appears to be a linear association. There does not appear to be a linear association. Let us compute r to see how this fact is reflected. For these data

$$\Sigma x = 672 \qquad \Sigma x^2 = 34{,}122 \qquad \Sigma xy = 2074.2$$
$$\Sigma y = 56.8 \qquad \Sigma y^2 = 177.8$$

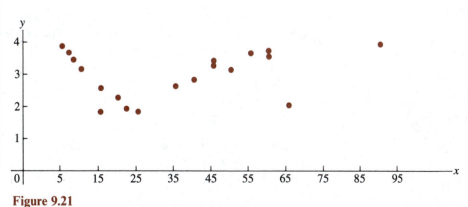

Figure 9.21
Scattergram of X, percentage of A's given, versus Y, average rating on student evaluations.

"Hello, FDA? I'd like to report research that directly links cheese with death in rats"

Figure 9.22
Copyrighted by the Chicago Tribune. Used with permission.

Hence,

$$r = \frac{19(2074.2) - (672)(56.8)}{\sqrt{[19(34122) - (672)^2][19(177.8) - (56.8)^2]}} = .23$$

This value indicates a weak positive correlation, as expected, since the scattergram indicates almost no linear association between X and Y.

A word of caution is in order. Strong linear correlation does not necessarily imply in any way a cause-and-effect relationship. For example, the variables X, the price of tea in China, and Y, the price of hot dogs in New York, may be strongly positively correlated. That does not mean that raising the price of tea in China causes the price of hot dogs to increase, or vice versa. It means that there is one force at work, probably worldwide inflation, that tends to have a similar effect on these variables. Figure 9.22 illustrates a result of misinterpreting a strong linear correlation.

Remember that in a regression study the variable x is not a random variable. In a correlation study, X must be a random variable. Thus, it does not really make sense to ask for the *correlation* between x and Y in a regression study. Even so, knowledge of the value of R can be useful in such a study. First, R tells us something concerning the nature of the regression line. To see why, note that

$$B = S_{xy}/S_{xx} \quad \text{and that} \quad R = S_{xy}/\sqrt{S_{xx}S_{yy}}$$

If we multiply B by $\sqrt{S_{xx}}/\sqrt{S_{yy}}$, we see that

$$B\sqrt{S_{xx}}/\sqrt{S_{yy}} = S_{xy}\sqrt{S_{xx}}/S_{xx}\sqrt{S_{yy}} = S_{xy}/\sqrt{S_{xx}S_{yy}} = R$$

The relationship between R and the slope of the regression line can be read from this equation. In particular, a positive value for R implies that the slope of the regression line is positive. This means that the graph of the regression line rises from left to right, and that the two variables x and Y tend to increase together; when one tends to be large, so does the other. Conversely, suppose that the value of R is negative. This forces the regression line to have a negative slope and thus to fall from left to right; when one variable is large in value, the other tends to be small. Furthermore, an R value of 0 implies that the estimated line of regression has slope 0 and is therefore horizontal. In this case, the estimated regression line is probably not of much use in predicting the value of Y.

The second use for R in a regression study is even more important than the first. As we will show, it is an indication of the adequacy of the simple linear regression model. To understand why this is true, recall that

$$S_{yy} = \Sigma(Y - \bar{Y})^2$$

measures the total variability exhibited by the random variable Y and that

$$\text{SSE} = S_{yy} - BS_{xy}$$

measures the random variability of the Y values about the estimated regression line. If we divide each side of this equation by S_{yy}, we see that

$$\text{SSE}/S_{yy} = 1 - BS_{xy}/S_{yy}$$

Since $B = S_{xy}/S_{xx}$, we have

$$\text{SSE}/S_{yy} = 1 - S_{xy}^2/S_{xx}S_{yy}$$

However, $R = S_{xy}/\sqrt{S_{xx}S_{yy}}$ and so $R^2 = S_{xy}^2/S_{xx}S_{yy}$. By substitution we see that

$$\text{SSE}/S_{yy} = 1 - R^2$$

or that $R^2 = [S_{yy} - \text{SSE}]/S_{yy}$.

Since S_{yy} measures the total variability in Y and SSE measures the random variability of Y about the estimated regression line, $S_{yy} - \text{SSE}$ measures the variability in Y explained by the regression line. The statistic R^2 measures the proportion of the variability in Y explained by the regression line. When this statistic is multiplied by 100% we obtain the percentage of the variability in Y that is explained by the simple linear regression model. The statistic R^2 is called the **coefficient of determination**. We shall interpret R^2 in a manner that parallels the interpretation of R. Figure 9.23 gives our interpretation of this statistic.

In Example 1, in Section 2, $r = .95$ and hence $r^2 = .90$. Multiplying by 100, we see that 90% of the variation in Y can be explained by the linear model. That is, 90% of the variation in weight loss can be accounted for because of similar variations in the number of hours of aerobic exercises performed per week by individuals in the study. Only 10% of the variation in Y is random or unex-

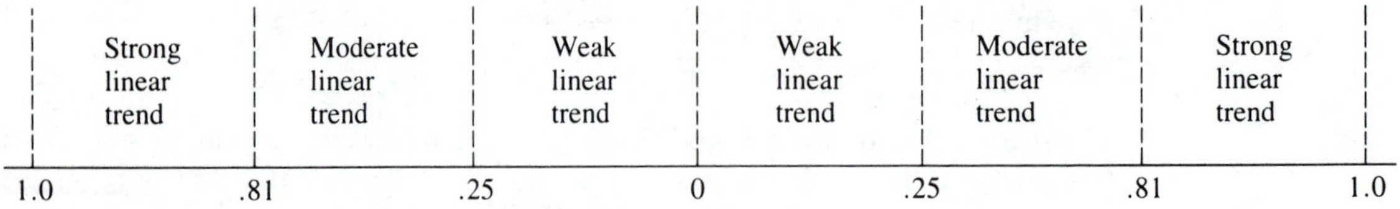

| 1.0 | .81 | .25 | 0 | .25 | .81 | 1.0 |

Figure 9.23
Guidelines for interpreting R^2.

plained. In this case, we can conclude that the data exhibits a strong linear trend. The linear regression model does an excellent job of explaining the variability in Y.

Exercises 9.7

In exercises 108–111, plot a scattergram for the given data set. Based on the scattergram, try to anticipate the nature of r. That is, try to anticipate its algebraic sign and magnitude. Compute the value of r in each case to test your intuition.

108.

x	y
1	3
2	4
3	8
4	9
5	10

109.

x	y
0	10
1	8
2	4
3	0
4	−1

110.

x	y
−3	9
−2	4
−1	0
1	1
2	5
3	8

111.

x	y
-3	3
-2	2
-1	3
1	2.5
2	3.5
3	2

112. A businessman is interested in studying the relationship between the amount of charges made by a customer the first month after the issuance of a credit card X and the total yearly charges made by the customer Y. The following data are obtained (reported in dollars):

x	y
10	120
5	75
25	250
100	800
30	360

   (a) Plot a scattergram for these data.
   (b) Do the random variables X and Y appear to be correlated? If so, is the correlation positive or negative?
   (c) Compute each of the following quantities:

$$\Sigma x \quad \Sigma y \quad \Sigma xy \quad \Sigma x^2 \quad \Sigma y^2$$

   (d) Find the value of r.

113. A researcher is interested in studying the relationship between the number of hours per day spent watching television, X, and the reading level in third-graders, Y. The following data were obtained for a sample of size 10 (higher scores imply a higher reading level):

x	y	x	y
3	4	1	5
5	3	3	3
6	2	5	2
2	4	6	1
0	5	2	5

   (a) Plot a scattergram for these data.
   (b) Do the random variables X and Y appear to be correlated? If so, is the correlation positive or negative?

(c) Compute the value of each of the following quantities:

$$\Sigma x \quad \Sigma y \quad \Sigma xy \quad \Sigma x^2 \quad \Sigma y^2$$

(d) Find the value of r.

114. A psychologist is interested in studying the relationship between a subject's age, X, and his or her attitude toward the role of older people in society, Y. A questionnaire is prepared to measure this attitude. Scores range from 0 (highly unfavorable attitude) to 10 (highly favorable attitude). The following data are obtained:

x	y	x	y
15	9	27	5
16	3	30	7
18	8	32	2
20	10	35	6
25	1	40	4

(a) Plot a scattergram for these data.
(b) Do the random variables X and Y appear to be correlated? If so, is the correlation positive or negative?
(c) For these data

$$\Sigma x = 258 \quad \Sigma y = 55 \quad \Sigma xy = 1{,}331$$

$$\Sigma x^2 = 7{,}308 \quad \Sigma y^2 = 385$$

Use this information to estimate ρ.

115. It can be shown that if X and Y are independent (the value assumed by X gives us no information concerning the value assumed by Y, and conversely), then $E[XY] = E[X]E[Y]$. Use this to show that if X and Y are independent, then $\rho = 0$.

116. An experiment consists of tossing a coin six times and observing the number of heads obtained, X, and then rolling a dice once and observing the number that appears, Y. Try this experiment for yourself ten times.
(a) Plot a scattergram for your observations.
(b) Would you expect r to be close to 1, -1, or 0?
(c) Compute r, thus estimating ρ.

117. Find and interpret the coefficient of determination for exercises 19–23.

COMPUTING SUPPLEMENT G

The SAS procedure PROC CORR estimates the Pearson coefficient of correlation. It also tests the null hypothesis that $\rho = 0$. The data of Example 2 is used to illustrate the use of the procedure.

Statement	*Purpose*
`DATA OBESITY;`	names data set
`INPUT X Y;`	names variables; each line contains the value of X and its corresponding Y value.
`LABEL X = LBS OVERWEIGHT;`	labels variable X
`LABEL Y = SYSTOLIC PRESSURE;`	labels variable Y
`CARDS;`	data follows
`5    115`	data
`10   118`	
`20   128`	
`25   130`	
`15   120`	
`28   135`	
`  ;`	signals end of data
`PROC PLOT;`	asks for a scattergram
`PLOT Y*X;`	first mentioned variable Y, will appear on the vertical axis
`TITLE 1 ARE X AND Y CORRELATED?;`	titles scattergram
`PROC CORR;`	calls for ρ to be estimated

The output of this program is shown below.

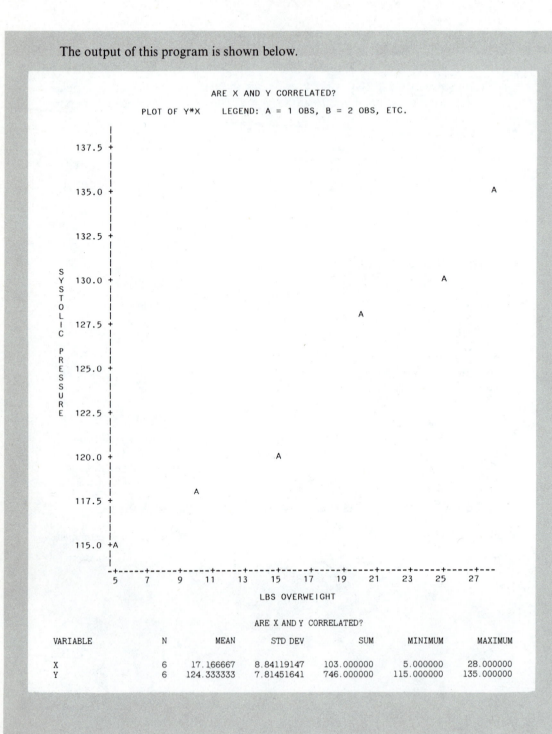

ARE X AND Y CORRELATED?

VARIABLE	N	MEAN	STD DEV	SUM	MINIMUM	MAXIMUM
X	6	17.166667	8.84119147	103.000000	5.000000	28.000000
Y	6	124.333333	7.81451641	746.000000	115.000000	135.000000

```
CORRELATION COEFFICIENTS/PROB > | R | UNDER H0 : RHO = 0 / N = 6

                                  X               Y
X                              1.00000         0.98037  ①
LBS OVERWEIGHT                 0.0000          0.0006   ②

Y                              0.98037         1.00000
SYSTOLIC PRESSURE              0.0006          0.0000
```

On the output, the estimated coefficient of correlation is given by ①. Note that, apart from roundoff error, this value agrees with that obtained by hand in Example 2. The P value for testing

$$H_0: \rho = 0$$

is given by ②. Since this P value is very small, .0006, we have strong statistical evidence that obesity and systolic blood pressure are correlated.

Vocabulary List and Key Concepts

simple linear regression

dependent variable

independent variable

designed regression study

curve of regression of Y on x

slope

Y intercept

method of least squares

residuals

SSE

model assumption in simple
    linear regression

uncorrelated

S_{yy}

S_{xx}

S_{xy}

confidence band

multiple linear regression

covariance

correlation

perfect positive correlation

perfect negative correlation

coefficient of determination

Review Exercises

118. In simple linear regression, we are assuming that

$$\mu_{Y|x} = \alpha + \beta x$$

Explain what this equation means in words. Which of the variables is the dependent variable? Is this variable a random variable? Which of the variables is the independent variable? Is this variable a random variable?

119. How can we get an idea of whether or not simple linear regression is applicable without having to do any computations?

120. To estimate α and β we use the method of least squares. Briefly explain the idea behind least squares and illustrate your discussion graphically.

121. What are the formulas for the least squares estimates for α and β?

122. What is a designed regression study?

Use the following data to answer exercises 123–129:

Let x denote the price, in dollars, of various computer games and Y the number, in thousands, of games sold.

x	y	x	y
10	50	30	32
10	45	30	31
10	52	40	20
20	35	40	26
20	37	50	19

123. Plot a scattergram for these data. Does simple linear regression seem appropriate? If so, would you expect the estimated slope to be positive or negative? Would you expect the computed value of R to be positive or negative? Would you expect R^2 to be close to 1?

124. For these data,

$$\Sigma x = 260 \qquad \Sigma y = 347 \qquad \Sigma xy = 7590$$

$$\Sigma x^2 = 8600 \qquad \Sigma y^2 = 13245$$

Use this information to find the value of the coefficient of determination. Based on this statistic, do you think that the simple linear regression model is appropriate? If so, estimate α and β.

125. Use the estimated regression equation found in exercise 124 to approximate the average demand for games that sell for $25.

126. Use the estimated regression equation found in exercise 124 to approximate the demand for a particular game that sells for $25.

*127. Estimate σ^2.

*128. Find a 95% confidence interval for the average demand for games that sell for $25.

*129. Find a 95% confidence interval for the demand for a particular game that sells for $25.

*130. What exactly are we trying to decide when we test $H_0: \beta = 0$ in a simple linear regression model?

*131. Explain briefly the difference between simple linear regression and multiple regression.

*132. Figures 9.24–9.26 (on pages 408–410) give the output for three models for relating the demand for seating on a particular airline flight to the price of the ticket and the season of the year. The seasons were coded as fall = 1, winter = 2, spring = 3, and summer = 4. Figure 9.24 is based on the multiple regression model that uses both variables. Do you think that this model is appropriate? If so, estimate the demand for space in the winter when the price of a ticket is $185. If not, continue to Figure 9.25. This printout is based on the simple linear regression model that includes only price. Does this model seem appropriate? If so, estimate the demand for space when tickets are priced at $185. If not, continue to Figure 9.26. This printout is based on the simple linear regression model that includes only the season of the year. Does this model seem appropriate? If so, estimate the demand for space during the winter months.

133. In a correlation study, we are studying two variables X and Y. In this case, is X a random variable? Is Y a random variable?

134. What does it mean to say that X and Y have perfect positive correlation? perfect negative correlation?

135. If X and Y are uncorrelated does this mean that there is no relationship at all between X and Y?

DEP VARIABLE: DEMAND

SAS

SOURCE	DF	SUM OF SQUARES	MEAN SQUARE	F VALUE	PROB>F
MODEL	2	1764.410	882.205	2.140	0.1737
ERROR	9	3710.507	412.279		
C TOTAL	11	5474.917			

ROOT MSE	20.304643	R-SQUARE	0.3223	
DEP MEAN	191.917	ADJ R-SQ	0.1717	
C.V.	10.57993			

VARIABLE	DF	PARAMETER ESTIMATE	STANDARD ERROR	T FOR H0: PARAMETER=0	PROB>|T|
INTERCEP	1	119.289	83.626011	1.426	0.1875
PRICE	1	0.317347	0.561711	0.565	0.5859
SEASON	1	7.894558	6.904960	1.143	0.2824

Figure 9.24
Multiple regression model for estimating the demand for seating based on ticket price and season.

DEP VARIABLE: DEMAND

SAS

SOURCE	DF	SUM OF SQUARES	MEAN SQUARE	F VALUE	PROB>F
MODEL	1	1225.490	1225.490	2.884	0.1203
ERROR	10	4249.426	424.943		
C TOTAL	11	5474.917			

ROOT MSE	20.614137	R-SQUARE	0.2238	
DEP MEAN	191.917	ADJ R-SQ	0.1462	
C.V.	10.74119			

VARIABLE	DF	PARAMETER ESTIMATE	STANDARD ERROR	T FOR H0: PARAMETER=0	PROB>|T|
INTERCEP	1	69.367647	72.408856	0.958	0.3607
PRICE	1	0.735294	0.432983	1.698	0.1203

Figure 9.25
Simple linear regression model for estimating the demand for seating based on price alone.

SAS

DEP VARIABLE: DEMAND

SOURCE	DF	SUM OF SQUARES	MEAN SQUARE	F VALUE	PROB>F
MODEL	1	1632.817	1632.817	4.250	0.0662
ERROR	10	3842.100	384.210		
C TOTAL	11	5474.917			

ROOT MSE		19.601275	R-SQUARE	0.2982	
DEP MEAN		191.917	ADJ R-SQ	0.2281	
C.V.		10.21343			

VARIABLE	DF	PARAMETER ESTIMATE	STANDARD ERROR	T FOR H0: PARAMETER=0	PROB> [T]
INTERCEP	1	165.833	13.860195	11.965	0.0001
SEASON	1	10.433333	5.061028	2.062	0.0662

Figure 9.26
Simple linear regression model for estimating the demand for seating based on season alone.

136. Discuss the practical significance of the algebraic sign of the covariance.

137. What is the estimator for the covariance? Is this estimator unbiased?

138. What is the estimator for ρ?

139. These data are obtained on the random variables X, the weight, in pounds, of a baby girl at birth, and Y, her weight, also in pounds, at age 21.

x	y	x	y
4.1	105	7.5	122
5.2	117	7.6	129
5.8	115	8.0	125
6.0	124	8.1	132
6.5	127	8.3	140
7.0	130	9.0	135

Plot a scattergram of these data. Would you expect the observed value of R to be positive or negative? Do you think that the correlation is strong, moderate, or weak? Evaluate R to test your intuition!

Categorical Data

We are concerned here with situations in which each observation can be classified as falling into exactly one of several mutually exclusive categories. Interest centers on the number of observations per category, and inferences are based on whether or not these numbers tend to refute a stated hypothesis. We will be concerned in particular with three problems.

1. Testing to see if a set of observations is drawn from a specified probability distribution
2. Testing to see if two or more variables used for classification purposes are independent
3. Comparing proportions

The statistical technique used for solving each problem has as its mathematical basis the multinomial distribution discussed briefly in the first section.

10.1 Multinomial Distribution

In Chapter 4 we considered the binomial model. In the binomial model, the experiment consists of a series of trials, each of which results in either a success or a failure. In this section we consider experiments in which each trial can result in two or more possible outcomes. We begin by defining what is meant by a *multinomial trial*.

Definition 10.1

Multinomial Trial

A **multinomial trial** with parameters p_1, p_2, ..., p_k is a trial that can result in exactly one of k possible outcomes. The probability that outcome i will occur on a given trial is p_i, for $i = 1, 2, ..., k$. Note that $0 \le p_i \le 1$ for each i and that

$$p_1 + p_2 + p_3 + \cdots + p_k = 1$$

Example 1

1. Ninety percent of the accounts handled by an accounting firm are paid to date, 1% show a credit balance, and the remainder show a credit debit. A single account is selected at random and the balance noted. This can be viewed as a multinomial trial with three possible outcomes—paid, credit, debit. In this case, $p_1 = .9$, $p_2 = .01$, and $p_3 = .09$.

2. Researchers are interested in ascertaining whether or not the blood-type distribution is the same among diabetics as it is among the general population in the United States. Empirical studies have fairly well established that the distribution in this population is

   type A: 41%
   type B: 9%
   type AB: 4%
   type O: 46%

   A single diabetic is selected, and the blood type noted. If blood types among diabetics are distributed the same as in the general population, this can be viewed as a multinomial trial with four possible outcomes: A, B, AB, and O. In this case, $p_1 = .41$, $p_2 = .09$, $p_3 = .04$, and $p_4 = .46$.

   We are interested in experiments that can be viewed as consisting of a series of n independent and identical multinomial trials. Such experiments naturally give rise to the *multinomial random variable* defined below.

Definition 10.2

Multinomial Random Variable

Let an experiment consist of n independent and identical multinomial trials with parameters p_1, p_2, ..., p_k. Let X_i denote the number of trials that result in outcome i for $i = 1, 2, ..., k$. The k-tuple $(X_1, X_2, ..., X_k)$ is called the **multinomial random variable** with parameters n and p_1, p_2, ..., p_k.

The word *k-tuple* may be new to you. In this context, the word is used to indicate k random variables considered in a specific order. Example 2 should make the idea clear.

Example 2

1. Assume that an officer of the accounting firm of Example 1 randomly samples ten accounts from among its extensive files. For all practical purposes, we may view this experiment as consisting of a series of $n = 10$ multinomial trials. The variable of interest is (X_1, X_2, X_3), where X_1 is the number of accounts paid to date, X_2 is the number showing a credit balance, and X_3 the number with a debit.

2. Assume that 125 diabetics are randomly selected and their blood type noted. If the distribution of blood types is the same among diabetics as it is in the general population, then this experiment can be viewed as consisting of a series of $n = 125$ multinomial trials. The variable of interest is (X_1, X_2, X_3, X_4), where X_1 is the number of patients observed with type A blood, X_2 the number with type B, X_3 the number with type AB, and X_4 the number with O.

Although it is easy to derive the probability function for a multinomial random variable, it is not necessary for our work here. However, it is necessary to determine the expected value of each of the random variables $X_1, X_2, \ldots, X_k$. To do so, consider a single multinomial trial and any fixed outcome i. In this particular trial outcome i either does or does not occur. If we consider the occurrence of outcome i to be a success, then the probability of success is p_i. In a series of n multinomial trials, X_i, the number of trials that result in outcome i, also counts the number of successes in the n trials. That is, X_i is binomial with parameters n and p_i. From the results of Chapter 4, we may conclude that $E[X_i] = n \cdot p_i$. That is, in a series of multinomial trials, the expected number of outcomes falling into category i is $n \cdot p_i$. This result is useful in that it makes the test statistic used in later sections intuitively appealing.

Example 3

If the distribution of blood types among diabetics is the same as that among the general population, then 41% are type A, 9% are type B, 4% are type AB and 46% are type O. In screening 125 diabetics, the expected number in each category respectively are

$$E[X_1] = 125(.41) = 51.25$$

$$E[X_2] = 125(.09) = 11.25$$

$$E[X_3] = 125(.04) = 5.00$$

$$E[X_4] = 125(.46) = 57.50$$

If we observe 53 individuals with type-A blood, 12 with type B, 5 with type AB and 55 with type O, can we conclude that there is an association between blood type and the presence of diabetes? That is, Is there enough difference between the observed number of individuals per category and the expected number to cause concern? This is the question that we will answer in Section 3. Figure 10.1 illustrates the fact that survey data is often categorical in nature.

"... and when asked about the Panama Canal treaty, 22% said 'no,' 17% said 'yes,' and 61% said 'what ever side it was John Wayne said he was on'"

Figure 10.1
Copyrighted by the Chicago Tribune. Used with permission.

Exercises 10.1

1. Seventy percent of the general public is thought to have average IQ, 10% slightly below average, and 10% slightly above. Five percent is classed as retarded and 5% as superior. A psychologist randomly selects 15 persons to participate in an experiment. If the above figures are correct, what is the expected number of individuals falling in each category?

2. It is felt that 60% of the general public favors an upcoming bond issue for higher education, 30% opposes the issue, and 10% is undecided. A poll is taken of 50 persons. Find the expected number in each category under the assumption that these figures are correct.

3. Find the expected number of times that a roulette wheel will land on green, red, or black in a series of 20 spins. (See Example 1 in Section 3.1.)

4. A city consultant feels that, for the accounting department, 70% of the jobs are productive, 20% are marginally productive, while 10% are nonproductive. If 15 jobs are selected at random from that department, what is the expected number of jobs in each category?

5. A recent survey reported that 18% of the American population feels that a great deal of unnecessary surgery is being done, 78% does not feel that this is the case, and 4% has no opinion. Twenty people are randomly selected and interviewed. Based on the above survey, how many people on the average

would you expect to find in each of the above categories? At the end of the interview, nobody was found with no opinion, 1 felt that too much unnecessary surgery is being performed, and 19 did not have that opinion. Do these results lead you to suspect that the stated percentages are in error? Explain.

6. A recent survey reported that 51% of the American public always trusts their doctor, 35% mostly does, 11% sometimes does, and 3% never does. If 50 people are randomly selected and questioned, how many on the average would you expect to fall into each of the above categories? If 15 are found to always trust their doctor, 25 mostly do, 5 sometimes do, and the rest never do, do you think that there is reason to suspect the accuracy of the given percentages? Explain.

10.2   Goodness-of-Fit Tests

The test we consider here is called the ***chi-square goodness-of-fit test***. The purpose of the procedure is to test the null hypothesis that a given set of observations is drawn from or fits a specified probability distribution against the alternative that it does not. The test statistic used in the test has an approximate chi-square distribution under the assumption that the null hypothesis is true. This procedure is particularly useful in testing to see if a given set of observations is drawn from a normal distribution, since this assumption underlies much of what has been said in previous chapters. The test is based on the following theorem, offered without proof.

Theorem 10.1

Let $(X_1, X_2, \ldots, X_k)$ be a multinomial random variable with parameters $n, p_1, p_2, \ldots, p_k$. For large n, the random variable

$$\sum \frac{(X_i - n \cdot p_i)^2}{n \cdot p_i}$$

follows an approximate chi-square distribution with $k - 1$ degrees of freedom.

Note that we can think of X_i as the actual or observed number of observations that fall into category i. Let us therefore denote X_i by O_i. Note also that $n \cdot p_i$ is the theoretical expected number of observations in category i. We let $E_i = n \cdot p_i$. The theorem thus says that

$$\sum \frac{(O_i - E_i)^2}{E_i}$$

is approximately chi-square with $k - 1$ degrees of freedom. It has been found that the approximation is acceptable if the categories are such that *no expected fre-*

quency is less than 1, and no more than 20% of the expected frequencies are less than 5.

The random variable

$$\Sigma \frac{(O_i - E_i)^2}{E_i}$$

serves as a test statistic for testing a null hypothesis that a given set of observations is drawn from a specified probability distribution. If H_0 is true, we know the value of p_i for each i and can compute E_i. The above statistic, then, is in effect simply comparing the observed number of observations per category with the number expected under H_0. If these figures agree fairly well (we have a good fit), then the term $(O_i - E_i)^2$ will be small and H_0 should not be rejected. If the observed and expected frequencies do not agree very well, then $(O_i - E_i)^2$ will be large, indicating a poor fit and calling for the rejection of H_0. Thus, our test is to reject H_0 if the observed value of the test statistic

$$\Sigma \frac{(O_i - E_i)^2}{E_i}$$

is too large based on the X_{k-1}^2 distribution.

Example 1

Suppose we suspect that the distribution of blood types is different among diabetics than it is among the general population. We wish to gain statistical support for this contention and therefore wish to test

> H_0: no difference in blood type distribution among diabetics and the distribution in the general United States population

> H_1: blood type distribution among diabetics differs from that of the general population

Mathematically, we are testing

> H_0: $p_1 = .41$, $p_2 = .09$, $p_3 = .04$, $p_4 = .46$

> H_1: at least one probability is not as stated

Since rejecting H_0, when in fact H_0 is true, leads to unnecessary and expensive research, let us test this hypothesis at the $\alpha = .01$ level, basing our decision on the data of Example 3 in Section 10.1. Note that the test statistic is

$$\Sigma \frac{(O_i - E_i)^2}{E_i}$$

which follows an approximate chi-square distribution with three degrees of freedom. From Table VII of Appendix A, the critical point for the test is 11.3. We reject H_0 if, and only if, the observed value of the test statistic falls on or above 11.3. The data upon which the test is based are given in Table 10.1.

Table 10.1

	Blood Type			
	A	B	AB	O
Observed O_i	53	12	5	55
Expected E_i under H_0	51.25	11.25	5.00	57.50
$O_i - E_i$	1.75	0.75	0	−2.5

A brief inspection of the last line in the table leads us to believe that H_0 will not be rejected. On an intuitive basis, the differences between the values expected under H_0 and those actually observed do not appear to be very great. To verify that this is so, we compute the observed value of the test statistic and compare it to the critical point of 11.3.

$$\Sigma \frac{(O_i - E_i)^2}{E_i} = \frac{(1.75)^2}{51.25} + \frac{(.75)^2}{11.25} + \frac{0^2}{5.00} + \frac{(-2.5)^2}{57.50} = .22$$

As expected, this value does not fall above the critical point. We are unable to reject H_0. We do not have sufficient statistical evidence to claim that the distribution of blood types among diabetics differs significantly from that of the general population of the United States.

Let us now turn our attention to a more commonly encountered problem involving the chi-square procedure, namely, that of testing for normality. We assume that we have available a random sample $Y_1, Y_2, \ldots, Y_n$ from the distribution of Y with unknown mean μ and unknown variance σ^2. We want to see if there is evidence to indicate that the distribution is not normal. The steps followed for this test are outlined below. They are carefully illustrated in the next example.

Testing for Normality

1. Break the real line into k mutually exclusive intervals or categories.

2. Estimate the population parameters μ and σ^2 from the data using these estimators:

$$\hat{\mu} = \bar{Y}$$

$$\hat{\sigma}^2 = \Sigma(Y_i - \bar{Y})^2/n = \Sigma Y_i^2/n - \bar{Y}^2$$

Note that we are estimating σ^2 slightly differently than we did in previous chapters. The theorem upon which the chi-square approximation rests does not require that the estimators for μ and σ^2 be unbiased. It does require that they be what are called *maximum likelihood estimators* [17]. It can be shown that $\bar{X}$ is the maximum likelihood estimator for μ. However, the maximum likelihood estimator for σ^2 is not S^2. Rather, it is obtained by dividing $\Sigma(Y_i - \bar{Y})^2$ by n rather than $n - 1$.

3. Approximate the probability of an observation falling into category i by $\hat{p}_i$. The method for doing so will be demonstrated shortly.

4. Approximate the expected number of observations falling into category i by $\hat{E}_i = n \cdot \hat{p}_i$.

5. Test H_0: Y is normally distributed
         H_1: Y is not normally distributed

using

$$\Sigma \frac{(O_i - \hat{E}_i)^2}{\hat{E}_i}$$

as the test statistic. It can be shown that in general the number of degrees of freedom (d.f.) associated with the above statistic is given by

d.f. = number of categories minus one minus the
        number of population parameters estimated from
        the data used in computing the expected
        category frequencies

In this case $d.f. = k - 1 - 2 = k - 3$ since we estimated two parameters μ and σ^2 from the data. The test is to reject the null hypothesis that Y is normally distributed if the observed value of the test statistic is too large.

Example 2   A new drug is being tested for possible use in the treatment of leukemia. The random variable is Y, the length of time of survival after the initial diagnosis of the disease. We wish to test Y for normality. The following observations were obtained for 40 patients who were treated with the drug (time is in years):

2.2	1.6	3.4	4.4	3.9	3.0	3.7	4.1
3.4	4.3	3.7	2.6	4.2	3.1	3.5	3.1
2.5	3.1	3.2	1.9	3.7	4.5	3.3	3.6
3.3	3.8	3.9	3.5	3.1	3.8	2.9	3.2
4.7	4.7	3.4	3.2	3.3	4.1	3.0	2.6

Step 1   We must first divide the real line into mutually exclusive categories such that the expected number of observations per category is at least 1, and such that no more than 20% are less than 5. Usually 5 to 20 categories are desirable, with each finite category being the same length. We begin by attempting to obtain 6 categories. Note that the largest observation is 4.7 and the smallest is 1.6. We must cover an interval of length $4.7 - 1.6 = 3.1$. To do so, the length of each finite category must be at least $(3.1)/6 = .52$. This length should be rounded *up* to the same number of decimal places as the data. Hence, we will use .6 as the actual length of each interval. The lower boundary for the first category should start 1/2 unit below the smallest observation. In this case, thinking of a unit as 1/10 of a year, the

Table 10.2

Category Number	Boundaries	Midpoint	Observed Number in Category O_i
1	1.55–2.15	1.85	2
2	2.15–2.75	2.45	4
3	2.75–3.35	3.05	13
4	3.35–3.95	3.65	13
5	3.95–4.55	4.25	6
6	4.55–5.15	4.85	2

lower boundary is 1.55. The remaining category boundaries are found by adding .6 successively to the preceding boundary value until all data points are covered. In this manner we obtain the 6 finite categories shown in Table 10.2. Note that this method guarantees that the boundary values have one more significant figure than the data points. Hence, no observation can fall on a boundary. In this sense, the categories are mutually exclusive, since each observation falls into one, and only one, of them. The above information can be summarized in a bar graph called a *frequency histogram* (Figure 10.2). In each case, we have constructed a rectangle centered at the midpoint of the category with height proportional to the observed number of observations in the category. It appears from the histogram that the observations do form a rough bell. We would not expect to be able to reject the hypothesis that Y is normal.

Step 2   Estimate μ and σ^2. For these data,

$$\bar{y} = 3.41 \qquad \Sigma y^2 = 485.07 \qquad \hat{\sigma}^2 = (\Sigma y^2)/40 - \bar{y}^2 = .50$$

$$\hat{\sigma} = .71$$

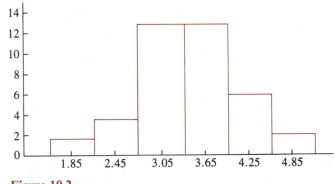

Figure 10.2
Histogram of patient survival times.

Table 10.3

Category Number	Boundaries	O_i	$\hat{p}_i$
1	$-\infty$–2.15	2	.0384
2	2.15–2.75	4	.1378
3	2.75–3.35	13	.2919
4	3.35–3.95	13	.3083
5	3.95–4.55	6	.1699
6	4.55–∞	2	.0537

Step 3   Approximate the cell probabilities. The first and last categories are in practice considered to be open-ended. Hence, to estimate the probability of an observation falling into category 1, we need to find

$$P[Y \leq 2.15]$$

This probability is found by standardizing and then making use of the standard normal table, Table II of Appendix A. We standardize by subtracting the estimated mean of 3.41 and dividing by the estimated standard deviation of .71.

$$P[Y \leq 2.15] = P\left[\frac{Y - 3.41}{.71} \leq \frac{2.15 - 3.41}{.71}\right] = P[Z \leq -1.77]$$

$$= .0384 = \hat{p}_1$$

Similarly, the probability of an observation falling into category 2 is given by

$$P[2.15 \leq Y \leq 2.75] = P[-1.77 \leq Z \leq -.93] = .1762 - .0384$$

$$= .1378 = \hat{p}_2$$

Other probabilities are found in the same manner. We summarize the results in Table 10.3.

Step 4   Approximate the expected cell frequencies. For each $i = 1, 2, 3, 4, 5, 6$; $\hat{E}_i = n \cdot \hat{p}_i = 40\hat{p}_i$. We thus obtain the expectations shown in Table 10.4.

Table 10.4

Category Number	Boundaries	O_i	$\hat{p}_i$	$\hat{E}_i = 40\hat{p}_i$
1	$-\infty$–2.15	2	.0384	1.54
2	2.15–2.75	4	.1378	5.51
3	2.75–3.35	13	.2919	11.68
4	3.35–3.95	13	.3083	12.33
5	3.95–4.55	6	.1699	6.80
6	4.55–∞	2	.0537	2.14

Table 10.5

Category Number	Boundaries	O_i	$\hat{p}_i$	$\hat{E}_i$
1	$-\infty$–2.75	6	.1762	7.05
2	2.75–3.35	13	.2919	11.68
3	3.35–3.95	13	.3083	12.33
4	3.95–4.55	6	.1699	6.80
5	4.55–∞	2	.0537	2.14

We want to satisfy the guideline that no expected cell frequency be less than 1, and no more than 20% be less than 5. Currently there are 2 cells in 6 or 33% of the cells with expected frequency less than 5. To satisfy the guideline, we combine the cell with the smaller expectation with the adjacent cell. We therefore base our actual test on the five categories shown in Table 10.5.

Step 5 We may now test

H_0: Y distributed normally

H_1: Y not distributed normally

The observed value of the test statistic is

$$\Sigma(O_i - \hat{E}_i)^2/\hat{E}_i = .45$$

The critical point for a size $\alpha = .10$ level test is found relative to the chi-square distribution with $k - 3 = 5 - 3 = 2$ degrees of freedom. Its value is 4.61. Since the observed value of the test statistic lies below the critical point, we are unable to reject the null hypothesis that Y is normally distributed.

Exercises 10.2

7. A sample of size 50 is drawn from a population that contains four types of objects: A, B, C, and D. Theory predicts that the proportion of these types of objects in the population is

Type	A	B	C	D
Proportion	.32	.25	.15	.28

In the sample, the observed number of objects of each type are

Type	A	B	C	D
Number	20	10	7	13

(a) Find the expected cell frequencies under the assumption that the theory is correct.

(b) Based on these expected frequencies, do you think that there is evidence that the theoretical values are incorrect?

(c) Test H_0: $p_A = .32$, $p_B = .25$, $p_C = .15$, $p_D = .28$ at the $\alpha = .05$ level. Did you anticipate the result?

8. A sample of size 100 is drawn from a population that contains five types of items: A, B, C, D, and E. Theory predicts that the proportions of these types of objects in the population are

Type	A	B	C	D	E
Proportion	.50	.10	.05	.25	.10

In the sample, the observed number of objects of each type is

Type	A	B	C	D	E
Number	30	18	12	20	20

(a) Find the expected cell frequencies under the assumption that the theory is correct.

(b) Based on these expected frequencies, do you think that there is strong evidence that the theoretical values are not correct?

(c) Test H_0: $p_A = .50$, $p_B = .10$, $p_C = .05$, $p_D = .25$, $p_E = .10$. Do you think that H_0 should be rejected? Explain, based on the P value of your test.

9. A die is being tested for fairness. In 120 tosses the following observations are obtained:

	1	2	3	4	5	6
O_i	15	19	26	20	18	22

Set up and test the appropriate hypothesis at the $\alpha = .10$ level using the chi-square procedure.

10. Test a coin for fairness by tossing it 100 times and observing the number of heads obtained. Use the chi-square procedure with $\alpha = .05$.

11. It is felt that 65% of the American public is in favor of government censorship of movies, 25% is opposed, and 10% is undecided. To test this distribution, a random sample of 500 persons is surveyed, and it is found that 315 are in favor, 140 are opposed, and 45 are undecided. Test at the $\alpha = .10$ level the null hypothesis

$$H_0: p_1 = .65, \qquad p_2 = .25, \qquad p_3 = .10$$

12. A psychologist is interested in color preferences among preschool children. Fifty preschoolers are shown identically shaped boards, each colored either

red, yellow, blue, or green. Each child is asked to pick the best-liked one. The following information is obtained:

	red	yellow	blue	green
O_i	18	15	12	5

Do these results indicate that there does exist a color preference among this age group? What is the P value of the test?

13. A federal directive outlines the following guidelines for a large company. The employees of this company must be 40% white males, 10% minority males, 40% white females, and 10% minority females. The company president selects 50 employees at random. It is noted that in this group 23 are white males, 12 are minority males, 10 are white females, and 5 are minority females. Test at the $\alpha = .05$ level the null hypothesis

$$H_0: p_1 = .40, \; p_2 = .10, \; p_3 = .40, \; p_4 = .10$$

14. A survey in 1977 reported that 50% of those surveyed felt that surgeons were too eager to operate, 34% did not feel this way, and 16% had no opinion. A recent survey was done to see if these figures have changed. In a sample of 200 persons interviewed, 120 felt that surgeons were too eager to operate, 60 did not feel this way, and 20 had no opinion. Do we have statistical evidence to reject the old percentages at the $\alpha = .05$ level?

15. A medical journal claimed that patients are, on the whole, happy with the care that they receive; that 92% have never considered suing for malpractice and only 7% have considered such action; and 1% had no opinion. To test this claim, a random sample of 1000 persons was polled. Fifty claimed that they had at one time or another considered bringing suit, 940 claimed that they had never considered such a move, and 10 had no opinion. At the $\alpha = .10$ level, do we have statistical evidence to refute the figures of the medical journal?

16. The following observations are obtained on the random variable Y, the number of pages in a large suburban daily newspaper.

24	43	41	40	42	40
40	20	30	26	28	28
40	42	41	26	41	32
36	34	30	36	28	36
34	43	28	20	30	42

(a) Construct a frequency histogram using originally 6 categories.
(b) Estimate μ and σ^2.
(c) Use the chi-square procedure to test for approximate normality.

17. The following observations are obtained on the random variable Y, the number of vehicles under warranty repaired per day by GM dealers throughout Virginia:

15	17	9	23	10	22	12	10
20	3	14	11	10	18	8	15
12	24	13	12	6	12	9	13
8	11	7	9	8	10	11	7
9	10	1	10	13	2	16	4

(a) Construct a frequency histogram for these data using originally 6 categories.

(b) Estimate μ and σ^2.

(c) Use the chi-square procedure to test for approximate normality.

18. The following observations are obtained on the random variable Y, the length of play (in minutes) of a popular tune recorded during the last ten years.

0.74	1.31	1.64	1.62	2.04	1.83	1.74
3.00	2.40	2.25	2.18	1.51	1.93	1.97
1.54	3.03	2.73	2.17	1.51	2.79	2.92
2.04	2.44	1.95	1.65	2.07	3.13	2.25
2.11	1.82	2.02	2.19	2.06	1.90	2.61
2.56	2.29	1.46	1.06	1.64	1.90	1.13

(a) Construct a frequency histogram for these data using originally 6 categories.

(b) Estimate μ and σ^2.

(c) Use the chi-square procedure to test for normality.

19. The following observations are obtained on the random variable Y, the length of time (in hours) that a carbon flashlight battery burns before burning out:

10.0	13.5	25.0	31.7	33.1	11.5	17.9
30.8	31.8	33.2	12.0	18.2	30.8	32.0
33.7	12.0	20.1	30.9	32.1	33.9	12.1
22.0	30.9	32.3	34.0	12.2	22.6	31.0
32.4	34.1	13.2	22.7	31.2	33.0	35.0

(a) Construct a frequency histogram for these data using originally 6 categories.

(b) Estimate μ and σ^2.

(c) Use the chi-square procedure to test for normality.

10.3   Testing for Independence

Many experiments, particularly in the social sciences, yield enumerative (count) data. Many are also characterized by the fact that each subject can be classified according to two or more criteria. It is often of interest to try to ascertain whether or not the variables used for classification are independent. Intuitively speaking, when these variables are **independent**, knowledge of the classification level relative to one variable has no bearing on the level assumed relative to another variable.

Example 1

We wish to determine whether or not the ethnic group to which an American belongs has any influence on income. If these variables are independent then knowledge of one's ethnic background, variable 1, will not help in predicting the level of income, variable 2. In conducting the experiment we identify three levels relative to ethnic background: black (B), Caucasian (C), and other (O). We identify five income levels: very low (VL), low (L), moderate (M), high (H), and very high (VH). These classifications when used together naturally generate the $3 \times 5 = 15$ mutually exclusive categories shown in Table 10.6. Since each individual falls into exactly one cell dependent or contingent upon its classification level relative to each variable, we call such a table a **contingency table**.

In general, we are interested in testing the hypothesis of independence by means of an $r \times c$ contingency table, where r represents the number of rows in the table, c the number of columns. Hence $r \cdot c$ is the number of categories generated. The above table would therefore be termed a 3×5 table with 15 categories or cells. We are interested in the following probabilities associated with the $r \times c$ table:

1. $p_{i\cdot}$, the probability that a subject is classed in level i relative to variable 1, $i = 1, 2, \ldots, r$. Note that the subscript i is being used to denote the row number.

2. $p_{\cdot j}$, the probability that a subject is classed in level j relative to variable 2 for $j = 1, 2, 3, \ldots, c$. Note that the subscript j is being used to denote the column number.

Table 10.6

	\multicolumn{5}{c}{*Variable 2*}
	VL	*L*	*M*	*H*	*VH*
Variable 1 B					
C					
O					

3. p_{ij}, the probability that a subject is classed in category or cell ij. Note that this refers to the ith level of variable 1 and the jth level of variable 2.

Note also that $\Sigma p_{i\cdot} = 1$ and $\Sigma p_{\cdot j} = 1$.

Example 2 illustrates this new notation in the context of the problem described in Example 1.

Example 2   Consider the contingency table shown in Table 10.6. Based on this table,

$p_1.$ = the probability that a randomly selected individual is a black American

$p_{\cdot 1}$ = the probability that a randomly selected individual is in the very low income classification

p_{11} = the probability that a randomly selected individual is a black American and in the very low income classification

Recall from Chapter 2 that two events A and B are independent if and only if $P[A \cap B] = P[A] \cdot P[B]$. For instance, if income is independent of ethnic background, then, referring to the notation of Example 1, we should have the following relationships:

$$P[B \cap VL] = P[B] \cdot P[VL] \quad \text{or} \quad p_{11} = (p_1.)(p_{\cdot 1})$$

$$P[B \cap L] = P[B] \cdot P[L] \quad \text{or} \quad p_{12} = (p_1.)(p_{\cdot 2})$$

$$P[B \cap M] = P[B] \cdot P[M] \quad \text{or} \quad p_{13} = (p_1.)(p_{\cdot 3})$$

$$\vdots$$

$$P[O \cap VH] = P[O] \cdot P[VH] \quad \text{or} \quad p_{35} = (p_3.)(p_{\cdot 5})$$

Generalizing the above, we can express the hypothesis

H_0: classification variables independent

H_1: classification variables not independent

mathematically by writing

$$H_0: p_{ij} = (p_{i\cdot})(p_{\cdot j}) \quad \text{for all } i = 1, 2, \ldots, r \quad \text{and} \quad j = 1, 2, \ldots, c$$

$$H_1: p_{ij} \neq (p_{i\cdot})(p_{\cdot j}) \quad \text{for some } i \text{ and } j$$

This hypothesis can be tested easily by means of the statistic

$$\sum_{i=1}^{r} \sum_{j=1}^{c} \frac{(O_{ij} - \hat{E}_{ij})^2}{\hat{E}_{ij}} = \sum_{\substack{\text{all} \\ \text{categories}}} \frac{\left(\begin{array}{c}\text{observed number} \\ \text{per category}\end{array} - \begin{array}{c}\text{estimated expected} \\ \text{number per category}\end{array}\right)^2}{\begin{array}{c}\text{estimated expected} \\ \text{number per category}\end{array}}$$

As we have noted previously, this statistic under H_0 has an approximate chi-square distribution with the number of degrees of freedom given by

$$\text{number of categories} - 1 - \begin{array}{c} \text{number of parameters estimated} \\ \text{from the data used in computing the} \\ \text{expected category frequencies} \end{array}$$

In the next example we will demonstrate that the number of degrees of freedom for the test for independence in an $r \times c$ contingency table is

---

Degrees of Freedom in Testing for Independence

$$r \cdot c - 1 - [(r - 1) + (c - 1)] = (r - 1) \cdot (c - 1)$$

Example 3

In studying the relationship between income and ethnic background, a random sample of size $n = 500$ persons is obtained. Table 10.7 summarizes the observed number of individuals falling into each of the 15 possible categories and into each classification level of the two variables. The numbers $n_{.1}, n_{.2}, \ldots, n_{.5}$ shown in the table are called *marginal column totals* and the numbers $n_{1.}, n_{2.},$ and $n_{3.}$ are called *marginal row totals*.

To apply the chi-square test for independence, we must use the observed data to estimate the expected number of observations in each category under the assumption that income is independent of ethnic background. We can consider falling into the $(i - j)$th cell a success. Since this occurs with probability p_{ij}, the expected number of successes, that is, the expected number of observations falling into the $(i - j)$th cell is

$$E_{ij} = p_{ij} \cdot n$$

If H_0 is true then $p_{ij} = (p_{i.})(p_{.j})$ and hence

$$E_{ij} = (p_{i.})(p_{.j})n$$

Table 10.7

	Variable 2					
	VL	L	M	H	VH	
Variable 1						
B	$O_{11} = 8$	$O_{12} = 16$	$O_{13} = 31$	$O_{14} = 4$	$O_{15} = 1$	$n_{1.} = 60$
C	$O_{21} = 23$	$O_{22} = 91$	$O_{23} = 210$	$O_{24} = 64$	$O_{25} = 22$	$n_{2.} = 410$
O	$O_{31} = 4$	$O_{32} = 8$	$O_{33} = 9$	$O_{34} = 7$	$O_{35} = 2$	$n_{3.} = 30$
	$n_{.1} = 35$	$n_{.2} = 115$	$n_{.3} = 250$	$n_{.4} = 75$	$n_{.5} = 25$	$n = 500$

Table 10.8

Variable 1	Variable 2					
	VL	L	M	H	VH	
B	8	16	31	4	1	60
	(4.2)	(13.8)	(30.0)	(9.0)	(3.0)	
C	23	91	210	64	22	410
	(28.7)	(94.3)	(205.0)	(61.5)	(20.5)	
O	4	8	9	7	2	30
	(2.1)	(6.9)	(15.0)	(4.5)	(1.5)	
	35	115	250	75	25	500

The above probabilities are theoretical. We do not know their true values. They can, however, be estimated from the data by

$$\hat{p}_{i\cdot} = n_{i\cdot}/n \qquad \hat{p}_{\cdot j} = n_{\cdot j}/n$$

and hence

> **Approximate Expected Cell Frequencies**
>
> $$\hat{E}_{ij} = (n_{i\cdot}/n)(n_{\cdot j}/n)(n) = (n_{i\cdot})(n_{\cdot j})/n$$
>
> $$\hat{E}_{ij} = \frac{\text{marginal row total} \cdot \text{marginal column total}}{\text{sample size}}$$

For example, for Table 10.7, we see that

$$\hat{E}_{11} = (n_{1\cdot})(n_{\cdot 1})/n = (60)(35)/500 = 4.2$$
$$\hat{E}_{21} = (n_{2\cdot})(n_{\cdot 2})/n = (410)(35)/500 = 28.7$$
$$\vdots$$
$$\hat{E}_{35} = (n_{3\cdot})(n_{\cdot 5})/n = (30)(25)/500 = 1.5$$

Following this procedure to estimate the expected number of observations in each of the 15 cells, we obtain the data shown in Table 10.8. Expected cell frequencies are given in parentheses. Note that since $p_{1\cdot} + p_{2\cdot} + p_{3\cdot} = 1$, we actually only need to estimate $r - 1 = 2$ of these probabilities from the data. The third is obtained by subtraction. Similarly, since $p_{\cdot 1} + p_{\cdot 2} + p_{\cdot 3} + p_{\cdot 4} + p_{\cdot 5} = 1$, we estimated only $c - 1 = 4$ of these five probabilities from the data, the fifth being obtained by subtraction. Thus, the number of parameters estimated from the data in computing the expected cell frequencies is $(r - 1) + (c - 1) = 6$. The number of degrees of freedom associated with the test statistic is therefore

$$r \cdot c - 1 - [(r - 1) + (c - 1)] = 15 - 1 - 6 = 8 = 2 \cdot 4 = (r - 1)(c - 1)$$

as claimed. For the problem at hand, the observed value of the test statistic

$$\sum_{i=1}^{3} \sum_{j=1}^{5} (O_{ij} - \hat{E}_{ij})^2 / \hat{E}_{ij} = 15.36$$

The critical point for a size $\alpha = .10$ level test based on the X_8^2 distribution is 13.4. Since the observed value of the test statistic is larger than this value, we can reject H_0 and can conclude that income and ethnic background are not independent.

It should be pointed out that in testing for independence, the *only value fixed by the researcher is the overall sample size n*. Both row and marginal totals are free to vary. Other types of contingency tables are discussed in the next section.

Exercises 10.3

20. Each object in a population can be classified by size as either large, medium, or small; and by color as either red, or not red. A random sample of size 445 is drawn and Table 10.9 results.
    (a) Find the row totals and column totals.
    (b) Find the expected cell frequencies under the assumption that size is independent of color.
    (c) Based on these expected frequencies, do you think that there is good evidence that size and color are not independent?
    (d) Test the null hypothesis of independence at the $\alpha = .05$ level.

Table 10.9

	Size			
	Large	Medium	Small	
Color				
Red	100	100	30	
Not Red	30	140	45	
				445

21. Each automobile in a population can be classified by weight as being either light, medium, or heavy; and by ease of handling as being either poor, good, or excellent. A random sample of size 80 is drawn and Table 10.10 results.
    (a) Find the row totals and column totals.
    (b) Find the expected cell frequencies under the assumption that weight and handling ease are independent.
    (c) Based on these expected frequencies, do you think that there is good evidence that weight and handling ease are not independent?
    (d) Test the null hypothesis of independence. Do you think that H_0 should be rejected? Explain, based on the approximate P value of your test.

Table 10.10

	Weight		
	Light	Medium	Heavy
Handling			
Poor	10	9	5
Good	12	11	7
Excellent	8	10	8

80

22. A study of a new flu vaccine is conducted. A random sample of 818 individuals is selected and each is classified according to inoculation status and the state of health. The results of the study are given in Table 10.11. State and test the hypothesis that one's state of health is independent of one's inoculation status at the $\alpha = .05$ level.

Table 10.11

	State of Health	
	Contracted Flu	Did Not Contract Flu
Inoculation Status		
Inoculated	276	3
Not Inoculated	473	66

818

23. A sociologist is interested in the relationship between religious affiliation and attitude toward government-funded social welfare programs. Religious affiliation is split into three levels: actively affiliated with a religious organization, inactive but claiming an affiliation, and claiming no religious affiliation. Four attitudes toward social welfare are identified: cut out welfare completely, maintain it at a reduced level, maintain it at the current level, increase it. One thousand persons are randomly selected and interviewed. The results of this survey are given in Table 10.12. State and test the null hypothesis that an individual's attitude toward social welfare programs is independent of his or her religious affiliation at the $\alpha = .05$ level.

Table 10.12

	Attitude Toward Social Welfare			
	Stop	Decrease	Remain the Same	Increase
Religious Affiliation				
Active	10	150	180	60
Inactive	36	141	158	15
None	28	98	115	9

1000

Table 10.13

			Opinion		
	Violently Opposed	Opposed	Undecided	Favor	Strongly Favor
Age					
18–30	76	81	60	62	21
31–40	76	74	36	51	13
41–50	61	63	27	41	8
51–60	48	46	23	28	5
Over 60	39	36	4	18	3

24. Political scientists are interested in determining whether or not one's opinion concerning a national guaranteed income is independent of age. A random sample of adults (18 and older) is obtained and the information gathered is shown in Table 10.13. Test for independence at the $\alpha = .10$ level.

25. Environmentalists claim that one's attitude about enforcement of air-pollution standards is independent of party affiliation. A random sample of 500 persons reveals the information given in Table 10.14. Is there evidence at the $\alpha = .05$ level that the environmentalists are wrong?

Table 10.14

	Party Affiliation		
	Democrat	Republican	Independent
Attitude Toward Enforcement			
Strictly Enforced Within 5 Years	61	24	15
Strict Enforcement Delayed			
For At Least 5 Years	121	52	27
Not Be Enforced At All	16	6	3
Don't Know	102	45	28

500

26. Researchers are interested in the relationship between obesity in children and their ability to function successfully in public school. A random sample of 2,000 sixth-graders reveals the information in Table 10.15. Is there evidence

Table 10.15

	Weight			
	Underweight	Normal	Slightly Overweight	Highly Overweight
School Performance				
Below Average	36	160	65	50
Average	180	840	300	185
Above Average	34	100	35	15

2000

that success in school is not independent of obesity? Explain based on the P value of your test statistic.

27. Recent studies have tended to support a relationship between eye color and sprinting speed in athletes. From a random sample of 100 male athletes the data of Table 10.16 are obtained. In the table fair is over 9.6 seconds, good is between 9.3 and 9.6 seconds, inclusive, excellent is under 9.3 seconds. Do these data support the contention that sprinting speed and eye color are not independent? Explain based on the P value of your test statistic.

Table 10.16

| | 100-Yard Dash Speed | | |
	Fair	Good	Excellent
Eye Color			
Blue	24	26	5
Brown	21	14	10

10.4   Comparing Proportions

In this section, we consider the use of contingency tables to compare proportions. Although the method of analysis is identical to that used in the last section, there is an important difference in the manner in which the experiment is conducted and in the statement of the null hypothesis. An example will illustrate this difference.

Example 1   Researchers are interested in the relationship between blood type and incidence of duodenal ulcer. Previous work suggests some association between blood group O and susceptibility to this type of ulcer. To investigate, 1,301 patients, persons diagnosed as having duodenal ulcer and 6,313 controls, persons who do not have this type of ulcer, are selected and their blood types determined. The results of this screening are given in Table 10.17. Note that there is an important difference

Table 10.17

| | Type of Subject | | |
	Patients	Controls	
Blood Group			
O	698	2,892	3,590
A	472	2,625	3,097
B	102	570	672
AB	29	226	255
	1,301	6,313	7,614
	(fixed)	(fixed)	

between this contingency table and those studied earlier. In particular the number of patients and the number of controls to be used in the study are *fixed* by the experimenter. They are not random variables.

In experiments such as that described in Example 1 in which column or row totals, but not both, are fixed by the experimenter, the hypothesis of "no association" between classification variables is expressed in terms of proportions. By fixing column totals we are selecting independent random samples from c populations. We want to see if the proportions of objects falling into various levels relative to the row classification variable are the same regardless of which population is involved. To express this idea mathematically, let p_{ij} ($i = 1, 2, \ldots, r$; $j = 1, 2, \ldots, c$) denote the proportion of objects in population j that lie in the ith level relative to the row classification variable. The null hypothesis is

$$H_0: p_{i1} = p_{i2} = \cdots = p_{ic} \qquad i = 1, 2, 3, \ldots, r$$

H_1: these proportions differ for at least one i

This looks difficult! It is not hard to understand in the context of a practical problem.

Example 2   In the study described in Example 1, we have a random sample of size 1,301 drawn from the population of individuals with duodenal ulcer and an independent random sample of size 6,313 drawn from the population of individuals that does not have this problem. We are interested in detecting any differences that exist in the way that blood types are distributed among patients and controls. That is, we want to answer these questions:

Is the proportion of ulcer patients with type-O blood the same as that of controls?

Is the proportion of ulcer patients with type-A blood the same as that of controls?

Is the proportion of ulcer patients with type-B blood the same as that of controls?

Is the proportion of ulcer patients with type-AB blood the same as that of controls?

Our null hypothesis is that the answer to each of these questions is yes. That is,

$$H_0: p_{11} = p_{12}, \; p_{21} = p_{22}, \; p_{31} = p_{32}, \; p_{41} = p_{42}$$

If we reject H_0, we are saying that at least one of these equalities fails to hold. Practically speaking, this suggests an association between blood type and the presence of this type of ulcer.

Mathematically we proceed in exactly the same manner as when testing for independence. However, we say in this case that we are testing for **homogeneity**— for uniformity of proportions within populations.

Table 10.18

	Type of Subject		
	Patients	Controls	
Blood Group			
O	698 (613.42)	2,892 (2,976.58)	3,590
A	472 (529.18)	2,625 (2,567.82)	3,097
B	102 (114.82)	570 (557.18)	672
AB	29 (43.57)	226 (211.43)	255
	1,301	6,313	7,614

Example 3

Table 10.18 presents the data of our previous study and gives the expected cell frequencies in parentheses. Inspection of the table leads one to suspect that there are some differences between patients and controls. In particular, note that the proportion of patients in blood group O is $698/1,301 = 54\%$; for controls this proportion is $2,892/6,313 = 46\%$. To see if this difference is great enough to cause rejection of H_0, we evaluate our X_3^2 test statistic. For these data,

$$\sum_{i=1}^{4} \sum_{j=1}^{3} (O_{ij} - \hat{E}_{ij})^2 / \hat{E}_{ij} = 29.12$$

This value allows rejection of H_0 with a P value less than .005. ($\chi^2_{.005} = 12.8$) We have established an association between blood type and duodenal ulcer.

It should be noted that saying that two variables are associated is not implying that one causes the other. They may both be related to some other variable or variables that cause the two to act as they do. The test for independence or homogeneity in a contingency table is a test of association, not a test of cause and effect. In the above example, we are saying that there appears to be a relationship between the incidence of duodenal ulcer and blood type. We are not saying that a particular blood type causes ulcers.

Exercises 10.4

28. Three populations are sampled independently and each object is classified by size as being either small, medium, or large. The data in Table 10.19 are obtained.
    (a) State the null hypothesis of no association in statistical terms. Express the null hypothesis verbally.

Table 10.19

	Population		
	I	II	III
Size			
Small	7	9	6
Medium	20	24	18
Large	38	42	36

(b) Find the row totals and column totals.
(c) Find the expected cell frequencies under the assumption that there is no association between population membership and size.
(d) Based on the expected cell frequencies, do you think that there is good statistical evidence that there is an association between population membership and size?
(e) Test the null hypothesis at the $\alpha = .10$ level.

Table 10.20

	Group	
	Male	*Female*
Oppose	60	80
Neutral	15	5
Favor		
	100	100

29. Samples of male and female students are selected and polled concerning their opinion about a renewal of the military draft. The results of the poll are given in Table 10.20.
   (a) Complete Table 10.20.
   (b) State the null hypothesis of no association both statistically and verbally.
   (c) Find the expected cell frequencies under the assumption that there is no association between an individual's sex and his or her opinion concerning the draft.
   (d) On the basis of these expected frequencies, do you think that there is an association between sex and opinion toward the draft?
   (e) Test H_0. Do you think that H_0 should be rejected? Explain based on the P value of your test.

30. A businessman is interested in determining the relationship between the day of the week selected for the midweek sale and the volume of sales. Over a two-year period, 20 midweek sales each were conducted Tuesdays, Wednesdays, and Thursdays. The businessman categorized the sales volume as being either light, moderate, or heavy. The results of the study are shown in Table 10.21.

Table 10.21

	Day of Sale			
	Tuesday	*Wednesday*	*Thursday*	
Volume of Sales				
Light	6	3	0	
Moderate	10	11	9	
Heavy	4	6	11	
	20	20	20	60

(a) State the appropriate null hypothesis.

(b) Is there evidence of an association between the day on which the sale was run and the volume of trade? Explain based on the P value of your test statistic.

31. A dietician is interested in studying food preferences among men and women. Random samples of 50 men and 50 women are selected and each subject is asked to indicate which of the green vegetables—green beans, peas, asparagus, or broccoli—is preferred. The results obtained are shown in Table 10.22. Is there evidence at the $\alpha = .05$ level that there is an association between vegetable preference and sex of the individual?

Table 10.22

	Sex		
	Men	Women	
Vegetable Type			
Green Beans	30	20	
Peas	15	20	
Asparagus	2	4	
Broccoli	3	6	
	50	50	100

32. A school psychologist is interested in the relationship between antisocial behavior among teenagers and self-image. A random sample of 50 teenagers who at some time during the school year have had behavioral problems is obtained. A random sample of 75 teenagers from the same schools who have not had any problems is selected. All students are asked to rate themselves on a self-image test. The survey results are given in Table 10.23. Do these data tend to support the contention that a negative self-image is found more often among students exhibiting behavioral problems than among other students? Explain.

Table 10.23

	Behavioral Status		
	Problems	No Problems	
Self-image			
Positive	4	13	
Neutral	20	32	
Negative	26	30	
	50	75	125

33. A study of the association between the amount of oxygen received by premature babies and the severity of vision defects among such babies is conducted. Records of 50 babies who received high-level oxygen for 4 to 7 weeks, treat-

Table 10.24

	Treatment		
	A	B	
Severity of Eye Damage			
Normal	20	42	
Mild Lesion	5	5	
Moderate Lesion	13	3	
Severe Lesion	4	0	
Blind	8	0	
	50	50	100

ment A, are obtained. Each baby is then classified as to vision status. Records of 50 babies who received low-level oxygen for 1 to 2 weeks, treatment B, are also obtained, and these babies are classified as to the status of their vision. The results of the study are given in Table 10.24. Is there evidence of an association between treatment used and severity of eye damage? Explain based on the P value of your test statistic. Practically speaking, which treatment appears to lessen the chances of eye damage?

Vocabulary List and Key Concepts

multinomial trial   independence

multinomial random variable   contingency table

chi-square goodness-of-fit test   homogeneity

Review Exercises

34. What is a multinomial trial?

35. Genetic theory predicts that when a carrier of color blindness and a normal male parent a child, the child will be either a color-blind male, a normal male, a normal female, or a carrier of color blindness. Each of these possibilities is equally likely. Forty babies are born to couples with this genetic makeup. These data are obtained:

    I   color-blind males: 8

    II   normal males: 12

    III   normal females: 11

    IV   carriers: 9

(a) Find the expected number in each category assuming that the genetic theory is correct.

(b) Test $H_0: p_1 = p_2 = p_3 = p_4$ at the $\alpha = .05$ level. Do you have evidence that the theory is incorrect?

36. The chi-square goodness-of-fit test is a large-sample procedure. It requires that expected cell frequencies not become too small. What guideline is suggested for these expectations? What should you do if an expected frequency is found to be too small?

37. Explain the logic behind a goodness-of-fit test.

38. In testing for normality, what estimator is used to estimate μ? to estimate σ^2? Is the estimator for σ^2 unbiased?

Use the following data to answer exercises 39–44:

Let X denote the time (in minutes) that it takes to complete a transaction at a bank's drive-in window. These data are obtained:

4.1	9.9	8.5	9.7	4.5	8.3
5.2	8.0	7.0	9.6	4.8	8.4
9.8	4.2	4.3	8.6	9.4	5.5
9.2	9.3	9.8	5.4	9.1	4.7
9.1	9.1	6.1	9.8	9.0	9.2

39. Construct a stem-and-leaf diagram for these data. On the basis of this diagram, do you think that the null hypothesis of normality can be rejected?

40. Break the data into five categories using the procedure discussed in this chapter.

41. Estimate μ and σ^2 using the maximum likelihood estimators.

42. Estimate the probability of an observation falling into each of the five categories found in exercise 40.

43. Estimate the expected cell frequencies and combine any categories that are too small.

44. Test the null hypothesis of normality at the $\alpha = .10$ level. Did you anticipate this result?

45. When all marginal totals in a contingency table are free to vary, what is the test of no association testing?

46. When row totals or column totals, but not both, are fixed by the experimenter, what is the test of no association testing?

47. Regardless of what type of contingency table is involved, the expected cell frequencies are computed in the same manner. What is the formula for these expected values?

Use the following data to answer exercises 48–51:

Jobs in the airline industry entail stress. It is thought that air-traffic controllers are particularly susceptible to stress-related disorders such as ulcers and high blood pressure. A sample of 500 air-traffic controllers is selected and observed; a

Table 10.25

	Population		
	Controllers	*Others*	
Stress Disorder Present			
Yes	115	125	
No	385	575	
	500	700	

sample of 700 workers in other areas of the airline industry is also studied. The results of the study are given in Table 10.25.

48. In testing for no association between job type and stress-related diseases, are we conducting a test of independence or a test of homogeneity?

49. State H_0 both statistically and verbally.

50. Find the expected cell frequencies and compare them with the observed frequencies. Based on this comparison, do you think that H_0 should be rejected?

51. Test H_0 at the $\alpha = .05$ level. Did you anticipate this result?

Use the following data to answer exercises 52–55:

A study is conducted to see if there is an association between age and the willingness to use a computerized banking system. A sample of 500 customers is selected and their opinion is sought. The data obtained are given in Table 10.26.

Table 10.26

	Use Computerized Banking		
	Yes	*No*	
Age			
Under 40	150	75	
40 or Over	150	125	
			500

52. In testing for no association between age and willingness to use the system, are we conducting a test of independence or a test of homogeneity?

53. State H_0 both statistically and verbally.

54. Find the expected cell frequencies and compare them with the observed frequencies. Based on this comparison, do you think that H_0 should be rejected?

55. Test H_0 at the $\alpha = .05$ level. Did you anticipate this result?

Analysis of Variance

In Chapter 8 we considered methods by which the means of two normal populations could be compared. In this chapter we extend the problem to the comparison of more than two means. The procedure used is called *analysis of variance* (ANOVA). We consider two problems, namely, the one-way classification and the randomized complete block design. The former is an extension of the pooled-T test presented in Chapter 8; the latter is an extension of the paired-T test.

11.1 One-Way Classification

The term **one-way classification** refers to the fact that there is only one **factor** or characteristic under direct study in the experiment. Each subject is classed as falling into exactly one level relative to this factor. Hence, we are dealing with one variable of classification.

There are two commonly encountered experimental situations in which the one-way classification analysis of variance comes into play. These situations are described in general below.

1. We have a single population and are interested in studying the effects of k different treatments. A random sample of N objects is selected and then randomly divided into k non-overlapping subgroups of sizes $n_1, n_2, \ldots, n_k$, as shown in Figure 11.1. Each subgroup receives a different treatment, and a response of some sort is noted. Any observed difference in response is attributed to the fact that the subgroups received different treatments. We want to test the hypothesis that the treatments have the same effect with a test that is especially likely to reject if the treatments differ in mean response. That is, we wish to test

$$H_0: \mu_1 = \mu_2 = \cdots = \mu_k$$

$$H_1: \mu_i \neq \mu_j \quad \text{for some } i \text{ and } j$$

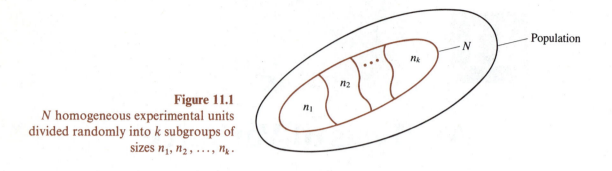

Figure 11.1
N homogeneous experimental units
divided randomly into k subgroups of
sizes $n_1, n_2, \ldots, n_k$.

where μ_i is the mean response for the ith treatment. This is a one-way classification problem and the single classification variable is the type of treatment received.

2. We have k populations, each population identified by some common characteristic to be studied in the experiment. Random samples of size $n_1, n_2, \ldots, n_k$ are selected from each of the populations, respectively, as shown in Figure 11.2. Each sample is given the same treatment, and a response of some sort is noted. Any observed difference is attributed to basic differences that exist among the k populations. We want to test the hypothesis that the groups are identical in response with a test that is especially likely to reject if the responses differ in mean value. Again, we test

$H_0: \mu_1 = \mu_2 = \cdots = \mu_k$

$H_1: \mu_i \neq \mu_j$ for some i and j

where μ_i is the mean response for the ith population. This is a one-way

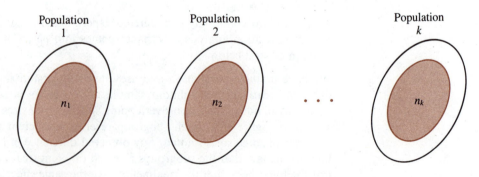

Figure 11.2
Independent random samples of sizes $n_1, n_2, \ldots, n_k$ are selected from k populations.

classification problem and the single classification variable is population membership.

Example 1 should clarify these ideas.

Example 1

1. In an experiment testing word fluency, 15 randomly selected students are divided randomly into three groups of five each. Each member of group I is asked individually to list orally as many words as possible beginning with the letter s. Each member of group II is asked individually to write as many words as possible beginning with the letter s. Members of group III are placed in a room together, and each is asked to list as many words as possible beginning with the letter s. The purpose of the experiment is to determine whether or not the manner in which the test is administered has an effect on the results.

   The one factor under study in the experiment is the mode of administering the test. Each subject is classed only according to the manner in which the test is taken. This is an example of the first situation described above, with

$$n_1 = n_2 = n_3 = 5 \quad \text{and} \quad N = 15$$

We are interested in testing

$$H_0: \mu_1 = \mu_2 = \mu_3$$

H_1: means are not all the same

Any difference in mean response will be attributed to the fact that the test was administered differently to the three groups. That is, they received different "treatments."

2. A sociologist is interested in the effect of family size on the independence of college freshmen. The one factor under study is family size. The sociologist is particularly interested in comparing students from families with 1, 2, 3, and more than 3 children, and has obtained random samples of sizes 15, 15, 14, and 13 from each of these four populations, respectively. A questionnaire that yields a score reflecting independence (on a scale from 0 to 100) is administered to each subject. This is an example of the second situation with

$$n_1 = 15, \quad n_2 = 15, \quad n_3 = 14, \quad n_4 = 13, \quad \text{and} \quad N = \Sigma n_i = 57$$

We are interested in testing

$$H_0: \mu_1 = \mu_2 = \mu_3 = \mu_4$$

H_1: means are not all the same

Any difference in mean scores will be attributed to the fact that the individuals involved come from families of different sizes.

Table 11.1

Data layout: one-way classification, completely random design

		Treatment		
1	2	3	...	k
X_{11}	X_{21}	X_{31}		X_{k1}
X_{12}	X_{22}	X_{32}		X_{k2}
X_{13}	X_{23}	X_{33}		X_{k3}
$\vdots$	$\vdots$	$\vdots$		$\vdots$
X_{1n_1}	X_{2n_2}	X_{3n_3}		X_{kn_k}
$T_1.$	$T_2.$	$T_3.$	...	$T_k.$ $T..$

The data collected in a single factor experiment are conveniently recorded in the format shown in Table 11.1. We use the following notation:

$$X_{ij} = \text{response of the } j\text{th experimental unit to the } i\text{th treatment}$$

$$i = 1, 2, \ldots, k \qquad j = 1, 2, \ldots, n_i$$

$$n_i = \text{sample size for the } i\text{th treatment}$$

$$\Sigma n_i = N = \text{total number of responses}$$

$$T_{i.} = \sum_{j=1}^{n_i} X_{ij} = \text{total of all responses to the } i\text{th treatment}$$

$$\bar{X}_{i.} = T_{i.}/n_i = \text{sample mean for the } i\text{th treatment}$$

$$T.. = \sum_{i=1}^{k} \sum_{j=1}^{n_i} X_{ij} = \sum_{i=1}^{k} T_{i.} = \text{total of all responses}$$

$$\bar{X}.. = T../N = \text{grand sample mean}$$

Note that, with the exception of sample sizes that are fixed by the experimenter, these are all statistics.

Example 2 demonstrates the use of this notation.

Example 2

When the word fluency experiment described in Example 1 is conducted, the data given in Table 11.2 results. For these data

$$T_1. = 105 \qquad\qquad T_2. = 100 \qquad\qquad T_3. = 119$$

$$\bar{x}_1. = 105/5 = 21 \qquad \bar{x}_2. = 100/5 = 20 \qquad \bar{x}_3. = 119/5 = 23.80$$

$$T.. = \sum_{i=1}^{3} T_{i.} = T_1. + T_2. + T_3. = 105 + 100 + 119 = 324$$

$$\bar{x}.. = T../N = 324/15 = 21.60$$

Table 11.2

Mode of Administering Test

Individual and Oral	*Individual and Written*	*Group and Written*
$x_{11} = 23$	$x_{21} = 20$	$x_{31} = 25$
$x_{12} = 26$	$x_{22} = 23$	$x_{32} = 29$
$x_{13} = 23$	$x_{23} = 21$	$x_{33} = 17$
$x_{14} = 14$	$x_{24} = 19$	$x_{34} = 21$
$x_{15} = 19$	$x_{25} = 17$	$x_{35} = 27$

To have some understanding of the computations involved in the analysis of variance procedure, it is necessary to consider briefly the underlying ideas. The analysis is based primarily on the assumption that a given response X_{ij} can be partitioned into components attributable to various logical and identifiable sources. Let

μ = theoretical overall average or expected response ignoring the actual treatment involved

μ_i = theoretical average or expected response for the ith treatment $i = 1, 2, \ldots, k$

The model for the one-way classification is written as follows:

Model: One-Way Classification

$$X_{ij} = \mu + (\mu_i - \mu) + (X_{ij} - \mu_i)$$

In words, we are claiming that

$$
\begin{array}{ccccc}
\text{response of} & & & & \text{random fluctuation} \\
\text{jth experimental} & \text{overall} & \text{effect due to} & & \text{within treatment i} \\
\text{unit to ith} & = \text{average} + & \text{treatment i} + & \text{due to individual} \\
\text{treatment} & \text{response} & & & \text{differences} \\
X_{ij} & \mu & (\mu_i - \mu) & & (X_{ij} - \mu_i)
\end{array}
$$

For instance, in our word fluency example, we are claiming that an individual's response can be partitioned into components corresponding to the overall average score that would be attained regardless of the method by which the test was administered. In addition there would be some effect caused by the method of administration plus a random effect because of individual differences.

To develop a test statistic for testing the null hypothesis of equal means, we must make some assumptions concerning the populations from which our samples are drawn. The assumptions made parallel those made in the two-sample pooled-T context. They are

---

Model Assumptions One-Way Classification

1. The k samples represent independent random samples drawn from k populations with means $\mu_1, \mu_2, \mu_3, \ldots, \mu_k$.
2. Each of the k populations is normally distributed.
3. Each of the k populations has the same variance σ^2.

To see the logic behind the test statistic, we first rewrite the model by subtracting μ from each side to obtain

$$X_{ij} - \mu = (\mu_i - \mu) + (X_{ij} - \mu_i)$$

Since $\mu, \mu_1, \mu_2, \ldots, \mu_k$ are the theoretical means, they must be estimated from the data. This can be done easily by using the corresponding sample means $\bar{X}.., \bar{X}_{1.}, \bar{X}_{2.}, \ldots, \bar{X}_k.$ defined previously. By replacing the theoretical model parameters by their estimators, we obtain

$$X_{ij} - \bar{X}.. = (\bar{X}_{i.} - \bar{X}..) + (X_{ij} - \bar{X}_{i.})$$

If each side of the above equation is squared and summed over all possible values of i and j, the following equation is obtained:

$$\sum_{i=1}^{k} \sum_{j=1}^{n_i} (X_{ij} - \bar{X}..)^2 = \sum_{i=1}^{k} n_i(\bar{X}_{i.} - \bar{X}..)^2 + \sum_{i=1}^{k} \sum_{j=1}^{n_i} (X_{ij} - \bar{X}_{i.})^2$$

It should be observed that

$$\sum_{i=1}^{k} \sum_{j=1}^{n_i} (X_{ij} - \bar{X}..)^2 = \text{sum of squares of deviations of observations from the grand mean}$$

$$= \text{measure of the total amount of variability in data}$$

$$= \text{total sum of squares} = SS_{\text{Total}}$$

$$\sum_{i=1}^{k} n_i(\bar{X}_{i.} - \bar{X}..)^2 = \text{weighted sum of squares of deviations of treatment means from the grand mean}$$

$$= \text{measure of amount of variation in data attributed to the fact that different treatments are used}$$

$$= \text{treatment sum of squares} = SS_{\text{Tr}}$$

$$\sum_{i=1}^{k} \sum_{j=1}^{n_i} (X_{ij} - \bar{X}_{i\cdot})^2 = \text{sum of squares of deviations of observations from the treatment mean associated with the observation}$$

$$= \text{measure of amount of variation in data attributed to random or unexplained sources}$$

$$= \text{residual or error sum of squares} = SS_E$$

Using this shorthand notation, we can write the previous equation as

$$SS_{\text{Total}} = SS_{\text{Tr}} + SS_E$$

The equation is referred to as a *sum of squares identity*. This identity expresses the idea that, in a one-way classification, the total variability in the data, SS_{Total}, can be partitioned into two components. The first, SS_{Tr}, measures the variability due to the fact that different levels of the classification variable are used; the second, SS_E, measures the random or unexplained variability in the data.

We now define two statistics called the *treatment mean square* and the *error mean square*. The former reflects the variability in response due to the fact that different levels of the classification variable are used; it is defined in terms of SS_{Tr}. The latter reflects the random or unexplained variability in response among experimental units; it is defined in terms of SS_E. These statistics are given by

Mean Squares

$$\text{treatment mean square} = MS_{\text{Tr}} = SS_{\text{Tr}}/(k-1)$$

$$\text{error mean square} = MS_E = SS_E/(N-k)$$

The test statistic used to test for equality of means is based upon the premise that if treatment means are unequal, then most of the variation in response should be attributed to differences in treatments; little of the variation should be unexplained. Hence, the error mean square should be small relative to the treatment mean square. This suggests that to test for equality of means, we need to compare MS_{Tr} to MS_E. We do so via the ratio

$$MS_{\text{Tr}}/MS_E$$

Remember that to test a null hypothesis, the distribution of the test statistic must be known under the assumption that the null hypothesis is true. It can be shown that when our model assumptions are satisfied and when $\mu_1 = \mu_2 = \cdots = \mu_k$, the statistic MS_{Tr}/MS_E follows an F distribution with $k-1$ and $N-k$ degrees of freedom. Thus the test statistic in a one-way classification is given by

Test Statistic One-Way Classification

$$\text{MS}_{\text{Tr}}/\text{MS}_{\text{E}} = F_{k-1,\,N-k}$$

It can also be shown that if H_0 is true the expected value of this statistic is 1; otherwise its expectation exceeds 1. Thus, to conduct our test, we evaluate the test statistic and reject

$$H_0: \mu_1 = \mu_2 = \ldots = \mu_k$$

for values of the F ratio that are too large to have occurred by chance. That is, our test is always a right-tailed test.

The reason for the name *analysis of variance* should now be clear. The analysis is accomplished by partitioning the variability observed in the data into components and then making use of their probability distributions to devise a test statistic.

The formulas presented for SS_{Total}, SS_{Tr}, and SS_{E} are a bit awkward to handle. In practice these computational formulas are used.

Computational Shortcuts

$$\text{SS}_{\text{Total}} = \sum_{i=1}^{k} \sum_{j=1}^{n_i} X_{ij}^2 - T_{..}^2/N$$

$$\text{SS}_{\text{Tr}} = \sum_{i=1}^{k} T_{i.}^2/n_i - T_{..}^2/N$$

$$\text{SS}_{\text{E}} = \sum_{i=1}^{k} \sum_{j=1}^{n_i} X_{ij}^2 - \sum_{i=1}^{k} T_{i.}^2/n_i = \text{SS}_{\text{Total}} - \text{SS}_{\text{Tr}}$$

Although these formulas look complicated, they are not difficult to use. To illustrate, we continue our analysis of the word fluency data begun in Examples 1 and 2.

Example 3

Let us test $H_0: \mu_1 = \mu_2 = \mu_3$ at the $\alpha = .10$ level. In the last example, we found these totals

$$T_1. = 105 \qquad T_2. = 100 \qquad T_3. = 119 \qquad T.. = 324$$

The only new total that is needed is the sum of the squares of the individual data points given in Table 11.2. For these data,

$$\sum_{i=1}^{3} \sum_{j=1}^{5} X_{ij}^2 = X_{11}^2 + X_{12}^2 + X_{13}^2 + \cdots + X_{35}^2$$

$$= 23^2 + 26^2 + 23^2 + \cdots + 27^2 = 7{,}236$$

$$T_{..}^2/N = (324)^2/15 = 6998.40$$

$$\sum_{i=1}^{3} T_{i.}^2/n_i = T_{1.}^2/5 + T_{2.}^2/5 + T_{3.}^2/5$$

$$= 105^2/5 + 100^2/5 + 119^2/5 = 7{,}037.20$$

$$SS_{Total} = \sum_{i=1}^{3} \sum_{j=1}^{5} X_{ij}^2 - T_{..}^2/N = 7{,}236 - 6{,}998.4 = 237.60$$

$$SS_{Tr} = \sum_{i=1}^{3} T_{i.}^2/n_i - T_{..}^2/N = 7{,}037.20 - 6{,}998.4 = 38.80$$

$$SS_E = SS_{Total} - SS_{Tr} = 237.60 - 38.80 = 198.80$$

$$MS_{Tr} = SS_{Tr}/(k-1) = 38.80/2 = 19.40$$

$$MS_E = SS_E/(N-k) = 198.80/(15-3) = 16.57$$

The observed value of the $F_{2,\,12}$ statistic is

$$MS_{Tr}/MS_E = 19.40/16.57 = 1.17$$

The critical point for an $\alpha = .10$ test from Table VIII in Appendix A is 2.81. Since 1.17 is not greater than this point, we are unable to reject H_0. We do not have sufficient statistical evidence to conclude that the mode of administering the test has an effect on the mean response.

Analysis of variance data are usually presented in summary form in an *analysis of variance table*. This is a table that summarizes all of the important statistics in easily readable form. It lists the source of variation in response, the observed values for SS_{Tr} and SS_E, the degrees of freedom or the denominators used to find the mean squares, the observed values for MS_{Tr} and MS_E, and the observed value of the F ratio MS_{Tr}/MS_E. The ANOVA table for the data of Example 3 is shown in Table 11.3.

Since this technique is rather involved, let us consider one other illustration

Table 11.3

ANOVA

Source of Variation	Sum of Squares	Degrees of Freedom	Mean Square	F Ratio
Treatment	38.80 (SS_{Tr})	2($k-1$)	19.40 $\left(\dfrac{SS_{Tr}}{k-1}\right)$	1.17 $\left(\dfrac{MS_{Tr}}{MS_E}\right)$
Error	198.80 (SS_E)	12($N-k$)	16.57 $\left(\dfrac{SS_E}{N-k}\right)$	
Total	237.60	14($N-1$)		

before leaving you on your own! You can benefit most from this example by trying the analysis before looking at the text. If your results agree with ours, you have mastered the technique.

Example 4

When the sociological experiment described in Example 1 is conducted, the data shown in Table 11.4 results. Let us test

$$H_0: \mu_1 = \mu_2 = \mu_3 = \mu_4$$

at the $\alpha = .05$ level. The approximate critical point for the test based on the $F_{3, 53}$ distribution is 2.84. For these data

$$\sum_{i=1}^{4} \sum_{j=1}^{n_i} X_{ij}^2 = (59.1)^2 + (84.4)^2 + (76.0)^{2} + \cdots + (95.6)^2 + (83.8)^2$$

$$= 277{,}845.9$$

$$T_{..}^2/N = (3{,}914.4)^2/57 = 268{,}816.27$$

$$SS_{Total} = \sum_{i=1}^{4} \sum_{j=1}^{n_i} X_{ij}^2 - T_{..}^2/N$$

$$= 277{,}845.9 - 268{,}816.27 = 9{,}029.63$$

$$\sum_{i=1}^{4} T_{i.}^2/n_i = (1{,}001.3)^2/15 + (933.1)^2/15 + (946.9)^2/14 + (1{,}033.1)^2/13$$

$$= 271{,}029.07$$

Table 11.4

	Family Size		
1 Child	*2 Children*	*3 Children*	*More than 3 Children*
59.1	61.2	73.4	73.1
84.4	71.0	69.3	95.7
76.0	46.6	64.9	91.1
59.5	54.0	48.7	49.7
60.1	66.6	67.7	94.9
73.4	56.6	72.5	65.8
64.1	70.5	68.8	75.8
69.4	72.8	79.9	77.2
56.4	58.5	77.7	86.2
67.1	48.7	79.2	61.1
97.6	63.3	56.7	83.1
58.5	74.8	60.1	95.6
70.7	53.1	69.8	83.8
51.8	69.9	58.2	
53.2	65.5		
$T_{1.} = 1{,}001.3$	$T_{2.} = 933.1$	$T_{3.} = 946.9$	$T_{4.} = 1{,}033.1$ $T_{..} = 3914.4$

Table 11.5

ANOVA

Source of Variation	Sum of Squares	Degrees of Freedom	Mean Square	F Ratio
Treatment	2,212.80	3	737.60	5.73
Error	6,816.83	53	128.62	
Total	9,029.63	56		

$$SS_{Tr} = \sum_{i=1}^{4} T_i^2./n_i - T^2../N$$

$$= 271{,}029.07 - 268{,}816.27 = 2{,}212.80$$

$$SS_E = SS_{Total} - SS_{Tr}$$

$$= 9{,}029.63 - 2{,}212.80 = 6{,}816.83$$

$$MS_{Tr} = SS_{Tr}/(k-1) = 2{,}212.80/3 = 737.60$$

$$MS_E = SS_E/(N-k) = 6{,}816.83/(57-4) = 128.62$$

$$F_{3,53} = MS_{Tr}/MS_E = 737.60/128.62 = 5.73$$

These results are summarized in the ANOVA table of Table 11.5. Since 5.73 > 2.84, we reject H_0 and conclude that family size does have an effect on mean response.

In a one-way classification, sample sizes may vary from sample to sample. However, as in the pooled-T test, it is assumed that each of the k normal population variances are the same. Experience shows that the effects of a deviation from this assumption are minimized if sample sizes are the same. For this reason, whenever possible, one should design experiments so that $n_1 = n_2 = \cdots = n_k$. However, in practice, actual sample sizes used are often dictated by the availability of subjects and by cost considerations.

Exercises 11.1

1. Three normal populations are sampled. Each sample is of size 10. Suppose that $SS_{Tr} = 9{,}324$ and $SS_{Total} = 51{,}282$. Use these data to complete Table 11.6.
   (a) Find the critical point for an $\alpha = .10$ level test of $H_0: \mu_1 = \mu_2 = \mu_3$.
   (b) Can H_0 be rejected at the $\alpha = .10$ level?
   (c) What assumption are you making concerning the variances of the sampled populations?

Table 11.6

ANOVA

Source of Variation	Sum of Squares	Degrees of Freedom	Mean Square	F Ratio
Treatment	9,324			
Error				
Total	51,282			

2. Five normal populations are sampled. Each sample is of size 13. Suppose that $SS_{Tr} = 7,500$ and $SS_{Total} = 65,000$. Use these data to complete Table 11.7.
   (a) Find the critical point for an $\alpha = .10$ level test of $H_0: \mu_1 = \mu_2 = \mu_3 = \mu_4 = \mu_5$.
   (b) Can H_0 be rejected at the $\alpha = .10$ level?
   (c) What assumption are you making concerning the variances of the sampled populations?

3. A small pilot study is run to compare the television viewing habits among the following four groups:

   Group I: women under 20   Group II: women 20 or older
   Group III: men under 20   Group IV: men 20 or older

   The variable of interest is the number of hours of television watched per day. These data are obtained:

I	II	III	IV
5.1	4.5	8.2	10.0
8.2	7.2	8.5	11.1
7.3	.0	7.6	8.0
.0	8.1	1.0	5.4
.0	9.6	1.2	1.5

Table 11.7

ANOVA

Source of Variation	Sum of Squares	Degrees of Freedom	Mean Square	F Ratio
Treatment	7,500			
Error				
Total	65,000			

(a) For these data, find each of the following:

$$T_1. \qquad T_2. \qquad T_3. \qquad T_4. \qquad T..$$

$$\sum_{i=1}^{4} \sum_{j=1}^{5} X_{ij}^2 \qquad T_{..}^2/20 \qquad \sum_{i=1}^{4} T_{i.}^2/5$$

$$SS_{Total} \qquad SS_{Tr} \qquad MS_{Tr} \qquad SS_E \qquad MS_E$$

(b) Use the data from part (a) to test at the $\alpha = .10$ level the hypothesis that there is no difference in mean number of hours of television watched per day among the four groups.

4. An experiment is run to study inflation over a two-month period. In particular, the inflation rates in the following categories of foodstuffs are of interest:

Category I: staples (sugar, flour, salt, and others)

Category II: fresh fruits and vegetables

Category III: canned fruits and vegetables

Category IV: meat

Category V: dairy products

Ten items are selected randomly from among each group, and the percentage increase in cost over the two-month period is obtained. These data result:

$$T_1. = 1 \qquad T_2. = 5 \qquad T_3. = 6 \qquad T_4. = 10 \qquad T_5. = 3 \qquad \sum_{i=1}^{5} \sum_{j=1}^{10} X_{ij}^2 = 40$$

(a) Find each of the following:

$$T.. \qquad \sum_{i=1}^{5} T_{i.}^2/10 \qquad MS_{Tr} \qquad SS_E$$

$$T_{..}^2/50 \qquad SS_{Tr} \qquad SS_{Total} \qquad MS_E$$

(b) Use the results of part (a) to test at the $\alpha = .10$ level the hypothesis that there is no difference in the mean inflation rates among the five categories.

5. There has been considerable interest recently in the development of microwave ovens. Three different companies manufacture such ovens. We are interested in the mean length of time that it takes to cook a two-pound roast with each of the brands of oven. Tests are run on ovens from each company chosen randomly from among warehouse stock, and the following results are obtained (time is in minutes):

Oven manufacturer

A		B		C	
2.1	1.9	2.7	2.0	2.0	3.5
1.3	2.0	2.8	3.0	2.2	2.5
2.5	2.6	3.5		2.7	2.6
				1.8	

Test at the $\alpha = .025$ level the hypothesis that the oven brand has no effect on the mean length of time required to cook the roast.

6. In a study to assess how satisfied various groups of women are with their role in today's society, random samples of size 25 are obtained from each of the following populations: housewives, unmarried career women, married career women. Each subject is asked to fill out a questionnaire designed to measure role satisfaction on a scale from 0 to 100, with higher scores indicating a higher degree of satisfaction. The following results are obtained:

Housewives	Unmarried career women	Married career women
$T_{1.} = 1575$	$T_{2.} = 1875$	$T_{3.} = 1750$

$$\sum_{i=1}^{3} \sum_{j=1}^{25} X_{ij}^2 = 384,150.04$$

Use this information to test at the $\alpha = .10$ level the hypothesis that a woman's status in society has no effect on her degree of role satisfaction.

7. An experimenter is interested in the effect of sleep deprivation on manual dexterity. Thirty-two subjects are selected and randomly divided into four groups of size 8. After differing amounts of sleep deprivation, all subjects are given a series of tasks to perform, each of which requires a high degree of manual dexterity. A score from 0 (poor performance) to 10 (excellent performance) is obtained for each subject.

Group 1 (16 hours deprivation)		Group 2 (20 hours deprivation)		Group 3 (24 hours deprivation)		Group 4 (28 hours deprivation)	
8.95	6.48	7.70	8.04	5.99	5.78	3.78	4.87
8.04	7.81	5.81	5.96	6.79	7.60	3.35	3.14
7.72	7.50	6.61	7.30	6.43	5.78	2.45	3.98
6.21	6.90	6.07	7.46	5.85	6.00	4.27	2.47

Test at the $\alpha = .05$ level the hypothesis that the degree of sleep deprivation has no effect on manual dexterity.

8. An industrial psychologist is investigating the effects of three types of work hours on absenteeism among production-line workers. Each of the three methods is tried over a year's time in three different plant branches in neighboring towns. Use the results given on page 457 to test at the $\alpha = .05$ level the hypothesis that the type of work hours used has no effect on the mean number of days a worker is absent from work.

4-day work week *10-hour day* *(7 A.M. to 6 P.M.)*	*5-day work week,* *flexible* *8-hour day*	*5-day work week,* *8-hour day* *(7 A.M. to 4 P.M.)*
$T_1. = 900$	$T_2. = 527$	$T_3. = 909$
$n_1 = 100$	$n_2 = 85$	$n_3 = 90$

$$\sum_{i=1}^{3} \sum_{j=1}^{n_i} X_{ij}^2 = 50{,}511$$

COMPUTING SUPPLEMENT H

The SAS procedure PROC ANOVA is used to test H_0: $\mu_1 = \mu_2 = \cdots = \mu_k$ via the analysis of variance procedure. To illustrate, we let SAS analyze the data of Example 4.

Statement	*Purpose*
DATA SOC;	names the data set
INPUT SIZE SCORE;	each line contains the value of a variable SIZE, which identifies the group to which an observation belongs, and the value of the variable, SCORE, the individual's score on the questionnaire
CARDS;	data follows
1 59.1 1 84.4 . . . . 4 95.6 4 83.8	data
;	signals the end of data
PROC ANOVA;	calls for an analysis of variance to be conducted
CLASSES SIZE;	the data are grouped according to the value of the variable SIZE
MODEL SCORE = SIZE;	identifies the variable SCORE as the dependent or response variable
TITLE ONE WAY ANOVA;	titles the output

```
                          ONE WAY ANOVA
                  ANALYSIS OF VARIANCE PROCEDURE
                    CLASS LEVEL INFORMATION
              CLASS        LEVELS      VALUES
              SIZE           4         1 2 3 4

              NUMBER OF OBSERVATIONS IN DATA SET = 57
                          ONE WAY ANOVA
                  ANALYSIS OF VARIANCE PROCEDURE
```

DEPENDENT VARIABLE: SCORE

SOURCE	DF	SUM OF SQUARES	MEAN SQUARE	F VALUE
MODEL	3 ①	2212.80402448 ③	737.60134149 ⑥	5.73 ⑧
ERROR	53 ②	6816.82650183 ④	128.61936796 ⑦	PR > F
CORRECTED TOTAL	56	9029.63052632 ⑤		0.0019 ⑨

R-SQUARE	C.V.	ROOT MSE	SCORE MEAN
0.245060	16.5144	11.34104792	68.67368421

SOURCE	DF	ANOVA SS	F VALUE	PR > F
SIZE	3	2212.80402448	5.73	0.0019

Note that the source of variation that we called *treatment* is termed *model* on the SAS printout. If you compare the numbers on this printout to those given in Table 11.5 you will see that they agree apart from roundoff error. These quantities are given:

① Degrees of freedom associated with treatments $(k - 1)$
② Degrees of freedom associated with unexplained or random sources $(N - k)$
③ SS_{Tr}
④ SS_E
⑤ SS_{Total}
⑥ $MS_{Tr} = SS_{Tr}/(k - 1)$
⑦ $MS_E = SS_E/(N - k)$
⑧ $F = MS_{Tr}/MS_E$

The *P* value of the *F* test is given by ⑨. Since this *P* value is small, .0019, we can, as expected, reject H_0 and conclude that the treatments differ in mean effect.

11.2 Multiple Comparisons

Once a one-way classification analysis of variance has been run, we are in one of two positions:

1. We have been unable to reject H_0. We therefore conclude that, based on the available data, the treatments involved do not appear to differ in average effect.

2. We have been able to reject H_0. We therefore conclude that at least two of the treatments produce different average effects.

In case 1, the analysis of the data is over. However, in case 2, we naturally want to continue our investigation to try to determine exactly what differences are involved. There are several ways to approach this problem. We use what is termed *Scheffé's method of multiple comparisons*. To apply this method we must first define the term *linear contrast*.

Definition 11.1

Linear Contrast

Let $\mu_1, \mu_2, \ldots, \mu_k$ be the means of k populations. Any linear function of the form

$$\sum_{i=1}^{k} a_i \mu_i \qquad \text{where } \sum_{i=1}^{k} a_i = 0$$

is called a **linear contrast**.

Note that $\sum_{i=1}^{k} a_i = 0$ simply requires that the sum of the coefficients of the means involved in the contrast be zero.

Example 1

$\mu_1 - \mu_2$; $(1/2)\mu_1 + (1/2)\mu_2 - \mu_3$; $2\mu_1 - \mu_2 - \mu_3$ are all linear contrasts, since the sum of the coefficients of the means involved in each case is zero.

The purpose of a linear contrast is to make comparisons among means. In an experimental situation, the experimenter must choose those contrasts that are physically meaningful. We are interested in developing a method for testing hypotheses that the value of the contrast is zero. That is, we want to test hypotheses of the form

$$H_0: \sum_{i=1}^{k} a_i \mu_i = 0$$

Example 2

A tennis instructor is interested in various methods of teaching beginning tennis, and wants to compare the relative merits of the following combinations of teaching aids:

Treatment 1: scrimmage

Treatment 2: scrimmage and lectures on technique

Treatment 3: scrimmage and use of video tape

Treatment 4: scrimmage, lectures and video tape

The two extreme treatments, 1 and 4, may be compared by testing

$$H_0: \mu_1 - \mu_4 = 0$$

The effect of video tape may be detected by testing

$$H_0: (\mu_1 + \mu_2)/2 - (\mu_3 + \mu_4)/2 = 0$$

Scheffé's method of multiple comparisons consists of forming what is termed a Scheffé-type confidence interval on the contrast

$$\sum_{i=1}^{k} a_i \mu_i$$

of interest. If the numerical interval so constructed does not contain 0, then the hypothesis

$$H_0: \sum_{i=1}^{k} a_i \mu_i = 0$$

is rejected at the appropriate α level. The test is based on the following theorem.

Theorem 11.1

Scheffé's Method of Multiple Comparisons

Let $\mu_1, \mu_2, \ldots, \mu_k$ be the means of k normal populations, each with variance σ^2. The probability is $1 - \alpha$ that all contrasts simultaneously satisfy the following:

$$\sum_{i=1}^{k} a_i \bar{X}_1. - L \le \sum_{i=1}^{k} a_i. \mu_i \le \sum_{i=1}^{k} a_i \bar{X}_i. + L$$

where

$$L^2 = (k-1)f_{\alpha, k-1, N-k} \, \mathrm{MS_E} \sum_{i=1}^{k} (a_i^2/n_i)$$

This theorem looks rather forbidding, but it is easy to apply. Keep in mind the fact that the theorem comes into play *after* the hypothesis of equal means has been rejected in an analysis of variance problem. Hence, most of the computational work in constructing a Scheffé-type confidence interval has already been done. In particular, remember that

k = number of treatments

$f_{\alpha, k-1, N-k}$ = critical point used in rejecting

$H_0: \mu_1 = \mu_2 = \cdots = \mu_k$ in the ANOVA

MS_E = error mean square from ANOVA

n_i = sample size for the ith treatment

$\bar{X}_{i\cdot}$ = sample mean for treatment i

The next example demonstrates the use of this theorem.

Example 3

In an experiment designed to compare the methods of teaching tennis described in Example 2, the instructor taught one class of beginners using treatment 1, one class with treatment 2, one with treatment 3, and one using treatment 4. At the end of the semester, a skills test was administered to each student with high scores representing high performance. The following results were obtained:

Treatment 1	Treatment 2	Treatment 3	Treatment 4
$T_{1\cdot} = 1,200$	$T_{2\cdot} = 1,353$	$T_{3\cdot} = 1,453$	$T_{4\cdot} = 1,336$
$n_1 = 16$	$n_2 = 18$	$n_3 = 17$	$n_4 = 17$

$$\sum_{i=1}^{4} \sum_{j=1}^{n_i} X_{ij}^2 = 427,089$$

$$T_{\cdot\cdot} = \sum_{i=1}^{4} T_{i\cdot} = 5,342 \qquad T_{\cdot\cdot}^2/N = (5,342)^2/68 = 419,661.24$$

$$\sum_{i=1}^{4} T_{i\cdot}^2/n_i = (1,200)^2/16 + (1,353)^2/18 + (1,453)^2/17 + (1,336)^2/17$$
$$= 420,883.15$$

$$SS_{\text{Total}} = \sum_{i=1}^{4} \sum_{j=1}^{n_i} X_{ij}^2 - T_{\cdot\cdot}^2/N$$
$$= 427,089 - 419,661.24 = 7,427.76$$

$$SS_{\text{Tr}} = \sum_{i=1}^{4} T_{i\cdot}^2/n_i - T_{\cdot\cdot}^2/N$$
$$= 420,883.15 - 419,661.24 = 1,221.91$$

$$SS_E = SS_{\text{Total}} - SS_{\text{Tr}}$$
$$= 7,427.76 - 1,221.91 = 6,205.85$$

$$MS_{\text{Tr}} = SS_{\text{Tr}}/(k-1) = 1,221.91/3 = 407.30$$

$$MS_E = SS_E/(N-k) = 6,205.85/(68-4) = 96.97$$

$$F_{3,64} = MS_{\text{Tr}}/MS_E = 407.30/96.97 = 4.20$$

Table 11.8

ANOVA

Source of Variation	Sum of Squares	Degrees of Freedom	Mean Square	F Ratio
Treatment	1,221.91	3	407.30	4.20
Error	6,205.85	64	96.97	
Total	7,427.76	67		

We first test at the $\alpha = .05$ level the hypothesis

$$H_0: \mu_1 = \mu_2 = \mu_3 = \mu_4$$

The critical point for such a test is the point

$$f_{\alpha, k-1, N-k} = f_{.05, 3, 64} = 2.76$$

Since the observed value of the F ratio (4.20) exceeds 2.76, we can reject H_0 at the $\alpha = .05$ level and conclude that the treatments do not produce the same average effect. The ANOVA table for these data is shown in Table 11.8. To compare treatments 1 and 4, we test

$$H_0: \mu_1 - \mu_4 = 0$$

This is done by constructing a Scheffé-type confidence interval on the contrast $\mu_1 - \mu_4$. This confidence interval is given by

$$(\bar{X}_{1.} - \bar{X}_{4.}) \pm L$$

Thus, to test this hypothesis, we need to evaluate $\bar{X}_{1.}$ and $\bar{X}_{4.}$, and L for this particular contrast. The values of these statistics are:

$$\bar{X}_{1.} = T_{1.}/n_1 = 75.0 \qquad \bar{X}_{4.} = T_{4.}/n_4 = 78.59$$

$$L^2 = (k-1)f_{\alpha, k-1, N-k} \, MS_E \sum_{i=1}^{4} (a_i^2/n_i)$$

$$= 3(2.76)(96.97)[(1)^2/16 + (-1)^2/17] = 97.41$$

$$L = \sqrt{L^2} = \sqrt{97.41} = 9.87$$

The 95% Scheffé-type confidence interval is

$$(\bar{X}_{1.} - \bar{X}_{4.}) \pm 9.87$$

or

$$(75.0 - 78.59) \pm 9.87$$

This interval is

$$[-13.46, 6.28]$$

Since the interval does contain 0, we are unable to reject H_0: $\mu_1 - \mu_4 = 0$. We cannot conclude that there is a significant difference between the results obtained using treatments 1 and 4. To determine the effect of video tape, we test

$$H_0: (\mu_1 + \mu_2)/2 - (\mu_3 + \mu_4)/2 = 0$$

To do so, we compute $\bar{X}_1., \bar{X}_2., \bar{X}_3., \bar{X}_4.$ and the value of L for this contrast.

$$\bar{X}_1. = 75.0 \qquad \bar{X}_2. = 75.17 \qquad \bar{X}_3. = 85.47 \qquad \bar{X}_4. = 78.59$$

$$L^2 = 3(2.76)(96.97)[(1/2)^2/16 + (1/2)^2/18 + (-1/2)^2/17 + (-1/2)^2/17]$$

$$= 47.31$$

$$L = 6.88$$

The 95% Scheffé-type confidence interval is

$$(\bar{X}_1. + \bar{X}_2.)/2 - (\bar{X}_3. + \bar{X}_4.)/2 \pm L$$

Substituting the appropriate values into this expression, we obtain

$$(75.0 + 75.17)/2 - (85.47 + 78.59)/2 \pm 6.88$$

or

$$-6.95 \pm 6.88$$

Numerically, this interval is $[-13.83, -.07]$. Since this interval does not contain 0, we may reject H_0 and conclude that the mean performance score obtained using video tape is in fact higher than that obtained without it.

The Scheffé method has several nice features.

1. It is known to be affected very little if the populations involved are not normal and do not have equal variances (the basic assumption underlying analysis of variance).

2. Contrasts to be tested can and usually are selected after the experiment is conducted. We test first for equality of means, and if that is rejected, we attempt to discover which contrasts are responsible for the rejection. For instance, in our example, after inspection of the data, a third contrast of interest is $\mu_3 - \mu_4$, which compares the two treatments that appear to be closest in mean effect.

3. Whenever the F ratio in the ANOVA allows us to reject the hypothesis of equal means, one or more of the Scheffé-type confidence intervals that could be formed will not contain 0. That is, one or more contrasts will be found to be significantly different from 0.

4. The α level involved refers to the entire experiment. That is, the α level refers to the probability of drawing one or more false conclusions by rejecting hypotheses of the form

$$H_0: \sum_{i=1}^{k} a_i \mu_i = 0$$

Thus,

$$\alpha = P[\text{at least one null hypothesis is incorrectly rejected}]$$

One further comment should be made concerning point 4. The α level referred to there is called the **experimentwise error rate**. It gives the probability that at least one of the conclusions that we have drawn by rejecting a null hypothesis on a contrast is incorrect. For example, suppose that we set $\alpha = .05$ and investigate six different contrasts. If, at the end of the experiment, we reject four null hypotheses we will draw four concrete conclusions from our experiment. The probability that one or more of these conclusions is incorrect is only .05.

Exercises 11.2

9. A businessman experimented over a period of several years with four types of advertising. The variable of interest is the number of potential customers entering the store per day.

   Treatment 1: newspaper

   Treatment 2: newspaper and radio

   Treatment 3: newspaper and television

   Treatment 4: newspaper, radio, and television

   (a) Select a contrast that allows the businessman to detect the effect of some sort of auxiliary mode of advertisement in addition to the basic newspaper ad.
   (b) Select a contrast that allows the businessman to detect the effect of radio on the number of potential customers attracted.
   (c) What is the contrast

   $$(\mu_1 + \mu_2)/2 - (\mu_3 + \mu_4)/2$$

   designed to detect?

10. A psychologist is interested in studying the methods for treating severe anxiety among alcoholics and drug addicts. The psychologist has devised a means of measuring anxiety on a numerical scale and wants to compare the effects of the following modes of treatment:

    Treatment 1: individual counseling

    Treatment 2: individual counseling and group therapy

    Treatment 3: individual counseling and mild sedation

    Treatment 4: individual counseling, group therapy, and mild sedation

    (a) Select a contrast that can be used to detect the effect of sedation.
    (b) Select a contrast that can be used to detect the effect of group therapy.

11. A researcher is interested in testing two drugs for effectiveness in reducing high blood pressure. Each drug is available in both pill and liquid form. There are therefore four possible treatments:

Treatment 1: drug A in pill form

Treatment 2: drug B in pill form

Treatment 3: drug A in liquid form

Treatment 4: drug B in liquid form

The measured variable in the experiment is the percentage decrease in blood pressure 15 minutes after administration of the drug. The researcher antici-pates finding differences among these treatments.

(a) Select a contrast that could be used to compare drug A with drug B.

(b) Select a contrast that could be used to compare liquid form with pill form.

(c) Twenty persons suffering from high blood pressure are selected randomly and randomly divided into four groups of five persons each. Each group is then assigned randomly a treatment. The following data are obtained:

$$T_1. = 35 \qquad T_2. = 30 \qquad T_3. = 51 \qquad T_4. = 32 \qquad MS_E = 5.48$$

Using the analysis of variance procedure, test at the $\alpha = .10$ level the contention that the treatments differ in mean effect.

(d) Use the Scheffé method to test each of the contrasts of parts (a) and (b). Explain the practical consequences of your findings.

(e) Is there a best treatment? That is, is there a treatment that is clearly superior to all other treatments?

12. In Example 4 in Section 1, we concluded that family size does have an effect on independence in college freshmen. (High scores indicate a high degree of independence.)

(a) Select a contrast that can be used to compare "only children" with children from families with two or more children. Test the hypothesis that this contrast has value 0.

(b) Compare the mean score for children from families of three or more children to that of the group whose mean score appears to be closest to that of this group. Can we conclude that there exists one group that exhibits a higher degree of independence than any other?

(c) Compare the mean scores between the highest and lowest groups.

13. Consider exercise 6. Use Scheffé's method of multiple comparisons to compare:

(a) all possible pairs of means

(b) the mean score for career women with that of housewives

14. Consider exercise 8. Use Scheffé's method of multiple comparisons to compare:

(a) all possible pairs of means

(b) the mean absenteeism rate using flexible hours with the rate using fixed hours

15. A tennis ball manufacturer is interested in the manner in which the newest type of ball performs on various court surfaces. A series of tests are run with

the variable of interest being the number of games played before the ball goes dead. Five surfaces are used: clay, grass, composition, wood, and asphalt. The following information is obtained, based on samples of size 25 each.

Clay	Grass	Composition	Wood	Asphalt
$T_1. = 1,550$	$T_2. = 1,700$	$T_3. = 1,600$	$T_4. = 1,250$	$T_5. = 1,100$
$MS_E = 734.0$				

(a) Test at the $\alpha = .05$ level $H_0: \mu_1 = \mu_2 = \mu_3 = \mu_4 = \mu_5$.
(b) Use Scheffé's method to test $H_0: \mu_2 - \mu_5 = 0$.
(c) Use Scheffé's method to select and test a contrast that would allow one to compare the results obtained on a soft surface (clay, grass, composition) with those obtained on a hard surface (wood, asphalt).

*16. **Duncan's Multiple Range Test:** Scheffé's method of multiple comparisons is powerful in that it can be used to test any contrast. If interest centers solely on contrasts of the form $\mu_i - \mu_j$, $i \neq j$ then Duncan's multiple range test, as described below, may be applied when sample sizes are equal.

Steps in Duncan's Test When Sample Sizes Are Equal

1. Order the observed sample means from smallest to largest.

2. Consider any subset of p sample means $2 \leq p \leq k$ where k is the number of populations sampled. In order for the means of any of the p populations involved to be considered different, the range (largest sample mean − smallest sample mean) must exceed a specific value called the shortest significant range $(SSR)_p$.

3. The shortest significant range is calculated by means of Table IX in Appendix A and the following formula:

$$(SSR)_p = r_p \sqrt{(MS_E/n)}$$

where

r_p = least significant studentized range obtained from Table IX in Appendix A.

MS_E = error mean square obtained from ANOVA

n = common sample size

v = degrees of freedom associated with MS_E in the original ANOVA

4. Results are summarized by underlining any subset of adjacent means that is not considered to be significantly different at the α level selected.

Example   In an ANOVA to compare five means, the hypothesis of equality of means was rejected at the $\alpha = .05$ level. The observed sample means, based on samples of $n = 25$ and computed values of MS_E with 120 degrees of freedom, are given below:

$$\bar{x}_1. = 2.7 \quad \bar{x}_2. = 4.1 \quad \bar{x}_3. = 3.2 \quad \bar{x}_4. = 8.6 \quad \bar{x}_5. = 2.5 \quad MS_E = 3.75$$

We first order the sample means

$\bar{x}_5.$	$\bar{x}_1.$	$\bar{x}_3.$	$\bar{x}_2.$	$\bar{x}_4.$
2.5	2.7	3.2	4.1	8.6

We next construct the following chart giving the values of $(SSR)_p$ for $\alpha = .05$.

p	2	3	4	5
r_p (Table IX)	2.8	2.947	3.045	3.116
$(SSR)_p = r_p\sqrt{(MS_E/n)}$	1.08	1.14	1.18	1.21

The differences in means may be tested by comparing in the following order: largest minus smallest, largest minus next smallest, and so forth, ending finally with second smallest minus smallest. For example, since $8.6 - 2.5 = 6.1 > 1.21 = (SSR)_5$, we conclude that μ_4 is significantly different from μ_5. Since $8.6 - 2.7 = 5.9 > 1.18 = (SSR)_4$, we can conclude that μ_4 is significantly different from μ_1. Proceeding in this manner, it can be seen that μ_4 is significantly different from each of the other means. We indicate this by underlining $\bar{x}_4. = 8.6$ separately.

$\bar{x}_5.$	$\bar{x}_1.$	$\bar{x}_3.$	$\bar{x}_2.$	$\bar{x}_4.$
2.5	2.7	3.2	4.1	<u>8.6</u>

We now compare 4.1 in turn to each of the remaining means and find that there is a significant difference between μ_2 and μ_5 and between μ_2 and μ_1. However, there is no significant difference between μ_2 and μ_3.

$\bar{x}_5.$	$\bar{x}_1.$	$\bar{x}_3.$	$\bar{x}_2.$	$\bar{x}_4.$
2.5	2.7	<u>3.2</u>	<u>4.1</u>	<u>8.6</u>

Continuing in this manner, we finally obtain:

$\bar{x}_5.$	$\bar{x}_1.$	$\bar{x}_3.$	$\bar{x}_2.$	$\bar{x}_4.$
2.5	2.7	<u>3.2</u>	<u>4.1</u>	<u>8.6</u>

In an ANOVA to compare six means, the hypothesis of equality of means was rejected at the $\alpha = .05$ level. The observed sample means, based on samples of size $n = 5$, are

$$\bar{x}_1. = 19.84 \qquad \bar{x}_4. = 23.20$$
$$\bar{x}_2. = 14.50 \qquad \bar{x}_5. = 16.75$$
$$\bar{x}_3. = 21.12 \qquad \bar{x}_6. = 22.90$$

The observed value of MS_E is 2.45. Use Duncan's procedure to pinpoint the differences among the corresponding population means.

*17. There has been some discussion concerning academic standards for male athletes among some of the larger athletic conferences. Random samples of size 100 each were obtained from among male varsity athletes in the SEC, ACC, Big Ten, PAC 8, and Ivy League. The variable of interest is the gpa of the individual athlete. The following information was obtained:

SEC	ACC	Big Ten	PAC 8	Ivy League
$T_1. = 220$	$T_2. = 223$	$T_3. = 200$	$T_4. = 227$	$T_5. = 320$
$MS_E = 7.57$				

Test at the $\alpha = .05$ level the hypothesis that the means are the same. If differences are discovered, use Duncan's test to summarize the situation.

*18. Psychologists are interested in the psychological differences that may exist among owners of various types of pets. Random samples of size 36 each were obtained from persons owning or expressing a desire to own dogs, cats, birds, or no pets. Each subject answered a series of questions designed to give a score reflecting personality type. High scores indicate a tendency toward being extroverted; low scores a tendency toward being introverted. These results were obtained:

Dog	Cat	Bird	None
$T_1. = 3132$	$T_2. = 1404$	$T_3. = 1548$	$T_4. = 2160$
$MS_E = 6123$			

Test at the $\alpha = .05$ level the hypothesis that the mean scores are identical for each group. Use Duncan's test to summarize the differences.

11.3   Randomized Complete Block Design

In comparing treatments via the one-way classification, we randomly divide the experimental units into k subgroups. Each subgroup receives a different treatment, and any differences in means detected later are attributed to the fact that different treatments were used. Suppose that there is some **extraneous variable**, a variable not under direct study in the experiment, present. This variable might interfere with our ability to detect true differences in treatment means. For example, suppose that we want to compare three paints for weather resistance. The primary variable under study is the time that elapses before the paint begins to fade or peel. Suppose that we have 90 experimental houses spread throughout the 50 states. In a one-way classification, we would randomly divide these houses into three subgroups and randomly assign paints to subgroups. This scheme ignores an obvious extraneous variable, namely, geographic location. Paint on a house in Florida will probably not fare as well as it would if it were on a house in a less hot and humid climate; paint on a house in Alaska is subjected to harsher conditions than that on a house in North Carolina. In a one-way classification, we assume that the process of randomization tends to balance out differences in the extraneous variable. If so, then differences detected later are indeed due to differences in treatment. However, it is possible that randomization does not entirely equalize the effects of the extraneous variable, and thus the true treatment effect is masked. When there is an obvious extraneous variable present, it is preferable to use a design called a **randomized complete block design**. The purpose to this design is to control the effects of the extraneous variable. The randomized block design calls for one to group experimental units into subgroups called **blocks**, with the units within each block being as nearly alike as possible with respect to the extraneous or *blocking variable*. Treatments are then assigned randomly within blocks. In effect, the pairing procedure used in conducting the paired T-test (Chapter 8) is similar. There we worked essentially with blocks of size two. The first example should clarify this idea.

Example 1

1. Four different kinds of paving are being studied by the highway department for possible use on interstate highways within the state. Since location within the state can influence the results because of differences in weather and traffic patterns, a randomized block design is appropriate. Three sections of highway in different parts of the state are chosen for experimentation. Each section constitutes a block. Each block is divided into four strips, and the paving types are assigned randomly to one strip within each block. A year after the paving is done, the amount of wear for each strip is ascertained. We want to compare the average amount of wear for each paving type. A typical layout for this experiment is shown in Figure 11.3.

2. Three different drugs are under study for use in controlling high blood pressure. Twelve patients are available for study. These patients are divided into four groups of size three, with members within groups having essentially the same initial blood pressure readings. Within each group drugs are assigned randomly to subjects. The measured variable is the decrease in blood pressure over a one-week period.

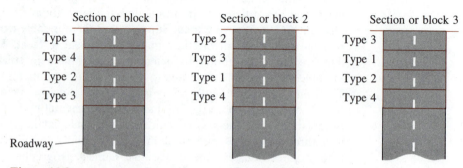

Figure 11.3
Typical layout for a randomized block experiment with three blocks and four treatments.

Table 11.9

Data Layout—Randomized Complete Block Treatment

		1	2	3	...	k	
Block	1	X_{11}	X_{21}	X_{31}	...	X_{k1}	$T_{\cdot 1}$
	2	X_{12}	X_{22}	X_{32}	...	X_{k2}	$T_{\cdot 2}$
	3	X_{13}	X_{23}	X_{33}	...	X_{k3}	$T_{\cdot 3}$
	⋮	⋮	⋮	⋮	...	⋮	⋮
	b	X_{1b}	X_{2b}	X_{3b}		X_{kb}	$T_{\cdot b}$
		$T_{1\cdot}$	$T_{2\cdot}$	$T_{3\cdot}$	...	$T_{k\cdot}$	$T_{\cdot\cdot}$

It should be noted that we are discussing in this section what are termed *complete blocks*. This refers to the fact that each treatment is used exactly once in each block. For instance, each paving type is used exactly once in each section, and each drug is used exactly once within each patient group.

The data layout for the randomized complete block design is given in Table 11.9. We use the following notation:

$$X_{ij} = \text{response to the } i\text{th treatment in } j\text{th block,}$$
$$i = 1, 2, \ldots, k \qquad j = 1, 2, \ldots, b$$

$$k = \text{number of treatments} = \text{number of observations per block}$$

$$b = \text{number of blocks} = \text{number of observations per treatment}$$

$$N = k \cdot b = \text{total number of observations}$$

$$T_{i\cdot} = \sum_{j=1}^{b} X_{ij} = \text{total of all responses to } i\text{th treatment}$$

$$\bar{X}_{i\cdot} = T_{i\cdot}/b = \text{sample mean for } i\text{th treatment}$$

$$T_{\cdot j} = \sum_{i=1}^{k} X_{ij} = \text{total of all responses in } j\text{th block}$$

$$\bar{X}_{\cdot j} = T_{\cdot j}/k = \text{sample mean for } j\text{th block}$$

$$T_{\cdot\cdot} = \sum_{i=1}^{k} \sum_{j=1}^{b} X_{ij} = \sum_{i=1}^{k} T_{i\cdot} = \sum_{j=1}^{b} T_{\cdot j} = \text{total of all responses}$$

$$\bar{X}_{\cdot\cdot} = T_{\cdot\cdot}/kb = \text{grand sample mean}$$

In Example 2 we illustrate this notation.

Example 2 When the paving experiment described in Example 1 is conducted, the data shown in Table 11.10 result. The scores obtained represent a rating given that takes into account such things as cracking, roughness, and change in the depth of paving.

Table 11.10

		\multicolumn{4}{c}{*Paving Type*}
		1	*2*	*3*	*4*	
Block	1	42.7	39.3	48.5	32.8	$T_{.1} = 163.3$
	2	50.0	38.0	49.7	40.2	$T_{.2} = 177.9$
	3	51.9	46.3	53.5	51.1	$T_{.3} = 202.8$
		$T_1. = 144.6$	$T_2. = 123.6$	$T_3. = 151.7$	$T_4. = 124.1$	$T.. = 544.0$
		$\bar{x}_1. = 48.2$	$\bar{x}_2. = 41.2$	$\bar{x}_3. = 50.6$	$\bar{x}_4. = 41.4$	$\bar{x}.. = 45.3$

The analysis proceeds along the same lines as in the one-way classification, with the exception that we now have one other component to consider, namely, the effect of blocking. We let

μ = theoretical overall average response ignoring the treatment or block involved

$\mu_i.$ = theoretical average response for the ith treatment, $i = 1, 2, \ldots, k$

$\mu_{.j}$ = theoretical average response for the jth block, $j = 1, 2, \ldots, b$

The model for the randomized complete block design is

Model: Randomized Complete Block Design

$$X_{ij} = \mu + (\mu_i. - \mu) + (\mu_{.j} - \mu) + (X_{ij} - \mu_i. - \mu_{.j} + \mu)$$

In words, we are claiming that

response to   overall   effect due to   effect due
ith treatment = average +   use of   + to use of +   random
in jth block   response ith treatment jth block   fluctuation
X_{ij} μ $(\mu_i. - \mu)$ $(\mu_{.j} - \mu)$ $(X_{ij} - \mu_i. - \mu_{.j} + \mu)$

We want to test

$H_0: \mu_1. = \mu_2. = \cdots = \mu_k.$ (treatment means are identical)

$H_1:$ (treatment means not all the same)

By replacing the theoretical means in the above model by their estimators we obtain

$$X_{ij} - \bar{X}.. = (\bar{X}_i. - \bar{X}..) + (\bar{X}_{.j} - \bar{X}..) + (X_{ij} - \bar{X}_i. - \bar{X}_{.j} + \bar{X}..)$$

If each side of this equation is squared, summed over all possible values of i and j, and simplified, the following equation results:

$$\sum_{i=1}^{k} \sum_{j=1}^{b} (X_{ij} - \bar{X}..)^2 = b \sum_{i=1}^{k} (\bar{X}_{i\cdot} - \bar{X}..)^2$$

$$+ k \sum_{j=1}^{b} (\bar{X}_{\cdot j} - \bar{X}..)^2$$

$$+ \sum_{i=1}^{k} \sum_{j=1}^{b} (X_{ij} - \bar{X}_{i\cdot} - \bar{X}_{\cdot j} + \bar{X}..)^2$$

This equation indicates that the total amount of variation present in the data, SS_{Total}, can be partitioned into three components: variation due to the fact that different treatments are involved, SS_{Tr}; variation due to the fact that different blocks are involved, SS_{Block}; and random fluctuation, SS_E. The sum of squares identity for this design is

$$SS_{Total} = SS_{Tr} + SS_{Block} + SS_E$$

These mean squares allow us to test H_0 by means of the F distribution:

Mean Squares

treatment mean square $= MS_{Tr} = SS_{Tr}/(k-1)$

block mean square $= MS_{Block} = SS_{Block}/(b-1)$

error mean square $= MS_E = SS_E/(k-1)(b-1)$

It can be shown that if H_0 is true,

$$MS_{Tr}/MS_E$$

is distributed as an F random variable with $k-1$ and $(k-1)(b-1)$ degrees of freedom. Thus our test satistic is

Test Statistic Randomized Complete Block

$$MS_{Tr}/MS_E = F_{k-1,\,(k-1)(b-1)}$$

Our test is to reject the hypothesis of equal treatment means if the observed value of the F statistic is too large. The computational formulas used in the analysis are

Computational Shortcuts

$$SS_{Tr} = \sum_{i=1}^{k} T_{i.}^2/b - T_{..}^2/kb$$

$$SS_{Block} = \sum_{j=1}^{b} T_{.j}^2/k - T_{..}^2/kb$$

$$SS_{Total} = \sum_{i=1}^{k} \sum_{j=1}^{b} X_{ij}^2 - T_{..}^2/kb$$

$$SS_E = SS_{Total} - SS_{Tr} - SS_{Block}$$

The use of F ratio is illustrated by completing the analysis of the data given in Table 11.10.

Example 3

To compare the four paving types based on the data of Table 11.10 we test

$$H_0: \mu_{1.} = \mu_{2.} = \mu_{3.} = \mu_{4.}.$$

The only new sum that must be computed from the data is the sum of the squares of the individual observations. For these data,

$$\sum_{i=1}^{4} \sum_{j=1}^{3} X_{ij}^2 = (42.7)^2 + (50.0)^2 + (51.9)^2 + \cdots + (51.1)^2$$
$$= 25,136.76$$

$$T_{..}^2/kb = (544.0)^2/12 = 24,661.33$$

$$\sum_{i=1}^{4} T_{i.}^2/b = T_{1.}^2/3 + T_{2.}^2/3 + T_{3.}^2/3 + T_{4.}^2/3$$
$$= (144.6)^2/3 + (123.6)^2/3 + (151.7)^2/3 + (124.1)^2/3$$
$$= 24,866.61$$

$$\sum_{j=1}^{3} T_{.j}^2/k = T_{.1}^2/4 + T_{.2}^2/4 + T_{.3}^2/4$$
$$= (163.3)^2/4 + (177.9)^2/4 + (202.8)^2/4$$
$$= 24,860.79$$

$$SS_{Tr} = \sum_{i=1}^{4} T_{i.}^2/b - T_{..}^2/kb$$
$$= 24,866.61 - 24,661.33 = 205.28$$

$$SS_{Block} = \sum_{j=1}^{3} T_{.j}^2/k - T_{..}^2/kb$$
$$= 24,860.79 - 24,661.33 = 199.46$$

Table 11.11

ANOVA

Source of Variation	Sum of Squares	Degrees of Freedom	Mean Square	F Ratio
Treatment	205.28 (SS_{Tr})	$3(k-1)$	$68.43\left(\dfrac{SS_{Tr}}{k-1}\right)$	$5.81\left(\dfrac{MS_{Tr}}{MS_E}\right)$
Block	199.46 (SS_{Block})	$2(b-1)$	$99.73\left(\dfrac{SS_{Block}}{b-1}\right)$	
Error	70.69 (SS_E)	$6(k-1)(b-1)$	$11.78\left(\dfrac{SS_E}{(k-1)(b-1)}\right)$	
Total	475.43 (SS_{Total})	$11(N-1)$		

$$SS_{Total} = \sum_{i=1}^{4} \sum_{j=1}^{3} X_{ij}^2 - T_{..}^2/kb$$
$$= 25{,}136.76 - 24{,}661.33 = 475.43$$
$$SS_E = SS_{Total} - SS_{Tr} - SS_{Block}$$
$$= 475.43 - 205.28 - 199.46 = 70.69$$

The ANOVA table, shown in Table 11.11, summarizes these computations. The critical point for a size $\alpha = .05$ test, based on the $F_{3,6}$ distribution, is 4.76. Since $5.81 > 4.76$, we reject

$$H_0: \mu_1. = \mu_2. = \mu_3. = \mu_4.$$

and conclude that there are differences among the mean scores for the paving types.

If the hypothesis of equal treatment means is rejected then differences can be investigated by means of Scheffé-type confidence intervals or by Duncan's procedure. These tests are performed exactly as before with

$$L^2 = (k-1) f_{\alpha, k-1, (k-1)(b-1)} \, MS_E \sum_{i=1}^{k} (a_i^2/b)$$

for the Scheffé intervals and

$$(SSR)_p = r_p \sqrt{(MS_E/b)}$$

for Duncan's test.

Before closing, let us mention that we can use the ratio

$$MS_{Block}/MS_E = F_{b-1, (k-1)(b-1)}$$

to test for equality among block means. This is not usually necessary, since we intentionally grouped experimental units into blocks via the extraneous variable. We would not have done this unless we were relatively certain that there were differences among blocks even at the outset of the experiment.

We have discussed in this chapter only the bare essentials in the broad field of analysis of variance. There are many other useful and interesting experimental designs of varying degrees of complexity. Texts on the subject that might be of use to you include the following: [14], [8], [9], [18].

Exercises 11.3

19. Four treatments are to be compared using five blocks. The block and treatment totals are

Treatment Totals	Block Totals
$T_1. = 61.8$	$T_{.1} = 58.8$
$T_2. = 100.0$	$T_{.2} = 78.0$
$T_3. = 119.3$	$T_{.3} = 80.1$
$T_4. = 73.9$	$T_{.4} = 64.7$
	$T_{.5} = 73.4$

Furthermore, $\sum_{i=1}^{4} \sum_{j=1}^{5} X_{ij}^2 = 6{,}810.28$

(a) Find the ANOVA table for these data.
(b) Test the null hypothesis of equality of treatment means at the $\alpha = .05$ level.
(c) Use Scheffé's method to test $H_0: \mu_1. = \mu_4.$ at the $\alpha = .05$ level.
(d) Use Scheffé's method to test $H_0: \mu_1. = \mu_2.$ at the $\alpha = .05$ level.

20. Three treatments are to be compared using eight blocks. The block and treatment totals are:

Treatment Totals	Block Totals	
$T_1. = 270.1$	$T_{.1} = 103.3$	$T_{.5} = 102.8$
$T_2. = 235.8$	$T_{.2} = 80.0$	$T_{.6} = 94.8$
$T_3. = 234.7$	$T_{.3} = 80.8$	$T_{.7} = 94.4$
	$T_{.4} = 95.6$	$T_{.8} = 88.9$

Furthermore, $\sum_{i=1}^{3} \sum_{j=1}^{8} X_{ij}^2 = 23{,}533.9$

(a) Find the ANOVA table for these data.
(b) Test the null hypothesis of equality of treatment means at the $\alpha = .05$ level.
(c) If appropriate, use Scheffé's method to test $H_0: \mu_1. = \mu_2.$ at the $\alpha = .05$ level.

(d) If appropriate, use Scheffé's method to test H_0: $\mu_{2\cdot} = \mu_{3\cdot}$ at the $\alpha = .05$ level.

21. In comparing three drugs for use in controlling high blood pressure as described in Example 1, these data are obtained (The response noted is the decrease in blood pressure over a one-week period.):

Degree of high blood pressure	Drug		
	1	2	3
Very high	5	4	2
High	4	5	4
Moderately high	6	8	8
Slightly high	3	6	6

Test at the $\alpha = .01$ level the hypothesis that $\mu_{1\cdot} = \mu_{2\cdot} = \mu_{3\cdot}$.

22. Company officials want to test three types of adhesive for possible use in modifying its tape. It is known that an adhesive's ability to hold is affected by the type of surface to which it is applied. Thus, it is necessary to test the adhesive on various types of surfaces. Six surface types are used. To each surface identical blocks of wood are affixed using each of the three adhesives. The variable of interest is the time in hours until the block falls. These results are observed:

	Adhesive type		
	1	2	3
Cinder	3.0	2.8	2.5
Smooth plaster	8.0	7.9	7.5
Rough plaster	6.5	6.6	6.3
Wood	6.9	7.2	7.0
Papered wall	9.2	9.0	8.5
Glass	11.0	10.8	10.5

Test at the $\alpha = .10$ level the null hypothesis that the average holding times are identical.

23. An experiment is conducted to study the average cost of insuring a midsize car with each of the four leading insurance firms in the United States. It is felt that the section of the country involved is an influencing factor, and hence this variable is controlled by blocking. The following observations are obtained:

	Company			
	A	B	C	D
South	140	138	170	129
Northeast	175	169	210	165
Midwest	150	150	173	135
Southwest	142	136	168	131
Far West	178	172	208	163

Test at the $\alpha = .05$ level the hypothesis that there is no difference among these companies with respect to the average cost of insurance.

24. Five detergents are being compared for cleaning power. It is felt that water temperature could be a factor, and hence each detergent is tested in warm, hot, and cold water. The cleanliness of the wash is rated on a scale from 0 to 10 by an impartial judge, with higher scores representing a cleaner wash. The following results are obtained:

	A (liquid)	B (liquid)	C (powder)	D (powder)	E (powder)
Warm	9.8	9.3	8.7	4.2	3.0
Hot	9.3	9.0	9.1	5.3	3.7
Cold	8.0	7.3	7.4	3.9	2.9

(a) Test at the $\alpha = .05$ level the hypothesis that the detergents have the same mean cleanliness rating.
(b) Use Scheffé's method of multiple comparison to compare the liquid detergents with the powdered detergents.
(c) Use Scheffé's method to determine whether or not one detergent is superior to all others.

25. In studying the increase in crime rates over the last 10 years criminologists gathered the following data.

	Type of crime			
	Murder	Rape	Robbery	Arson
Rural	.5%	1%	3%	.5%
Small town (under 50,000)	5%	6%	18%	4%
Medium cities (50,000–200,000)	15%	18%	42%	10%
Large cities	50%	60%	110%	45%

(a) Test at the $\alpha = .05$ level the hypothesis that the average rate of increase is the same for all four types of crime.
(b) Use Scheffé's method to determine whether there is one type of crime that shows an increase higher than any other.
(c) Use Scheffé's method to compare the rates of crimes against persons (murder, rape) with crimes against property (robbery, arson).

*26. Consider Example 3. Use Duncan's multiple range test to summarize the differences among paving types.

*27. Use Duncan's multiple range test to summarize the differences among the insurance companies of exercise 23.

COMPUTING SUPPLEMENT I

To illustrate the use of SAS in analyzing data from a randomized block experiment, we use the data of Example 2.

Statement	*Purpose*

```
DATA PAVING;
```
names data set

```
INPUT BLOCK TYPE SCORE;
```
each line will contain the block number, type of paving used, and wear score

```
CARDS;
```
signals that data follows

```
1  1  42.7
2  1  50.0
3  1  51.9
.  .   .
.  .   .
.  .   .
3  4  51.1
;
```
data

signals end of data

```
PROC ANOVA;
```
calls on analysis of variance procedure

```
CLASSES BLOCK TYPE;
```
indicates that observations are identified by both a block number and a paving type number

```
MODEL SCORE = BLOCK TYPE;
```
identifies variable on left, SCORE, as the response or measured variable

```
MEANS TYPE/DUNCAN;
```
asks for the Duncan's multiple range test to be run to compare the means of the paving types

```
TITLE RANDOMIZED BLOCKS;
```
titles output

```
                          RANDOMIZED BLOCKS
                  ANALYSIS OF VARIANCE PROCEDURE
                    CLASS LEVEL INFORMATION
                 CLASS      LEVELS      VALUES
                 BLOCK        3          123
                 TYPE         4         1234

            NUMBER OF OBSERVATIONS IN DATA SET = 12
                       RANDOMIZED BLOCKS
                  ANALYSIS OF VARIANCE PROCEDURE
```

DEPENDENT VARIABLE: SCORE

SOURCE	DF	SUM OF SQUARES	MEAN SQUARE	F VALUE
MODEL	5	404.72500000	80.94500000	6.87
ERROR	6	70.70166667 ③	11.78361111 ⑤	PR > F
CORRECTED TOTAL	11	475.42666667 ④		0.0181

R-SQUARE	C.V.	ROOT MSE	SCORE MEAN
0.851288	7.5722	3.43272648	45.33333333

SOURCE	DF	ANOVA SS	F VALUE	PR > F
BLOCK	2	199.45166667 ②	8.46 ⑧	0.0179 ⑨
TYPE	3	205.27333333 ①	5.81 ⑥	0.0330 ⑦

```
                       RANDOMIZED BLOCKS
                  ANALYSIS OF VARIANCE PROCEDURE
```

DUNCAN'S MULTIPLE RANGE TEST FOR VARIABLE: SCORE
NOTE: THIS TEST CONTROLS THE TYPE 1 COMPARISONWISE ERROR RATE,
      NOT THE EXPERIMENTWISE ERROR RATE.

ALPHA = 0.05 DF = 6 MSE = 11.7836

MEANS WITH THE SAME LETTER ARE NOT SIGNIFICANTLY DIFFERENT.

⑩DUNCAN	GROUPING	MEAN	N	TYPE
	A	50.567	3	3
	A			
B	A	48.200	3	1
B				
B		41.367	3	4
B				
B		41.200	3	2

On the printout shown, these quantities are given:

① SS_{Tr}

② SS_{Block}

③ SS_E

④ SS_{Total}

⑤ MS_E

Note that, apart from roundoff error, these figures agree with those given in Table 11.11.

The F ratio for testing for equality of treatment means is given by ⑥ and its P value by ⑦. Since P is small, .033, we reject H_0 and conclude that there are differences in the mean scores received by the four paving types.

The F ratio for testing for equality of block means is given by ⑧ and its P value by ⑨. Since P is small, we conclude that there are differences among localities, as expected.

The results of the Duncan's test are given by ⑩. To group these means we match those means labeled with the same letter. Thus the first two means listed, $\bar{x}_{1\cdot}$ and $\bar{x}_{3\cdot}$ are grouped together; also, the last three means listed, $\bar{x}_{2\cdot}$, $\bar{x}_{4\cdot}$ and $\bar{x}_{1\cdot}$ are grouped together. Using our previous notation, we report these groupings as

$$\underline{\bar{x}_{2\cdot} \qquad \bar{x}_{4\cdot} \qquad \bar{x}_{1\cdot} \qquad \bar{x}_{3\cdot}}$$

Vocabulary List and Key Concepts

one-way classification	experimentwise error rate
factor	extraneous variable
linear contrast	randomized complete block design
Scheffé's method of multiple comparisons	blocks

Review Exercises

28. What is the null hypothesis being tested in a one-way classification?

29. What is the model for the one-way classification?

30. What are the model assumptions for the one-way classification?

31. Explain briefly the overall purpose of the statistical technique known as analysis of variance.

32. What is the sum of squares identity for the one-way classification? Explain what each component in this identity measures.

33. Explain briefly the logic behind the test statistic used in the one-way classification analysis of variance.

34. The one-way classification is an extension to more than two samples of which T test?

Use the following data to answer exercises 35–39:

Officials of a large fast-food chain are experimenting with various promotional gimmicks. Forty franchises in the chain are randomly selected and split into subgroups. These treatments are then randomly assigned to subgroups:

I   no gimmick
II   specials on selected items
III games with food prizes
IV   games with money prizes

These data are obtained on sales during the one-month experimental period (sales are in $100,000 units):

I	II	III	IV		I	II	III	IV
.8	1.6	1.8	2.4		1.0	.7	1.5	2.6
.2	1.1	.3	2.6		1.1	.5	1.3	2.7
1.4	.2	1.1	2.8		1.4	1.3	2.9	2.1
.9	1.3	2.0	2.9		1.2	1.3	1.7	2.1
.9	1.5	2.8	1.7		1.0	.4	.8	2.3

35. State the null hypothesis to be tested, both statistically and in words.

36. Find the ANOVA table for these data and test H_0 at the $\alpha = .10$ level. Do these treatments seem to differ with respect to average sales produced?

37. Select a contrast that can be used to determine the effect of games in sales. Test this null hypothesis at the $\alpha = .10$ level. What practical conclusion can you draw based on the results of this test?

38. Select a contrast that can be used to determine the effect of some sort of promotional gimmick on sales. Test this contrast at the $\alpha = .10$ level. Based on this test, do you think that a promotional gimmick of some sort is useful?

*39. Use Duncan's test to see if there is a best treatment.

40. What is the purpose of blocking?

41. What is the primary null hypothesis being tested in a randomized complete block design?

42. What is the model for the randomized complete block design?

43. What is the sum of squares identity in the randomized complete block design? Explain what each component in this identity measures.

44. The randomized complete block design is an extension to more than two samples of which T test?

Use the following data to answer exercises 45–49:

An industrial psychologist wants to compare four different training methods in order to determine the most effective way to train newly hired employees. The company employs five training instructors. To remove the instructor effect from the experiment, she uses the instructors as blocks. Thus each instructor teaches the course once using each of the four methods. These results are obtained on the average score made on the final examination for each class: (high scores are desirable)

	Method			
	I	II	III	IV
	new	new	new	current
A	97	74	68	70
B	102	63	60	49
C	95	68	51	53
D	93	60	54	70
E	89	64	48	47

45. State the primary null hypothesis both statistically and in words.

46. Find the ANOVA table for these data and test H_0 at the $\alpha = .05$ level. Do these training methods seem to produce different average examination scores?

47. Select a contrast that can be used to compare the current method with the most promising new method. Test this contrast at the $\alpha = .05$ level. Assuming that all methods cost about the same, would you advise the company to use this new method rather than the one currently in use?

*48. Use Duncan's test to see if there is a best training method.

*49. We blocked to remove the instructor effect. Was blocking necessary? Explain by testing

$$H_0: \mu_{\cdot 1} = \mu_{\cdot 2} = \mu_{\cdot 3} = \mu_{\cdot 4} = \mu_{\cdot 5}$$

at the $\alpha = .10$ level.

Some Distribution-Free Alternatives

We have pointed out in previous chapters that most of the techniques employed have an underlying assumption of normality. That is, we are usually assuming that the samples with which we are dealing are drawn from distributions that are normal, or at least approximately so. For this reason, the procedures discussed that employed the Z, T, X^2, and F distributions are often referred to as constituting the body of *normal-theory statistics*. If the normality assumption is violated, then the normal-theory statistical tests are at best approximations whose accuracy is suspect. Unfortunately, in many instances, researchers are faced with situations in which it is apparent that the data are not normal. What can be done in this case? Statisticians began to tackle this problem around 1936, when a formal study of **distribution-free tests**, tests that presuppose little about the sampled population's shape and whose validity does not depend on it, was begun. We present here some well-known distribution-free tests. The tests chosen have been carefully selected to parallel the normal-theory tests presented earlier. Hence, if you are faced with nonnormal data, you will have a viable alternative method of analysis readily available.

12.1 One-Sample Procedures

Recall that in the case of normal-theory procedures the most commonly used measure of location or central tendency is the mean or theoretical average value of the random variable X. In most distribution-free tests the center of location is measured by using the median of the distribution. We assume throughout this discussion that the random variable under study is continuous. Recall that in the

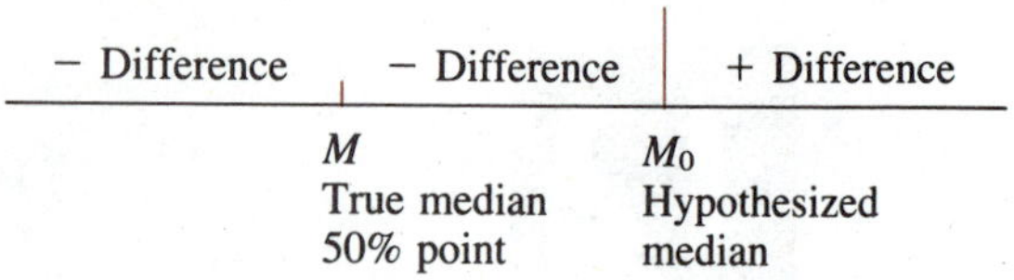

− Difference	− Difference	+ Difference
M		M_0
True median		Hypothesized
50% point		median

Figure 12.1

If $M < M_0$, then fewer than half of the observed differences should be positive.

case of a continuous random variable X, the **median** M is the point such that X is just as likely to fall above M as it is to fall below it. That is,

---

Median of a Continuous Random Variable

$$P[X \leq M] = P[X \geq M] = 1/2$$

In this section we present two procedures for testing hypotheses concerning the value of the median. The first, called the **sign test for median**, is based upon the binomial distribution studied in Chapter 4.

Sign Test for Median   Suppose that we have available a random sample X_1, X_2, ..., X_n from a continuous distribution with median M, and we want to test a hypothesis concerning the value of M. For example, suppose we wish to test

$$H_0: M = M_0$$

$$H_1: M < M_0$$

Note that if H_0 is true, then each of the variables X_i has the same chance of falling below M_0 as it does of falling above M_0. Hence, if H_0 is true, each of the differences, $X_i - M_0$, has the same chance of being positive as it does of being negative. Furthermore, the probability of a zero difference is theoretically zero due to the fact that for continuous variables $P[X = M_0] = 0$. Let Q_+ denote the number of positive signs obtained. If H_0 is true, then Q_+ is a binomial random variable with parameters n and $1/2$. This means that if H_0 is true, the average or expected number of positive signs is $n \cdot p = n \cdot (1/2)$. Thus, if H_0 is tue, we should observe about as many positive signs as we do negative ones. Consider the diagram of Figure 12.1 which represents the relationship between M and M_0 under H_0. If H_1 is true, we should observe too many negative signs and therefore too few positive signs. Thus Q_+ can be used as a test statistic for testing the above hypothesis. We simply reject H_0 in favor of H_1 at the appropriate α level or P value if the observed number of positive signs is too small. Example 1 demonstrates the use of this statistic.

Example 1   Past experience indicates that the median age for a first date for girls was 14. Sociologists feel that this value is too large. To test their theory, 15 randomly selected high school senior girls are questioned concerning this matter. The following responses are obtained:

13.0	12.5	13.5	14.2	11.5
12.5	15.0	15.5	13.5	13.0
16.0	15.5	13.7	12.0	14.5

We are testing

$$H_0: M = 14$$

$$H_1: M < 14$$

When 14 is subtracted from each observation above, we obtain 9 negative signs and 6 positive signs. Hence, the observed value of the test statistic Q_+ is 6. We use Table I of Appendix A to find the P value for the test. From the binomial table with $n = 15$ and $p = 1/2$,

$$P \text{ value} = P[Q_+ \leq 6 \,|\, p = 1/2] = .3036$$

Since this value is fairly large, we do not reject H_0. We do not have conclusive evidence that $M < 14$.

The logic used to test

$$H_0: M = M_0$$

$$H_1: M > M_0$$

is similar to that used to conduct a left-tailed test. However, in this case the test statistic is Q_-, the number of negative signs obtained. We reject H_0 if the observed value of Q_- is too small to have occurred by chance. The test statistic used to conduct a two-tailed test is the smaller of Q_+ and Q_-.

Before considering our next example, let us mentioned the problem of zero differences. Theoretically, a zero difference should not occur because $P[X = M_0] = 0$. However, zero differences do occur in practice due to the difficulty of obtaining precise measurements on continuous random variables. Various suggestions have been made for dealing with this problem. We take a conservative approach. Since a zero difference lends support to H_0, we suggest that zero differences be assigned the algebraic sign least conducive to the rejection of H_0. This means the following:

Handling Zero Differences

1. In a left-tailed test, a zero difference is considered to be positive.

2. In a right-tailed test, a zero difference is considered to be negative.

3. In a two-tailed test, a zero difference assumes the sign of the least-frequently occurring difference.

These guidelines are conservative in the sense that if we can still reject H_0 even after having given H_0 the benefit of the doubt, we can be very sure that our conclusions are correct.

As a second illustration of the sign test, let us consider a two-tailed hypothesis.

Example 2

Executives of a trucking firm are considering the possibility of rerouting one of its runs along a newly opened stretch of highway. Due to differing traffic patterns it is not known whether this move will increase or decrease the old median travel time of 4 hours. To decide, we test

$$H_0: M = 4$$

$$H_1: M \neq 4$$

These twenty observations are obtained on X, the time in hours required to make the run using the new route.

| 4.4 | 3.9 | 5.2 | 4.6 | 4.1 | 4.6 | 4.3 | 3.6 | 4.4 | 5.5 |
| 3.9 | 4.7 | 4.1 | 3.8 | 4.8 | 4.5 | 5.6 | 4.1 | 4.5 | 4.0 |

When we subtract 4 from each observation, we obtain 15 positive signs, 4 negative signs, and one zero. Following the guidelines on handling zero scores, we consider the zero to be negative. We conduct the test as though we have 15 positive signs and 5 negative ones. From the binomial table with $n = 20$ and $p = 1/2$, we see that

$$P[Q_- \leq 5] = .0207$$

Since the test is two-tailed, the P value is double this value or .0414. Since this probability seems small, we reject H_0 and conclude that the median time using the new route is not 4 hours.

Note that to apply the sign test we assume only that the random variable involved is continuous. We assume nothing more concerning its distribution. In particular, we do not assume normality. If, in fact, the population is normal, then the median of the population and the mean of the population are identical. In this case, the sign test is testing the same hypothesis as that being tested in normal theory by a one-sample T test.

The second test that we consider is the **Wilcoxon signed-rank test**. It is a test of location that can be used for symmetric distributions that are not necessarily normally distributed. The test makes use of the algebraic sign of the difference between an observation and the proposed median, and also takes into consideration the magnitude of this difference.

Wilcoxon Signed-Rank Test for Median (F. Wilcoxon, 1945) (Optional)
Consider a random sample $X_1, X_2, \ldots, X_n$ from a continuous distribution that is

symmetric about some median M. We wish to test the null hypothesis that the value of this median is M_0. We first form a set of n difference scores:

$$D_1 = X_1 - M_0$$

$$D_2 = X_2 - M_0$$

$$\vdots$$

$$D_n = X_n - M_0$$

The null hypothesis can now be thought of as being that the median of the population of differences is 0. To run the test, we first form the n absolute differences $|D_1|, |D_2|, ..., |D_n|$. We then order these differences from smallest to largest and rank them from 1 to n. We next assign to each rank R_i, the algebraic sign of the difference score that generated the rank. Note that, because of the assumption of continuity, no zeros and no ties should occur. Note also that, because of the hypothesized symmetry about 0, a priori, each rank is just as likely to be accompanied by a positive sign as it is by a negative sign. This means that under H_0, each of the 2^n distinguishably different sets of signed ranks should occur with equal probability of $1/2^n$. We can calculate the value of the following statistics:

$$W_+ = \Sigma R_i \qquad \text{and} \qquad |W_-| = |\Sigma R_i|$$
$$\text{(all positive ranks)} \qquad\qquad \text{(all negative ranks)}$$

Under H_0 each of these should be moderate in size. If a large number of observations lies above M_0, indicating that the true median is larger than the hypothesized value of M_0, then we obtain too many positive ranks and W_+ is inflated, forcing $|W_-|$ to be too small. Similarly, if a large number of observations lie below M_0, indicating that the true median is below M_0, then we obtain too many negative ranks and thus $|W_-|$ is inflated and W_+ is too small. Thus, we can use as our test statistic, W, *the one expected to be the smaller of the two under the assumption that the alternative is true.* This means that to test for the median of a symmetric distribution, we use these statistics:

Wilcox on Statistics

1. For a right-tailed test of $H_0: M = M_0$, we reject H_0 if $W = |W_-|$ is too small.

2. For a left-tailed test of $H_0: M = M_0$, we reject H_0 if $W = W_+$ is too small.

3. For a two-tailed test of $H_0: M = M_0$, we reject H_0 if $W = \min\{|W_-|, W_+\}$ is too small.

The distribution of this statistic under H_0 is in Table X in Appendix A for various approximate α levels and samples of size 5 to 50. The test is always to reject if the observed value of the test statistic W falls on or below the critical value listed in the table.

Example 3

The illiteracy rate nationally for the year 1969 among whites, age 14 and over, was reported to be .7% (*source:* United States Bureau of the Census). It is felt that the median rate in large metropolitan areas (cities with a population of 500,000 or more) was higher than this. Assuming symmetry, we want to test

$$H_0: M = .7$$

$$H_1: M > .7$$

To test this hypothesis, 20 cities were selected randomly and the illiteracy rate among the group of interest was obtained. The following data were collected: (data represent percentages so that $.6 = .6\%$)

.6	.5	.62	1.7	.75
1.0	.69	.8	.8	.57
.9	1.5	.95	.53	1.1
1.2	2.0	.65	.79	.61

We first form the set of differences by subtracting .7, the hypothesized median rate, from each observation to obtain these differences:

−.1	−.20	−.08	1.0	.05
.3	−.01	.10	.1	−.13
.2	.8	.25	−.17	.4
.5	1.3	−.05	.09	−.09

We now order the absolute differences from smallest to largest, rank them from 1 to 20, and assign to each rank the algebraic sign of the difference score that generated the rank. Note that theoretically we should have no ties. However, in practice, ties do occur. For example, we obtain an absolute difference of .1 three times. When this occurs, all tied scores are assigned the *average* group rank. This idea is illustrated in Table 12.1:

Table 12.1

D_i	−.01	.05	−.05	−.08	.09	−.09	.1	.1	−.1	−.13
$|D_i|$	.01	.05	.05	.08	.09	.09	.1	.1	.1	.13
Rank R_i	1	2.5	2.5	4	5.5	5.5	8	8	8	10
Signed Rank	−1	2.5	−2.5	−4	5.5	−5.5	8	8	−8	−10

D_i	−.17	.2	−.2	.25	.3	.4	.5	.8	1.0	1.3
$|D_i|$	.17	.2	.2	.25	.3	.4	.5	.8	1.0	1.3
Rank R_i	11	12.5	12.5	14	15	16	17	18	19	20
Signed Rank	−11	12.5	−12.5	14	15	16	17	18	19	20

For these data,

$$W_+ = \Sigma R_i = 2.5 + 5.5 + 8 + \cdots + 19 + 20 = 155.5$$
$$\text{(all positive ranks)}$$

$$|W_-| = |\Sigma R_i| = 54.5$$
$$\text{(all negative ranks)}$$

If the illiteracy rate is higher than .7, then we expect too many positive signs and too few negative ones. The observed value of the test statistic is

$$W = |W_-| = 54.5$$

We see from Table X in Appendix A, that, for a sample of size 20, W must be less than or equal to 60 to reject H_0 at the largest α level listed, namely $\alpha = .05$. Since $54.5 < 60$, we are able to conclude that the median illiteracy rate among whites, aged 14 or over, in a large metropolitan area is higher than that on the national scene.

Note that in using the Wilcoxon test to test a hypothesis concerning the value of the median, we are assuming symmetry but we are not assuming normality. If, in fact, the population is normal, then once again we are testing the same hypothesis as that being tested in normal theory by the one-sample T test. If from past experience, we know that the population is symmetric, then this test is a valid test both of the mean and the median. If we do not have evidence of symmetry, then the sign test is preferred as a test of location.

As we have said before, statistics is an art as well as a science. Analyzing data requires you to make some subjective judgments. The chart shown in Figure 12.2 should be of help when testing for location based on a single sample drawn from a continuous distribution.

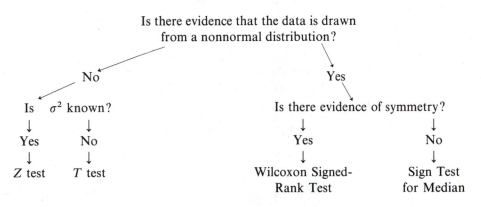

Figure 12.2
Strategy for testing for location based on a single sample.

Exercises 12.1

1. In each case, find the P value of the sign test based on the given data:

	Alternative Hypothesis	n	Number of Positive Signs	Number of Negative Signs	Number of Zeros
(a)	$M > M_0$	15	12	3	0
(b)	$M > M_0$	15	12	1	2
(c)	$M > M_0$	20	14	5	1
(d)	$M < M_0$	10	2	8	0
(e)	$M < M_0$	10	2	7	1
(f)	$M < M_0$	20	3	15	2
(g)	$M \neq M_0$	20	2	18	0
(h)	$M \neq M_0$	20	2	17	1
(i)	$M \neq M_0$	20	15	4	1

2. It is hypothesized that the median length of a telephone call is three minutes. Officials think that this value is too small. To test this hypothesis they observe 10 randomly selected calls and record the following observations on the length of each call in minutes:

   2.7   10.5   3.8   15.2   5.7   3.5   2.1   4.0   3.7   3.2

   Set up and test the appropriate hypothesis using the sign test. What is the P value of your test?

3. A sociologist is interested in testing the hypothesis that the median age for marriage among Alaskan Eskimo women is 18. The sociologist thinks that this figure is too low and wants statistical support for the contention. A random sample of size 20 generates these data:

   18.2   15.0   14.0   15.2   27.0   16.0   18.5   17.6   16.7   25.7
   17.4   18.1   22.0   18.3   17.8   21.6   23.0   18.3   20.0   19.6

   Can we conclude at the $\alpha = .05$ level that the sociologist is correct using the sign test?

4. A small clothing store wants to advertise that 50% of its items are priced under $25.00. In order to get evidence that the claim is justified, the manager randomly sampled 20 items from the listed inventory and obtained the following observations:

   25.98   18.75   16.50   29.95   2.00
   13.25   16.31   20.00   22.50   1.00
   5.98   32.50   25.00   50.00   37.00
   6.15   12.98   17.50   16.95   24.98

   Is this the evidence the manager wants? Explain on the basis of the P value of the sign test.

5. An insurance company is interested in ascertaining the median age of its

policyholders. The executives think that the median age is greater than 45.0. If their thoughts are substantiated, they intend to launch a massive compaign to enroll new and younger clients. To confirm their suspicions, they randomly select 75 client names from their files and obtain these data on their ages:

37.1	36.1	21.8	52.6	31.1
47.2	25.2	22.7	59.5	45.0
50.6	47.8	49.5	33.1	49.3
51.0	49.9	48.3	71.2	52.7
49.9	53.3	39.9	78.9	40.6
35.8	65.6	62.7	63.8	71.5
60.3	29.7	57.9	43.3	41.5
63.7	71.1	58.8	32.0	90.4
62.1	80.0	30.3	72.0	42.9
51.2	50.2	73.2	81.6	85.8
20.6	28.3	72.0	39.0	25.7
48.1	53.7	28.1	48.2	69.6
47.2	23.6	31.8	49.3	74.3
41.0	40.1	70.2	36.9	26.2
26.3	61.2	38.9	53.8	73.1

Calculate $P[Q_- \leq q | p = 1/2]$, where q is the number of negative signs obtained, by using the normal approximation to the binomial distribution. Is this sufficient evidence to confirm their suspicion and warrant an enrollment drive? Explain on the basis of the P value involved.

*6. Recall from high-school algebra that $\sum_{i=1}^{n} i = n(n + 1)/2$. That is, the sum of the first n integers is given by $n(n + 1)/2$. By applying this to the Wilcoxon procedure, we see that $W_+ + |W_-| = n(n + 1)/2$. This can be used as a check on the accuracy of your ranking and computation of the value of W_+ and $|W_-|$. Verify this result for the figures of Example 3.

*7. The median life expectancy nationally for white males born in 1900 was 46.6 years. A random sample of 15 white males born in that year in a small town revealed the following data on age at death:

51.1	60.0	42.0	55.8	25.0	37.0	49.5	29.5
45.1	alive	15.0	53.1	47.5	alive	57.6	

Is there evidence at the $\alpha = .05$ level that in this town the median life expectancy is higher among the group involved than the national figure of 46.6? Use the Wilcoxon procedure and assign living individuals the highest positive ranks. What are you assuming concerning the distribution of X, the age at death of individuals in this town?

*8. In 1975 it was reported that the average price of pork chops nationally was $1.87 per pound, and that, nationally, eggs sold for an average of $.78 per dozen. A random sample of 10 outlets for these products was selected and studied during the year 1975 in a specific Southern city. These data were collected on the average cost of each of the above items over the year:

Pork chops (X)				Eggs (Y)			
1.86	1.85	1.88	1.83	.75	.79	.77	.73
1.90	1.89	1.88	1.92	.69	.77	.72	.74
1.79	1.91			.80	.78		

(a) Use Wilcoxon's procedure to see if it can be claimed that the median cost of pork chops in this city was above the national average of $1.87 per pound.
(b) Use Wilcoxon's procedure to see if it can be claimed that the median cost of eggs in this city was below the national average of $.78 per dozen.
(c) What are you assuming concerning the distribution of the random variables X and Y? If this assumption does not seem to be reasonable, what distribution-free test can be used to test medians?

*9. A large importer of expensive sports cars assumes that the median age for a buyer of his model 20ZZ is under 35 years. A random sample of 18 buyers of this model yields the following data:

42.2	40.3	35.6	29.9	33.1	30.5	31.3	36.7	36.5
60.3	55.6	29.2	38.9	42.7	36.3	29.8	39.0	36.2

Use whatever test you think is appropriate to test the correctness of the importer's assumption. Be ready to defend your choice.

12.2   Two-Sample Procedures – Matched Data

We consider here two distribution-free tests involving paired observations. The first, the **sign test for median difference**, is a simple extension of the sign test to a set of difference scores. The second, the Wilcoxon signed-rank test for matched data, is a special application of the Wilcoxon procedure and is one of the most widely used methods of handling matched data.

Sign Test for Median Difference   Consider any collection of n paired observations $(X_1, Y_1), (X_2, Y_2), \ldots, (X_n, Y_n)$, with the random variables X and Y considered to be continuous. These observations generate a collection of n continuous difference scores

$$D_1 = X_1 - Y_1$$
$$D_2 = X_2 - Y_2$$
$$\vdots$$
$$D_n = X_n - Y_n$$

We wish to test the null hypothesis that the median difference score is 0. That is, we want to test

$$H_0: M_D = M_{X-Y} = 0$$

If H_0 is true, an observed difference is just as likely to be positive as it is to be negative. If H_0 is not true, then either the number of positive signs or the number of negative signs observed will be too small. Thus our test procedure is

---

Sign Test for Median Difference

1. We reject $H_0: M_D = 0$ in favor of $H_1: M_D < 0$ if Q_+, the number of positive signs observed, is too small.

2. We reject $H_0: M_D = 0$ in favor of $H_1: M_D > 0$ if Q_-, the number of negative signs observed, is too small.

3. We reject $H_0: M_D = 0$ in favor of $H_1: M_D \neq 0$ if $\min\{Q_+, Q_-\}$ is too small.

Example 1 illustrates the use of this test.

Example 1   A soap manufacturer wants to advertise that its product "creams not dries" the skin. To get support for this claim, the manufacturer selects 10 women and asks each of them to wash one side of her face with its product and the other side with the soap of the nearest competitor twice each day for two weeks. At the end of this period, an impartial judge rates the skin for dryness on a scale from 1 to 10, with low scores indicating a high degree of dryness. The following pairs of observations are obtained:

Company brand (X)	5.0	4.3	7.3	2.1	9.8	6.9	10.0	1.5	8.2	7.3
Competitor (Y)	6.1	4.5	6.0	2.0	7.5	8.0	9.2	1.0	8.0	6.9
Sign of $D = X - Y$	−	−	+	+	+	−	+	+	+	+

We want to test

$$H_0: M_D = M_{X-Y} = 0 \qquad \text{(company brand as likely to beat competitor as to lose)}$$

$$H_1: M_D = M_{X-Y} > 0 \qquad \text{(company brand more likely to beat competitor than to lose)}$$

The observed value of the test statistic Q_-, the number of negative signs obtained, is 3. We are to reject H_0 if this number is too small to have occurred reasonably by chance under H_0. From Table I of Appendix A, the binomial table with $n = 10$ and $p = 1/2$, we see that

$$P[Q_- \leq 3 \,|\, p = 1/2] = .1719$$

Since .1719 is moderate in size, and making a false advertising claim may have serious consequences, it would be best not to reject H_0. There is not sufficient evidence to support the claim that the company brand causes less dryness to the skin than does the soap of the nearest competitor.

Note that as in the previous sign test we are assuming only continuity. We are not assuming normality. However, if the random variables X and Y are normally distributed, then $X - Y$ is also normally distributed. In this case, the median difference and the mean difference are equal. Hence,

$$M_{X-Y} = \mu_{X-Y} = \mu_X - \mu_Y$$

That is, if X and Y are both normally distributed, then the null hypothesis $M_{X-Y} = 0$ is equivalent to the null hypothesis that the means of the two populations are identical. In this case, the sign test for median difference is testing the same hypothesis as that tested by the normal-theory paired-T test.

Wilcoxon Signed-Rank Test—Matched Pairs (Optional)   Consider any collection $(X_1, Y_1), (X_2, Y_2), \ldots, (X_n, Y_n)$ of ordered pairs of observations on continuous random variables X and Y. These can be used to generate a collection of n continuous difference scores

$$D_1 = X_1 - Y_1$$
$$D_2 = X_2 - Y_2$$
$$\vdots$$
$$D_n = X_n - Y_n$$

We assume that these differences are drawn from a symmetric population. We wish to test the null hypothesis that the population median is zero. This problem is a special case of the problem handled previously by the Wilcoxon signed-rank test for median with $M_0 = 0$. It can therefore be solved using this procedure. Example 2 illustrates how this is done.

Example 2

A pharmaceutical company has two different methods available for analyzing potency in its drug used for the treatment of bee stings. It is suspected that method A yields consistently smaller results than B. Assuming symmetry, we want to test

$$H_0: M_{A-B} = 0$$

$$H_1: M_{A-B} < 0$$

To test this hypothesis the following data are obtained:

Table 12.2

Method A, (X_i)	1.5	1.4	1.4	1.0	1.1	.9	1.3	1.2	1.1	.9	.7	1.8
Method B, (Y_i)	2.0	1.8	.7	1.3	1.2	1.5	1.1	.9	1.5	1.7	.9	.9
$D_i = X_i - Y_i$	−.5	−.4	.7	−.3	−.1	−.6	.2	.3	−.4	−.8	−.2	.9

Ordering the absolute observed differences from smallest to largest, we obtain these ranks and signed ranks:

Table 12.3

$|D_i|$	.1	.2	.2	.3	.3	.4	.4	.5	.6	.7	.8	.9
Rank, R_i	1	2.5	2.5	4.5	4.5	6.5	6.5	8	9	10	11	12
Signed Rank	−1	−2.5	2.5	−4.5	4.5	−6.5	−6.5	−8	−9	10	−11	12

The observed values of the Wilcoxon statistics W_+ and $|W_-|$ are given by

$$W_+ = \Sigma R_i = 29 \qquad |W_-| = |\Sigma R_i| = 49$$
$$\text{all positive} \qquad\qquad \text{all negative}$$
$$\text{ranks} \qquad\qquad\quad \text{ranks}$$

If method A tends to yield smaller readings than B, then we expect an excess of negative signs. Our test statistic is that statistic which is expected to be the smaller of the two if H_1 is true. In this case, the test statistic is W_+ and its observed value is $W_+ = 29$. From Table X of Appendix A, we see that the critical

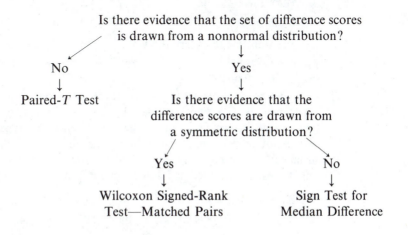

Figure 12.3
Strategy for testing for location based on paired data.

point for an $\alpha = .05$ test for a sample of size 12 is 17. Since $29 > 17$ we are unable to reject H_0. We do not have sufficient evidence to claim that the potency reading obtained using method A tends to be smaller than that obtained with method B.

Note that, as in the previous Wilcoxon procedure, we are assuming symmetry, but not normality. However, if the readings obtained using each method are normally distributed, then once again this procedure is testing the same hypothesis as that tested by means of the normal-theory paired-T test. If you have no prior evidence of symmetry, then the sign test for median difference is preferred as a means of testing for the median difference.

Figure 12.3 on page 497 gives a suggested strategy for handling paired data.

Exercises 12.2

10. A new type of car wax, Wax Ease, which is in semiliquid form, has been developed. The company marketing the new wax wishes to advertise that it takes less time to wax the average car with the new wax than it does with the leading competitor. Since it intends to name the competitor in the ad, the manufacturer wishes to be able to back its claim statistically. To do so, 15 matched pairs of cars are obtained. One car is waxed with Wax Ease and the other with the competitor's brand. Each pair is waxed by the same individual, and the order of the type of wax used is random. The time required to do the job is noted in each case. The following data are obtained (time is in hours):

Wax Ease	2.1	1.0	3.6	2.5	4.0	1.7	2.9	3.0
Competitor	2.7	1.3	2.0	2.3	3.9	1.8	3.0	3.0

Wax Ease	4.5	3.1	3.1	1.5	1.6	1.0	.8
Competitor	4.6	3.0	3.4	1.7	1.8	1.3	1.0

Use the sign test to determine whether or not the makers of Wax Ease have the statistical evidence that they want. Support your opinion on the basis of the P value involved.

11. In trying to develop a quick-drying, oil-based paint for artists, a company experimented with two formulas. It was felt that under similar conditions, type II tends to dry faster than type I, and hence that type II is preferable. To test this theory, 10 different painting surfaces were obtained. Half of each surface was painted using type I and the other half with type II. The drying time for each was noted as follows (time is in hours):

Type I	1.0	1.5	.75	.90	2.1	2.7	.50	.78	1.50	2.2
Type II	.75	1.25	.80	.85	2.0	2.7	.45	.75	1.25	2.1

Do these data tend to support the contention that type II is faster drying than type I? Use the sign test and support your answer on the basis of the P value involved.

12. A medical experiment involving the measurement of the thickness of the wall of the coronary artery in millimeters is performed. At each point of interest several measurements are taken using special calipers, and the results are averaged to yield a single reading. Two observers, A and B, take part in the experiment. These results are obtained:

A	.52	.38	.27	.40	.45	.72	.66	.31	.33	.52	.33	.36
B	.52	.33	.26	.40	.51	.68	.62	.30	.36	.60	.36	.37
A	.35	.43	.60	.43	.30	.36	.51	.40	.72	.70	.61	.52
B	.31	.40	.57	.40	.37	.40	.50	.39	.82	.72	.53	.47

Use the sign test to test at the $\alpha = .05$ level the hypothesis that observer A tends to give higher readings than does observer B. Use the normal approximation to the binomial distribution to calculate $P[Q_- \leq q \,|\, p = 1/2]$ where q is the number of negative signs obtained.

13. A researcher is interested in testing the effects of taste in toothpaste. A group of 10 subjects is selected at random. Each person is asked to brush one side of the mouth with a baking soda paste, labeled X, and the other side with a baking soda paste with mint flavoring, labeled Y. When they have completed brushing on each side, a dentist is asked to rate the sides on a scale of 0 to 3, with 0 indicating not very clean teeth and 3 indicating extremely clean teeth. The responses were as follows:

X	1.2	1.4	2.0	2.2	2.1	1.7	2.1	1.6	3.0	1.0
Y	2.3	3.0	2.1	1.0	3.0	3.0	2.0	1.0	3.0	3.0

Use the sign test to test at the $\alpha = .05$ level the contention that the side brushed with mint-flavored paste tends to be cleaner than that brushed with unflavored paste.

*14. An insurance company wishes to gain statistical support for its contention that its yearly premiums tend to be lower than those of its nearest competitor. To support this claim, a random sample of 15 large metropolitan areas is selected, and the yearly premiums for its policy and that for a comparable policy offered by its competitor are obtained. The following data are gathered:

Company	500	498	505	495	490	498	480	501
Competitor	515	495	500	510	500	502	483	500
Company	496	478	520	513	506	497	482	
Competitor	498	490	526	515	513	499	490	

Use the Wilcoxon test to determine whether or not the company has the evidence that it needs. What are you assuming concerning the distribution of the population from which the difference scores are drawn?

*15. A psychologist thinks that women tend to be judged more on their looks
than do men. To test this hypothesis, pictures of what are considered to be
an unattractive male and an unattractive female are shown to a random
sample of 12 persons who work as interviewers for major corporations. The
interviewers are asked to select from among a given list of adjectives those
they think best describe the individuals pictured. Based on the adjectives, a
score from −5 to 5 is obtained, with negative scores indicating negative
feelings toward the subject. The same is done for pairs of pictures of what are
considered to be a male and a female of average attractiveness and a male
and a female of above-average attractiveness. These data are gathered:

Unattractive

| Male | −3 | .0 | −1.0 | 1.0 | 1.5 | 1.7 | 1.3 | −4.0 | .1 | −3.1 | .12 | −.75 |
| Female | −5 | −3.5 | −2.1 | .0 | −.25 | 1.5 | −.26 | −1.6 | .3 | −3.7 | .15 | −1.3 |

Average

| Male | 2.0 | 2.7 | −1.0 | 1.3 | 1.5 | .0 | 2.3 | 2.6 | 1.35 | 1.95 | 1.65 | 2.5 |
| Female | 1.8 | 2.3 | −.8 | 1.2 | 1.25 | .2 | 2.0 | 2.55 | 1.2 | 1.03 | 1.63 | 1.0 |

Above average

| Male | 3.0 | 4.5 | 4.2 | 3.9 | 4.7 | 3.9 | 3.7 | 4.5 | 4.75 | 4.35 | 4.2 | 4.95 |
| Female | 3.5 | 4.8 | 4.1 | 4.0 | 4.5 | 5.0 | 4.2 | 4.25 | 4.30 | 4.30 | 4.5 | 4.80 |

Assuming symmetry, use the Wilcoxon test to analyze each of the above data
sets. Be prepared to comment on the results obtained.

*16. Sociologists think that females tend to be less inclined toward harsh punish-
ment of criminal defendants than are males. To test this contention, 25
couples are selected. Each couple receives a brief written description of the
crime committed and the background of the defendant. Without consultation
with one another each member of the couple is asked to indicate the
minimum active sentence (in years) that should be given the defendant. The
data collected are:

| Male | .5 | 1.75 | 5.0 | 20 | 1.5 | 6.0 | 7.5 | 25 | 23 | 25 | 29 | 25.5 |
| Female | .0 | 1.0 | 7.0 | 18 | 2.0 | 3.0 | 8.5 | 20 | 22 | 25 | 25 | 31.5 |

| Male | .75 | 3.5 | 10.0 | 8.25 | 8.30 | 10.5 | 35 | 15.7 | 11.0 | 15.2 | 10.7 | 1.6 | 9.5 |
| Female | .50 | 3.1 | 6.2 | 9.6 | 8.15 | 11.8 | 25 | 12.3 | 12.6 | 15.1 | 9.3 | 2.0 | 5.5 |

Assuming symmetry, use the Wilcoxon procedure to test the above contention.

17. It is thought that, because of dangers involved in both jobs, fire fighters in
large metropolitan areas should be paid salaries comparable to police officers. To push their case, the fire fighters wish to gain statistical support for
their contention that they are currently being underpaid. They obtain the
following data concerning the average salaries paid to fire fighters and police

officers in 10 randomly selected metropolitan areas (salaries in thousands· of dollars per year):

| Fire fighters | 10.5 | 9.3 | 10.1 | 8.7 | 12.0 | 13.0 | 7.3 | 9.50 | 9.7 | 11.6 |
| Police officers | 11.1 | 10.2 | 10.0 | 8.5 | 12.5 | 12.6 | 8.0 | 9.25 | 9.8 | 11.2 |

Do the fire fighters have the evidence they want? Use whatever test you think is appropriate, but be ready to defend your choice.

*18. Assume that we are dealing with six difference scores drawn from a symmetric distribution and that there are no ties and no zero scores, so that the set of unsigned ranks with which we are dealing is $\{1, 2, 3, 4, 5, 6\}$.
   (a) If the null hypothesis that the median difference is zero is true, then, before the experiment is run, each of these ranks is just as likely to have a plus sign associated with it as it is a minus sign. How many equally likely sets of signed ranks can be generated from the set $\{1, 2, 3, 4, 5, 6\}$? What is the probability of the occurrence of each?
   (b) Find the exact probability of each of the following:

$$P[W_+ \leq 0] \qquad P[W_+ \leq 1] \qquad P[W_+ \leq 2] \qquad P[W_+ \leq 3]$$

This problem in essence verifies the numbers given in your Wilcoxon table for $n = 6$.

*19. **Normal Approximation to W:** It has been found that for large values of n, W_+ (or $|W_-|$) is approximately normally distributed with mean $\mu = n(n + 1)/4$ and the variance $\sigma^2 = n(n + 1)(2n + 1)/24$. This fact can be used to determine α levels when n is not in Table X of Appendix A. We need only standardize W_+ (or $|W_-|$) and determine α by means of the standard normal table.
   (a) For a sample of size 60, find $P[W_+ \leq 600]$.
   (b) For a sample of size 100, find $P[W_+ \leq 2{,}700]$.

12.3   Two-Sample Procedures – Unmatched Data

We consider here a distribution-free test to use when we have independent random samples of sizes m and n drawn from continuous populations X and Y. We wish to test the null hypothesis that the two populations are identical with a test that is likely to reject when they differ in center of location, mean, or median. The test is called the **Wilcoxon rank-sum test.**

Wilcoxon Rank-Sum Test   Consider a random sample of size m from an X population and an independent random sample of size n from a Y population. Assume that $m \leq n$. If the null hypothesis of identical populations is true, then the $n + m$ observations constitute a random sample of size $n + m$ from a common parent population. These observations can be ordered smallest to largest and ranked from 1 to $n + m$. There are $\binom{m + n}{m}$ ways to assign ranks to the m observations that constitute the smaller-sized sample. Furthermore, if the null hypothesis is true, then, a priori, each of these assignments is equally likely. For each combination, we can compute the value of the Wilcoxon statistic, W_m, the sum of the ranks associated with the m observations that constitute the smaller-sized sample.

The frequency distribution of W_m under H_0 is in Table XI in Appendix A. The most extreme scores lead to the rejection of H_0 where the meaning of the word *extreme* depends upon the form of the alternative.

Example 1

The personnel director of a large insurance firm claims that insurance agents trained in personal-social relations make more favorable impressions on prospective clients than do those without such training. To test this hypothesis, 22 individuals are randomly selected from those most recently hired, and 10 are assigned randomly to the personal-social relations course (experimental group). The remaining 12 are used as a control. Following the training period, each of 22 individuals is observed in three simulated interviews with clients and is rated on a 10-point scale for his or her ease in establishing personal relationships. The reported score is the average rating for the three interviews. The higher the score, the better the rating. We want to test

H_0: the two populations are identical

H_1: the experimental group tends to score
higher than the control group

This calls for a right-tailed test. We will reject H_0 if the observed value of the test statistic W_m is too large to have occurred by chance. Do these data tend to support the personnel director's claim?

Experimental (X)	8.1	7.9	9.0	4.3	7.0	9.1	7.2	8.0	9.0	3.1		
Control (Y)	9.1	6.3	2.5	6.0	0.0	2.0	7.0	5.5	1.0	9.0	9.7	5.1

We first order these observations from smallest to largest, retaining their group identity, then rank them from 1 to 22, treating ties as we have in the preceding Wilcoxon procedures.

Observed value	.0	1.0	2.0	2.5	3.1	4.3	5.1	5.5	6.0	6.3	7.0
Group identity	Y	Y	Y	Y	X	X	Y	Y	Y	Y	Y
Rank	1	2	3	4	5	6	7	8	9	10	11.5

Observed value	7.0	7.2	7.9	8.0	8.1	9.0	9.0	9.0	9.1	9.1	9.7
Group identity	X	X	X	X	X	X	X	Y	X	Y	Y
Rank	11.5	13	14	15	16	18	18	18	20.5	20.5	22

The observed value of the Wilcoxon statistic W_m is found by summing the ranks associated with the smaller sized X sample. In this case,

$$W_m = 5 + 6 + 11.5 + \cdots + 20.5 = 137$$

Since the claim is that the course increases the observed scores, we reject H_0 if the above value is too large. From Table XI in Appendix A, the critical value for an $\alpha = .05$ level one-sided test with $m = 10$ and $n = 12$ is 141. Since $137 < 141$, we are unable to reject H_0. We do not have sufficient evidence to support the claim.

If the samples are drawn from normal populations, then the Wilcoxon procedure tests the same hypothesis as that of the normal theory two-sample T procedures.

Exercises 12.3

20. To find out whether a new serum arrests leukemia, 9 mice that have all reached an advanced stage of the disease are selected. Five mice receive treatment and four do not. The survival times (in months) from the time the experiment commenced are as follows:

Treated	2.1	5.3	1.4	4.6	.9
Untreated	1.9	0.5	2.8	3.1	

    At the .05 level of significance, can the serum be said to be effective? Use the Wilcoxon rank-sum test.

21. Sociologists think that females tend to view the performance of government officials a bit more optimistically than do males. To test this contention, random samples of 20 females and 25 males are obtained. Each individual is asked to complete a questionnaire concerning the performance of the President of the United States during his first six months in office. From this questionnaire, a score is obtained for each individual with low scores representing high optimism. The following data are obtained:

Female				Male				
23.18	19.49	16.00	30.08	22.01	26.43	22.51	27.26	27.97
26.34	25.11	18.35	25.18	19.54	11.09	21.28	29.11	30.19
24.03	25.59	10.39	20.66	28.08	19.16	37.07	30.77	23.79
22.00	25.81	14.73	19.21	28.33	28.34	25.14	31.73	24.38
17.10	21.90	16.44	15.67	18.94	35.77	15.61	20.03	26.23

    Test the appropriate hypothesis at the $\alpha = .05$ level, using the Wilcoxon procedure.

22. It is thought that the percentage of convicted defendants given active sentences in United States district courts is higher in the South than it is in other parts of the country. To gain statistical support for this contention, random samples of 10 courts in the South and 20 throughout the remainder of the country are selected. For each, the percentage of defendants receiving active sentences is obtained. The following data result:

South		Other sections			
49.95	49.09	47.09	51.04	42.95	51.01
45.41	49.21	47.30	45.97	42.49	44.10
50.83	52.99	47.79	46.62	44.93	44.62
54.28	49.81	43.58	49.76	43.90	40.27
52.31	49.49	44.12	41.58	51.71	44.52

Use the Wilcoxon rank-sum test to test the appropriate hypothesis at the $\alpha = .025$ level.

*23. Let $m = 3$ and $n = 4$ and consider the set of ranks

$$\{1, 2, 3, 4, 5, 6, 7\}$$

(a) How many possible combinations of 3 ranks can be formed?

(b) Since, under H_0, each of the above combinations occurs with probability $1/35$, the two most extreme sets in either direction would yield critical regions for size $\alpha \approx .05$ tests ($2/35 = .0571$). Verify that the critical values for left-tail and right-tail tests based on W_m are 7 and 17, respectively, as given in Table XI in Appendix A.

*24. **Normal Approximation to W_m:** It has been found that for large sample sizes, W_m is distributed approximately normally with mean $\mu = m(m + n + 1)/2$ and variance $\sigma^2 = mn(m + n + 1)/12$. This fact can be used to test hypotheses that call for the use of the Wilcoxon procedure in situations in which the sample sizes involved are not in Table XI in Appendix A. We need only standardize W_m and then use the standard normal table to determine the α level involved in the test.

(a) Let $m = 50$ and $n = 75$; find $P[W_m \leq 3,000]$.

(b) Let $m = 25$ and $n = 60$; find $P[W_m \leq 1,200]$.

12.4   Correlation Procedures

We considered the problem of correlation in Chapter 9. The measure of linearity used there was the Pearson coefficient of correlation r. We consider here another measure of linear association, namely Spearman's rank-correlation coefficient.

Spearman's rank-correlation coefficient (C. Spearman, 1904)   The **Spearman rank-correlation coefficient** r_S is an adaptation of the Pearson coefficient r for use with ranked data. The ranks may be natural, such as the class ranking assigned to students according to their grade averages, or subjective, such as the ranks assigned by a judge to a group of paintings in an art exhibit. Basically, all that is done is to order separately the observations on the two random variables X and Y. Each observation is then replaced by its respective rank in the ordering. Tied scores are again assigned the average of the tied ranks. The Pearson coefficient

$$r = \frac{n\Sigma xy - \Sigma x\Sigma y}{\sqrt{[n\Sigma x^2 - (\Sigma x)^2][n\Sigma y^2 - (\Sigma y)^2]}}$$

is then calculated, making use not of the original observations, but of the ranks of these observations. This procedure yields results that are slightly different from those obtained by the Pearson method, with the agreement improving for large samples. The Spearman approach is computationally easier than Pearson's, since it can be shown that for ranked data with no ties

$$r \approx r_s = 1 - \frac{6 \sum_{i=1}^{n} d_i^2}{n(n^2 - 1)}$$

where d_i = rank of x_i − rank of y_i. The Spearman method should not be used if there is a large number of ties involved.

Example 1

Consider these data on the grades on tests for six students taking both algebra and trigonometry:

Actual Scores			Ranked Scores			
Student	Algebra	Trigonometry	Student	Alegebra	Trigonometry	d
1	78	70	1	3	3	0
2	60	64	2	1.5	2	−.5
3	87	81	3	5	4	1
4	85	83	4	4	5	−1
5	60	58	5	1.5	1	.5
6	90	100	6	6	6	0

For these observations

$$r = .8991 \text{ (using actual calculations)}$$

$$r_s = .9276 \text{ (using Pearson's } r \text{ with observations replaced by ranks)}$$

$$r_s \approx .9286 \text{ (using computational shortcut to approximate } r_s)$$

Exercises 12.4

25. In a carefully monitored experiment to study the effects of caffeine on sleep, the following observations are obtained on the number of ounces of coffee drunk in an 8-hour period X and the length (in minutes) of the deep sleep cycle of the subject Y:

x	y	x	y
80	20	8	48
76	22	75	27
50	35	15	45
32	42	24	40

Compute r and r_s and compare the values obtained.

26. The following observations are obtained on the variables X, the length of a Sunday sermon, and Y, the number of persons in the church the following Sunday (time is in minutes):

x	y	x	y
3	150	7	139
5	150	25	90
12	130	30	85
2.5	147	10	125
15	110	20	100

Compute the value of r_s for these data.

27. A psychological experiment is conducted that involves eight sets of identical twins. The point of the experiment is to consider the correlation, if any, between growth rate during the first year of the first-born and the second-born twin. The following data are obtained on the height (in inches) of each twin on the first birthday:

Twin set	1	2	3	4	5	6	7	8
First born	24.0	27.0	26.5	23.2	28.0	21.0	26.1	25.0
Second born	23.0	25.1	27.0	22.9	27.4	20.0	27.2	25.0

Compute the value of r_s.

28. A husband and wife who go bowling together keep their scores for ten lines to see if there is a correlation between their scores. The scores are:

Line	1	2	3	4	5	6	7	8	9	10
Husband	147	158	131	142	183	151	196	129	155	158
Wife	122	128	125	123	115	120	108	143	124	123

Compute the value of r_s.

12.5   K-Sample Procedures

Recall that, in the case of normal theory statistics, we studied two problems, the one-way classification analysis of variance and the randomized complete block design. Each is intended to test the hypothesis that k normal populations have the same center of location, based on random samples drawn from these populations. If we deal with data for which the normality assumption fails, then distribution-free methods should be used. We consider here two distribution-free techniques for testing k populations for identical centers of location. The first, the **Kruskal-Wallis k-sample test**, is the distribution-free analog of the normal theory one-way classification analysis of variance; the second, the **Friedman test**, is used in situations that ordinarily call for a randomized complete block design.

Kruskal-Wallis *k*-Sample Test (W. H. Kruskal and W. A. Wallis, 1952)   Assume that we have available random samples of sizes $n_1, n_2, \ldots, n_k$ drawn from k populations. We wish to determine whether or not these populations are identical with a test that is especially likely to reject if they differ in center of location. If H_0 is true, then the collection of observations can be considered as being a random sample of size $N = n_1 + n_2 + \cdots + n_k$ from a common parent population. We may rank these observations, smallest to largest, from 1 to N and replace each observation by its respective rank. Let R_i, represent the sum of the ranks associated with the ith sample. If H_0 is true, then higher ranks should be scattered throughout the k samples producing moderate rank sums. If H_0 is not true, then the higher ranks tend to fall within the sample drawn from the population with a higher center of location. Thus, we need a test statistic that detects a situation in which one or more rank sums is too large to have occurred by chance. One such statistic is the Kruskal-Wallis H given by:

$$H = \left[12/(N)(N+1) \sum_{i=1}^{k} R_i^2/n_i \right] - 3(N+1)$$

The exact distribution of this statistic can be found using elementary counting techniques. A typical example is given in exercise 29. However, obtaining this distribution is time consuming. We can approximate the needed probabilities by using the chi-square distribution with $k - 1$ degrees of freedom. Since an inflated value of R_i will be emphasized by squaring when evaluating H, we reject H_0 for observed values of H that are too large to have occurred by chance. Example 1 demonstrates the use of the Kruskal-Wallis test.

Example 1   Random samples of size five yield these data on the pollution level of three major rivers. High scores indicate a high level of pollution.

First river		Second river		Third river	
2.7	(13)	2.9	(14)	.6	(1)
1.4	(4)	2.4	(11.5)	1.2	(2.5)
2.0	(8)	3.7	(15)	1.5	(5)
1.2	(2.5)	1.6	(6)	1.7	(7)
2.1	(9.5)	2.4	(11.5)	2.1	(9.5)

The ranks of the scores are given in parentheses, with tied scores receiving the average group rank. Summing these ranks we see that

$$R_1 = 37 \qquad R_2 = 58 \qquad R_3 = 25$$

$$H = \left[12/(N)(N+1) \sum_{i=1}^{3} R_i^2/n_i \right] - 3(N+1)$$

$$= 12/(15)(16)[(37)^2/5 + (58)^2/5 + (25)^2/5] - 3(16) = 5.58$$

Comparing this value with the critical points of the X_2^2 distribution, we see that the null hypothesis of identical populations can be rejected at the $\alpha = .10$ level because the observed value of 5.58 is greater than the critical point of 4.61. There is some evidence that the pollution levels in the three rivers are not the same.

The Kruskal-Wallis test deals with independent samples drawn from k populations. When observations are matched across samples, we use the Friedman test to compare population locations.

Friedman Test (M. Friedman, 1937)   Assume that we are concerned with k treatments but we think that, by randomly assigning these treatments to experimental units, we may introduce extraneous factors that obscure our results. To overcome this difficulty, we draw k units from each of b possibly different populations. In this way we obtain b blocks of size k. We then randomly assign the k treatments within each block. We wish to test the hypothesis that the k treatments have identical effects with a test that is especially likely to reject if the treatments differ in location. To do so, we record the observations in a table of k columns representing treatments and b rows representing blocks. The observations within each row (block) are then ranked from 1 to k with the rank of 1 going to the smallest observation and the rank of k to the largest. If the treatments have identical effects under each matching condition, then the observation within each block may be considered to be a random sample of size k from a common parent population. In this case, large ranks should be scattered throughout the k columns in every row. If H_0 is not true, then large ranks have a tendency to cluster in the column representing the treatment with the higher center of location. Let R_i be the sum of the ranks in the ith treatment column. We need a statistic that detects a situation in which one or more of these rank sums has become too large to have occurred by chance. One such statistic is the Friedman statistic S given by

$$S = \sum_{i=1}^{k} [R_i - b(k + 1)/2]^2$$

As in the case of the Kruskal-Wallis statistic, this statistic looks formidable. It is, in fact, not feasible to attempt to table the probability distribution of S directly, However, it has been shown that the statistic

$$12S/bk(k + 1)$$

follows an approximate chi-square distribution, with $k - 1$ degrees of freedom. This fact allows us to use the chi-square table, rejecting H_0 for values of $12S/bk(k + 1)$ that are too large to have occurred by chance. As you will see, this statistic is not hard to use.

Example 2   A consumer-research group has as its major responsibility the testing of various products that are to be marketed to the public. A study is conducted to compare three types of bicycle braking systems. It is thought that the brand of bicycle involved in the test could influence the results, and, hence, six of the better-known bicycle brands are used in the test. We thus obtain six blocks. Each type of brake system is tested once on each type of bicycle, with the variable of interest being the length of service (in weeks) until a major repair is required. These results are obtained:

Bike type	Brake type		
	A	B	C
S	5.2 (2)	7.3 (3)	3.0 (1)
V	6.8 (1)	8.9 (3)	7.5 (2)
JH	6.3 (2.5)	6.3 (2.5)	6.0 (1)
R	13.0 (1.5)	14.8 (3)	13.0 (1.5)
C	12.8 (2.5)	12.8 (2.5)	11.0 (1)
Ra	15.0 (2)	15.2 (3)	14.5 (1)

To see if there are significant differences among brake types, we first rank the observations within each row. These ranks are given above in parentheses. From the above, we may now compute the values of R_1, R_2, R_3, S, and the test statistic.

$$R_1 = 11.5 \qquad R_2 = 17.0 \qquad R_3 = 7.5$$

$$S = \sum_{i=1}^{3} [R_i - b(k+1)/2]^2 = \sum_{i=1}^{3} [R_i - 6(4)/2]^2$$

$$= (11.5 - 12)^2 + (17 - 12)^2 + (7.5 - 12)^2 = 45.50$$

The observed value of the test statistic is therefore

$$12S/bk(k+1) = 12(45.50)/(6)(3)(4) = 7.58$$

From the chi-square table we see that this value is significant at the $\alpha = .025$ level, since it is larger than the tabled value of 7.38. Hence, we conclude that there are differences among the braking systems. Furthermore, it appears upon inspection that system B is preferable.

We have given in this chapter only the briefest of introductions to the field of distribution-free statistics. This is one of the most interesting and fastest growing areas of statistics. You are referred to any of the following excellent references for further information on the subject: [2], [4], [5].

Exercises 12.5

29. Consider a situation in which $n_1 = 2$, $n_2 = 1$, $n_3 = 1$ and hence $N = 4$. If H_0 is true, and these are samples from three identical populations, then each of the $4!/2! 1! 1! = 12$ ways to split the ranks 1, 2, 3, 4 into subsets of sizes 2, 1,

1 are equally likely, and each occurs with probability 1/12. These 12 splits are given below:

Treatment 1, R_1	Treatment 2, R_2	Treatment 3, R_3	H
{1, 2}	{3}	{4}	
{1, 2}	{4}	{3}	
{1, 3}	{2}	{4}	
{1, 3}	{4}	{2}	
{1, 4}	{2}	{3}	
{1, 4}	{3}	{2}	
{2, 3}	{1}	{4}	
{2, 3}	{4}	{1}	
{2, 4}	{1}	{3}	
{2, 4}	{3}	{1}	
{3, 4}	{1}	{2}	
{3, 4}	{2}	{1}	

(a) For each of the above splits, compute the values of R_1, R_2, R_3, and H.
(b) Use this information to complete the following probability table for H when $n_1 = 2$, $n_2 = n_3 = 1$.

h	.3	1.8	2.7
$P[H = h] = f(h)$			

30. In a psychological experiment three types of rats—well-trained, partially-trained, and untrained—are required to pass through a maze. The time in minutes required to complete the maze is recorded for each rat. The following data are obtained.

Well-trained		Partially-trained		Untrained	
.7	1.6	1.9	1.3	1.3	2.8
1.2	.4	1.4	2.1	1.8	1.3
1.3	.8	1.8	2.7	2.4	1.9
.9	1.8	2.7	2.6	1.9	1.6

Use the Kruskal-Wallis test to determine whether or not there are differences among the times involved. Let $\alpha = .05$.

31. Three new methods of teaching golf are being tested by a physical educator. Twenty subjects, all beginners, are to be used. The subjects are divided randomly into four groups of five each. One group is taught by the usual method, while each of the other three groups is taught by one of the new techniques. A skills tests is administered at the end of the learning period and the scores recorded. These data result (high scores represent high skill levels):

New Techniques			Usual Method
I	*II*	*III*	*IV*
63.0	85.0	90.2	65.0
47.0	80.1	70.7	45.2
51.0	79.0	86.0	50.9
74.0	67.0	62.3	75.0
60.0	82.3	72.3	58.8

   Use the Kruskal-Wallis test to see if there is a significant difference among the scores obtained. Let $\alpha = .10$.

32. At a driving scnool, a new driver's education course is instituted. A change in the roster of regular driving instructors is also suggested. To compare results, 15 beginning students are randomly placed into three groups of five students each. At the end of the course, the standard 100-point driving skills test is administered. The following results are obtained:

Regular course (regular instructors)	New course (regular instructors)	New course (race-car driver instructors)
55	78	81
90	85	71
76	70	89
73	65	69
79	80	84

   Use the Kruskal-Wallis test to see if there is a significant difference among the scores obtained. What is the *P* value of the test?

33. An insurance company in a recent ad compared its rates with those of two of its leading competitors. Since the location of the policyholder has an effect on rates, this extraneous factor is controlled by blocking. Twenty cities (blocks) are used, and the yearly rate for comparable policies offered by the three companies are obtained. The following information is collected:

	Company	Competitor I	Competitor II
Buffalo, NY	560	597	563
Cincinnati, OH	398	409	381
Miami, FL	480	511	564
Minneapolis, MN	454	522	425
New Orleans, LA	527	573	519
Akron, OH	452	439	438
Gary, IN	598	587	516
Nashville, TN	322	430	424
Pasadena, CA	429	483	448
Rochester, NY	457	480	462
San Mateo, CA	476	580	510
Tacoma, WA	410	465	383
Appleton, WI	358	371	336
Bismarck, ND	320	317	259
Farmington, NM	429	390	352
Huntsville, AL	344	341	295
New Castle, PA	358	428	402
Pocatello, ID	271	396	358
Rock Hill, SC	410	265	292
Rutland, VT	407	500	393

Use the Friedman test to see if there are significant differences among the rates of the three companies. If so, be ready to comment on an intuitive level on the results. Let $\alpha = .05$.

34. A sportsman is attempting to find a faster horse than his current entry in the racing circuit. He is considering five other horses for purchase. Because a jockey can influence the performance of a horse, the sportsman decides to try each horse using four different jockeys. The data given below represent the time in seconds required to cover a specified distance. Is there sufficient evidence at the $\alpha = .10$ level to indicate that the horses do not perform identically? If so, which one would you recommend that he buy?

Jockey	Horse 1	Horse 2	Horse 3	Horse 4	Horse 5	Current entry
A	13.0	13.9	17.0	19.0	18.3	19.1
B	12.1	13.7	13.7	18.3	17.9	13.8
C	17.0	17.0	17.1	18.7	18.3	17.1
D	15.3	15.5	18.2	18.2	18.1	15.7

35. An advertising executive is interested in the effect of color on ad appeal. Ten persons are selected randomly and asked to rate each of five magazine layouts in order of eye appeal, with a rank of 1 being given to the most appealing ad. The layouts are identical except for the color scheme involved. The following data are obtained:

Person	I, no color	II, red-predominant	III, yellow-predominant	IV, blue-predominant	V, mixed color
1	1	2	4	3	5
2	1	2	5	3	4
3	3	4	5	1	2
4	1	2	3	5	4
5	4	5	1	3	2
6	1	2	3	4	5
7	2	1	5	4	3
8	1	2	4	5	3
9	3	5	1	2	4
10	1	2	5	4	3

Is there evidence of a color effect? Explain, based on the P value of your test.

Vocabulary List and Key Concepts

distribution-free tests

median

sign test for median

Wilcoxon signed-rank test

sign test for median difference

Wilcoxon rank-sum test

Kruskal-Wallis k-sample test

Spearman rank-correlation coefficient

Friedman test

Review Exercises

36. The sign test for median is a distribution-free alternative to what normal theory test? Under what circumstances would the sign test be preferred to the normal theory test?

37. In the sign test, we assume that no zeros will occur. What is the theoretical basis for this assumption? Who do zeros scores occur occasionally anyway? How are these zeros handled?

Use the following data to answer exercises 38–40:

A new 24-hour card operated banking system is installed. Bank executives want to advertise that the median time required to complete a transaction is less than 5 minutes.

| 2.1 | 3.6 | 2.3 | 4.9 | 2.5 | 2.8 | 7.7 | 2.0 | 5.1 | 3.4 |
| 3.1 | 3.7 | 6.0 | 7.2 | 2.1 | 8.2 | 4.1 | 8.2 | 2.0 | 3.4 |

38. Construct a stem-and-leaf diagram for these data. Do you think that there is reason to suspect that X, the median time required to complete a transaction, is not normally distributed?

39. State and test the null hypothesis using the sign test. Do you think that H_0 should be rejected? Explain, based on the P value of the test.

40. What assumption underlies the Wilcoxon signed-rank test for median? Would you feel comfortable analyzing this data set via the Wilcoxon procedure? Explain.

41. The sign test for median difference is a distribution-free alternative to what normal-theory test?

42. A restaurant manager claims that men tend to tip better than women. These data are obtained on the amount of tip left by men and women on bills of equivalent sizes:

Men	Women	Men	Women
$3.50	$2.75	$2.75	$2.60
3.25	3.10	3.10	2.70
2.70	2.85	1.50	1.75
4.25	4.50	4.20	4.15
4.50	4.40	3.25	3.10
5.10	5.50	3.00	3.00
4.00	4.25	5.00	4.90
8.00	6.00		

State and test the null hypothesis needed to support the manager's claim. Use whatever procedure you deem appropriate. Be ready to defend your choice.

43. The Wilcoxon rank-sum test is a distribution-free alternative to what normal-theory test or tests?

44. It is thought that people who drive red cars tend to drive faster than those who drive cars that are not red. To test this contention, the speeds of 15 randomly selected cars are monitored. Do these data support the contention at the $\alpha = .05$ level?

Red	Not Red
55.1	62.0
50.5	55.0
65.3	57.3
69.2	63.0
58.0	75.0
60.0	48.0
	51.8
	60.0
	45.0

45. A criminologist claims that the active sentences given to men tend to be longer than those given to women convicted of similar crimes. When ordered smallest to largest, the length of sentences for a random sample of 10 women and 15 men yielded these data: (M = men, W = women)

   W W M W M M W W W M M M W W M W W M M M M M M M M

   Do these data support the contention of the criminologist at the $\alpha = .05$ level?

46. The Kruskal-Wallis test is a distribution-free alternative to what normal theory procedure?

47. The Friedman test is a distribution-free alternative to what normal theory procedure?

48. Five different computer keyboards are being tested. Ten experienced operators are allowed to use each keyboard for a one-week period. At the end of the experimental period each individual ranks the keyboards in order of preference with a rank of 1 being the highest rank possible. These data result:

	Keyboard				
Operator	A	B	C	D	E
1	5	2	1	3	4
2	5	3	2	1	4
3	3	2	1	4	5
4	4	2	1	3	5
5	5	1	3	2	4
6	3	2	1	5	4
7	4	1	2	3	5
8	2	3	1	5	4
9	5	2	1	3	4
10	3	4	1	2	5

Test the null hypothesis that these keyboards are liked equally well at the $\alpha = .10$ level.

49. A study of absenteeism among assembly line workers is conducted. Workers are randomly selected from three shifts of a major automobile manufacturer. These data are obtained on the percentage of working-hours absent:

Morning shift	Afternoon shift	Evening shift
.25	1.00	4.00
1.00	1.85	1.50
.00	2.25	.50
1.75	1.95	3.25
2.00	3.60	
	3.30	
	3.95	

At the $\alpha = .10$ level, test the null hypothesis that there are no differences in the absentee rates among these three groups.

50. A sportswriter claims that in NFL football games the visiting teams tend to receive more yards in penalties than does the home team. At the $\alpha = .10$ level, do these data support this contention?

Game	Home	Visitor
1	15	25
2	65	60
3	75	75
4	20	25
5	10	50
6	65	70
7	40	80
8	65	100
9	55	50
10	60	65

51. In Olympic diving, the scores assigned by judges from two countries with different political philosophies are being studied. These data are obtained on scores assigned to 10 dives:

Dive	Judge A	Judge B
1	9.8	4.5
2	9.6	5.9
3	5.1	8.2
4	4.5	9.1
5	10.0	7.2
6	9.3	4.8
7	9.7	5.8
8	6.2	5.5
9	7.3	6.0
10	8.9	4.3

Find the Spearman correlation coefficient. Comment on the practical implications of this result.

52. A study is conducted to compare the speed with which four private carriers can deliver packages. Identical packages are shipped from a particular location to 8 different destinations. These data are obtained on the time in hours required for delivery.

	Carrier			
Package	I	II	III	IV
1	8	10	9	11
2	12	11	9	14
3	3	5	4	6
4	24	26	23	25
5	9	8	10	7
6	6	7	5	9
7	16	17	16	18
8	18	19	18	20

Test the null hypothesis that these carriers are identical with respect to delivery time at the $\alpha = .05$ level.

Appendix A

Table I
Cumulative binomial distribution

$$P[X \leqslant t] = \sum_{x=0}^{[t]} \binom{n}{x} p^x (1-p)^{n-x}$$

n	t	0.10	0.20	0.25	0.30	0.40	0.50	0.60	0.70	0.80	0.90
5	0	0.5905	0.3277	0.2373	0.1681	0.0778	0.0312	0.0102	0.0024	0.0003	0.0000
	1	0.9185	0.7373	0.6328	0.5282	0.3370	0.1875	0.0870	0.0308	0.0067	0.0005
	2	0.9914	0.9421	0.8965	0.8369	0.6826	0.5000	0.3174	0.1631	0.0579	0.0086
	3	0.9995	0.9933	0.9844	0.9692	0.9130	0.8125	0.6630	0.4718	0.2627	0.0815
	4	1.0000	0.9997	0.9990	0.9976	0.9898	0.9688	0.9222	0.8319	0.6723	0.4095
	5	1.0000	1.0000	1.0000	1.0000	1.0000	1.0000	1.0000	1.0000	1.0000	1.0000
10	0	0.3487	0.1074	0.0563	0.0282	0.0060	0.0010	0.0001	0.0000	0.0000	0.0000
	1	0.7361	0.3758	0.2440	0.1493	0.0464	0.0107	0.0017	0.0001	0.0000	0.0000
	2	0.9298	0.6778	0.5256	0.3828	0.1673	0.0547	0.0123	0.0016	0.0001	0.0000
	3	0.9872	0.8791	0.7759	0.6496	0.3823	0.1719	0.0548	0.0106	0.0009	0.0000
	4	0.9984	0.9672	0.9219	0.8497	0.6331	0.3770	0.1662	0.0474	0.0064	0.0002
	5	0.9999	0.9936	0.9803	0.9527	0.8338	0.6230	0.3669	0.1503	0.0328	0.0016
	6	1.0000	0.9991	0.9965	0.9894	0.9452	0.8281	0.6177	0.3504	0.1209	0.0128
	7	1.0000	0.9999	0.9996	0.9984	0.9877	0.9453	0.8327	0.6172	0.3222	0.0702
	8	1.0000	1.0000	1.0000	0.9999	0.9983	0.9893	0.9536	0.8507	0.6242	0.2639
	9	1.0000	1.0000	1.0000	1.0000	0.9999	0.9990	0.9940	0.9718	0.8926	0.6513
	10	1.0000	1.0000	1.0000	1.0000	1.0000	1.0000	1.0000	1.0000	1.0000	1.0000
15	0	0.2059	0.0352	0.0134	0.0047	0.0005	0.0000	0.0000	0.0000	0.0000	0.0000
	1	0.5490	0.1671	0.0802	0.0353	0.0052	0.0005	0.0000	0.0000	0.0000	0.0000
	2	0.8159	0.3980	0.2361	0.1268	0.0271	0.0037	0.0003	0.0000	0.0000	0.0000
	3	0.9444	0.6482	0.4613	0.2969	0.0905	0.0176	0.0019	0.0001	0.0000	0.0000
	4	0.9873	0.8358	0.6865	0.5155	0.2173	0.0592	0.0094	0.0007	0.0000	0.0000
	5	0.9978	0.9389	0.8516	0.7216	0.4032	0.1509	0.0338	0.0037	0.0001	0.0000
	6	0.9997	0.9819	0.9434	0.8689	0.6098	0.3036	0.0951	0.0152	0.0008	0.0000
	7	1.0000	0.9958	0.9827	0.9500	0.7869	0.5000	0.2131	0.0500	0.0042	0.0000
	8	1.0000	0.9992	0.9958	0.9848	0.9050	0.6964	0.3902	0.1311	0.0181	0.0003
	9	1.0000	0.9999	0.9992	0.9963	0.9662	0.8491	0.5968	0.2784	0.0611	0.0023
	10	1.0000	1.0000	0.9999	0.9993	0.9907	0.9408	0.7827	0.4845	0.1642	0.0127

(Table continues on following page.)

A1

Table I

Cumulative binomial distribution *(continued)*

$$P[X \leqslant t] = \sum_{x=0}^{[t]} \binom{n}{x} p^x (1-p)^{n-x}$$

n	t	0.10	0.20	0.25	0.30	0.40	0.50	0.60	0.70	0.80	0.90
	11	1.0000	1.0000	1.0000	0.9999	0.9981	0.9824	0.9095	0.7031	0.3518	0.0556
	12	1.0000	1.0000	1.0000	1.0000	0.9997	0.9963	0.9729	0.8732	0.6020	0.1841
	13	1.0000	1.0000	1.0000	1.0000	1.0000	0.9995	0.9948	0.9647	0.8329	0.4510
	14	1.0000	1.0000	1.0000	1.0000	1.0000	1.0000	0.9995	0.9953	0.9648	0.7941
	15	1.0000	1.0000	1.0000	1.0000	1.0000	1.0000	1.0000	1.0000	1.0000	1.0000
20	0	0.1216	0.0115	0.0032	0.0008	0.0000	0.0000	0.0000	0.0000	0.0000	0.0000
	1	0.3917	0.0692	0.0243	0.0076	0.0005	0.0000	0.0000	0.0000	0.0000	0.0000
	2	0.6769	0.2061	0.0913	0.0355	0.0036	0.0002	0.0000	0.0000	0.0000	0.0000
	3	0.8670	0.4114	0.2252	0.1071	0.0160	0.0013	0.0001	0.0000	0.0000	0.0000
	4	0.9568	0.6296	0.4148	0.2375	0.0510	0.0059	0.0003	0.0000	0.0000	0.0000
	5	0.9887	0.8042	0.6172	0.4164	0.1256	0.0207	0.0016	0.0000	0.0000	0.0000
	6	0.9976	0.9133	0.7858	0.6080	0.2500	0.0577	0.0065	0.0003	0.0000	0.0000
	7	0.9996	0.9679	0.8982	0.7723	0.4159	0.1316	0.0210	0.0013	0.0000	0.0000
	8	0.9999	0.9900	0.9591	0.8867	0.5956	0.2517	0.0565	0.0051	0.0001	0.0000
	9	1.0000	0.9974	0.9861	0.9520	0.7553	0.4119	0.1275	0.0171	0.0006	0.0000
	10	1.0000	0.9994	0.9961	0.9829	0.8725	0.5881	0.2447	0.0480	0.0026	0.0000
	11	1.0000	0.9999	0.9991	0.9949	0.9435	0.7483	0.4044	0.1133	0.0100	0.0001
	12	1.0000	1.0000	0.9998	0.9987	0.9790	0.8684	0.5841	0.2277	0.0321	0.0004
	13	1.0000	1.0000	1.0000	0.9997	0.9935	0.9423	0.7500	0.3920	0.0867	0.0024
	14	1.0000	1.0000	1.0000	1.0000	0.9984	0.9793	0.8744	0.5836	0.1958	0.0113
	15	1.0000	1.0000	1.0000	1.0000	0.9997	0.9941	0.9490	0.7625	0.3704	0.0432
	16	1.0000	1.0000	1.0000	1.0000	1.0000	0.9987	0.9840	0.8929	0.5886	0.1330
	17	1.0000	1.0000	1.0000	1.0000	1.0000	0.9998	0.9964	0.9645	0.7939	0.3231
	18	1.0000	1.0000	1.0000	1.0000	1.0000	1.0000	0.9995	0.9924	0.9308	0.6083
	19	1.0000	1.0000	1.0000	1.0000	1.0000	1.0000	1.0000	0.9992	0.9885	0.8784
	20	1.0000	1.0000	1.0000	1.0000	1.0000	1.0000	1.0000	1.0000	1.0000	1.0000

Table II
Cumulative distribution: Standard normal

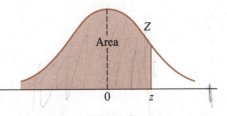

$P[Z \leqslant z]$

z	0.00	0.01	0.02	0.03	0.04	0.05	0.06	0.07	0.08	0.09
−3.4	0.0003	0.0003	0.0003	0.0003	0.0003	0.0003	0.0003	0.0003	0.0003	0.0002
−3.3	0.0005	0.0005	0.0005	0.0004	0.0004	0.0004	0.0004	0.0004	0.0004	0.0003
−3.2	0.0007	0.0007	0.0006	0.0006	0.0006	0.0006	0.0006	0.0005	0.0005	0.0005
−3.1	0.0010	0.0009	0.0009	0.0009	0.0008	0.0008	0.0008	0.0008	0.0007	0.0007
−3.0	0.0013	0.0013	0.0013	0.0012	0.0012	0.0011	0.0011	0.0011	0.0010	0.0010
−2.9	0.0019	0.0018	0.0017	0.0017	0.0016	0.0016	0.0015	0.0015	0.0014	0.0014
−2.8	0.0026	0.0025	0.0024	0.0023	0.0023	0.0022	0.0021	0.0021	0.0020	0.0019
−2.7	0.0035	0.0034	0.0033	0.0032	0.0031	0.0030	0.0029	0.0028	0.0027	0.0026
−2.6	0.0047	0.0045	0.0044	0.0043	0.0041	0.0040	0.0039	0.0038	0.0037	0.0036
−2.5	0.0062	0.0060	0.0059	0.0057	0.0055	0.0054	0.0052	0.0051	0.0049	0.0048
−2.4	0.0082	0.0080	0.0078	0.0075	0.0073	0.0071	0.0069	0.0068	0.0066	0.0064
−2.3	0.0107	0.0104	0.0102	0.0099	0.0096	0.0094	0.0091	0.0089	0.0087	0.0084
−2.2	0.0139	0.0136	0.0132	0.0129	0.0125	0.0122	0.0119	0.0116	0.0113	0.0110
−2.1	0.0179	0.0174	0.0170	0.0166	0.0162	0.0158	0.0154	0.0150	0.0146	0.0143
−2.0	0.0228	0.0222	0.0217	0.0212	0.0207	0.0202	0.0197	0.0192	0.0188	0.0183
−1.9	0.0287	0.0281	0.0274	0.0268	0.0262	0.0256	0.0250	0.0244	0.0239	0.0233
−1.8	0.0359	0.0352	0.0344	0.0336	0.0329	0.0322	0.0314	0.0307	0.0301	0.0294
−1.7	0.0446	0.0436	0.0427	0.0418	0.0409	0.0401	0.0392	0.0384	0.0375	0.0367
−1.6	0.0548	0.0537	0.0526	0.0516	0.0505	0.0495	0.0485	0.0475	0.0465	0.0455
−1.5	0.0668	0.0655	0.0643	0.0630	0.0618	0.0606	0.0594	0.0582	0.0571	0.0559
−1.4	0.0808	0.0793	0.0778	0.0764	0.0749	0.0735	0.0722	0.0708	0.0694	0.0681
−1.3	0.0968	0.0951	0.0934	0.0918	0.0901	0.0885	0.0869	0.0853	0.0838	0.0823
−1.2	0.1151	0.1131	0.1112	0.1093	0.1075	0.1056	0.1038	0.1020	0.1003	0.0985
−1.1	0.1357	0.1335	0.1314	0.1292	0.1271	0.1251	0.1230	0.1210	0.1190	0.1170
−1.0	0.1587	0.1562	0.1539	0.1515	0.1492	0.1469	0.1446	0.1423	0.1401	0.1379
−0.9	0.1841	0.1814	0.1788	0.1762	0.1736	0.1711	0.1685	0.1660	0.1635	0.1611
−0.8	0.2119	0.2090	0.2061	0.2033	0.2005	0.1977	0.1949	0.1922	0.1894	0.1867
−0.7	0.2420	0.2389	0.2358	0.2327	0.2296	0.2266	0.2236	0.2206	0.2177	0.2148
−0.6	0.2743	0.2709	0.2676	0.2643	0.2611	0.2578	0.2546	0.2514	0.2483	0.2451
−0.5	0.3085	0.3050	0.3015	0.2981	0.2946	0.2912	0.2877	0.2843	0.2810	0.2776

(Table continues on following page.)

Table II
Cumulative distribution: Standard normal *(continued)*

$$P[Z \leqslant z]$$

z	0.00	0.01	0.02	0.03	0.04	0.05	0.06	0.07	0.08	0.09
−0.4	0.3446	0.3409	0.3372	0.3336	0.3300	0.3264	0.3228	0.3192	0.3156	0.3121
−0.3	0.3821	0.3783	0.3745	0.3707	0.3669	0.3632	0.3594	0.3557	0.3520	0.3483
−0.2	0.4207	0.4168	0.4129	0.4090	0.4052	0.4013	0.3974	0.3936	0.3897	0.3859
−0.1	0.4602	0.4562	0.4522	0.4483	0.4443	0.4404	0.4364	0.4325	0.4286	0.4247
−0.0	0.5000	0.4960	0.4920	0.4880	0.4840	0.4801	0.4761	0.4721	0.4681	0.4641
0.0	0.5000	0.5040	0.5080	0.5120	0.5160	0.5199	0.5239	0.5279	0.5319	0.5359
0.1	0.5398	0.5438	0.5478	0.5517	0.5557	0.5596	0.5636	0.5675	0.5714	0.5753
0.2	0.5793	0.5832	0.5871	0.5910	0.5948	0.5987	0.6026	0.6064	0.6103	0.6141
0.3	0.6179	0.6217	0.6255	0.6293	0.6331	0.6368	0.6406	0.6443	0.6480	0.6517
0.4	0.6554	0.6591	0.6628	0.6664	0.6700	0.6736	0.6772	0.6808	0.6844	0.6879
0.5	0.6915	0.6950	0.6985	0.7019	0.7054	0.7088	0.7123	0.7157	0.7190	0.7224
0.6	0.7257	0.7291	0.7324	0.7357	0.7389	0.7422	0.7454	0.7486	0.7517	0.7549
0.7	0.7580	0.7611	0.7642	0.7673	0.7704	0.7734	0.7764	0.7794	0.7823	0.7852
0.8	0.7881	0.7910	0.7939	0.7967	0.7995	0.8023	0.8051	0.8078	0.8106	0.8133
0.9	0.8159	0.8186	0.8212	0.8238	0.8264	0.8289	0.8315	0.8340	0.8365	0.8389
1.0	0.8413	0.8438	0.8461	0.8485	0.8508	0.8531	0.8554	0.8577	0.8599	0.8621
1.1	0.8643	0.8665	0.8686	0.8708	0.8729	0.8749	0.8770	0.8790	0.8810	0.8830
1.2	0.8849	0.8869	0.8888	0.8907	0.8925	0.8944	0.8962	0.8980	0.8997	0.9015
1.3	0.9032	0.9049	0.9066	0.9082	0.9099	0.9115	0.9131	0.9147	0.9162	0.9177
1.4	0.9192	0.9207	0.9222	0.9236	0.9251	0.9265	0.9278	0.9292	0.9306	0.9319
1.5	0.9332	0.9345	0.9357	0.9370	0.9382	0.9394	0.9406	0.9418	0.9429	0.9441
1.6	0.9452	0.9463	0.9474	0.9484	0.9495	0.9505	0.9515	0.9525	0.9535	0.9545
1.7	0.9554	0.9564	0.9573	0.9582	0.9591	0.9599	0.9608	0.9616	0.9625	0.9633
1.8	0.9641	0.9649	0.9656	0.9664	0.9671	0.9678	0.9686	0.9693	0.9699	0.9706
1.9	0.9713	0.9719	0.9726	0.9732	0.9738	0.9744	0.9750	0.9756	0.9761	0.9767
2.0	0.9772	0.9778	0.9783	0.9788	0.9793	0.9798	0.9803	0.9808	0.9812	0.9817
2.1	0.9821	0.9826	0.9830	0.9834	0.9838	0.9842	0.9846	0.9850	0.9854	0.9857
2.2	0.9861	0.9864	0.9868	0.9871	0.9875	0.9878	0.9881	0.9884	0.9887	0.9890
2.3	0.9893	0.9896	0.9898	0.9901	0.9904	0.9906	0.9909	0.9911	0.9913	0.9916
2.4	0.9918	0.9920	0.9922	0.9925	0.9927	0.9929	0.9931	0.9932	0.9934	0.9936
2.5	0.9938	0.9940	0.9941	0.9943	0.9945	0.9946	0.9948	0.9949	0.9951	0.9952
2.6	0.9953	0.9955	0.9956	0.9957	0.9959	0.9960	0.9961	0.9962	0.9963	0.9964
2.7	0.9965	0.9966	0.9967	0.9968	0.9969	0.9970	0.9971	0.9972	0.9973	0.9974
2.8	0.9974	0.9975	0.9976	0.9977	0.9977	0.9978	0.9979	0.9979	0.9980	0.9981
2.9	0.9981	0.9982	0.9982	0.9983	0.9984	0.9984	0.9985	0.9985	0.9986	0.9986
3.0	0.9987	0.9987	0.9987	0.9988	0.9988	0.9989	0.9989	0.9989	0.9990	0.9990
3.1	0.9990	0.9991	0.9991	0.9991	0.9992	0.9992	0.9992	0.9992	0.9993	0.9993
3.2	0.9993	0.9993	0.9994	0.9994	0.9994	0.9994	0.9994	0.9995	0.9995	0.9995
3.3	0.9995	0.9995	0.9995	0.9996	0.9996	0.9996	0.9996	0.9996	0.9996	0.9997
3.4	0.9997	0.9997	0.9997	0.9997	0.9997	0.9997	0.9997	0.9997	0.9997	0.9998

Table III
Poisson distribution function

$$F(t) = P[X \leq t]$$

[t]	.50	1.0	2.0	3.0	4.0	5.0	6.0	7.0	8.0	9.0
0	.607	.368	.135	.050	.018	.007	.002	.001	.000	.000
1	.910	.736	.406	.199	.092	.040	.017	.007	.003	.001
2	.986	.920	.677	.423	.238	.125	.062	.030	.014	.006
3	.998	.981	.857	.647	.433	.265	.151	.082	.042	.021
4	1.000	.996	.947	.815	.629	.440	.285	.173	.100	.055
5	1.000	.999	.983	.961	.785	.616	.446	.301	.191	.116
6	1.000	1.000	.995	.966	.889	.762	.606	.450	.313	.207
7	1.000	1.000	.999	.988	.949	.867	.744	.599	.453	.324
8	1.000	1.000	1.000	.996	.979	.932	.847	.729	.593	.456
9	1.000	1.000	1.000	.999	.992	.968	.916	.830	.717	.587
10	1.000	1.000	1.000	1.000	.997	.986	.957	.901	.816	.706
11	1.000	1.000	1.000	1.000	.999	.995	.980	.947	.888	.803
12	1.000	1.000	1.000	1.000	1.000	.998	.991	.973	.936	.876
13	1.000	1.000	1.000	1.000	1.000	.999	.996	.987	.966	.926
14	1.000	1.000	1.000	1.000	1.000	1.000	.999	.994	.983	.959
15	1.000	1.000	1.000	1.000	1.000	1.000	.999	.998	.992	.978
16	1.000	1.000	1.000	1.000	1.000	1.000	1.000	.999	.996	.989
17	1.000	1.000	1.000	1.000	1.000	1.000	1.000	1.000	.998	.995
18	1.000	1.000	1.000	1.000	1.000	1.000	1.000	1.000	.999	.998
19	1.000	1.000	1.000	1.000	1.000	1.000	1.000	1.000	1.000	.999
20	1.000	1.000	1.000	1.000	1.000	1.000	1.000	1.000	1.000	1.000

(Table continues on following page.)

Table III
Poisson distribution function *(continued)*

$$F(t) = P[X \leqslant t]$$

[*t*]	10.0	11.0	12.0	13.0	14.0	15.0
				λ*s*		
2	.003	.001	.001	.000	.000	.000
3	.010	.005	.002	.001	.000	.000
4	.029	.015	.008	.004	.002	.001
5	.067	.038	.020	.011	.006	.003
6	.130	.079	.046	.026	.014	.008
7	.220	.143	.090	.054	.032	.018
8	.333	.232	.155	.100	.062	.037
9	.458	.341	.242	.166	.109	.070
10	.583	.460	.347	.252	.176	.118
11	.697	.579	.462	.353	.260	.185
12	.792	.689	.576	.463	.358	.268
13	.864	.781	.682	.573	.464	.363
14	.917	.854	.772	.675	.570	.466
15	.951	.907	.844	.764	.669	.568
16	.973	.944	.899	.835	.756	.664
17	.986	.968	.937	.890	.827	.749
18	.993	.982	.963	.930	.883	.819
19	.997	.991	.979	.957	.923	.875
20	.998	.995	.988	.975	.952	.917
21	.999	.998	.994	.986	.971	.947
22	1.000	.999	.997	.992	.983	.967
23	1.000	1.000	.999	.996	.991	.981
24	1.000	1.000	.999	.998	.995	.989
25	1.000	1.000	1.000	.999	.997	.994
26	1.000	1.000	1.000	1.000	.999	.997
27	1.000	1.000	1.000	1.000	.999	.998
28	1.000	1.000	1.000	1.000	1.000	.999
29	1.000	1.000	1.000	1.000	1.000	1.000

Table IV
Random digits

Lines	(1)	(2)	(3)	(4)	(5)	(6)	(7)	(8)	(9)	(10)	(11)	(12)	(13)	(14)
1	10480	15011	01536	02011	81647	91646	69179	14194	62590	36207	20969	99570	91291	90700
2	22368	46573	25595	85393	30995	89198	27982	53402	93965	34095	52666	19174	39615	99505
3	24130	48360	22527	97265	76393	64809	15179	24830	49340	32081	30680	19655	63348	58629
4	42167	93093	06243	61680	07856	16376	39440	53537	71341	57004	00849	74917	97758	16379
5	37570	39975	81837	16656	06121	91782	60468	81305	49684	60672	14110	06927	01263	54613
6	77921	06907	11008	42751	27756	53498	18602	70659	90655	15053	21916	81825	44394	42880
7	99562	72905	56420	69994	98872	31016	71194	18738	44013	48840	63213	21069	10634	12952
8	96301	91977	05463	07972	18876	20922	94595	56869	69014	60045	18425	84903	42508	32307
9	89579	14342	63661	10281	17453	18103	57740	84378	25331	12566	58678	44947	05585	56941
10	85475	36857	43342	53988	53060	59533	38867	62300	08158	17983	16439	11458	18593	64952
11	28918	69578	88231	33276	70997	79936	56865	05859	90106	31595	01547	85590	91610	78188
12	63553	40961	48235	03427	49626	69445	18663	72695	52180	20847	12234	90511	33703	90322
13	09429	93969	52636	92737	88974	33488	36320	17617	30015	08272	84115	27156	30613	74952
14	10365	61129	87529	85689	48237	52267	67689	93394	01511	26358	85104	20285	29975	89868
15	07119	97336	71048	08178	77233	13916	47564	81056	97735	85977	29372	74461	28551	90707
16	51085	12765	51821	51259	77452	16308	60756	92144	49442	53900	70960	63990	75601	40719
17	02368	21382	52404	60268	89368	19885	55322	44819	01188	65255	64835	44919	05944	55157
18	01011	54092	33362	94904	31273	04146	18594	29852	71585	85030	51132	01915	92747	64951
19	52162	53916	46369	58586	23216	14513	83149	98736	23495	64350	94738	17752	35156	35749
20	07056	97628	33787	09998	42698	06691	76988	13602	51851	46104	88916	19509	25625	58104
21	48663	91245	85828	14346	09172	30168	90229	04734	59193	22178	30421	61666	99904	32812
22	54164	58492	22421	74103	47070	25306	76468	26384	58151	06646	21524	15227	96909	44592
23	32639	32363	05597	24200	13363	38005	94342	28728	35806	06912	17012	64161	18296	22851
24	29334	27001	87637	87308	58731	00256	45834	15398	46557	41135	10367	07684	36188	18510
25	02488	33062	28834	07351	19731	92420	60952	61280	50001	67658	32586	86679	50720	94953
26	81525	72295	04839	96423	24878	82651	66566	14778	76797	14780	13300	87074	79666	95725
27	29676	20591	68086	26432	46901	20849	89768	81536	86645	12659	92259	57102	80428	25280
28	00742	57392	39064	66432	84673	40027	32832	61362	98947	96067	64760	64584	96096	98253
29	05366	04213	25669	26422	44407	44048	37937	63904	45766	66134	75470	63520	34693	90449
30	91921	26418	64117	94305	26766	25940	39972	22209	71500	64568	91402	42416	07844	69618
31	00582	04711	87917	77341	42206	35126	74087	99547	81817	42607	43808	76655	62028	76630
32	00725	69884	62797	56170	86324	88072	76222	36086	84637	93161	76038	65855	77919	88006
33	69011	65797	95876	55293	18988	27354	26575	08625	40801	59920	29841	80150	12777	48501
34	25976	57948	29888	88604	67917	48708	18912	82271	65424	69774	33611	54262	85963	03547
35	09763	83473	73577	12908	30883	18317	28290	35797	05998	41688	34952	37888	38917	88050
36	91567	42595	27958	30134	04024	86385	29880	99730	55536	84855	29080	09250	79656	73211
37	17955	56349	90999	49127	20044	59931	06115	20542	18059	02008	73708	83517	36103	42791
38	46503	18584	18845	49618	02304	51038	20655	58727	28168	15475	56942	53389	20562	87338
39	92157	89634	94824	78171	84610	82834	09922	25417	44137	48413	25555	21246	35509	20468
40	14577	62765	35605	81263	39667	47358	56873	56307	61607	49518	89656	20103	77490	18062

(Table continues on following page.)

Table IV
Random digits *(continued)*

Lines	(1)	(2)	(3)	(4)	(5)	(6)	(7)	(8)	(9)	(10)	(11)	(12)	(13)	(14)
						Columns								
41	98427	07523	33362	64270	01638	92477	66969	98420	04880	45585	46565	04102	46880	45709
42	34914	63976	88720	82765	34476	17032	87589	40836	32427	70002	70683	88863	77775	69348
43	70060	28277	39475	46473	23219	53416	94970	25832	69975	94884	19661	72828	00102	66794
44	53976	54914	06990	67245	68350	82948	11398	42878	80287	88267	47363	46634	06541	97809
45	76072	29515	40980	07391	58745	25774	22987	80059	39911	96189	41151	14222	60697	59583
46	90725	52210	83974	29992	65831	38857	50490	83765	55657	14361	31720	57375	56228	41546
47	64364	67412	33339	31926	14883	24413	59744	92351	97473	89286	35931	04110	23726	51900
48	08962	00358	31662	25388	61642	34072	81249	35648	56891	69352	48373	45578	78547	81788
49	95012	68379	93526	70765	10593	04542	76463	54328	02349	17247	28865	14777	62730	92277
50	15664	10493	20492	38391	91132	21999	59516	81652	27195	48223	46751	22923	32261	85653

Reprinted with permission from Beyer, W.H., Ed., in *CRC Handbook of Tables for Probability and Statistics*, 2nd ed. Copyright © CRC Press, Inc., Boca Raton, Florida.

Table V

Patient	Sex	Systolic Pressure	Diastolic Pressure	Patient	Sex	Systolic Pressure	Diastolic Pressure
1	M	112	80	31	M	120	80
2	F	112	80	32	M	110	84
3	M	130	76	33	M	115	70
4	M	157	86	34	F	120	100
5	F	152	80	35	M	142	94
6	F	130	80	36	F	120	85
7	M	140	90	37	M	102	64
8	M	120	74	38	M	110	78
9	F	100	82	39	F	118	80
10	M	118	100	40	M	120	80
11	F	120	90	41	F	120	78
12	M	122	72	42	M	126	76
13	F	130	80	43	M	144	100
14	M	96	65	44	F	140	95
15	F	90	56	45	M	116	80
16	M	102	70	46	M	108	80
17	F	115	80	47	F	136	98
18	M	130	80	48	M	135	85
19	M	136	94	49	M	114	76
20	F	130	80	50	F	150	90
21	M	122	80	51	F	110	50
22	M	118	84	52	M	120	70
23	M	108	76	53	M	120	80
24	M	110	86	54	F	122	80
25	F	140	88	55	M	120	80
26	M	130	80	56	M	165	70
27	F	120	80	57	M	120	80
28	M	120	76	58	F	136	90
29	M	125	80	59	M	140	70
30	F	130	80	60	M	110	66

(Table continues on following page.)

Table V (continued)

Patient	Sex	Systolic Pressure	Diastolic Pressure	Patient	Sex	Systolic Pressure	Diastolic Pressure
61	F	142	80	91	F	110	72
62	F	120	80	92	M	120	75
63	M	136	86	93	F	110	80
64	M	142	80	94	M	120	80
65	F	95	80	95	M	116	80
66	M	110	80	96	F	122	95
67	F	118	80	97	M	130	90
68	M	97	75	98	M	110	70
69	M	120	84	99	F	150	80
70	F	110	60	100	M	122	80
71	M	110	80	101	M	108	58
72	F	118	88	102	F	132	88
73	M	117	76	103	M	110	70
74	M	100	78	104	M	122	82
75	F	152	88	105	F	140	100
76	M	112	80	106	M	138	96
77	F	160	88	107	M	127	90
78	M	110	75	108	F	130	80
79	M	128	85	109	M	130	80
80	F	118	90	110	M	110	70
81	M	140	100	111	F	130	90
82	F	90	58	112	M	130	82
83	M	110	70	113	M	160	105
84	F	102	60	114	F	132	80
85	M	140	90	115	M	120	80
86	F	110	80	116	M	123	73
87	M	116	74	117	F	110	80
88	M	118	81	118	M	112	80
89	F	140	85	119	F	118	72
90	M	106	70	120	M	140	78

Table VI
Cumulative T distribution

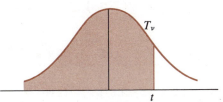

$$P[T_\nu \leq t]$$

ν	F							
	0.60	0.75	0.90	0.95	0.975	0.99	0.995	0.9995
1	0.325	1.000	3.078	6.314	12.706	31.821	63.657	636.619
2	0.289	0.816	1.886	2.920	4.303	6.965	9.925	31.598
3	0.277	0.765	1.638	2.353	3.182	4.541	5.841	12.924
4	0.271	0.741	1.533	2.132	2.776	3.747	4.604	8.610
5	0.267	0.727	1.476	2.015	2.571	3.365	4.032	6.869
6	0.265	0.718	1.440	1.943	2.447	3.143	3.707	5.959
7	0.263	0.711	1.415	1.895	2.365	2.998	3.499	5.408
8	0.262	0.706	1.397	1.860	2.306	2.896	3.355	5.041
9	0.261	0.703	1.383	1.833	2.262	2.821	3.250	4.781
10	0.260	0.700	1.372	1.812	2.228	2.764	3.169	4.587
11	0.260	0.697	1.363	1.796	2.201	2.718	3.106	4.437
12	0.259	0.695	1.356	1.782	2.179	2.681	3.055	4.318
13	0.259	0.694	1.350	1.771	2.160	2.650	3.012	4.221
14	0.258	0.692	1.345	1.761	2.145	2.624	2.977	4.140
15	0.258	0.691	1.341	1.753	2.131	2.602	2.947	4.073
16	0.258	0.690	1.337	1.746	2.120	2.583	2.921	4.015
17	0.257	0.689	1.333	1.740	2.110	2.567	2.898	3.965
18	0.257	0.688	1.330	1.734	2.101	2.552	2.878	3.922
19	0.257	0.688	1.328	1.729	2.093	2.539	2.861	3.883
20	0.257	0.687	1.325	1.725	2.086	2.528	2.845	3.850
21	0.257	0.686	1.323	1.721	2.080	2.518	2.831	3.819
22	0.256	0.686	1.321	1.717	2.074	2.508	2.819	3.792
23	0.256	0.685	1.319	1.714	2.069	2.500	2.807	3.767
24	0.256	0.685	1.318	1.711	2.064	2.492	2.797	3.745
25	0.256	0.684	1.316	1.708	2.060	2.485	2.787	3.725
26	0.256	0.684	1.315	1.706	2.056	2.479	2.779	3.707
27	0.256	0.684	1.314	1.703	2.052	2.473	2.771	3.690
28	0.256	0.683	1.313	1.701	2.048	2.467	2.763	3.674
29	0.256	0.683	1.311	1.699	2.045	2.462	2.756	3.659
30	0.256	0.683	1.310	1.697	2.042	2.457	2.750	3.646

(Table continues on following page.)

Table VI

Cumulative T distribution *(continued)*

$$P[T_\nu \leq t]$$

ν	0.60	0.75	0.90	0.95	0.975	0.99	0.995	0.9995
40	0.255	0.681	1.303	1.684	2.021	2.423	2.704	3.551
60	0.254	0.679	1.296	1.671	2.000	2.390	2.660	3.460
120	0.254	0.677	1.289	1.658	1.980	2.358	2.617	3.373
∞	0.253	0.674	1.282	1.645	1.960	2.326	2.576	3.291

Reprinted with permission from Beyer, W. H., Ed., in *CRC Handbook of Tables for Probability and Statistics*, 2nd ed. Copyright © CRC Press, Inc., Boca Raton, Florida.

Table VII
Cumulative distribution—Chi square

X_v^2

t

F

$P(X_v^2 \le t)$

v	0.005	0.010	0.025	0.050	0.100	0.250	0.500	0.750	0.900	0.950	0.975	0.990	0.995
1	0.0000393	0.000157	0.000982	0.00393	0.0158	0.102	0.455	1.32	2.71	3.84	5.02	6.63	7.88
2	0.0100	0.0201	0.0506	0.103	0.211	0.575	1.39	2.77	4.61	5.99	7.38	9.21	10.6
3	0.0717	0.115	0.216	0.352	0.584	1.21	2.37	4.11	6.25	7.81	9.35	11.3	12.8
4	0.207	0.297	0.484	0.711	1.06	1.92	3.36	5.39	7.78	9.49	11.1	13.3	14.9
5	0.412	0.554	0.831	1.15	1.61	2.67	4.35	6.63	9.24	11.1	12.8	15.1	16.7
6	0.676	0.872	1.24	1.64	2.20	3.45	5.35	7.84	10.6	12.6	14.4	16.8	18.5
7	0.989	1.24	1.69	2.17	2.83	4.25	6.35	9.04	12.0	14.1	16.0	18.5	20.3
8	1.34	1.65	2.18	2.73	3.49	5.07	7.34	10.2	13.4	15.5	17.5	20.1	22.0
9	1.73	2.09	2.70	3.33	4.17	5.90	8.34	11.4	14.7	16.9	19.0	21.7	23.6
10	2.16	2.56	3.25	3.94	4.87	6.74	9.34	12.5	16.0	18.3	20.5	23.2	25.2
11	2.60	3.05	3.82	4.57	5.58	7.58	10.3	13.7	17.3	19.7	21.9	24.7	26.8
12	3.07	3.57	4.40	5.23	6.30	8.44	11.3	14.8	18.5	21.0	23.3	26.2	28.3
13	3.57	4.11	5.01	5.89	7.04	9.30	12.3	16.0	19.8	22.4	24.7	27.7	29.8
14	4.07	4.66	5.63	6.57	7.79	10.2	13.3	17.1	21.1	23.7	26.1	29.1	31.3
15	4.60	5.23	6.26	7.26	8.55	11.0	14.3	18.2	22.3	25.0	27.5	30.6	32.8
16	5.14	5.81	6.91	7.96	9.31	11.9	15.3	19.4	23.5	26.3	28.8	32.0	34.3
17	5.70	6.41	7.56	8.67	10.1	12.8	16.3	20.5	24.8	27.6	30.2	33.4	35.7
18	6.26	7.01	8.23	9.39	10.9	13.7	17.3	21.6	26.0	28.9	31.5	34.8	37.2
19	6.84	7.63	8.91	10.1	11.7	14.6	18.3	22.7	27.2	30.1	32.9	36.2	38.6
20	7.43	8.26	9.59	10.9	12.4	15.5	19.3	23.8	28.4	31.4	34.2	37.6	40.0

(Table continues on following page.)

Table VII
Cumulative distribution—Chi square (*continued*)

$$P(X_\nu^2 \le t)$$

ν	0.005	0.010	0.025	0.050	0.100	0.250	0.500	0.750	0.900	0.950	0.975	0.990	0.995
							F						
21	8.03	8.90	10.3	11.6	13.2	16.3	20.3	24.9	29.6	32.7	35.5	38.9	41.4
22	8.64	9.54	11.0	12.3	14.0	17.2	21.3	26.0	30.8	33.9	36.8	40.3	42.8
23	9.26	10.2	11.7	13.1	14.8	18.1	22.3	27.1	32.0	35.2	38.1	41.6	44.2
24	9.89	10.9	12.4	13.8	15.7	19.0	23.3	28.2	33.2	36.4	39.4	43.0	45.6
25	10.5	11.5	13.1	14.6	16.5	19.9	24.3	29.3	34.4	37.7	40.6	44.3	46.9
26	11.2	12.2	13.8	15.4	17.3	20.8	25.3	30.4	35.6	38.9	41.9	45.6	48.3
27	11.8	12.9	14.6	16.2	18.1	21.7	26.3	31.5	36.7	40.1	43.2	47.0	49.6
28	12.5	13.6	15.3	16.9	18.9	22.7	27.3	32.6	37.9	41.3	44.5	48.3	51.0
29	13.1	14.3	16.0	17.7	19.8	23.6	28.3	33.7	39.1	42.6	45.7	49.6	52.3
30	13.8	15.0	16.8	18.5	20.6	24.5	29.3	34.8	40.3	43.8	47.0	50.9	53.7

Table VIII
Cumulative distribution: F

$$P[F_{\nu_1, \nu_2} \leq t] = .9$$

ν_1

ν_2	1	2	3	4	5	6	7	8	9	10	12	15	20	24	30	40	60	120	∞
1	39.86	49.50	53.59	55.83	57.24	58.20	58.91	59.44	59.86	60.19	60.71	61.22	61.74	62.00	62.26	62.53	62.79	63.06	63.33
2	8.53	9.00	9.16	9.24	9.29	9.33	9.35	9.37	9.38	9.39	9.41	9.42	9.44	9.45	9.46	9.47	9.47	9.48	9.49
3	5.54	5.46	5.39	5.34	5.31	5.28	5.27	5.25	5.24	5.23	5.22	5.20	5.18	5.18	5.17	5.16	5.15	5.14	5.13
4	4.54	4.32	4.19	4.11	4.05	4.01	3.98	3.95	3.94	3.92	3.90	3.87	3.84	3.83	3.82	3.80	3.79	3.78	3.76
5	4.06	3.78	3.62	3.52	3.45	3.40	3.37	3.34	3.32	3.30	3.27	3.24	3.21	3.19	3.17	3.16	3.14	3.12	3.10
6	3.78	3.46	3.29	3.18	3.11	3.05	3.01	2.98	2.96	2.94	2.90	2.87	2.84	2.82	2.80	2.78	2.76	2.74	2.72
7	3.59	3.26	3.07	2.96	2.88	2.83	2.78	2.75	2.72	2.70	2.67	2.63	2.59	2.58	2.56	2.54	2.51	2.49	2.47
8	3.46	3.11	2.92	2.81	2.73	2.67	2.62	2.59	2.56	2.54	2.50	2.46	2.42	2.40	2.38	2.36	2.34	2.32	2.29
9	3.36	3.01	2.81	2.69	2.61	2.55	2.51	2.47	2.44	2.42	2.38	2.34	2.30	2.28	2.25	2.23	2.21	2.18	2.16
10	3.29	2.92	2.73	2.61	2.52	2.46	2.41	2.38	2.35	2.32	2.28	2.24	2.20	2.18	2.16	2.13	2.11	2.08	2.06
11	3.23	2.86	2.66	2.54	2.45	2.39	2.34	2.30	2.27	2.25	2.21	2.17	2.12	2.10	2.08	2.05	2.03	2.00	1.97
12	3.18	2.81	2.61	2.48	2.39	2.33	2.28	2.24	2.21	2.19	2.15	2.10	2.06	2.04	2.01	1.99	1.96	1.93	1.90
13	3.14	2.76	2.56	2.43	2.35	2.28	2.23	2.20	2.16	2.14	2.10	2.05	2.01	1.98	1.96	1.93	1.90	1.88	1.85
14	3.10	2.73	2.52	2.39	2.31	2.24	2.19	2.15	2.12	2.10	2.05	2.01	1.96	1.94	1.91	1.89	1.86	1.83	1.80
15	3.07	2.70	2.49	2.36	2.27	2.21	2.16	2.12	2.09	2.06	2.02	1.97	1.92	1.90	1.87	1.85	1.82	1.79	1.76
16	3.05	2.67	2.46	2.33	2.24	2.18	2.13	2.09	2.06	2.03	1.99	1.94	1.89	1.87	1.84	1.81	1.78	1.75	1.72
17	3.03	2.64	2.44	2.31	2.22	2.15	2.10	2.06	2.03	2.00	1.96	1.91	1.86	1.84	1.81	1.78	1.75	1.72	1.69
18	3.01	2.62	2.42	2.29	2.20	2.13	2.08	2.04	2.00	1.98	1.93	1.89	1.84	1.81	1.78	1.75	1.72	1.69	1.66
19	2.99	2.61	2.40	2.27	2.18	2.11	2.06	2.02	1.98	1.96	1.91	1.86	1.81	1.79	1.76	1.73	1.70	1.67	1.63

(Table continues on following page.)

Table VIII
Cumulative distribution: F (continued)

$$P[F_{v_1, v_2} \leq t] = .9$$

v_1

v_2	1	2	3	4	5	6	7	8	9	10	12	15	20	24	30	40	60	120	∞
20	2.97	2.59	2.38	2.25	2.16	2.09	2.04	2.00	1.96	1.94	1.89	1.84	1.79	1.77	1.74	1.71	1.68	1.64	1.61
21	2.96	2.57	2.36	2.23	2.14	2.08	2.02	1.98	1.95	1.92	1.87	1.83	1.78	1.75	1.72	1.69	1.66	1.62	1.59
22	2.95	2.56	2.35	2.22	2.13	2.06	2.01	1.97	1.93	1.90	1.86	1.81	1.76	1.73	1.70	1.67	1.64	1.60	1.57
23	2.94	2.55	2.34	2.21	2.11	2.05	1.99	1.95	1.92	1.89	1.84	1.80	1.74	1.72	1.69	1.66	1.62	1.59	1.53
24	2.93	2.54	2.33	2.19	2.10	2.04	1.98	1.94	1.91	1.88	1.83	1.78	1.73	1.70	1.67	1.64	1.61	1.57	1.55
25	2.92	2.53	2.32	2.18	2.09	2.02	1.97	1.93	1.89	1.87	1.82	1.77	1.72	1.69	1.66	1.63	1.59	1.56	1.52
26	2.91	2.52	2.31	2.17	2.08	2.01	1.96	1.92	1.88	1.86	1.81	1.76	1.71	1.68	1.65	1.61	1.58	1.54	1.50
27	2.90	2.51	2.30	2.17	2.07	2.00	1.95	1.91	1.87	1.85	1.80	1.75	1.70	1.67	1.64	1.60	1.57	1.53	1.49
28	2.89	2.50	2.29	2.16	2.06	2.00	1.94	1.90	1.87	1.84	1.79	1.74	1.69	1.66	1.63	1.59	1.56	1.52	1.48
29	2.89	2.50	2.28	2.15	2.06	1.99	1.93	1.89	1.86	1.83	1.78	1.73	1.68	1.65	1.62	1.58	1.55	1.51	1.47
30	2.88	2.49	2.28	2.14	2.05	1.98	1.93	1.88	1.85	1.82	1.77	1.72	1.67	1.64	1.61	1.57	1.54	1.50	1.46
40	2.84	2.44	2.23	2.09	2.00	1.93	1.87	1.83	1.79	1.76	1.71	1.66	1.61	1.57	1.54	1.51	1.47	1.42	1.38
60	2.79	2.39	2.18	2.04	1.95	1.87	1.82	1.77	1.74	1.71	1.66	1.60	1.54	1.51	1.48	1.44	1.40	1.35	1.29
120	2.75	2.35	2.13	1.99	1.90	1.82	1.77	1.72	1.68	1.65	1.60	1.55	1.48	1.45	1.41	1.37	1.32	1.26	1.19
∞	2.71	2.30	2.08	1.94	1.85	1.77	1.72	1.67	1.63	1.60	1.55	1.49	1.42	1.38	1.34	1.30	1.24	1.17	1.00

Table VIII
Cumulative distribution: F (continued)

$$P[F_{\nu_1, \nu_2} \leq t] = .95$$

ν_2	1	2	3	4	5	6	7	8	9	10	12	15	20	24	30	40	60	120	∞
1	161.4	199.5	215.7	224.6	230.2	234.0	236.8	238.9	240.5	241.9	243.9	245.9	248.0	249.1	250.1	251.1	252.2	253.3	254.3
2	18.51	19.00	19.16	19.25	19.30	19.33	19.35	19.37	19.38	19.40	19.41	19.43	19.45	19.45	19.46	19.47	19.48	19.49	19.50
3	10.13	9.55	9.28	9.12	9.01	8.94	8.89	8.85	8.81	8.79	8.74	8.70	8.66	8.64	8.62	8.59	8.57	8.55	8.53
4	7.71	6.94	6.59	6.39	6.26	6.16	6.09	6.04	6.00	5.96	5.91	5.86	5.80	5.77	5.75	5.72	5.69	5.66	5.63
5	6.61	5.79	5.41	5.19	5.05	4.95	4.88	4.82	4.77	4.74	4.68	4.62	4.56	4.53	4.50	4.46	4.43	4.40	4.36
6	5.99	5.14	4.76	4.53	4.39	4.28	4.21	4.15	4.10	4.06	4.00	3.94	3.87	3.84	3.81	3.77	3.74	3.70	3.67
7	5.59	4.74	4.35	4.12	3.97	3.87	3.79	3.73	3.68	3.64	3.57	3.51	3.44	3.41	3.38	3.34	3.30	3.27	3.23
8	5.32	4.46	4.07	3.84	3.69	3.58	3.50	3.44	3.39	3.35	3.28	3.22	3.15	3.12	3.08	3.04	3.01	2.97	2.93
9	5.12	4.26	3.86	3.63	3.48	3.37	3.29	3.23	3.18	3.14	3.07	3.01	2.94	2.90	2.86	2.83	2.79	2.75	2.71
10	4.96	4.10	3.71	3.48	3.33	3.22	3.14	3.07	3.02	2.98	2.91	2.85	2.77	2.74	2.70	2.66	2.62	2.58	2.54
11	4.84	3.98	3.59	3.36	3.20	3.09	3.01	2.95	2.90	2.85	2.79	2.72	2.65	2.61	2.57	2.53	2.49	2.45	2.40
12	4.75	3.89	3.49	3.26	3.11	3.00	2.91	2.85	2.80	2.75	2.69	2.62	2.54	2.51	2.47	2.43	2.38	2.34	2.30
13	4.67	3.81	3.41	3.18	3.03	2.92	2.83	2.77	2.71	2.67	2.60	2.53	2.46	2.42	2.38	2.34	2.30	2.25	2.21
14	4.60	3.74	3.34	3.11	2.96	2.85	2.76	2.70	2.65	2.60	2.53	2.46	2.39	2.35	2.31	2.27	2.22	2.18	2.13
15	4.54	3.68	3.29	3.06	2.90	2.79	2.71	2.64	2.59	2.54	2.48	2.40	2.33	2.29	2.25	2.20	2.16	2.11	2.07
16	4.49	3.63	3.24	3.01	2.85	2.74	2.66	2.59	2.54	2.49	2.42	2.35	2.28	2.24	2.19	2.15	2.11	2.06	2.01
17	4.45	3.59	3.20	2.96	2.81	2.70	2.61	2.55	2.49	2.45	2.38	2.31	2.23	2.19	2.15	2.10	2.06	2.01	1.96
18	4.41	3.55	3.16	2.93	2.77	2.66	2.58	2.51	2.46	2.41	2.34	2.27	2.19	2.15	2.11	2.06	2.02	1.97	1.92
19	4.38	3.52	3.13	2.90	2.74	2.63	2.54	2.48	2.42	2.38	2.31	2.23	2.16	2.11	2.07	2.03	1.98	1.93	1.88

(Table continues on following page.)

Table VIII
Cumulative distribution: F *(continued)*

$$P[F_{\nu_1, \nu_2} \le t] = .95$$

ν_2	ν_1																		
	1	2	3	4	5	6	7	8	9	10	12	15	20	24	30	40	60	120	∞
20	4.35	3.49	3.10	2.87	2.71	2.60	2.51	2.45	2.39	2.35	2.28	2.20	2.12	2.08	2.04	1.99	1.95	1.90	1.84
21	4.32	3.47	3.07	2.84	2.68	2.57	2.49	2.42	2.37	2.32	2.25	2.18	2.10	2.05	2.01	1.96	1.92	1.87	1.81
22	4.30	3.44	3.05	2.82	2.66	2.55	2.46	2.40	2.34	2.30	2.23	2.15	2.07	2.03	1.98	1.94	1.89	1.84	1.78
23	4.28	3.42	3.03	2.80	2.64	2.53	2.44	2.37	2.32	2.27	2.20	2.13	2.05	2.01	1.96	1.91	1.86	1.81	1.76
24	4.26	3.40	3.01	2.78	2.62	2.51	2.42	2.36	2.30	2.25	2.18	2.11	2.03	1.98	1.94	1.89	1.84	1.79	1.73
25	4.24	3.39	2.99	2.76	2.60	2.49	2.40	2.34	2.28	2.24	2.16	2.09	2.01	1.96	1.92	1.87	1.82	1.77	1.71
26	4.23	3.37	2.98	2.74	2.59	2.47	2.39	2.32	2.27	2.22	2.15	2.07	1.99	1.95	1.90	1.85	1.80	1.75	1.69
27	4.21	3.35	2.96	2.73	2.57	2.46	2.37	2.31	2.25	2.20	2.13	2.06	1.97	1.93	1.88	1.84	1.79	1.73	1.67
28	4.20	3.34	2.95	2.71	2.56	2.45	2.36	2.29	2.24	2.19	2.12	2.04	1.96	1.91	1.87	1.82	1.77	1.71	1.65
29	4.18	3.33	2.93	2.70	2.55	2.43	2.35	2.28	2.22	2.18	2.10	2.03	1.94	1.90	1.85	1.81	1.75	1.70	1.64
30	4.17	3.32	2.92	2.69	2.53	2.42	2.33	2.27	2.21	2.16	2.09	2.01	1.93	1.89	1.84	1.79	1.74	1.68	1.62
40	4.08	3.23	2.84	2.61	2.45	2.34	2.25	2.18	2.12	2.08	2.00	1.92	1.84	1.79	1.74	1.69	1.64	1.58	1.51
60	4.00	3.15	2.76	2.53	2.37	2.25	2.17	2.10	2.04	1.99	1.92	1.84	1.75	1.70	1.65	1.59	1.53	1.47	1.39
120	3.92	3.07	2.68	2.45	2.29	2.17	2.09	2.02	1.96	1.91	1.83	1.75	1.66	1.61	1.55	1.50	1.43	1.35	1.25
∞	3.84	3.00	2.60	2.37	2.21	2.10	2.01	1.94	1.88	1.83	1.75	1.67	1.57	1.52	1.46	1.39	1.32	1.22	1.00

Table VIII
Cumulative distribution: F (continued)

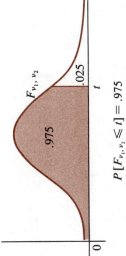

$$F_{\nu_1, \nu_2}$$

.975

.025

$P[F_{\nu_1, \nu_2} \leq t] = .975$

ν_2	1	2	3	4	5	6	7	8	9	10	12	15	20	24	30	40	60	120	∞
1	647.8	799.5	864.2	899.6	921.8	937.1	948.2	956.7	963.3	968.6	976.7	984.9	993.1	997.2	1001	1006	1010	1014	1018
2	38.51	39.00	39.17	39.25	39.30	39.33	39.36	39.37	39.39	39.40	39.41	39.43	39.45	39.46	39.46	39.47	39.48	39.49	39.50
3	17.44	16.04	15.44	15.10	14.88	14.73	14.62	14.54	14.47	14.42	14.34	14.25	14.17	14.12	14.08	14.04	13.99	13.95	13.90
4	12.22	10.65	9.98	9.60	9.36	9.20	9.07	8.98	8.90	8.84	8.75	8.66	8.56	8.51	8.46	8.41	8.36	8.31	8.26
5	10.01	8.43	7.76	7.39	7.15	6.98	6.85	6.76	6.68	6.62	6.52	6.43	6.33	6.28	6.23	6.18	6.12	6.07	6.02
6	8.81	7.26	6.60	6.23	5.99	5.82	5.70	5.60	5.52	5.46	5.37	5.27	5.17	5.12	5.07	5.01	4.96	4.90	4.85
7	8.07	6.54	5.89	5.52	5.29	5.12	4.99	4.90	4.82	4.76	4.67	4.57	4.47	4.42	4.36	4.31	4.25	4.20	4.14
8	7.57	6.06	5.42	5.05	4.82	4.65	4.53	4.43	4.36	4.30	4.20	4.10	4.00	3.95	3.89	3.84	3.78	3.73	3.67
9	7.21	5.71	5.08	4.72	4.48	4.32	4.20	4.10	4.03	3.96	3.87	3.77	3.67	3.61	3.56	3.51	3.45	3.39	3.33
10	6.94	5.46	4.83	4.47	4.24	4.07	3.95	3.85	3.78	3.72	3.62	3.52	3.42	3.37	3.31	3.26	3.20	3.14	3.08
11	6.72	5.26	4.63	4.28	4.04	3.88	3.76	3.66	3.59	3.53	3.43	3.33	3.23	3.17	3.12	3.06	3.00	2.94	2.88
12	6.55	5.10	4.47	4.12	3.89	3.73	3.61	3.51	3.44	3.37	3.28	3.18	3.07	3.02	2.96	2.91	2.85	2.79	2.72
13	6.41	4.97	4.35	4.00	3.77	3.60	3.48	3.39	3.31	3.25	3.15	3.05	2.95	2.89	2.84	2.78	2.72	2.66	2.60
14	6.30	4.86	4.24	3.89	3.66	3.50	3.38	3.29	3.21	3.15	3.05	2.95	2.84	2.79	2.73	2.67	2.61	2.55	2.49
15	6.20	4.77	4.15	3.80	3.58	3.41	3.29	3.20	3.12	3.06	2.96	2.86	2.76	2.70	2.64	2.59	2.52	2.46	2.40
16	6.12	4.69	4.08	3.73	3.50	3.34	3.22	3.12	3.05	2.99	2.89	2.79	2.68	2.63	2.57	2.51	2.45	2.38	2.32
17	6.04	4.62	4.01	3.66	3.44	3.28	3.16	3.06	2.98	2.92	2.82	2.72	2.62	2.56	2.50	2.44	2.38	2.32	2.25
18	5.98	4.56	3.95	3.61	3.38	3.22	3.10	3.01	2.93	2.87	2.77	2.67	2.56	2.50	2.44	2.38	2.32	2.26	2.19
19	5.92	4.51	3.90	3.56	3.33	3.17	3.05	2.96	2.88	2.82	2.72	2.62	2.51	2.45	2.39	2.33	2.27	2.20	2.13

ν_1

(Table continues on following page.)

A19

Table VIII
Cumulative distribution: *F (continued)*

$$P[F_{v_1, v_2} \leq t] = .975$$

v_2 \ v_1	1	2	3	4	5	6	7	8	9	10	12	15	20	24	30	40	60	120	∞
20	5.87	4.46	3.86	3.51	3.29	3.13	3.01	2.91	2.84	2.77	2.68	2.57	2.46	2.41	2.35	2.29	2.22	2.16	2.09
21	5.83	4.42	3.82	3.48	3.25	3.09	2.97	2.87	2.80	2.73	2.64	2.53	2.42	2.37	2.31	2.25	2.18	2.11	2.04
22	5.79	4.38	3.78	3.44	3.22	3.05	2.93	2.84	2.76	2.70	2.60	2.50	2.39	2.33	2.27	2.21	2.14	2.08	2.00
23	5.75	4.35	3.75	3.41	3.18	3.02	2.90	2.81	2.73	2.67	2.57	2.47	2.36	2.30	2.24	2.18	2.11	2.04	1.97
24	5.72	4.32	3.72	3.38	3.15	2.99	2.87	2.78	2.70	2.64	2.54	2.44	2.33	2.27	2.21	2.15	2.08	2.01	1.94
25	5.69	4.29	3.69	3.35	3.13	2.97	2.85	2.75	2.68	2.61	2.51	2.41	2.30	2.24	2.18	2.12	2.05	1.98	1.91
26	5.66	4.27	3.67	3.33	3.10	2.94	2.82	2.73	2.65	2.59	2.49	2.39	2.28	2.22	2.16	2.09	2.03	1.95	1.88
27	5.63	4.24	3.65	3.31	3.08	2.92	2.80	2.71	2.63	2.57	2.47	2.36	2.25	2.19	2.13	2.07	2.00	1.93	1.85
28	5.61	4.22	3.63	3.29	3.06	2.90	2.78	2.69	2.61	2.55	2.45	2.34	2.23	2.17	2.11	2.05	1.98	1.91	1.83
29	5.59	4.20	3.61	3.27	3.04	2.88	2.76	2.67	2.59	2.53	2.43	2.32	2.21	2.15	2.09	2.03	1.96	1.89	1.81
30	5.57	4.18	3.59	3.25	3.03	2.87	2.75	2.65	2.57	2.51	2.41	2.31	2.20	2.14	2.07	2.01	1.94	1.87	1.79
40	5.42	4.05	3.46	3.13	2.90	2.74	2.62	2.53	2.45	2.39	2.29	2.18	2.07	2.01	1.94	1.88	1.80	1.72	1.64
60	5.29	3.93	3.34	3.01	2.79	2.63	2.51	2.41	2.33	2.27	2.17	2.06	1.94	1.88	1.82	1.74	1.67	1.58	1.48
120	5.15	3.80	3.23	2.89	2.67	2.52	2.39	2.30	2.22	2.16	2.05	1.94	1.82	1.76	1.69	1.61	1.53	1.43	1.31
∞	5.02	3.69	3.12	2.79	2.57	2.41	2.29	2.19	2.11	2.05	1.94	1.83	1.71	1.64	1.57	1.48	1.39	1.27	1.00

Table VIII
Cumulative distribution: F (continued)

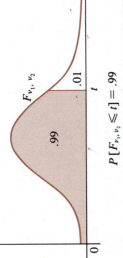

F_{ν_1, ν_2}

$.99$

$.01$

t

$P[F_{\nu_1, \nu_2} \leq t] = .99$

ν_2	1	2	3	4	5	6	7	8	9	10	12	15	20	24	30	40	60	120	∞
1	4052	4999.5	5403	5625	5764	5859	5928	5982	6022	6056	6106	6157	6209	6235	6261	6287	6313	6339	6366
2	98.50	99.00	99.17	99.25	99.30	99.33	99.36	99.37	99.39	99.40	99.42	99.43	99.45	99.46	99.47	99.47	99.48	99.49	99.50
3	34.12	30.82	29.46	28.71	28.24	27.91	27.67	27.49	27.35	27.23	27.05	26.87	26.69	26.60	26.50	26.41	26.32	26.22	26.13
4	21.20	18.00	16.69	15.98	15.52	15.21	14.98	14.80	14.66	14.55	14.37	14.20	14.02	13.93	13.84	13.75	13.65	13.56	13.46
5	16.26	13.27	12.06	11.39	10.97	10.67	10.46	10.29	10.16	10.05	9.89	9.72	9.55	9.47	9.38	9.29	9.20	9.11	9.02
6	13.75	10.92	9.78	9.15	8.75	8.47	8.26	8.10	7.98	7.87	7.72	7.56	7.40	7.31	7.23	7.14	7.06	6.97	6.88
7	12.25	9.55	8.45	7.85	7.46	7.19	6.99	6.84	6.72	6.62	6.47	6.31	6.16	6.07	5.99	5.91	5.82	5.74	5.65
8	11.26	8.65	7.59	7.01	6.63	6.37	6.18	6.03	5.91	5.81	5.67	5.52	5.36	5.28	5.20	5.12	5.03	4.95	4.86
9	10.56	8.02	6.99	6.42	6.06	5.80	5.61	5.47	5.35	5.26	5.11	4.96	4.81	4.73	4.65	4.57	4.48	4.40	4.31
10	10.04	7.56	6.55	5.99	5.64	5.39	5.20	5.06	4.94	4.85	4.71	4.56	4.41	4.33	4.25	4.17	4.08	4.00	3.91
11	9.65	7.21	6.22	5.67	5.32	5.07	4.89	4.74	4.63	4.54	4.40	4.25	4.10	4.02	3.94	3.86	3.78	3.69	3.60
12	9.33	6.93	5.95	5.41	5.06	4.82	4.64	4.50	4.39	4.30	4.16	4.01	3.86	3.78	3.70	3.62	3.54	3.45	3.36
13	9.07	6.70	5.74	5.21	4.86	4.62	4.44	4.30	4.19	4.10	3.96	382	3.66	3.59	3.51	3.43	3.34	3.25	3.17
14	8.86	6.51	5.56	5.04	4.69	4.46	4.28	4.14	4.03	3.94	3.80	3.66	3.51	3.43	3.35	3.27	3.18	3.09	3.00
15	8.68	6.36	5.42	4.89	4.56	4.32	4.14	4.00	3.89	3.80	3.67	3.52	3.37	3.29	3.21	3.13	3.05	2.96	2.87
16	8.53	6.23	5.29	4.77	4.44	4.20	4.03	3.89	3.78	3.69	3.55	3.41	3.26	3.18	3.10	3.02	2.93	2.84	2.75
17	8.40	6.11	5.18	4.67	4.34	4.10	3.93	3.79	3.68	3.59	3.46	3.31	3.16	3.08	3.00	2.92	2.83	2.75	2.65
18	8.29	6.01	5.09	4.58	4.25	4.01	3.84	3.71	3.60	3.51	3.37	3.23	3.08	3.00	2.92	2.84	2.75	2.66	2.57
19	8.18	5.93	5.01	4.50	4.17	3.94	3.77	3.63	3.52	3.43	3.30	3.15	3.00	2.92	2.84	2.76	2.67	2.58	2.49

ν_1

(Table continues on following page.)

Table VIII
Cumulative distribution: F *(continued)*

$$P[F_{\nu_1, \nu_2} \leqslant t] = .99$$

ν_2	ν_1 1	2	3	4	5	6	7	8	9	10	12	15	20	24	30	40	60	120	∞
20	8.10	5.85	4.94	4.43	4.10	3.87	3.70	3.56	3.46	3.37	3.23	3.09	2.94	2.86	2.78	2.69	2.61	2.52	2.42
21	8.02	5.78	4.87	4.37	4.04	3.81	3.64	3.51	3.40	3.31	3.17	3.03	2.88	2.80	2.72	2.64	2.55	2.46	2.36
22	7.95	5.72	4.82	4.31	3.99	3.76	3.59	3.45	3.35	3.26	3.12	2.98	2.83	2.75	2.67	2.58	2.50	2.40	2.31
23	7.88	5.66	4.76	4.26	3.94	3.71	3.54	3.41	3.30	3.21	3.07	2.93	2.78	2.70	2.62	2.54	2.45	2.35	2.26
24	7.82	5.61	4.72	4.22	3.90	3.67	3.50	3.36	3.26	3.17	3.03	2.89	2.74	2.66	2.58	2.49	2.40	2.31	2.21
25	7.77	5.57	4.68	4.18	3.85	3.63	3.46	3.32	3.22	3.13	2.99	2.85	2.70	2.62	2.54	2.45	2.36	2.27	2.17
26	7.72	5.53	4.64	4.14	3.82	3.59	3.42	3.29	3.18	3.09	2.96	2.81	2.66	2.58	2.50	2.42	2.33	2.23	2.13
27	7.68	5.49	4.60	4.11	3.78	3.56	3.39	3.26	3.15	3.06	2.93	2.78	2.63	2.55	2.47	2.38	2.29	2.20	2.10
28	7.64	5.45	4.57	4.07	3.75	3.53	3.36	3.23	3.12	3.03	2.90	2.75	2.60	2.52	2.44	2.35	2.26	2.17	2.06
29	7.60	5.42	4.54	4.04	3.73	3.50	3.33	3.20	3.09	3.00	2.87	2.73	2.57	2.49	2.41	2.33	2.23	2.14	2.03
30	7.56	5.39	4.51	4.02	3.70	3.47	3.30	3.17	3.07	2.98	2.84	2.70	2.55	2.47	2.39	2.30	2.21	2.11	2.01
40	7.31	5.18	4.31	3.83	3.51	3.29	3.12	2.99	2.89	2.80	2.66	2.52	2.37	2.29	2.20	2.11	2.02	1.92	1.80
60	7.08	4.98	4.13	3.65	3.34	3.12	2.95	2.82	2.72	2.63	2.50	2.35	2.20	2.12	2.03	1.94	1.84	1.73	1.60
120	6.85	4.79	3.95	3.48	3.17	2.96	2.79	2.66	2.56	2.47	2.34	2.19	2.03	1.95	1.86	1.76	1.66	1.53	1.38
∞	6.63	4.61	3.78	3.32	3.02	2.80	2.64	2.51	2.41	2.32	2.18	2.04	1.88	1.79	1.70	1.59	1.47	1.32	1.00

Reprinted with permission from Beyer, W. H., Ed., in *CRC Handbook of Tables for Probability and Statistics*, 2nd ed. Copyright © CRC Press, Inc., Boca Raton, Florida.

Table IX

	Least significant studentized ranges r_p $\alpha = .05$ P						Least significant studentized ranges r_p $\alpha = 0.01$ P				
r	2	3	4	5	6	r	2	3	4	5	6
1	17.97	17.97	17.97	17.97	17.97	1	90.03	90.03	90.03	90.03	90.03
2	6.085	6.085	6.085	6.085	6.085	2	14.04	14.04	14.04	14.04	14.04
3	4.501	4.516	4.516	4.516	4.516	3	8.261	8.321	8.321	8.321	8.321
4	3.927	4.013	4.033	4.033	4.033	4	6.512	6.677	6.740	6.756	6.756
5	3.635	3.749	3.797	3.814	3.814	5	5.702	5.893	5.898	6.040	6.065
6	3.461	3.587	3.649	3.680	3.694	6	5.243	5.439	5.549	5.614	5.655
7	3.344	3.477	3.548	3.588	3.611	7	4.949	5.145	5.260	5.334	5.383
8	3.261	3.399	3.475	3.521	3.549	8	4.746	4.939	5.057	5.135	5.189
9	3.199	3.339	3.420	3.470	3.502	9	4.596	4.787	4.906	4.986	5.043
10	3.151	3.293	3.376	3.430	3.465	10	4.482	4.671	4.790	4.871	4.931
11	3.113	3.256	3.342	3.397	3.435	11	4.392	4.579	4.697	4.780	4.841
12	3.082	3.225	3.313	3.370	3.410	12	4.320	4.504	4.622	4.706	4.767
13	3.055	3.200	3.289	3.348	3.389	13	4.260	4.442	4.560	4.644	4.705
14	3.033	3.178	3.268	3.329	3.372	14	4.210	4.391	4.508	4.591	4.654
15	3.014	3.160	3.250	3.312	3.356	15	4.168	4.347	4.463	4.547	4.610
16	2.998	3.144	3.235	3.298	3.343	16	4.131	4.309	4.425	4.509	4.572
17	2.984	3.130	3.222	3.285	3.331	17	4.099	4.275	4.391	4.475	4.539
18	2.971	3.118	3.210	3.274	3.321	18	4.071	4.246	4.362	4.445	4.509
19	2.960	3.107	3.199	3.264	3.311	19	4.046	4.220	4.335	4.419	4.483
20	2.950	3.097	3.190	3.255	3.303	20	4.024	4.197	4.312	4.395	4.459
24	2.919	3.066	3.160	3.226	3.276	24	3.956	4.126	4.239	4.322	4.386
30	2.888	3.035	3.131	3.199	3.250	30	3.889	4.506	4.168	4.250	4.314
40	2.858	3.006	3.102	3.171	3.224	40	3.825	3.988	4.098	4.180	4.244
60	2.829	2.976	3.073	3.143	3.198	60	3.762	3.922	4.031	4.111	4.174
120	2.800	2.947	3.045	3.116	3.172	120	3.702	3.858	3.965	4.044	4.107
∞	2.772	2.918	3.017	3.089	3.146	∞	3.643	3.796	3.900	3.978	4.040

Abridgement of H. L. Harter's "Critical Values for Duncan's New Multiple Range Test," *Biometrics*, Vol. 16, No. 4 (1960).

Table X
Critical values of W in the Wilcoxon Signed Rank Test

$n = 5(1)50$

One-sided	Two-sided	$n = 5$	$n = 6$	$n = 7$	$n = 8$	$n = 9$	$n = 10$
$P = .05$	$P = .10$	1	2	4	6	8	11
$P = .025$	$P = .05$		1	2	4	6	8
$P = .01$	$P = .02$			0	2	3	5
$P = .005$	$P = .01$				0	2	3

One-sided	Two-sided	$n = 11$	$n = 12$	$n = 13$	$n = 14$	$n = 15$	$n = 16$
$P = .05$	$P = .10$	14	17	21	26	30	36
$P = .025$	$P = .05$	11	14	17	21	25	30
$P = .01$	$P = .02$	7	10	13	16	20	24
$P = .005$	$P = .01$	5	7	10	13	16	19

One-sided	Two-sided	$n = 17$	$n = 18$	$n = 19$	$n = 20$	$n = 21$	$n = 22$
$P = .05$	$P = .10$	41	47	54	60	68	75
$P = .025$	$P = .05$	35	40	46	52	59	66
$P = .01$	$P = .02$	28	33	38	43	49	56
$P = .005$	$P = .01$	23	28	32	37	43	49

One-sided	Two-sided	$n = 23$	$n = 24$	$n = 25$	$n = 26$	$n = 27$	$n = 28$
$P = .05$	$P = .10$	83	92	101	110	120	130
$P = .025$	$P = .05$	73	81	90	98	107	117
$P = .01$	$P = .02$	62	69	77	85	93	102
$P = .005$	$P = .01$	55	61	68	76	84	92

One-sided	Two-sided	$n = 29$	$n = 30$	$n = 31$	$n = 32$	$n = 33$	$n = 34$
$P = .05$	$P = .10$	141	152	163	175	188	201
$P = .025$	$P = .05$	127	137	148	159	171	183
$P = .01$	$P = .02$	111	120	130	141	151	162
$P = .005$	$P = .01$	100	109	118	128	138	149

One-sided	Two-sided	$n = 35$	$n = 36$	$n = 37$	$n = 38$	$n = 39$	
$P = .05$	$P = .10$	214	228	242	256	271	
$P = .025$	$P = .05$	195	208	222	235	250	
$P = .01$	$P = .02$	174	186	198	211	224	
$P = .005$	$P = .01$	160	171	183	195	208	

One-sided	Two-sided	$n = 40$	$n = 41$	$n = 42$	$n = 43$	$n = 44$	$n = 45$
$P = .05$	$P = .10$	287	303	319	336	353	371
$P = .025$	$P = .05$	264	279	295	311	327	344
$P = .01$	$P = .02$	238	252	267	281	297	313
$P = .005$	$P = .01$	221	234	248	262	277	292

(Table continues on following page.)

Table X
Critical values of W in the Wilcoxon Signed Rank Test *(continued)*

$n = 5(1)50$

One-sided	Two-sided	$n = 46$	$n = 47$	$n = 48$	$n = 49$	$n = 50$
$P = .05$	$P = .10$	389	408	427	446	466
$P = .025$	$P = .05$	361	379	397	415	434
$P = .01$	$P = .02$	329	345	362	380	398
$P = .005$	$P = .01$	307	323	339	356	373

Table XI
Critical values for the Wilcoxon Rank Sum Test

$m = 3(1)25$ and $n = m(1)m + 25$
$P = .05$ one-sided; $P = .10$ two-sided

n	$m = 3$	$m = 4$	$m = 5$	$m = 6$	$m = 7$	$m = 8$	$m = 9$	$m = 10$	$m = 11$	$m = 12$	$m = 13$	$m = 14$
$n = m$	6,15	12,24	19,36	28,50	39,66	52,84	66,105	83,127	101,152	121,179	143,208	167,239
$n = m + 1$	7,17	13,27	20,40	30,54	41,71	54,90	69,111	86,134	105,159	125,187	148,216	172,248
$n = m + 2$	7,20	14,30	22,43	32,58	43,76	57,95	72,117	89,141	109,166	129,195	152,225	177,257
$n = m + 3$	8,22	15,33	24,46	33,63	46,80	60,100	75,123	93,147	112,174	134,202	157,233	182,266
$n = m + 4$	9,24	16,36	25,50	35,67	48,85	62,106	78,129	96,154	116,181	138,210	162,241	187,275
$n = m + 5$	9,27	17,39	26,54	37,71	50,90	65,111	81,135	100,160	120,188	142,218	166,250	192,284
$n = m + 6$	10,29	18,42	27,58	39,75	52,95	67,117	84,141	103,167	124,195	147,225	171,258	197,293
$n = m + 7$	11,31	19,45	29,61	41,79	54,100	70,122	87,147	107,173	128,202	151,233	176,266	203,301
$n = m + 8$	11,34	20,48	30,65	42,84	57,104	73,127	90,153	110,180	132,209	155,241	181,274	208,310
$n = m + 9$	12,36	21,51	32,68	44,88	59,109	75,133	93,159	114,186	136,216	159,249	185,283	213,319
$n = m + 10$	13,38	22,54	33,72	46,92	61,114	78,138	96,165	117,193	139,224	164,256	190,291	218,328
$n = m + 11$	13,41	23,57	34,76	48,96	63,119	80,144	100,170	120,200	143,231	168,264	195,299	223,337
$n = m + 12$	14,43	24,60	36,79	50,100	65,124	83,149	103,176	124,206	147,238	172,272	199,308	228,346
$n = m + 13$	15,45	25,63	37,83	52,104	68,128	86,154	106,182	127,213	151,245	177,279	204,316	234,354
$n = m + 14$	15,48	26,66	39,86	53,109	70,133	88,160	109,188	131,219	155,252	181,287	209,324	239,363
$n = m + 15$	16,50	27,69	40,90	55,113	72,138	91,165	112,194	134,226	159,259	185,295	214,332	244,372
$n = m + 16$	17,52	28,72	42,93	57,117	74,143	94,170	115,200	138,232	163,266	190,302	218,341	249,381
$n = m + 17$	17,55	29,75	43,97	59,121	77,147	96,176	118,206	141,239	167,273	194,310	223,349	254,390
$n = m + 18$	18,57	30,78	44,101	61,125	79,152	99,181	121,212	145,245	171,280	198,318	228,357	260,398
$n = m + 19$	19,59	31,81	46,104	62,130	81,157	102,186	124,218	148,252	175,287	203,325	233,365	265,407
$n = m + 20$	19,62	32,84	47,108	64,134	83,162	104,192	127,224	152,258	178,295	207,333	237,374	270,416
$n = m + 21$	20,64	33,87	49,111	66,138	86,166	107,197	130,230	155,265	182,302	211,341	242,382	275,425
$n = m + 22$	21,66	34,90	50,115	68,142	88,171	109,203	133,236	159,271	186,309	216,348	247,390	280,434
$n = m + 23$	21,69	35,93	52,118	70,146	90,176	112,208	136,242	162,278	190,316	220,356	252,398	285,443
$n = m + 24$	22,71	37,95	53,122	72,150	92,181	115,213	139,248	166,284	194,323	224,364	257,406	291,451
$n = m + 25$	23,73	38,98	54,126	73,155	94,186	117,219	142,254	169,291	198,330	229,371	261,415	296,460

Table XI
Critical values for the Wilcoxon Rank Sum Test (*continued*)

$m = 3(1)25$ and $n = m(1)m + 25$
$P = .05$ one-sided; $P = .10$ two-sided

n	m = 15	m = 16	m = 17	m = 18	m = 19	m = 20	m = 21	m = 22	m = 23	m = 24	m = 25
n = m	192,273	220,308	249,346	280,386	314,427	349,471	386,517	424,566	465,616	508,668	552,723
n = m + 1	198,282	226,318	256,356	287,397	321,439	356,484	394,530	433,579	474,630	517,683	562,738
n = m + 2	203,292	232,328	262,367	294,408	328,451	364,496	402,543	442,592	483,644	527,697	572,753
n = m + 3	209,301	238,338	268,378	301,419	336,462	372,508	410,556	450,606	492,658	536,712	582,768
n = m + 4	215,310	244,348	275,388	308,430	343,474	380,520	418,569	459,619	501,672	546,726	592,783
n = m + 5	220,320	250,358	281,399	315,441	350,486	387,533	427,581	468,632	511,685	555,741	602,798
n = m + 6	226,329	256,368	288,409	322,452	358,497	395,545	435,594	476,646	520,699	565,755	612,813
n = m + 7	231,339	262,378	294,420	329,463	365,509	403,557	443,607	485,659	529,713	574,770	622,828
n = m + 8	237,348	268,388	301,430	336,474	372,521	411,569	451,620	494,672	538,727	584,784	632,843
n = m + 9	242,358	274,398	307,441	342,486	380,532	419,581	459,633	502,686	547,741	594,798	642,858
n = m + 10	248,367	280,408	314,451	349,497	387,544	426,594	468,645	511,699	556,755	603,813	652,873
n = m + 11	254,376	286,418	320,462	356,508	394,556	434,606	476,658	520,712	565,769	613,827	662,888
n = m + 12	259,386	292,428	327,472	363,519	402,567	442,618	484,671	528,726	574,783	622,842	672,903
n = m + 13	265,395	298,438	333,483	370,530	409,579	450,630	492,684	537,739	584,796	632,856	682,918
n = m + 14	270,405	304,448	340,493	377,541	416,591	458,642	501,696	546,752	593,810	642,870	692,933
n = m + 15	276,414	310,458	346,504	384,552	424,602	465,655	509,709	554,766	602,824	651,885	702,948
n = m + 16	282,423	316,468	353,514	391,563	431,614	473,667	517,722	563,779	611,838	661,899	712,963
n = m + 17	287,433	322,478	359,525	398,574	438,626	481,679	526,734	572,792	620,852	670,914	723,977
n = m + 18	293,442	328,488	366,535	405,585	446,637	489,691	534,747	581,805	629,866	680,928	733,992
n = m + 19	299,451	334,498	372,546	412,596	453,649	497,703	542,760	589,819	639,879	690,942	743,1007
n = m + 20	304,461	340,508	379,556	419,607	461,660	505,715	550,773	598,832	648,893	699,957	753,1022
n = m + 21	310,470	347,517	385,568	426,618	468,672	512,728	559,785	607,845	657,907	709,971	763,1037
n = m + 22	315,480	353,527	392,577	433,629	475,684	520,740	567,798	615,859	666,921	718,986	773,1052
n = m + 23	321,489	359,537	398,588	439,641	483,695	528,752	575,811	624,872	675,935	728,100	783,1067
n = m + 24	327,498	365,547	405,598	446,652	490,707	536,764	583,824	633,885	684,949	738,1014	793,1082
n = m + 25	332,508	371,557	411,609	453,663	498,718	544,776	592,836	642,898	694,962	747,1029	803,1097

(*Table continues on following page.*)

Table XI
Critical values for the Wilcoxon Rank Sum Test *(continued)*

$m = 3(1)25$ and $n = m(1)m + 25$
$P = .05$ one-sided; $P = .10$ two-sided

n	m = 3	m = 4	m = 5	m = 6	m = 7	m = 8	m = 9	m = 10	m = 11	m = 12	m = 13	m = 14
n = m	5,16	11,25	18,37	26,52	37,68	49,87	63,108	79,131	96,157	116,184	137,214	160,246
n = m + 1	6,18	12,28	19,41	28,56	39,73	51,93	66,114	82,138	100,164	120,192	141,223	165,255
n = m + 2	6,21	12,32	20,45	29,61	41,78	54,98	68,121	85,145	103,172	124,200	146,231	170,264
n = m + 3	7,23	13,35	21,49	31,65	43,83	56,104	71,127	88,152	107,179	128,208	150,240	174,274
n = m + 4	7,26	14,38	22,53	32,70	45,88	58,110	74,133	91,159	110,187	131,217	154,249	179,283
n = m + 5	8,28	15,41	24,56	34,74	46,94	61,115	77,139	94,166	114,194	135,225	159,257	184,292
n = m + 6	8,31	16,44	25,60	36,78	48,99	63,121	79,146	97,173	118,201	139,233	163,266	189,301
n = m + 7	9,33	17,47	26,64	37,83	50,104	65,127	82,152	101,179	121,209	143,241	168,274	194,310
n = m + 8	10,35	17,51	27,68	39,87	52,109	68,132	85,158	104,186	125,216	147,249	172,283	198,320
n = m + 9	10,38	18,54	29,71	41,91	54,114	70,138	88,164	107,193	128,224	151,257	176,292	203,329
n = m + 10	11,40	19,57	30,75	42,96	56,119	72,144	90,171	110,200	132,231	155,265	181,300	208,338
n = m + 11	11,43	20,60	31,79	44,100	58,124	75,149	93,177	113,207	135,239	159,273	185,309	213,347
n = m + 12	12,45	21,63	32,83	45,105	60,129	77,155	96,183	117,213	139,246	163,281	190,317	218,356
n = m + 13	12,48	22,66	33,87	47,109	62,134	80,160	99,189	120,220	143,253	167,289	194,326	222,366
n = m + 14	13,50	23,69	35,90	49,113	64,139	82,166	101,196	123,227	146,261	171,297	198,335	227,375
n = m + 15	13,53	24,72	36,94	50,118	66,144	84,172	104,202	126,234	150,268	175,305	203,343	232,384
n = m + 16	14,55	24,76	37,98	52,122	68,149	87,177	107,208	129,241	153,276	179,313	207,352	237,393
n = m + 17	14,58	25,79	38,102	53,127	70,154	89,183	110,214	132,248	157,283	183,321	212,360	242,402
n = m + 18	15,60	26,82	40,105	55,131	72,159	92,188	113,220	136,254	161,290	187,329	216,369	247,411
n = m + 19	15,63	27,85	41,109	57,135	74,164	94,194	115,227	139,261	164,298	191,337	221,377	252,420
n = m + 20	16,65	28,88	42,113	58,140	76,169	96,200	118,233	142,268	168,305	195,345	225,386	256,430
n = m + 21	16,68	29,91	43,117	60,144	78,174	99,205	121,239	145,275	171,313	199,353	229,395	261,439
n = m + 22	17,70	30,94	45,120	61,149	80,179	101,211	124,245	148,282	175,320	203,361	234,403	266,448
n = m + 23	17,73	31,97	46,124	63,153	82,184	103,217	127,251	152,288	179,327	207,369	238,412	271,457
n = m + 24	18,75	31,101	47,128	65,157	84,189	106,222	129,258	155,295	182,335	211,377	243,420	276,466
n = m + 25	18,78	32,104	48,132	66,162	86,194	108,228	132,264	158,302	186,342	216,384	247,429	281,475

Table XI
Critical values for the Wilcoxon Rank Sum Test *(continued)*

$m = 3(1)25$ and $n = m(1)m + 25$
$P = .025$ one-sided; $P = .05$ two-sided

n	m = 15	m = 16	m = 17	m = 18	m = 19	m = 20	m = 21	m = 22	m = 23	m = 24	m = 25
n = m	185,280	212,316	240,355	271,395	303,438	337,483	373,530	411,579	451,630	493,683	536,739
n = m + 1	190,290	217,327	246,366	277,407	310,450	345,495	381,543	419,593	460,644	502,698	546,754
n = m + 2	195,300	223,337	252,377	284,418	317,462	352,508	389,556	428,606	468,659	511,713	555,770
n = m + 3	201,309	229,347	258,388	290,430	324,474	359,521	397,569	436,620	477,673	520,728	565,785
n = m + 4	206,319	234,358	264,399	297,441	331,486	367,533	404,583	444,634	486,687	529,743	574,801
n = m + 5	211,329	240,368	271,409	303,453	338,498	374,546	412,596	452,648	494,702	538,758	584,816
n = m + 6	216,339	245,379	277,420	310,464	345,510	381,559	420,609	460,662	503,716	547,773	593,832
n = m + 7	221,349	251,389	283,431	316,476	351,523	389,571	428,622	469,675	512,730	556,788	603,847
n = m + 8	227,358	257,399	289,442	323,487	358,535	396,584	436,635	477,689	520,745	565,803	612,863
n = m + 9	232,368	262,410	295,453	329,499	365,547	403,597	443,649	485,703	529,759	575,817	622,878
n = m + 10	237,378	268,420	301,464	336,510	372,559	411,609	451,662	493,717	538,773	584,832	632,893
n = m + 11	242,388	274,430	307,475	342,522	379,571	418,622	459,675	502,730	546,788	593,847	641,909
n = m + 12	248,397	279,441	313,486	349,533	386,583	426,634	467,688	510,744	555,802	602,862	651,924
n = m + 13	253,407	285,451	319,497	355,545	393,595	433,647	475,701	518,758	564,816	611,877	660,940
n = m + 14	258,417	291,461	325,508	362,556	400,607	440,660	482,715	526,772	572,831	620,892	670,955
n = m + 15	263,427	296,472	331,519	368,568	407,619	448,672	490,728	535,785	581,845	629,907	679,971
n = m + 16	269,436	302,482	338,529	375,579	414,631	455,685	498,741	543,799	590,859	638,922	689,986
n = m + 17	274,446	308,492	344,540	381,591	421,643	463,697	506,754	551,813	599,873	648,936	699,1001
n = m + 18	279,456	314,502	350,551	388,602	428,655	470,710	514,767	560,826	607,888	657,951	708,1017
n = m + 19	284,466	319,513	356,562	395,613	435,667	477,723	522,780	568,840	616,902	666,966	718,1032
n = m + 20	290,475	325,523	362,573	401,625	442,679	485,735	530,793	576,854	625,916	675,981	727,1048
n = m + 21	295,485	331,533	368,584	408,636	449,691	492,748	537,807	584,868	633,931	684,996	737,1063
n = m + 22	300,495	336,544	374,595	414,648	456,703	500,760	545,820	593,881	642,945	693,1011	747,1078
n = m + 23	306,504	342,554	380,606	421,659	463,715	507,773	553,833	601,895	651,959	703,1025	756,1094
n = m + 24	311,514	348,564	387,616	427,671	470,727	515,785	561,846	609,909	660,973	712,1040	766,1109
n = m + 25	316,524	353,575	393,627	434,682	477,739	522,798	569,859	618,922	668,988	721,1055	775,1125

References

[1] William Beyer, ed. *Handbook of Tables for Probability and Statistics*, 2nd ed. Cleveland: The Chemical Rubber Company, 1968.

[2] James V. Bradley. *Distribution-Free Statistical Tests*. Englewood Cliffs, N. J.: Prentice-Hall, 1968.

[3] Cathy Campbell and Brian Joiner. "How to Get the Answer without Being Sure You've Asked the Question." *The American Statistician*, Vol. 27, 1973.

[4] W. J. Conover. *Practical Nonparametric Statistics*. New York: John Wiley and Sons, 1971.

[5] Wayne Daniel. *Applied Nonparametric Statistics*. Boston: Houghton Mifflin Company, 1978.

[6] J. L. Freedman, J. M. Carlsmith, and David Sears. *Social Psychology*, 2nd ed. Englewood Cliffs, N. J.: Prentice-Hall, 1974.

[7] Vivian Gourevitch. *Statistical Methods: A Problem Solving Approach*. Boston: Allyn and Bacon, 1966.

[8] W. C. Guenther. *Analysis of Variance*. Englewood Cliffs, N. J.: Prentice-Hall, 1964.

[9] Charles Hicks. *Fundamental Concepts in the Design of Experiments*. New York: Holt, Rinehart and Winston, 1964.

[10] Paul Hoel. *Introduction to Mathematical Statistics*. New York: John Wiley and Sons, 1971.

[11] J. Kemeny, Laurie Snell, and G. L. Thompson. *Introduction to Finite Mathematics*. Englewood Cliffs, N. J.: Prentice-Hall, 1964.

[12] David Kleinbaum and L. Kupper. *Applied Regression Analysis and Other Multivariate Methods*. North Scituate, Mass.: Duxbury Press, 1978.

[13] Harold Larson. *Introduction to Probability Theory and Statistical Inference*. New York: John Wiley and Sons, 1969.

[14] J. S. Milton and J. Arnold. *Probability and Statistics in the Engineering and Computing Sciences*. New York: McGraw-Hill, 1986.

[15] J. S. Milton and C. P. Tsokos. *Probability Theory with the Essential Analysis*. Reading, Mass.: Addison-Wesley, 1976.

[16] J. S. Milton and J. O. Tsokos. *Statistical Methods in the Biological and Health Sciences*. New York: McGraw-Hill, 1983.

[17] Alexander Mood, F. Graybill, and D. Boes. *Introduction to the Theory of Statistics.* New York: McGraw-Hill, 1974.

[18] George Snedecor and C. Cochran. *Statistical Methods.* Ames, Iowa: Iowa State University Press, 1980.

[19] Mark Stephens. *Three Mile Island.* New York: Random House, 1981.

[20] Judith Tanur et al. *Statistics: A Guide to the Unknown.* San Francisco, California: Holden-Day, 1972.

[21] C. P. Tsokos. *Probability Distributions: An Introduction to Probability Theory with Applications.* Belmont, California: Duxbury Press, 1972.

[22] John Tukey. *Exploratory Data Analysis.* Reading, Mass.: Addison-Wesley, 1977.

[23] Zalman Usiskin. *Algebra Through Applications with Probability and Statistics.* Chicago: University of Chicago Press, 1976.

[24] Ronald Walpole and Raymond Myers. *Probability and Statistics for Engineers and Scientists.* New York: The Macmillan Company, 1972.

[25] Mary Sue Younger. *A Handbook for Linear Regression.* North Scituate, Mass.: Duxbury Press, 1979.

Answers to Odd-numbered Exercises

Introduction   Review Exercises

1. (a) The 1000 loans granted by the bank

   (b) X: record a 1 for a loan that exceeds $10,000 and a 0 otherwise
       Y: the amount of a loan for an individual
       Both X, Y are discrete.

   (c) The proportion of loans that exceed $10,000; the average of the loans

   (d) It exists.

   (e) No; depending on how the sample was selected and other factors such as chance, the largest loans (or smallest) could have been chosen.

3. X: record a 1 for a citizen who favors the service and a 0 otherwise
   Y: the number of blocks an individual is willing to walk to catch the bus
   Both X, Y are discrete.

5. Population parameters

Chapter 1   Section 1.1

1. Sample data; no

3. No; some individuals evade the census.

5.

0	14321
0	6687595
1	1301
1	6557
2	4103
2	75
3	2
3	5
4	0
4	6

No; yes; data are clustered below 2.5.

7. (a)

Nonsmokers			Smokers
15	1	15	173029
16	29	16	8103
17	2564	17	729
18	9813	18	103
19	8579823	19	499
20	06154127	20	5
21	218143	21	236
22	6114	22	841
23	063	23	2180
24	19	24	31198
25	0	25	8712201

(b) Yes

Section 1.2

9. (a) 3.5

(b) .58

(c) .6

	(d) Class boundaries	(e) Class mark
Class 1	14.05–14.65	14.35
Class 2	14.65–15.25	14.95
Class 3	15.25–15.85	15.55
Class 4	15.85–16.45	16.15
Class 5	16.45–17.05	16.75
Class 6	17.05–17.65	17.35

11.

0	2
0	97596
1	4210324
1	5565769658
2	1302
2	7
3	01
3	

Yes

13. (a)

	Class boundaries	Class mark	Frequency
Class 1	−5.5–4.5	−.5	3
Class 2	4.5–14.5	9.5	9
Class 3	14.5–24.5	19.5	8
Class 4	24.5–34.5	29.5	21
Class 5	34.5–44.5	39.5	8
Class 6	44.5–54.5	49.5	8
Class 7	54.5–64.5	59.5	3

(c) The doctor saw the patient early.

(d) Yes

(e) Yes

15. (a)

1.2	232140443432001214
1.2	8597667
1.3	4244
1.3	7656599895
1.4	4322012
1.4	69588799775655

(b)

	Class boundaries	Class mark	Frequency
Class 1	1.195–1.245	1.22	18
Class 2	1.245–1.295	1.27	7
Class 3	1.295–1.345	1.32	4
Class 4	1.345–1.395	1.37	10
Class 5	1.395–1.445	1.42	7
Class 6	1.445–1.495	1.47	14
Class 7	1.495–1.545	1.52	0

(c) No

17. (a)

	Class boundaries	Class mark	Frequency
Class 1	13.5–624.5	319	23
Class 2	624.5–1235.5	930	9
Class 3	1235.5–1846.5	1541	7
Class 4	1846.5–2457.5	2152	4
Class 5	2457.5–3068.5	2763	4
Class 6	3068.5–3679.5	3374	2
Class 7	3679.5–4290.5	3985	1

(b) Yes; skewed right

(c) No

Section 1.3

19. (a) Average of 355th and 356th observations on the ordered list

(b) 407th observation on the ordered list

(c) 526th observation on the ordered list

(d) Average of the 1000th and the 1001st observations on the ordered list

21. $350; no, this is only a sample.

23. (a) $\bar{x}_1 = 30.19$
$\bar{x}_2 = 17.41$
$\bar{x}_3 = 50.42$
$\bar{x}_4 = 13.59$

(b) 24.29; no

(c) 32.01; yes; sample sizes are equal.

(d) 28.29

(e) 28.29; same as (d)

(f) Yes

(g) 18.10

25. (a) Society A 18.9; Society B 23.9
Yes

(b) Society A 16.5; Society B 25
No

(c)

	Society A		Society B
1	33444444	1	4
1	5566789	1	5889
2	01244	2	00124
2	56	2	5566677
3	0	3	012
3	5	3	5

Yes; Society A skewed right

Section 1.4

27. (a) $\Sigma x = 40$; $\bar{x} = 3.6364$

(b) $s^2 = 6.6545$

(c) $\Sigma x^2 = 212$; $s^2 = 6.6545$

(d) $s = 2.5796$; number of jobs

29. (a)

	Men (X)		Women (Y)
0	0	0	5005
1	0557	1	0575580
2	555055	2	005
3	05	3	0
4	05	4	
5	55	5	
6	05	6	
7	0	7	

(b) Men, men, men, men

Men (X)	Women (Y)
$\bar{x} = 3.285$	$\bar{y} = 1.367$
median $= 2.5$	median $= 1.5$
$s^2 = 3.846$	$s^2 = .758$
$s = 1.961$	$s = .871$

(c) Yes

31. The second group, fall (Y)

spring (X)	fall (Y)
$s^2 = 372.73$	$s^2 = 72.32$

33. (b) Society A
    (c)

Society A	Society B
$s^2 = 34.897$	$s^2 = 31.229$

Section 1.5

35.

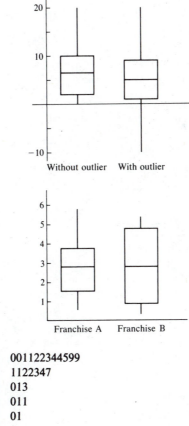

39.

1	001122344599
2	1122347
3	013
4	011
5	01
6	3
7	7
8	3

Section 1.6

41. (a)
Class	Midpoint
1	2
2	5
3	8
4	11
5	14

Same midpoints for both branches

(c) No

(d) Branch A skewed to the left

(e) $\bar{x}$ (branch A) $\approx$ 9.08
$\bar{x}$ (branch B) $\approx$ 8.00

(f) s^2 (branch A) $\approx$ 17.0036
s^2 (branch B) $\approx$ 7.2727

(g) s_A = 4.1235
s_B = 2.6968

43. $\bar{x} \approx$ 1284.38
$s \approx$ 1222.5610

Chapter 1 Review Exercises

45. 16.25; no

47. Dollars; s = 1.13; No

49. Yes; 17.22

51. No

53. s^2 = 90.0074

55. $\bar{x} \approx$ 43.97; no

57. $s \approx$ 9.07

59.
Smokers	Nonsmokers
lh = 28.8	lh = 26.4
uh = 54.6	uh = 36.0

Chapter 2

Section 2.1

1. Personal

3. Relative frequency

5. Classical

7. 3/6; 3/6; 2/6; classical

9. 20/65; relative frequency

11. 4/50; classical

13. 1000/10000; classical

15. 10/150; classical

17. 26/52; 26/52; 13/52; 4/52; 12/52; classical

19. This assumes that each state delegation is of the same size.

Section 2.2

21. (a) $\{(1, 1), (1, 2), (2, 1), (2, 2), (2, 3), (3, 2)\}$

   (b) $\{(1, 1), (2, 2), (3, 3), (4, 4), (5, 5), (6, 6)\}$

   (c) $\{(1,1), (1, 2), (2, 1), (2, 2), (2, 3), (3, 2), (3, 3), (4, 4), (5, 5), (6, 6)\}$

   (d) $\{(1, 1), (2, 2)\}$

   (e) $\{(3, 3), (2, 4), (4, 2), (1, 5), (5, 1), (5, 2), (2, 5), (4, 3), (3, 4), (1, 6), (6, 1), (4, 4), (5, 3), (3, 5), (6, 2), (2, 6), (6, 3), (3, 6), (5, 4), (4, 5), (5, 5), (6, 4), (4, 6), (6, 5), (5, 6), (6, 6)\}$

   (f) $\{(3, 3), (4, 4), (5, 5), (6, 6)\}$

   (g) No; $A \cap B$ contains 2 elements.

23. (a) $P \cup C$: The parents of at least one member of the couple were divorced or the couple divorces.

   $P \cap C$: The parents of at least one member of the couple were divorced and the couple divorces.

   $P \cap C'$: The parents of at least one member of the couple were divorced and the couple did not divorce.

   $P' \cap C'$: The parents of neither member of the couple were divorced and the couple did not divorce.

   (b) $P' \cap C$

   (c) No

25. (a) P': A child is not premature.
   $P \cap M \cap B'$: A child is premature and the child's mother smokes but the child does not have a birth defect.
   $P' \cap M' \cap B$: A child is not premature and the child's mother does not smoke and the child has a birth defect.
   $(P \cup B) \cap M'$: A child is premature or has a birth defect and the child's mother does not smoke.

   (b) M'

   (c) $M' \cap P$

   (d) $P \cap M \cap B$

   (e) $P' \cap M' \cap B'$

   (f) No

Section 2.3

27. 1/6; 1/18; 8/36; property 3

29. 3/4

31. (a) .4

   (b) .7

   (c) .8

   (d) .2

   (e) .5

   (f) .2

33. (a) .3

   (b) .3

35. (a) .9

   (b) .1

   (c) .1

   (d) .5

   No; $V \cap C \neq \emptyset$

37. (a) .1

   (b) .4

   (c) .5

39. (a) .3

   (b) .07

   (c) .23

Section 2.4

43. (a) 5/13

   (b) 1/2

   (c) 5/26

   (d) 5/13

   (e) 1/2

   (f) No

45. (a) 1/2

   (b) 1/9

   (c) 5/18

   (d) 2/5

   (e) 2/9

   (f) 1/6

   (g) 1/6

(h) 1

(i) 1/3

(j) 1/18

(k) 1/3

(l) 1/5

47. (a) .01

(b) .20

(c) .50

(d) .50

49. $P[H \mid N] = 6/75$; $P[N' \mid H] = 3/5$

51. (a) 4/5

(b) 1/5

(c) 2/3

53. (a) 3/7

(b) 3/13

Section 2.5

55. (a) .8

(b) 1/2

(c) Yes

57. 9/16; 1/16; 3/8

59. .0000015; independence

61. P[individual is carrier] = .0110
P[pool is virus-free] = .989

63. .10; .10

65. 15/85

67. .9

Section 2.6

69. .4412; .4118

71. .8444

73. (a) .3651

(b) .1923

75. .80

Section 2.7

77. {*MYD, MYB, MYND, MND, MNB, MNND, MUD, MUB, MUND, FYD, FYB, FYND, FND, FNB, FNND, FUD, FUB, FUND*}

79. (b) {*AAAA, AAAU, AAUA, AAUU, AUAA, AUAU, AUUA, AUUU, UAAA, UAAU, UAUA, UAUU, UUAA, UUAU, UUUA, UUUU*}

    (c) *I* = {*AAAA, AAAU, AAUA, AUAA, UAAA*}

    (d) *O* = {*AAUU, AUAU, AUUA, AUUU, UAAU, UAUA, UAUU, UUAA, UUAU, UUUA, UUUU*}

    (e) Yes

81. (a) 20

    (b) 6

    (c) 3/20

83. (a) 1000

    (b) 1/1000

    (c) 216

    (d) 1/216

85. 16; 256

87. (a) 32

    (b) 1/16

89. (a) 64

    (b) 24

    (c) 40

    (d) 6/64

Section 2.8

91. Permutation

93. Combination

95. Combination

97. 24; 24

99. (a) 40320

    (b) 336; 7/8

101. 1680

Section 2.9

103. (a) 126

(b) 1

(c) 1

(d) 28

(e) 20

(f) 1

105. 1365

107. $\binom{90}{5} / \binom{100}{5}$

109. $\binom{90}{5} / \binom{100}{5}$; $\binom{90}{3} \binom{10}{2} / \binom{100}{5}$

Chapter 2 Review Exercises

111. 1/1000

113. .70

115. No

117. 1/3

119. 1/38

121. 18/38

123. $(18/38) \cdot (18/38)$

125. Yes

131. .37

133. (a) 35

(b) 20; 15

(c) 15/35

135. 120

Chapter 3

Section 3.1

1. Discrete; values are 0, 1, 2, 3, 4, 5

3. Continuous

5. Continuous

7. Discrete; values 0, 1, 2, 3, ..., K

9. Continuous

11. Discrete; values 0, 1, 2, 3, ..., K

13. Continuous

Section 3.2

15.

Y	1	2	3	4	5	6
f_Y	1/6	1/6	1/6	1/6	1/6	1/6

$P[Y \leqslant 3] = 1/2$
$P[Y < 3] = 1/3$; no

17.

D	−5	−4	−3	−2	−1	0	1	2	3	4	5
f_D	1/36	2/36	3/36	4/36	5/36	6/36	5/36	4/36	3/36	2/36	1/36

15/36; 1/6; 1/12

19. (a) .18

(b) .52; .27

(c) .72

21.

W	1	2	3	4	···
f_W	2/38	36/38 · 2/38	$(36/38)^2$ 2/38	$(36/38)^3$ 2/38	···

$$P[W > 3] = 1 - \left[\frac{2}{38} + \frac{36}{38} \cdot \frac{2}{38} + \left(\frac{36}{38}\right)^2 \frac{2}{38} \right]$$

23. (a) Figure (a) $P[L \leqslant 1]$
       Figure (b) $P[1 \leqslant L \leqslant 3]$
       Figure (c) $P[3 \leqslant L \leqslant 4]$
       Figure (d)   no area shaded

(c) 1/2

(d) Figure (a)   1/8
     Figure (b)   6/8
     Figure (c)   1/8
     Figure (d)   0

25. (a) 9/10

(b) 1/10

(c) 0

(d) 8/10

(e) 40 weeks
Minimum = 12.5
Maximum = 37.5

Section 3.3

27. (a) 4.89

(b) 26.02

(c) $\sigma^2 = 2.1079$
   $\sigma = 1.4519$

(d) Number of cases

29. (a) 1.35

    (b) 3.55

    (c) $\sigma^2 = 1.73$; $\sigma = 1.32$

    (d) (number of checks)^2; no; number of checks

31. No

33. $-\$.014$; No

35. (a) Y

    (b) X

37. (a) 18

    (b) 5

    (c) 13

    (d) -6

    (e) 7

    (f) 9

    (g) 21

    (h) 25

39. (a) 0; 0

    (b) 1; 1

    (c) $E\left[\dfrac{X - \mu_X}{\sigma_X}\right] = 0$

       $\text{Var}\left[\dfrac{X - \mu_X}{\sigma_X}\right] = 1$

Chapter 3 Review Exercises

41. Continuous; 0

43. (a) $b \cdot P[A] - a \cdot P[A']$

    (c) $1/5$; \$5; \$10

    (d) \$1000

    (e) 1:37; \$37

    (f) 1:5

45. No; $E[X] = -\$1.15$

47. No; $f(0) < 0$

Chapter 4

Section 4.1

1. Binomial; $n = 10$; $p = 8/36$; $E[X] = 2.222$; Var $X = 1.728$

3. Binomial; $n = 15$; $p = 1/2$; $E[X] = 7.5$; Var $X = 3.75$

5. Neither

7. (a) $E[X] = 1.25$; Var $X = 0.938$; $\sigma = .968$

   (b) $f(x) = \dbinom{5}{x}\left(\dfrac{1}{4}\right)^x\left(\dfrac{3}{4}\right)^{5-x}$

   (c) $P[X = 0] = \left(\dfrac{3}{4}\right)^5 \approx .237$; $P[X > 4] = P[X = 5] = \left(\dfrac{1}{4}\right)^5 \approx .001$

   (d) $P[X = 4 \text{ or } X = 5] = P[X \geqslant 4] = P[X = 4] + P[X = 5] \approx .015 + .001 = .016$

   Yes, if the student were guessing then only 1.6% of the time would he get 4 or 5 correct.

9. (a) $E[X] = 10 \cdot (.2) = 2$

   (b) Var $X = 2 \cdot (.8) = 1.6$; $\sigma = 1.2649$

   (c) $f(x) = \binom{10}{x}(.2)^x(.8)^{10-x}$

   (d) $P[X = 0] = (.8)^{10} \approx .1074$

   (e) $P[X \geqslant 1] = 1 - P[X < 1] = 1 - P[X = 0] = 1 - .1074 = .8926$

Section 4.2

11. (a) $E[X] = .1$; Var $X = .099$

    (c) .4048

13. (a) $1/4$

    (b) 2.5

    (c) .0197

15. .0207

17. X: number of children with blue eyes;

    $P[X = 0] = \binom{10}{0}(1/4)^0(3/4)^{10} \approx .0563$

    If the parents are both heterogeneous

Section 4.3

19. (a) 8; 4; 2

    (b) $f(x) = \dfrac{1}{2\sqrt{2\pi}}\, e^{\{-(1/2)\,[(x - 8)/2]^2\}}$

    (c) $8 \pm 2 = 6, 10$

21. (a)   2.55
    (b)   .96
    (c) −1.28
    (d)   1.28
    (e)   1.645
    (f)   1.96
    (g)   2.575

Section 4.4

23. (a) .0004
    (b) .0475
    (c) .5934
25. Counter Y
27. $8.48 \leqslant \mathrm{pH} \leqslant 9.52$

Section 4.5

31. (a) .0207
    (b) .0222
    (c) .0148
    (d) .0153
33. .0526
35. .9119
37. (a) $E[X] = 50$
    (b) $P[x \leqslant 40 \mid p = 1/3] = .0495$
    (c) $P[x > 40 \mid p = 1/4] = .2843$

Section 4.6

39. (a) .453
    (b) .313
    (c) .687
    (d) .140
    (e) .675
    (f) .836

41. (a) 14

(b) 14

(c) .032; .642; .419

(d) $P[X > 23] = .009$

43. 2; .135; .865

45. 8; .313; yes

47. .0110

Chapter 4 Review Exercises

49. $n = 100$
$p = .9$

51. $E[X] = 90$
$\text{Var } X = 9$
$\sigma = 3$

53. $P[X > 95] = .0336$

55. $Pf[X \le 81] = .0035$

57. $n = 5; p = .25$

59. $f(x) = \binom{5}{x} (.25)^x (.75)^{5-x}$

61. $P[X \ge 1] = 1 - .2373 = .7627$

63. $f(x) = \dfrac{1}{\sqrt{2\pi}\,(.25)}\, e^{\{-(1/2)\,[(x\,-\,9)/\,.25)]^2\}}$

65. $P[X < 8.7] = .1151$

67. $P[8.72 < X < 9.35] = .7878$

69. 9.4113

71. $P[X \ge 15 \mid p = .5] = 1 - .9793$
$= .0307$

Chapter 5

Section 5.1

1. A statistical study is not called for.

3. A statistical study is called for. Population is all patients; hypothesis testing

5. A statistical study is called for. Population is all chips; estimation

7. A statistical study is not called for.

9. A statistical study is not called for.

17. $\binom{100}{10}$; $\dfrac{1}{\binom{100}{10}}$; $\dfrac{\binom{99}{9}}{\binom{100}{10}}$

Section 5.2

19. (a) 51; 369

    (b) 5.1

23. .2

Section 5.3

25. (a) H_0: $p \leqslant .5$
        H_1: $p > .5$

27. (a) H_0: $\mu \geqslant 1$
        H_1: $\mu < 1$

29. (a) H_0: $p \geqslant .2$
        H_1: $p < .2$

    (b) $C = \{0\}$

    (c) $\alpha = P[X = 0 \mid p = .2] \approx .0352$

    (d) $\beta = P[X \geqslant 1 \mid p = .1] \approx .7941$; no

31. (a) $\alpha = P[X \geqslant 8 \mid p = .5] = .0547$; $\beta = P[X \leqslant 7 \mid p = .6] = .8327$;
        power $= .1673$

    (b) $\alpha = P[X \geqslant 11 \mid p = .5] = .0592$; $\beta = P[X \leqslant 10 \mid p = .6] = .7827$;
        power $= .2173$

    (c) $C = \{14, 15, 16, 17, 18, 19, 20\}$; $\beta = P[X \leqslant 13 \mid p = .6] = .7500$;
        power $= .2500$

33. (a) H_0: $p \geqslant .6$
        H_1: $p < .6$

    (b) $P[X \leqslant 10 \mid p = .6] = .2447$

Chapter 5 Review Exercises

37. (a) H_0: $p \leqslant .1$
        H_1: $p > .1$

    (b) $P[X \geqslant 3 \mid p = .1] = .1841$; no

39. Yes; no

Chapter 6

Section 6.1

1. 8.8

3. $13.92

9. (a) .87; 1.07; .81

   (b) .92

Section 6.2

11. (a) (9.02, 10.98)

    (b) (7.342, 8.658)

    (c) (−17.3032, −12.6968)

13. (a) (10.89, 11.51)

15. (41.41, 48.59)

17. (a) 25

    (b) 3.6

19. (a) No

    (b) 2.5; 6.75

    (c) .50

Section 6.3

21. (a) 113

    (b) 7.5333

    (c) 1075

    (d) 15.9809; 3.9976

23. (a) (6.0640, 7.9360)

    (b) (9.3156, 10.6844)

    (c) (−5.5032, −4.4968)

    (d) (99.1235, 100.8765)

25. (b) 64.2; 296.26

    (c) 4.0125; 2.5772; 1.6054

    (d) (3.1572, 4.8678)

    (e) $243.39

27. (a) (.14, .26)

    (b) Yes

29. (a) 62

    (b) 27

    (c) 601

Section 6.4

31. (a) 1.729

    (b) 1.311

    (c) 2.326

    (d) -1.734

    (e) -2.492

    (f) $-2.064, 2.064$

    (g) $-1.699, 1.699$

    (h) $-2.576, 2.576$

33. (a) $H_0: \mu = 1.9$
      $H_1: \mu > 1.9$

    (c) 1.753

    (d) Test statistic $= 2.5077$; reject H_0

35. $H_0: \mu = .10$
    $H_1: \mu > .10$
    Test statistic $= 1.07$; critical point 1.645; unable to reject H_0; type II error

37. (a) $-2.064, 2.064$

    (b) Test statistic $= -1.333$; unable to reject H_0; type II error

39. P value of the test is between .005 and .0005; yes

41. P value of the test is between .10 and .20; no

43. (a) $H_0: \mu = 30$
      $H_1: \mu > 30$

    (c) Test statistic $= 6.6776$; P value of the test is smaller than .0005; yes

45. (a) $H_0: \mu = 80$
      $H_1: \mu > 80$

    (b) $s = 3.2617$
      $\bar{x} = 84.8685$

    (c) Test statistic $= 6.6776$; P value of the test is smaller than .0005; yes

47. $H_0: \mu = 5$

    $H_1: \mu > 5$
    Test statistic $= 3.16$; P value is smaller than .005.

Section 6.5

49. (a) (4.6201, 17.0595)

    (b) (2.1494, 4.1303)

    (c) (8.7538, 32.3232)

    (d) (2.9587, 5.6854)

51. (72.9483, 269.3603) confidence interval on σ^2; (8.5410, 16.4122) confidence interval on σ

53. (15.6250, 46.2329) confidence interval on σ^2; (3.9528, 6.7995) confidence interval on σ

55. (a) (177.3704, 182.6296)

    (b) (768.6723, 1024.0962) confidence interval on σ^2; (27.7249, 32.0015) confidence interval on σ

Section 6.6

59. (a) Test statistic = 26.125; critical point = 30.1; fail to reject H_0

    (b) Test statistic = 39.875; critical point = 39.1; reject H_0

    (c) Test statistic = 14.4; critical point = 13.8; fail to reject H_0

    (d) Test statistic = 13.8; critical point = 11.7; fail to reject H_0

    (e) Test statistic = 24.9; critical points = 14.6, 43.2; fail to reject H_0

    (f) Test statistic = 13.8333; critical points = 7.26, 25.0; fail to reject H_0

61. (a) H_0: $\sigma = 2$ (or $\sigma^2 = 4$)
        H_1: $\sigma > 2$ (or $\sigma^2 > 4$)

    (b) Test statistic = 29.04; critical point = 94.87; fail to reject H_0

63. (a) Test statistic = 23.75; P value is between .25 and .10.

    (b) Test statistic = 17.4; P value is between .05 and .10.

    (c) Test statistic = 21.6; P value is between .50 and 1.00.

65. H_0: $\sigma^2 = .09$
    H_1: $\sigma^2 < .09$
    Test statistic = 6.6667; P value is between .025 and .05. We can reject H_0 at the .05 level.

67. (a) H_0: $\sigma^2 = 0.25$
        H_1: $\sigma^2 > 0.25$

    (b) Critical point $\approx$ 61.93

    (c) Test statistic = 70.56; reject H_0

Chapter 6 Review Exercises

69. $\bar{X}$

71. Confidence interval on μ with σ^2 unknown

73. Yes

75. Normal distribution

79. $s^2 = 208.6625$
    $s = 14.4452$

81. (10.6685, 22.3605)

83. When σ^2 is unknown

85. $\bar{x} = 2.164$

87. $H_0: \mu = 2$ Right-tailed test
$H_1: \mu > 2$

89. Test statistic $= .8458$; H_0 cannot be rejected.

91. Critical point $= 13.8$

95. Test statistic $= -3.3638$; P value is between .005 and .0005.

Chapter 7

Section 7.1

1. .2015; about 60

3. .2; about 4000

5. (a) (.0482, .2376)

(b) longer; (.0258, .2600)

7. (.0430, .2298)

9. (.7723, .8777) method I; (.7557, .8943) method II
(11585, 13116) method I; (11336, 13415) method II

11. (a) 1691

(b) 384

(c) 457

13. $n \approx 1068$

Section 7.2

17. (a) -1.282

(b) -1.645

(c) 1.96

(d) 2.326

(e) $-1.645, 1.645$

(f) $-1.96, 1.96$

19. $H_0: p = .5$
$H_1: p > .5$ critical point is 1.28
Test statistic is 2.2; reject H_0; Type I

21. $H_0: p = .5$
$H_1: p > .5$ critical point is 2.326
Test statistic is .9494; fail to reject H_0; Type II error

23. $H_0: p = .4$
$H_1: p < .4$
Test statistic is -1.44; H_0 could be rejected at the $\alpha \approx .075$ level.

25. $H_0: p = .01$
$H_1: p > .01$
Test statistic is 1.12; P value of test $\approx .13$

Section 7.3

27. (0.11, 0.19)
    We can be 96% sure that the true difference in proportions lies between 11% and 19%, with the percentage of clergy favoring the proposal being the higher.

29. (a) $(-.2480, .1480)$

    (b) $(-.1097, .3097)$

    (c) Longer

31. (a) (.0156, .0844)

    (b) Yes; use supplier II

Section 7.4

35. (a) H_0: $p_1 - p_2 = .02$
    H_1: $p_1 - p_2 > .02$

    (b) Test statistic $= .5774$; critical point $= 1.645$; fail to reject H_0

37. (a) Test statistic $= .7782$; critical point $= 1.645$; fail to reject H_0

    (b) Test statistic $= -.2786$; critical point $= -1.282$; fail to reject H_0

    (c) Test statistic $= -.6427$; critical points $= -1.96, 1.96$; fail to reject H_0

39. (a) H_0: $p_1 = p_2$
    H_1: $p_1 < p_2$

    (b) Critical point $= -1.645$; test statistic $= -.3417$; unable to reject H_0

    (c) Type II

41. To test H_0: $p_1 = p_2$
    H_1: $p_1 > p_2$
    Test statistic $= 1.2660$; H_0 could be rejected at $\alpha = .1020$ level.

43. (a) H_0: $p_1 = p_2$
    H_1: $p_1 \neq p_2$

    (b) Test statistic $= 3.27$; H_0 could be rejected at the .001 level.

Chapter 7 Review Exercises

45. $$\left(\frac{X}{n} - z_{\alpha/2} \sqrt{\frac{(X/n)\,[1 - (X/n)]}{n}} \;,\; \frac{X}{n} = z_{\alpha/2} \sqrt{\frac{(X/n)\,[1 - (X/n)]}{n}} \right)$$
    $$\left(\frac{X}{n} - z_{\alpha/2} \sqrt{\frac{1}{4n}} \;,\; \frac{X}{n} + z_{\alpha/2} \sqrt{\frac{1}{4n}} \right)$$
    The second type when $X/n = 1/2$

47. (a) $n \approx 271$

    (b) (.2565, .3435)

    (c) No

49. $\dfrac{X/n - p_0}{\sqrt{p_0\,(1 - p_0)/n}}$; approximately normal

51. $\dfrac{X_1}{n_1} - \dfrac{X_2}{n_2}$; yes

53. $(-.0388, .0788)$; no, confidence interval includes zero.

55. $\hat{p} = \dfrac{X_1 + X_2}{n_1 + n_2}$

57. (a) $H_0\!: p_1 = p_2$
   $H_1\!: p_1 > p_2$

   (b) Critical point is 1.645.

   (c) Test statistic $= .6667$; cannot reject H_0

Chapter 8   Section 8.1

1. (a) 2.28

   (b) 1.00

   (c) 4.07

   (d) 3.67

   (e) .0517

   (f) .1908

   (g) .4310

   (h) $a = .4255$; $b = 2.42$

   (i) $a = .3788$; $b = 3.22$

3. $H_0\!: \sigma_1^2 = \sigma_2^2$
   $H_1\!: \sigma_1^2 > \sigma_2^2$; the critical point is 3.18.
   Test statistic is 7.5; reject H_0; Type I

7. (a) $H_0\!: \sigma_1^2 = \sigma_2^2$
   $H_1\!: \sigma_1^2 \neq \sigma_2^2$

   (b) Test statistic is .78; unable to reject H_0 at the $\alpha = .10$ level; can reject H_0 at the $\alpha = .20$ level.

9. $H_0\!: \sigma_A^2 = \sigma_B^2$
   $H_1\!: \sigma_A^2 < \sigma_B^2$
   Test statistic is .47; can reject H_0 at $\alpha = .05$; cannot reject H_0 at $\alpha = .025$.

11. Confidence interval is $(.4848, 6.8928)$; contains 1

13. Confidence interval is $(0.48, 1.90)$; no; one is contained in the interval.

15. Confidence interval is $(.8271, 13.4332)$; no; confidence interval contains one.

Section 8.2

17. (a) $\dfrac{s_1^2}{s_2^2} = 1.0714$; critical points are .5076, 1.97; fail to reject $H_0\!: \sigma_1^2 = \sigma_2^2$

   (b) $s_p^2 = 29$; $s_p = 5.3852$

   (c) Because $n_1 = n_2$

19. (a) $\bar{x}_1 - \bar{x}_2 = 9.58$
    Population I liberal arts

   (b) $s_1^2 = 108.10$
       $s_1^2 = 102.52$
       Test statistic is 1.0544; critical points are .4098, 2.44; cannot reject H_0 at the $\alpha = .20$ level.

   (c) Yes, $s_p^2 = 105.31$; $s_p = 10.26$

   (d) Confidence interval (1.62, 17.54); yes; interval does not contain 0; population I higher than population II

21. (a) Test statistic $= .8409$; critical points are .7937 and 1.26; unable to reject H_0

   (b) $s_p^2 = .405$; $s_p = .6364$

   (c) Confidence interval is $(-.412, -.088)$; 1960 scores appear to be lower; confidence interval does not contain 0.

Section 8.3

25. (a) $s_p = 3.3166$; test statistic $= 1.7056$; critical point $= 1.697$; reject H_0.

   (b) $s_p = 1.8207$; test statistic $= -3.8439$; critical point $= -2.021$; reject H_0.

   (c) $s_p = 3.4178$; test statistic $= 3.6528$; critical points are -1.671 and 1.671; reject H_0.

27. (a) H_0: $\mu_1 = \mu_2$
       H_1: $\mu_1 < \mu_2$

   (b) Test statistic $= .7143$; critical points are .5882 and 1.7; fail to reject H_0: $\sigma_1^2 = \sigma_2^2$.

   (c) $s_p = .2449$;
       Test statistic $= -.4331$; critical point $= -1.684$; fail to reject H_0: $\mu_1 = \mu_2$; make cigarettes using filter I.

29. H_0: $\mu_{\text{FEMALE}} = \mu_{\text{MALE}}$
    H_1: $\mu_{\text{FEMALE}} > \mu_{\text{MALE}}$
    Pool; $s_p^2 = 20.01$; test statistic $= 1.71$; H_0 can be rejected at the $\alpha = .05$ level.

31. Unable to reject H_0: $\sigma_1^2 = \sigma_2^2$ at $\alpha = .1$ level; $s_p^2 = 80.17$;

    test statistic $\dfrac{-8.4}{(8.9538)(.3162)} = -2.9667$;

    can reject H_0: $\mu_1 = \mu_2$ at the $\alpha = .005$ level.

Section 8.4

33. (a) (.6186, 9.3814)

   (b) (.8851, 7.1149)

   (c) $(-4.0772, .0772)$

35. (a) Test statistic is 1.78; conclude that $\sigma_1^2 \neq \sigma_2^2$.

   (b) Confidence interval is (2.1367, 5.0633); μ_1 is larger than μ_2; confidence interval does not contain 0.

37. (a) $\bar{x}_1 - \bar{x}_2 = -.3544$; yes

    (b) Test statistic $= 3.3182$; reject H_0: $\sigma_1^2 = \sigma_2^2$

    (c) H_0: $\mu_1 = \mu_2$
       H_1: $\mu_1 < \mu_2$
       Test statistic $= -2.1217$; using T_{24}, we can reject H_0 at the $\alpha = .025$ level.

39. H_0: $\mu_1 = \mu_2$
    H_1: $\mu_1 > \mu_2$
    There is evidence at the $\alpha = .1$ level that $\sigma_2 \neq \sigma_2^2$; do not pool; $\nu \approx 43$; test statistic is 1.76; H_0 can be rejected at the $\alpha = .05$ level.

Section 8.5

43. (a) $\widehat{\mu_X - \mu_Y} = 8.55$

    (b) $(7.52, 9.58)$; the three-step approach appears to be superior to the four-step approach.

45. (a) $(.0799, .1681)$

    (b) BBQ tends to be rated higher.

47. H_0: $\mu_X = \mu_Y$
    H_1: $\mu_X < \mu_Y$
    The critical point is -1.383; test statistic $= -2.33$; reject H_0; conclude that the bathroom scales tend to give lower readings than the office scale.

49. H_0: $\mu_X = \mu_Y$ $\bar{d} = .55$
    H_1: $\mu_X > \mu_Y$ $s_d^2 = 6.84$
    Critical point is 1.303; test statistic $= 1.33$; Reject H_0; yes; The probability that the advertising is false is only about 10%.

51. Test statistic is 26.66; can reject H_0 at the .0005 level.

Chapter 8 Review Exercises

53. A test of hypothesis to see if the population variances are equal; set the α level of the test fairly large; make the sample sizes the same.

55. Pooling; by a weighted average

57. By a normal distribution

59. Test statistic $= .3147$; critical points are .3984 and 2.72; reject H_0: $\sigma_1^2 = \sigma_2^2$

61. $\bar{x}_1 - \bar{x}_2 = .2635$

63. H_0: $\mu_1 = \mu_2$
    H_1: $\mu_1 > \mu_2$

65. Test statistic $= 6.3595$;
    critical point $= 1.697$; reject H_0.

67. $\bar{x}_1 - \bar{x}_2 = 24.0909$

69. H_0: $\mu_1 - \mu_2 \neq 10$
    H_1: $\mu_1 - \mu_2 > 10$

71. Test statistic = 5.6369; critical points are .4608 and 2.10; reject H_0: $\sigma_1^2 = \sigma_2^2$; cannot pool

     Test statistic = 2.5614; T-distribution with $\nu \neq 14$

     P value is between .01 and .005 therefore reject H_0

73. Fail to reject H_0: $\sigma_1^2 = \sigma_2^2$

     Pool variances; $s_p = 48131.89$

     Confidence interval $(-44598.52, 45850.52)$; no difference indicated; 0 is in the confidence interval.

Chapter 9

Section 9.1

1. Linear regression is applicable.

3. Linear regression does not appear to be applicable.

5. Linear regression is applicable.

Section 9.2

7. $b = 3$; $a = 6.4$

9. 13.9

11. No; 10 is beyond the range of the given data points.

13. $b = -4.086$; $a = 2.8$

15. -11.501

17. (a) $b = -.08$; $a = 13.64$

    (b) Decrease; slope is < 0; the regression line falls from left to right.

    (c) $\hat{y} = 9.64$

19. (b) $\Sigma x = 88$, $\Sigma y = 49$

    $\Sigma x^2 = 1614$, $\Sigma xy = 899$

    (c) $\hat{\alpha} = -.04$

    $\hat{\beta} = .56$

    (d) $\hat{\mu}_{Y/x} = -.04 + .56x$

    (e) 10 (approximately)

21. (b) $\Sigma x = 302$, $\Sigma y = 66$

    $\Sigma x^2 = 9424$, $\Sigma xy = 2086.5$

    (c) $\hat{\mu}_{Y/x} = -2.76 + .31x$

    (d) 8.1

23. (b) $\hat{\mu}_{Y/x} = 8.17 + 2.6x$

    (c) 12.07

25. $\hat{\mu}_{Y/x} = 43.8 - 0.47x$;

    24 miles per gallon

27. (a) $\hat{\mu}_{Y/x} = 260.83 - 0.42x$

  (b) The year 2000 would have an x value of 100.

  (c) Yes, it is impossible to run the mile in zero time.

Section 9.3

29. $\Sigma_{xy} = 102$; $\Sigma y = 36$; $S_{xy} = 12$

31. $a = 3$

33. $\hat{y}_1 = 5.4$
   $\hat{y}_2 = 7.8$
   $\hat{y}_3 = 10.2$
   $\hat{y}_4 = 12.6$

35. $\Sigma y^2 = 354$; $S_{yy} = 30$

37. $\hat{\sigma}^2 = .6$

39. $\Sigma x = 42$; $\Sigma x^2 = 288$
   $S_{xx} = 36$

41. $b = 2.1389$

45. $e_1 = -.1547$
   $e_2 = -.2936$
   $e_3 = .5675$
   $e_4 = 1.873$
   $e_5 = 1.5675$
   $e_6 = -.7103$
   $e_7 = -2.8492$

47. $\Sigma y^2 = 1877$
   $S_{yy} = 179.7143$

49. $\hat{\sigma}^2 = 3.016$

Section 9.4

51. $S_{xx} = .5838$ $S_{xy} = -2.34$
   $S_{yy} = 10.46$ $\bar{x} = 1.675$

53. (15.2481, 16.1467)

55. $\Sigma x = 464$ $S_{xx} = 5048$
   $\Sigma x^2 = 31960$ $\bar{x} = 61.75$

57. $\Sigma xy = 3194.4$; $S_{xy} = 503.2$

59. (4.3068, 4.9502)

61.

x values	confidence interval
25	(1.6196, 2.6524)
45	(3.7798, 4.4802)
58	(5.1315, 5.7207)
90	(8.178, 9.055)

Yes

63.

x values	confidence interval
4	(6.2935, 16.2937)
6	(10.7979, 20.3449)
8	(14.8491, 24.8493)
10	(18.5014, 29.7526)

Yes

65. (7.7022, 8.4778)

Section 9.5

71. $\Sigma x = 47$ $S_{xx} = 44.1$
    $\Sigma x^2 = 265$

73. $\Sigma xy = 505$; $S_{xy} = 86.7$

75. $\hat{\sigma}^2 = .306$; $\hat{\sigma} = .5532$

77. No

79. $\Sigma x = 47$ $S_{xx} = 44.1$
    $\Sigma x^2 = 265$

81. $\Sigma xy = 124$ $S_{xy} = -2.9$

83. The observed value of the test statistic is $-.2640$; H_0 cannot be rejected even at the $\alpha = .50$ level ; no

85. $S_{xx} = 716.9$ $S_{xy} = 615.5$
    $S_{yy} = 586.5$ $SSE = 58.0317$
    $\hat{\sigma}^2 = 7.254$ $\hat{\sigma} = 2.6933$

87. $\hat{y}|_{x=28} = 29.0305 \approx 29$

89. Test statistic $= 8.0929$; critical points $= -2.0447, 2.0447$
    Reject H_0; $\hat{\mu}_{Y/x} = -1.4744 + 4.67x$
    $\hat{y}|_{x=1.75} = 6.7$

Section 9.6

91. $\Sigma y = 254$ $S_{yy} = 334.8$
    $\bar{y} = 50.8$

93. $\hat{\beta}_1 = .9181$

95. Test statistic $= 1.8796$; can reject H_0: $\beta_1 = 0$ at $\alpha = .20$ level

97. $\hat{\beta}_2 = 9.2143$

99. Test statistic $= 2.7106$; can reject H_0: $\beta_2 = 0$ at the .10 level

101. $\hat{Y}|_{x=1.8} = 50.8000$

103. $\mu_{Y|x_1, x_2, x_3} = 9.8343 + .1703x_1 - .5044x_2 - .2652x_3$

105. No

107. Yes; $\hat{\mu}_{Y/x_1} = 70.6769 + .4410x_1$
     $\hat{\mu}_{Y|40} = 88.3169$

Section 9.7

109. $r = -.9848$

111. $r = -.0698$

113. (b) Yes; negative

(c) $\Sigma x = 33$; $\Sigma y = 34$;
$\Sigma x^2 = 149$; $\Sigma y^2 = 134$;
$\Sigma xy = 87$

(d) $r = -.93$

115. $\rho = \dfrac{\text{Cov }(X, Y)}{\sqrt{\text{Var }X \text{Var }Y}}$

If X and Y are independent, then

$\text{Cov }(X, Y) = E[XY] - E[X]E[Y] = 0$

and hence $\rho = \dfrac{0}{\sqrt{\text{Var }X \text{Var }Y}} = 0$

117. For exercise 19 $r^2 = .9414$; 94% of the variation in Y can be attributed to a linear association between X and Y.
For exercise 20 $r^2 = .4983$; 50% of the variation in Y can be attributed to a linear association between X, Y
For exercise 21 $r^2 = .9279$; 93% of the variation in Y can be attributed to a linear association between X, Y
For exercise 22 $r^2 = .6734$; 67% of the variation in Y can be attributed to a linear association between X and Y
For exercise 23 $r^2 = .7643$; 76% of the variation in Y can be attributed to a linear association between X, Y.

Chapter 9 Review Exercises

119. Plot a scattergram

121. $\hat{\beta} = b = \dfrac{n\Sigma xy - \Sigma x \Sigma y}{n\Sigma x^2 - (\Sigma x)^2}$

$\hat{\alpha} = a = \bar{y} - b\bar{x}$

123. Yes; negative; negative; yes

125. 35.4773

127. $\hat{\sigma}^2 = 11.1968$

129. (27.3822, 43.5724)

133. Yes; yes

135. No

137. $\widehat{\text{Cov}(X, Y)} = \Sigma(X - \bar{X})(Y - \bar{Y})/(n - 1)$
Yes

139. Positive; strong; $r = .8815$

Chapter 10   **Section 10.1**

1. $E[X_1] = 10.5$; $E[X_2] = 1.5 = E[X_3]$; $E[X_4] = .75 = E[X_5]$

3. $E[\text{green}] = 1.0526$; $E[\text{red}] = E[\text{black}] = 9.4737$

5. 3.6; 15.6; .8
   The probability of these results is .0320, which is fairly small.

Section 10.2

7. (a)

Type	A	B	C	D
Expected no.	16	12.5	7.5	14

   (b) No

   (c) Test statistic is 1.6047; critical point $= 7.81$; we are unable to reject H_0.

9. H_0: $p_i = 1/6$ $i = 1, 2, 3, 4, 5, 6$ (die is fair)
   H_1: $p_i \neq 1/6$ for some i (die is not fair)
   $E_i = 20$ for each i; critical point is 9.24;
   Test statistic is 3.5; unable to reject H_0.

11. $E_1 = 325, E_2 = 125, E_3 = 50$
    The critical point is 4.61; test statistic is 2.61; unable to reject H_0

13. $E_1 = 20, E_2 = 5, E_3 = 20, E_4 = 5$
    The critical point is 7.81; the observed value of the test statistic is 15.25; reject H_0

15. H_0: $p_1 = .92$ $E_1 = 920$
    $p_2 = .07$ $E_2 = 70$
    $p_3 = .01$ $E_3 = 10$
    The critical point is 4.61; the observed value of the test statistic is 2.50; unable to reject H_0

17. (a)

Category number	Boundaries	O_i	P_i	$E_i = 40P_i$
1	$-\infty$ to 3.5	3	.0655	2.62 ⎫
2	3.5 to 6.5	2	.1107	4.43 ⎭
3	6.5 to 9.5	9	.1832	7.33
4	9.5 to 12.5	13	.2277	9.11
5	12.5 to 15.5	6	.2010	8.04
6	15.5 to 18.5	3	.1281	5.12 ⎫
7	18.5 to 21.50	1	.0588	2.35 ⎬
8	21.5 to ∞	3	.0250	.100 ⎭

   (b) $\bar{y} = 11.35$
   $\Sigma y^2 = 6230$
   $\hat{\sigma}^2 = 26.93$

   (c) The observed value of the test statistic is 3.41; the critical point is 7.78; unable to reject the null hypothesis of normality.

19. (a)

Category number	Boundaries	O_i	P_i	E_i
1	$-\infty$ to 14.15	8	.0934	3.27
2	14.15 to 18.35	2	.1099	3.85
3	18.35 to 22.55	2	.1636	5.73
4	22.55 to 26.75	3	.1888	6.61
5	26.75 to 30.95	4	.1800	6.3
6	30.95 to ∞	16	.2643	9.25

(b) $\bar{y} = 25.51$; $\Sigma y^2 = 25383.19$
$\sigma^2 = 74.40$

(c) The observed value of the X_4^2 statistic is 11.33; the critical point for an $\alpha = .025$ test is 11.3; reject the null hypothesis of normality.

Section 10.3

21.

	Light		Medium		Heavy		
Poor	(9) 10		(9) 9		(6) 5		24
(a) Good	(11.25) 12		(11.25) 11		(7.5) 7		30
(b) Excellent	(9.75) 8		(9.75) 10		(6.5) 8		26
	30		30		20		80

(c) No

(d) $\Sigma\Sigma(O - \hat{E})^2/\hat{E} = 1.0334$ is the test statistic; P value of the test is between .05 and .10; reject H_0.

23. H_0: $p_{ij} = p_{i\cdot} \cdot p_{\cdot j}$; $i = 1, 2, 3$
$\qquad\qquad\qquad\qquad j = 1, 2, 3, 4$
Attitude is independent of religious affiliation
H_1: $p_{ij} \neq p_{i\cdot} \cdot p_{\cdot j}$ for some i, j
The observed value of the test statistic is 56.87; we can reject H_0 at the $\alpha = .005$ level.

25. The critical point is 12.6; the observed value of the test statistic is .84; unable to reject H_0.

27. The observed value of the test statistic is 4.51; P value is between .25 and .1; thus we can reject H_0: independence only at the $\alpha = .25$ level

Section 10.4

29. (b) Is the proportion of males that oppose the draft the same as females?
Is the proportion of males that are neutral about the draft the same as females?
Is the proportion of males that favor the draft the same as females?
H_0: $p_{11} = p_{12}$
$\quad\ \ p_{21} = p_{22}$
$\quad\ \ p_{31} = p_{32}$

(a)

	Male		Female		
Oppose	(70) 60		(70) 80		140
(c) Neutral	(10) 15		(10) 5		20
Favor	(20) 25		(20) 15		40
	100		100		200

(d) Yes

(e) The observed value of the test statistic is 10.3572; the P value is between .01 and .005; reject H_0.

31. The critical point is 7.81; the observed value of the test statistic is 4.38; unable to reject H_0.

33. The observed value of the test statistic is 26.06; H_0 can be rejected at the $\alpha = .005$ level (critical point is 14.9); treatment B

Chapter 10 Review Exercises

35. (a) E[each category] = 10

(b) The observed value of the test statistic is 1.0; critical point is 7.81; fail to reject H_0.

41. $\hat{\mu} = 7.65$; $\hat{\sigma}^2 = 2.08$

43.

Category number	$\hat{E}$
1	1.42
2	4.67
3	8.90
4	8.90
5	6.10

45. Independence

47. $\hat{E}_{ij} = (n_{i.})(n_{.j})/n$

49. There is no difference between controllers and other people in terms of stress disorders.
H_0: $p_{11} = p_{12}$
$p_{21} = p_{22}$

51. The observed value of the test statistic is 4.82; critical point is 3.84; reject H_0; yes.

53. There is no difference between those under 40 and those 40 and older in the use of computerized banking.
H_0: $p_{11} = p_{12}$
$p_{21} = p_{22}$

55. The observed value of the test statistic is 7.58; the critical point is 3.84; reject H_0; yes.

Chapter 11

Section 11.1

1.

Source of Variation	Sum of Squares	Degrees of Freedom	Mean Square	F ratio
Treatment	9324	2	4662	3
Error	41958	27	1554	
Total	51282	29		

(a) 2.51

(b) Yes

(c) That they are the same

3. (a) $T_{1.} = 20.6$, $T_{2.} = 29.4$, $T_{3.} = 26.5$, $T_{4.} = 36$

$T_{..} = 112.5$ $\quad \sum_{i=1}^{4} \sum_{j=1}^{5} X_{ij}^2 = 894.71$

$T_{..}^2/20 = 632.8125$

$\sum_{i=1}^{4} T_{i.}^2/5 = 657.394$

$SS_{total} = 261.8975 \qquad SS_{Tr} = 24.5815$

$MS_{Tr} = 8.1938 \qquad SS_E = 237.316$

$\cdot \quad MS_E = 14.8322$

(b) Test statistic $= .5524$;

Critical point $= 2.46$, $\alpha = .10$; cannot reject H_0

5.

Source of Variation	Sum of Squares	Degrees of Freedom	Mean Square	Ratio
Treatment	1.49	2	.75	2.68
Error	4.15	15	.28	
Total	5.64	17		

H_0: $\mu_A = \mu_B = \mu_C$

The critical point is 4.77; unable to reject H_0

7.

Source of Variation	Sum of Squares	Degrees of Freedom	Mean Square	Ratio
Treatment	71.92	3	23.97	35.78
Error	18.79	28	.67	
Total	90.71	31		

H_0: $\mu_1 = \mu_2 = \mu_3 = \mu_4$

The critical point is 2.95; reject H_0.

Section 11.2

9. (a) $\mu_1 - \dfrac{\mu_2 + \mu_3 + \mu_4}{3}$

(b) $\dfrac{\mu_1 + \mu_3}{2} - \dfrac{\mu_2 + \mu_4}{2}$

(c) The effect of television

11. (a) $\dfrac{\mu_1 + \mu_3}{2} - \dfrac{\mu_2 + \mu_4}{2}$

(b) $\dfrac{\mu_1 + \mu_2}{2} - \dfrac{\mu_3 + \mu_4}{2}$

(c)

Source of Variation	Sum of Squares	Degrees of Freedom	Mean Square	F ratio
Treatment	54.80	3	18.27	3.33
Error	87.68	16	5.48	
Total	142.48	19		

Critical point is 2.46; reject H_0

(d) $L^2 = 8.09$

$$\dfrac{\bar{x}_1 + \bar{x}_3 - \bar{x}_2 - \bar{x}_4}{2} \pm L = (-.44, 5.24);$$

There is no difference on the average between A and B; $(-4.64, 1.04)$ there is no difference on the average between pill and liquid form.

(e) Test $H_0: \mu_1 - \mu_3 = 0$
$L^2 = 16.18; (-0.82, 7.22);$ There is no best treatment.

13. (a) $H_0: \mu_1 - \mu_2 = 0; L = 10.76; (-22.7602, -1.24);$ reject H_0; conclude that unmarried career women scored higher on the satisfaction scale than housewives;
$H_0: \mu_1 - \mu_3 = 0; L = 10.76; (-17.76, 3.76);$ unable to reject H_0; no evidence of differences between married career women and housewives;
$H_0: \mu_2 - \mu_3 = 0; L = 10.76; (-5.76, 15.76);$ unable to reject H_0; no evidence of differences between married and unmarried career women

(b) $H_0: \mu_1 - \dfrac{\mu_2 + \mu_3}{2} = 0;$

$L = 9.32; (-18.82, -.18);$ reject H_0; conclude that career women score higher on the satisfaction scale than housewives

15. (a)   *ANOVA*

Source of Variation	Sum of Squares	Degrees of Freedom	Mean Square	F ratio
Treatment	10280	4	2570	3.50
Error	88080	120	734	
Total	92360	124		

The critical point is 2.45; reject H_0

(b) $H_0: \mu_2 - \mu_5 = 0; L = 23.99; (0.01, 47.99);$ reject H_0 and conclude that balls last longer on grass than on asphalt.

(c) H_0: $\dfrac{\mu_1 + \mu_2 + \mu_3}{3} - \dfrac{\mu_4 + \mu_5}{2}$

$L = 15.48$; $(2.19, 33.15)$; reject H_0; conclude that balls last longer on soft surfaces.

17.

ANOVA

Source of Variation	Sum of Squares	Degrees of Freedom	Mean Square	F ratio
Treatment	88.38	4	22.10	2.92
Error	3747.15	495	7.57	
Total	3835.53	499		

The critical point is 2.37; reject H_0;

$\bar{x}_{3.}$	$\bar{x}_{1.}$	$\bar{x}_{2.}$	$\bar{x}_{4.}$	$\bar{x}_{5.}$
2.0	2.2	2.23	2.27	3.2

p	2	3	4	5
r_p	2.772	2.918	3.017	3.089
$(SSR)_p$	.763	.803	.83	.85

Section 11.3

19. (a)

ANOVA

Source of Variation	Sum of Squares	Degrees of Freedom	Mean Square	F ratio
Treatment	401.338	3	133.7793	61.3498
Block	81.525	4	20.3812	
Error	26.167	12	2.1806	
Total	509.03	19		

(b) The critical point is 3.49; reject H_0: $\mu_{1.} = \mu_{2.} = \mu_{3.} = \mu_{4.}$

(c) H_0: $\mu_{1.} = \mu_{4.}$
$L^2 = 9.1324$
$(-5.442, .602)$; unable to reject H_0

(d) $(-10.662, -4.618)$; reject H_0

21.

ANOVA

Source of Variation	Sum of Squares	Degrees of Freedom	Mean Square	F ratio
Treatment	3.17	2	1.59	.88
Block	22.92	3		
Error	10.83	6	1.81	
Total	36.92	11		

The critical point is 10.92; unable to reject H_0: $\mu_{1.} = \mu_{2.} = \mu_{3.}$

23.

ANOVA

Source of Variation	Sum of Squares	Degrees of Freedom	Mean Square	F ratio
Treatment	4803.8	3	1601.27	172.74
Block	5440.8	4		
Error	111.2	12	9.27	
Total	10355.8	19		

The critical point is 3.49; reject H_0: $\mu_{A.} = \mu_{B.} = \mu_{C.} = \mu_{D.}$

25. (a)

ANOVA

Source of Variation	Sum of Squares	Degrees of Freedom	Mean Square	F ratio
Treatment	2007.13	3	669.04	4.3
Block	10232	3		
Error	1401.37	9	155.71	
Total	13640.5	15		

The critical point is 3.86; reject H_0: $\mu_{1.} = \mu_{2.} = \mu_{3.} = \mu_{4.}$

(b) H_0: $\mu_{3.} - \mu_{2.} = 0$ (robbery-rape)

$L = 30.03$; $(-8.03, 52.03)$; unable to reject H_0; cannot conclude that one type of crime showed a higher increase than any other.

(c) H_0: $\dfrac{\mu_{1.} + \mu_{2.}}{2} - \dfrac{\mu_{3.} + \mu_{4.}}{2} = 0$

$L = 21.23$; $(-30.86, 11.60)$; unable to reject H_0

27.

$\bar{x}_{4.}$	$\bar{x}_{2.}$	$\bar{x}_{1.}$	$\bar{x}_{3.}$
144.6	153	157	185.8

p	2	3	4
r_p	3.082	3.225	3.313
$(SSR)_p$	4.19	4.39	4.50

Chapter 11 Review Exercises

35. H_0: $\mu_{I.} = \mu_{II.} = \mu_{III.} = \mu_{IV.}$

The various treatments (or none) have no effect on the sales.

37. H_0: $\dfrac{\mu_1 + \mu_2}{2} - \dfrac{\mu_3 + \mu_4}{2} = 0$

$L^2 = .2008$

Confidence interval $(-1.4781, -.5819)$ does not contain 0; reject H_0; can conclude that games make a difference in sales

39.

$\bar{x}_{1.}$	$\bar{x}_{2.}$	$\bar{x}_{3.}$	$\bar{x}_{4.}$	test at $\alpha = .05$
.99	.99	1.62	2.42	

p	2	3	4
r_p	2.858	3.006	3.102
$(SSR)_p$	.4897	.5149	.5314

The best treatment is games with money prizes.

41. $H_0: \mu_1 = \mu_2 = \mu_3 = \cdots = \mu_K$

43. $SS_{TOTAL} = SS_{Tr} + SS_{BLOCK} + SS_E$
    SS_{Tr}: variation due to the fact that different treatments are involved
    SS_{BLOCK}: variation due to the fact that different blocks are involved
    SS_E: random fluctuation

45. $H_0: \mu_{1.} = \mu_{2.} = \mu_{3.} = \mu_{4.}$
    The various training methods have no effect on the student scores.

47. $H_0: \mu_1 - \mu_4 = 0$
    $L^2 = 170.5563$
    Confidence interval is (24.3403, 50.4597); reject H_0; conclude there is a difference between I and IV; yes

49.

Source of Variation	Sum of Squares	Degrees of Freedom	Mean Square	F ratio
Treatment	488.5	4	122.125	.3382
Error	5417.25	15	361.15	
Total	5905.75	19		

$f_{.05, 4, 15} = 2.9$
Fail to reject H_0; instructor effect not significant; blocking is not necessary.

Chapter 12   Section 12.1

1. (a) .0176
   (b) .0176
   (c) .0577
   (d) .0547
   (e) .1719
   (f) .0207
   (g) .0004
   (h) .0026
   (i) .0414

3. $H_0: M = 18$
   $H_1: M > 18$
   $Q_- = 8$
   $P[Q_- \leqslant 8 \mid p = 1/2] = .2517$; no

5. $H_0: M = 45$
   $H_1: M > 45$

$$Q_- = 30$$
$$P[Q_- \leqslant 30 \mid p = 1/2]$$
$$\approx P\left[Z \leqslant \frac{30.5 - 37.5}{\sqrt{18.75}}\right]$$
$$= .0526$$
Yes; reject H_0.

7. H_0: $M = 46.6$
   H_1: $M > 46.6$
   The observed value of the test statistic is $W = \mid W_- \mid = 51$; the critical point is 30; unable to reject H_0; assume the distribution of X is symmetric

9. Use the Wilcoxon procedure;
   H_0: $\mu = 35$
   H_1: $\mu < 35$
   The observed value of the test statistic is $W = \mid W_- \mid = 60$; the critical point for an $\alpha = .05$ test is 47; unable to reject H_0

Section 12.2

11. $X =$ drying time for type I
    $Y =$ drying time for type II
    H_0: $\mu_{X-Y} = 0$
    H_1: $\mu_{X-Y} > 0$
    The number of negative signs is 2; $P[Q_- \leqslant 2 \mid p = 1/2] = .0547$; reject H_0

13. H_0: $\mu_{X-Y} = 0$
    H_1: $\mu_{X-Y} > 0$
    The number of positive signs is 4; $P[Q_+ \leqslant 4 \mid p = 1/2] = .3770$; we cannot reject H_0.

15. H_0: $M_{M-F} = 0$
    H_1: $M_{M-F} > 0$
    Unattractive: $W = \mid W_- \mid = 14.5$; the critical point for an $\alpha = .05$ test is 17; reject H_0; the unattractive male fared better than the unattractive female.
    Average: $W = \mid W_- \mid = 17$; reject H_0
    Above average: $W = W_+ = 27.5$; unable to reject H_0

17. H_0: $M_{F-P} = 0$
    H_1: $M_{F-P} < 0$
    $W = W_+ = 19.5$; the critical point for an $\alpha = .05$ test is 11; unable to reject H_0

19. (a) $\dfrac{n(n + 1)}{4} = 915$ $\dfrac{n(n + 1)(2n + 1)}{24} = 18452.5$

    $$P[W_+ \leqslant 600] \approx P\left[Z \leqslant \frac{600.5 - 915}{135.84}\right] \approx .0102$$

    (b) $\dfrac{n(n + 1)}{4} = 2525$ $\dfrac{n(n + 1)(2n + 1)}{24} = 84587.5$

    $$P[W_+ \leqslant 2700] \approx P\left[Z \leqslant \frac{2700.5 - 2525}{290.84}\right] \approx .7257$$

Section 12.3

21. If the assertion is true, then W_m should be small. $W_m = 348$; the critical point is 387; conclude that females tend to view the performance of government officials more optimistically than males.

23. (a) $\binom{7}{3} = 35$

    (b) The lower 2 rank sums would occur by observing the sets $\{1, 2, 3\}$, for which $W_m = 6$ and $\{1, 2, 4\}$ $W_m = 7$. Hence 7 is the lower critical value. The higher 2 rank sums occur for the sets $\{5, 6, 7\}$, for which $W_m = 18$ and $\{4, 6, 7\}$, for which $W_m = 17$ as claimed.

Section 12.4

25. $r = -0.98$

$$r_s = 1 - \frac{6(166)}{8(63)} = -0.98$$

27. $\Sigma d^2 = 8$; $r_s = 0.90$

Section 12.5

29. (a) $R_1 = R_2 = 3$; $R_3 = 4$; $H = 2.7$
      $R_1 = R_3 = 3$; $R_2 = 4$; $H = 2.7$
      $R_1 = R_3 = 4$; $R_2 = 2$; $H = 1.8$
      $R_1 = R_2 = 4$; $R_3 = 2$; $H = 1.8$
      $R_1 = 5$; $R_2 = 2$; $R_3 = 3$; $H = .3$
      $R_1 = 5$; $R_2 = 3$; $R_3 = 2$; $H = .3$
      $R_1 = 5$; $R_2 = 1$; $R_3 = 4$; $H = 2.7$
      $R_1 = 5$; $R_2 = 4$; $R_3 = 1$; $H = 2.7$
      $R_1 = 6$; $R_2 = 1$; $R_3 = 3$; $H = 1.8$
      $R_1 = 6$; $R_2 = 3$; $R_3 = 1$; $H = 1.8$
      $R_1 = 7$; $R_2 = 1$; $R_3 = 2$; $H = 2.7$
      $R_1 = 7$; $R_2 = 2$; $R_3 = 1$; $H = 2.7$

    (b)

h	.3	1.8	2.7
$f(h)$	2/12	4/12	6/12

31. $R_1 = 33$; $R_2 = 76$; $R_3 = 69$; $R_4 = 32$; $H = 9.29$; reject H_0 at the $\alpha = .10$ level

33. $R_1 = 38$; $R_2 = 52$; $R_3 = 30$; $S = 248$;

$$\frac{12S}{bk(k+1)} = 12.4; \text{ reject } H_0 \text{ at the .05 level}$$

35. $R_1 = 18$; $R_2 = 27$; $R_3 = 36$; $R_4 = 34$; $R_5 = 35$; $S = 230$

$$\frac{12S}{bk(k+1)} = 9.20; \text{ can reject } H_0 \text{ at the } \alpha = .1 \text{ level}$$

Chapter 12 Review Exercises

37. For continuous variables $P[X_i = M_0] = 0$; zero differences do occur in practice due to the difficulty of obtaining precise measurements on continuous random variables; zero differences are assigned the algebraic sign least conducive to the rejection of H_0.

39. H_0: $M = 5$ $Q+ = 6$
    $P[Q+ \leqslant 6 \mid p = 1/2] = .0577$
    Yes; reject H_0

41. Paired T test

43. Two-sample T test

45. Wilcoxon Rank Sum Test
    $W_m = 87$; critical point is 100; reject H_0

47. Randomized complete block design

49. Kruskal-Wallis k Sample Test
    $R_m = 24.5$; $R_A = 74.5$; $R_E = 37$
    $H = 4.376$
    The critical point for X_2^2 (.1) is 4.61; unable to reject H_0

51. $\sum\limits_{i=1}^{10} d_i^2 = 228$

    $r_s = -.3818$

Index

V Test Statistics

H_0: $\mu = \mu_0$

$$T_{n-1} = \frac{\overline{X} - \mu_0}{S/\sqrt{n}}$$

H_0: $p = p_0$

$$Z = \frac{\hat{p} - p_0}{\sqrt{p_0(1 - p_0)/n}}$$

H_0: $\sigma^2 = \sigma_0^2$

$$X_{n-1}^2 = (n - 1)S^2/\sigma_0^2$$

H_0: $\mu_1 = \mu_2$
$(\sigma_1^2 = \sigma_2^2)$

$$T_{n_1+n_2-2} = \frac{\overline{X}_1 - \overline{X}_2}{S_p\sqrt{1/n_1 + 1/n_2}}$$

H_0: $\mu_1 = \mu_2$
$(\sigma_1^2 \neq \sigma_2^2)$

$$T_\nu = \frac{\overline{X}_1 - \overline{X}_2}{\sqrt{S_1^2/n_1 + S_2^2/n_2}} \quad \text{where}$$

$$\nu \approx \frac{(S_1^2/n_1 + S_2^2/n_2)^2}{\dfrac{(S_1^2/n_1)^2}{n_1 - 1} + \dfrac{(S_2^2/n_2)^2}{n_2 - 1}}$$

H_0: $\mu_1 = \mu_2$
(Paired data)

$$T_{n-1} = \frac{\overline{D}}{S_D/\sqrt{n}}$$

H_0: $\sigma_1^2 = \sigma_2^2$

$$F_{n_1-1,n_2-1} = S_1^2/S_2^2$$

H_0: $p_1 = p_2$

$$Z = \frac{\hat{p}_1 - \hat{p}_2}{\sqrt{\hat{p}(1 - \hat{p})(1/n_1 + 1/n_2)}} \quad \text{where}$$

$$\hat{p} = \frac{X_1 + X_2}{n_1 + n_2}$$